Distinctive Techniques for Organic Synthesis

A Practical Guide

Distinctive Techniques for Organic Synthesis

A Practical Guide

Tse-Lok Ho
National Chiao Tung University, Taiwan

Published by

World Scientific Publishing Co. Pte. Ltd.

P O Box 128, Farrer Road, Singapore 912805

USA office: Suite 1B, 1060 Main Street, River Edge, NJ 07661

UK office: 57 Shelton Street, Covent Garden, London WC2H 9HE

Library of Congress Cataloging-in-Publication Data

Ho, Tse-Lok.

Distinctive techniques for organic synthesis : a practical guide / Tse-Lok Ho.

p. cm.

Includes bibliographical references and index.

ISBN 9810232527

1. Organic compounds -- Synthesis. I. Title.

QD262.H588 1998

547'.2--dc21 98-8006

CIP

British Library Cataloguing-in-Publication Data

A catalogue record for this book is available from the British Library.

Printed in Singapore by Eurasia Press Pte Ltd

CONTENTS

PREFACE

This book is intended to survey, in a very succinct style, several valuable and distinctive techniques for organic synthesis which have not found daily use in the laboratory. The relatively low degree of technical demands of these processes should not have discouraged their routine employment by preparative chemists, yet, unfortunately, the frequency of applications is incommensurate with their synthetic potentials. It is hoped that this volume will somewhat remedy the neglect. In order to facilitate the readers to adopt the methods in their work, experimental procedures are included where appropriate. Because my prime objective is to gather all the essential information within a single cover, only representative examples are used to illustrate the methodologies. The enormous amounts of data cannot be included, even in a tabular form. For the same reason I basically refrain from describing detailed theories underlying the reactions of which source materials are available elsewhere.

The idea of grouping all these techniques in a single volume was many years old, and a somewhat abbreviated Chinese version of this text was published in 1993. Immediately afterwards I presented a proposal for an English edition under the title of "Organic Synthesis Under Extreme Conditions" to an American publisher and was rejected. Curiously, a multiauthor volume with virtually the same title and coverage appeared in the market in 1997. In light of this situation the readers are urged to compare and judge the merits of this and the other book. Of course the proposed readership of this work is the synthetic community.

Tse-Lok Ho

PREFACE

This book is intended [illegible] a very [illegible] distinctive techniques for organic synthesis [illegible] daily use in the laboratory. [illegible] relatively low degree of technical demands of these processes [illegible] not [illegible] used [illegible] routine [illegible] chemists, [illegible] the [illegible] of applications [illegible] potential. It is hoped that [illegible] will [illegible] facilitate the readers to [illegible] procedures [illegible] included where [illegible] information [illegible] used to illustrate the methodology [illegible] included [illegible] tabular [illegible] for the same [illegible] describing [illegible] underlying the [illegible] of which [illegible] materials are available elsewhere.

The idea of [illegible] many years ago [illegible] version of this text [illegible] published in [illegible] Immediately afterwards [illegible] English edition under the title of "Organic Synthesis Under [illegible] Conditions" [illegible] published and [illegible]. Curiously, [illegible] appeared in [illegible] 1997 [illegible] light of [illegible] to compare and judge the merits of this and the other book. Of course, [illegible] readership of this book [illegible] synthetic community.

[illegible]

General Abbreviations

Ac	acetyl
acac	acetylacetonate
AIBN	2,2'-azobisisobutyronitrile
aq	aqueous
Ar	aryl
Bn	benzyl
Bu	*n*-butyl
Bz	benzoyl
18-c-6	18-crown-6
c-	cyclo
cat	catalytic
Cp	cyclopentadienyl
DABCO	1,4-diazabicyclo[2.2.2]octane
DBN	1,5-diazabicyclo[4.3.0]non-5-ene
DDQ	2,3-dichloro-5,6-dicyano-1,4-benzoquinone
de	diastereomer excess
DMAP	4-(dimethylamino)pyridine
DME	1,2-dimethoxyethane
DMF	dimethylformamide
DMSO	dimethyl sulfoxide
E	COOMe
ee	enantiomer excess
Et	ethyl
HMPA	hexamethylphosphoric triamide
hv	light
iPr	isopropyl
kbar	kilobar
L	ligand
LAH	lithium aluminum hydride
LDA	lithium diisopropylamide
lut	2,6-lutidine
M	metal (alkali)
Me	methyl
Ms	mesyl (methanesulfonyl)
Nu	nucleophile
Oc	octyl
PEG	polyethylene glycol
Ph	phenyl
Pr	*n*-propyl
py	pyridine
Q^+	quaternary onium ion
RaNi	Raney nickel
R^f	perfluoroalkyl
(s)	solid
sens.	photosensitizer
TBS	=TBDMS, *t*-butyldimethylsilyl
THF	tetrahydrofuran
TMEDA	*N,N,N',N'*-tetramethylethylenediamnine
TMS	trimethylsilyl
Ts	tosyl (*p*-toluenesulfonyl)
Δ	heat

1 PHASE TRANSFER REACTIONS

Chemical reactions involving two or more molecules usually proceed at higher rates in one phase. However, it is not always feasible or expedient to render some reactions under homogeneous conditions. In such cases the phase transfer technique should be considered.

Improvement of reactions by phase transfer catalysis (PTC) represents a breakthrough in chemical technology. Although sporadic reports appeared before the 1960s, consolidation of the concept through a series of publications on the theme of extractive alkylation [Markosza, 1965], a patent describing catalysis of heterogeneous reactions [Starks, 1968], and the observations concerning reactions of quaternary ammonium salts in nonpolar media [Brandström, 1969, 1972] initiated an avalanche of research in the methodology.

While the most common phase transfer technique deals with reactants in two immiscible liquids, its scope encompasses gas-liquid, gas-solid, liquid-solid, and presumably solid-solid systems. In comparison with conventional methods, PTC holds considerable advantages in terms of cost (which avoids expensive anhydrous dipolar aprotic solvents), time, mildness, and simplicity. It is often observed that reactivities and selectivities are enhanced under such conditions.

From an electrochemical study [Tan, 1994] of the Williamson ether synthesis under phase transfer catalysis, it is concluded that the role of the catalyst is to establish a Galvani distribution potential difference between the two phases which acts as the driving force for transferring the reactive ions from the aqueous medium to the organic layer.

The general features of a typical PTC substitution reaction carried out in a system of organic and aqueous phases specifies reaction taking place in the organic layer. The phase transfer agent is present in catalytic amount and capable of moving one reactant in its resident solvent into that of the other reactant or the interfacial region. Only then the two species can react with reasonable speed. Needless to say that strong agitation is a most crucial to maximize reaction rates.

Actually there are two mechanistic models for the PTC reaction. In the Starks extraction mechanism the catalyst is biphilic, thus well qualified by a quaternary ammonium or phosphonium salt, either of which having finite solubility in both phases on account of its charge and organophilicity, capable of ferrying the counter anion from the aqueous phase to the organic layer by the partition phenomenon. The anion generated from reaction will pair with the quaternary onium species and enter the aqueous phase and a second catalytic cycle starts at this point. The interfacial mechanism (proposed by Makosza and modified by Brandström and Montanari) describes cases in which the catalyst is highly organophilic and it functions by anion exchange in the interfacial region. In other words, the cation of such a catalyst virtually never stays in the aqueous phase.

$$R\text{-}X + M^{+}Nu^{-} \xrightarrow{Q^{+}} R\text{-}Nu + M^{+}X^{-} \qquad R\text{-}H + E^{+}X^{-} \xrightarrow{B^{-}} R\text{-}E + H^{+}X^{-}$$

R-X → R-Nu; $Q^{+}Nu^{-}$ $Q^{+}X^{-}$; org. phase; aq. (solid) phase; $M^{+}X^{-}$ $M^{+}Nu^{-}$

R-H → R-E; $Q^{+}Nu^{-}$ $Q^{+}X^{-}$; $M^{+}X^{-}$ $M^{+}Nu^{-}$

cation-catalyzed phase transfer reaction

anion-catalyzed phase transfer reaction

It should be recognized that a PTC reaction consists of anion transfer and intrinsic reaction steps and the determining factors for each may be different, and high rates of both steps of the process are necessary. It is also important that the transferred anion be highly reactive, and likely be when pairing with a quaternary ammonium or phosphonium ion. The greater distance separating the cation and the anion in comparison with the that in the corresponding alkali metal salts increases the nucleophilicity of the anion. Actually the same principle applies when a crown ether with proper cavity to accommodate the cation is added to complex the metal ion in the alternative PTC system. Interestingly, tris(3,6-dioxaheptyl)amine is an excellent solid-liquid PTC [Soula, 1985] because it is inexpensive, relatively nontoxic and capable solubilizing alkali metal ions as well as transition metal ions by virtue of molecular flexibility., although the anion activation power is lower than cryptands.

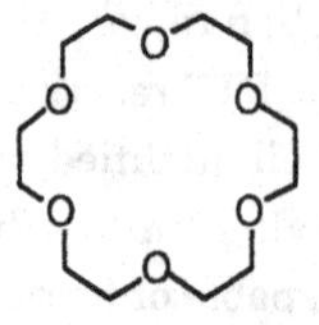

18-crown-6

tris{2-(2-methoxyethoxy)ethyl}amine

A recently developed method for expedient preparation of cyclophosphazenic polypodands [Gobbi, 1994], which have high catalytic activities, should render these very competitve in future utility.

Micelles and PTC have similarities and differences. Micelles are formed by addition of surfactants to aqueous-organic biphasic systems, but PTC are not necessarily surfactants and vice versa. Generally micelle systems have limited synthetic value owing to problems associated with emulsions.

Common anions are hydrophilic, their transfer into an organic phase must overcome the energy of hydration, which is particularly important for the smaller ions and those having a greater charge. The association of the anion to quaternary onium ions has a positive effect. Divalent anions have practically no affinity for cations in nonpolar organic solvents and therefore do not function at all. On the other hand, overly strong association renders ion pairs highly lipophilic and the catalyst regeneration becomes difficult. Thus quaternary iodide salts are rather poor catalysts and reactions that generate iodide ion are usually shut down at relatively low conversions.

It should be reiterated that a useful PTC should be soluble in the organic solvent because typical catalyst regeneration rate in the aqueous phase is 5-10 orders of magnitude greater than reaction rate in the organic layer. Quaternary salts having 10-30 carbon atoms are usually suitable, but the cation structure is also critical. Thus a cation containing one long alkyl chain and three methyl or ethyl groups are poor catalysts, for they tend to form micelles and stay in the aqueous phase. A general conclusion is that quaternary onium ions possessing all butyl or larger groups have superior anion transfer capability. Aryl and secondary alkyl group substituents present, the latter tend to undergo elimination.

While both quaternary ammonium and phosphonium salts have been successfully used as phase transfer catalysts, different situations may demand the choice of one or the other. Phosphonium salts are thermally more stable than ammonium analogues (which are prone to undergo internal displacement at elevated temperatures), but both are attacked by hydroxide ion to give phosphine oxides and Hofmann elimination products, respectively.

Other conditions that affect a phase transfer reaction include pertinent ion-pairing and the polarity of the organic solvent. Generally, saturation of the aqueous phase with an inorganic salt improves phase transfer by salting out (reducing the degree of hydration) the phase transferrent (anion). Sometimes the presence of a solvating compound such as an alcohol in small quantities to modify the structure of the anion can have dramatic benefit. Regarding organic solvent polarity, nonpolar solvents tend to magnify the partition differences of the anions while polar solvents such as CH_2Cl_2 have a leveling effect.

The choice of PTC is also based on criteria such as selectivity, stability under reaction conditions, availability and cost, and ease of removal or recovery. For example, a series of pyridinium salts [Dehmlow, 1989] are excellent catalysts for anhydrous alkylation of sodium phenoxides, but they are useless in the presence of aqueous bases because of their susceptibility to under decomposition to pyridones (and thiopyridines in the presence of Na_2S).

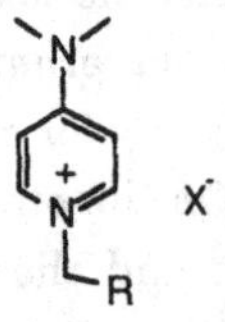

The diverse demands provide a constant stimulus to the development of new catalysts. For example, triphenylsulfonium and -selenonium salts [Kondo, 1989, 1992] meet the criteria of effective PTC, and the simple admixture of an acyl chloride with a tertiary amine furnishes acyltrialkylammonium chloride [Bhalerao, 1992] which serves well in solid-liquid phase transfer reactions. Chiral quaternary ammonium salts have shown great promise is promoting asymmetric reactions.

While most catalysts serve to transfer anions to the organic phase in which reactions take place, a few others can ferry cationic species across the interfacial boundary. Salts of tetrakis{3,5-bis(trifluoromethyl)phenyl}borate have proven highly satisfactory for catalyzing reactions of diazonium and oxonium species, including diazo coupling, Friedel-Crafts alkylation, nitrosation, acid-mediated hydrolysis of esters [Ichikawa, 1988]. The bulky anion is very stable to acid and oxidants.

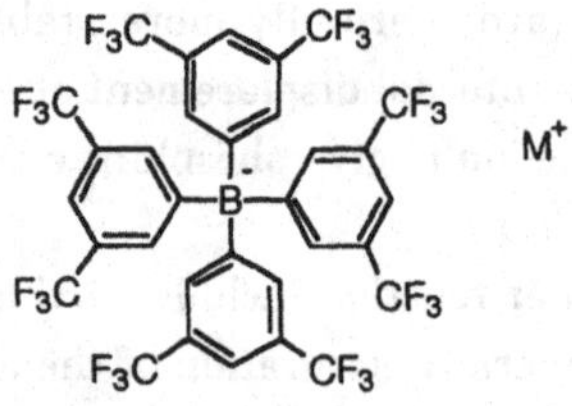

Regarding catalyst removal, those anchored to an insoluble resin are the easiest, for they require simple filtration at the end of the reaction. These catalysts have both advantages and defects: reactions are slower, and their activity may be more sensitive to environment, e.g., the degree of cross-linking of the resin.

New biphasic systems composing of a hydrocarbon and a fluorocarbon (perfluorinated alkanes, ethers, or tertiary amines) which are larglely immiscible are valuable for conducting water-sensitive reactions. Successful hydroformylation of alkenes has been demonstrated [Horvath, 1994] using a rhodium catalyst bound to a phosphine ligand which is soluble in the fluorous phase. Reaction occurs when reactants come into contact with the catalyst in that phase or at the interphase, with assistance of a PTC.

Triphase reactions [Regen, 1982a] have received attention from a few investigators. In the most common systems the catalyst is present in the solid phase (insolubilized ammonium and phosphonium salts, crown ethers, cryptands on resins, silica gel or alumina) together with two immiscible liquids. In some situations the triphase reaction has unique attractiveness. Thus macrolide formation in good to excellent yields has been achieved without resorting to high-dilution technique by stirring mesyloxy carboxylic acids with a resin-bound phosphonium mesylate in toluene/aqeous potasium bicarbonate [Regen, 1982b].

In phase transfer reactions, particularly those involving a solid, mass transfer is critical. Accordingly many more experimental parameters influence the triphasic reactions.

1.1. SUBSTITUTIONS

Many polar organic reactions were improved when conducted in polar aprotic solvents such as DMF and DMSO. The advent of phase transfer techniques then brought about an operational option which is often superior. In practically every case, common substitution reactions can profit by adopting this inexpensive but expedient protocol. Among the major advantages are convenience of execution including workup, remarkable absence of side reactions related to solvolysis, applicability to industrial scale reactions where cost is also critical.

Generally, the structure-reactivity relationship of substrate (typically alkyl halides) in simple substitution reactions under PTC conditions follows that established for S_N2 process in dipolar aprotic solvents [Smiley, 1960]. Thus in alkyl halides the trend of primary > secondary > tertiary halides is observed. Variation of the leaving group brings about gradation of reactivity in the rank of R-OMs > R-Br > R-Cl, while R-I and R-OTs behave poorly, due to strong association of I^- and TsO^- ions with quaternary ammonium and phosphonium ions. Characteristic of the S_N2 mechanism for PTC substitution reactions inversion of configuration is usually observed .

1.1.1. With Cyanide Ion

One of the most thoroughly studied aliphatic substitution reactions concerns the preparation of nitriles from alkyl halides (except iodides) and sulfonates (tosylates excluded). Quaternary onium salts and crown ethers and cryptands have been used successfully in the liquid-liquid PTC reaction [Liotta, 1974a; Cinquini, 1975].

If under anhydrous conditions using equimolar quantity of an onium cyanide, the reaction proceeds almost instantaneously and explosively even at room temperature [Simchen, 1975]. However, the so-called anhydrous conditions is not rigorously free of water, a trace of which must be present. However, as water content is increased further, the rate is depressed.

$$R\text{-}X + Q^+CN^- \longrightarrow R\text{-}CN + Q^+X^-$$

The solid-liquid PTC reaction proceeds better with 18-crown-6 than tetra-*n*-butylammonium bromide, probably indicating the differential efficiencies for transporting the anion from the solid to the organic phase. At least in some cases evidence adduced the formation of a ternary substrate-catalyst-solid reagent complex.

For industrial applications such as the preparation of nitrile intermediates of ibuprofen [Hampl, 1990] and Nylon-8 [Nanba, 1988] the cost excludes crown ethers as catalysts.

Treatment of clays with quaternary ammonium salts forms suitable catalysts which are easily recovered and reused [Lin, 1991].

Attention must be paid to avoid side reactions arising from alkylation of highly reactive nitriles. For example, phenylacetonitrile initially formed by the reaction of benzyl chloride and sodium cyanide is susceptible to benzylation when the pH of the aqueous medium rises (by reaction of NaCN with water to produce NaOH and HCN). Slow addition of NaCN minimizes the undesired alkylation [Coates, 1973].

$$PhCH_2Cl \xrightarrow{Q^+CN^-} PhCH_2CN \xrightarrow[Ph_2CH_2Cl]{Q^+CN^- / H_2O} PhCH(CN)CH_2Ph$$

Naked cyanide (from solid MCN and catalytic amount of a crown ether) in acetonitrile or benzene is an excellent nucleophile, capable of displacing the chlorine atom from chlorotrimethylsilane [Zubrick, 1975].

Glutaronitrile. [Liotta, 1974]

A magnetically stirred mixture of 1,3-dichloropropane (5.08g, 45 mmol), dry KCN (11.7 g, 180 mmol), 18-crown-6 (1.01 g, 3.8 mmol) in acetonitrile (25 mL) is refluxed. After 1.5 h the completed reaction (GC monitoring) is cooled, filtered, and reduced to ca. one-third volume. Addition of water, extraction with dichloromethane, drying over $MgSO_4$, evaporation in vacuo, and distillation gives glutaronitrile (4.10 g, 96.8%).

Displacement of vinyl halides becomes possible in the KCN/18-crown-6 system in the presence of $(Ph_3P)_4Pd$ [Yamamura, 1977].

Sensitive nitriles such as cyanoformates are obtainable in 62-94% yields from the chloroformates [Childs, 1976]. Aroyl cyanides are accessible more conveniently by PTC reaction of the aroyl chlorides with NaCN [Koenig, 1974]. The moderate yields, when the substrates are more electron-deficient, are due to dimer formation.

$$ArCOCl + CN^- \xrightarrow{Q^+X^-} ArCOCN \xrightarrow{CN^-} Ar\text{-}C(CN)_2\text{-}O^- \xrightarrow{ArCOCl} Ar\text{-}C(CN)_2\text{-}OCOAr$$

In conjunction with acetone cyanohydrin, 1,4-addition of the naked cyanide ion to enones (including cholestenone) can be effected.

1.1.2. With Halide Ions

The conversion of one alkyl halide to another or a mesylate to an alkyl halide is relatively facile with both liquid-liquid and solid-liquid phase transfer catalysis. For alkyl fluoride preparation (RCl being better substrates than RBr) it requires a higher temperature therefore quaternary phosphonium salts are employed [Landini, 1974a]. An ion exchange resin with residential quaternary ammonium ions (e.g., Amberlyst A-26) may be converted to the fluoride and used in the PTC reaction [Cainelli, 1976].

The displacement using concentrated hydrohalogenic acids is feasible in the presence of a quaternary ammonium salt [Landini, 1992] which has a great accelerating effect.

Olefins and alcohols are formed as side products but alcohols are absent when the reaction is conducted with a KF/crown ether system [Liotta, 1974b]. Secondary halides are prone to elimination because naked fluoride ion is also a strong base.

Water content in the reaction media is critical for fluoride displacement [Dermeik, 1985; Landini, 1991]. When quaternary ammonium salts are used as catalysts and KF as reagent, 0.33 mole of water per mole of KF would be optimal. This ratio is a compromise between fluoride transfer and catalyst stability.

Alkyl mesylates are better substrates because of mesyloxide ion is a better leaving group than halide ions and the weaker association with the cation makes catalyst regeneration more efficient.

Phase transfer conditions are also suitable for the conversion of sulfonyl chlorides to the corresponding fluorides [Bianchi, 1977].

Aromatic fluorides such as 2,4-dinitrofluorobenzene can be prepared by halogen exchange with the KF/crown ether system [Bram, 1988].

Cl, NO_2, NO_2 + KF —18-crown-6→ F, NO_2, NO_2 + KCl

Exchange between any other pair of halogen atoms in alkyl halides is perhaps of less significant values in the preparative sense. However, equilibria in favor of one direction are definitely more easily controlled in the two-phase system, for example by using an aqueous solution containing an excess metal halide to overwhelm the phase transfer catalyst. Alternatively, setting up the equilibrium with equimolar amount of a salt in water and replace the aqueous phase with fresh aliquot. After several cycles the organic phase should contain the exchange product as the major component. The nature of the cation in the added salt must also be considered, as equilibrium is dependent on the unequal (saturated) aqueous solubilities of M^+X^- and M^+Y^-.

A study comparing the catalytic activity for halogen exchange indicates cryptands > a quaternary phosphonium bromide > crown ethers [Cinquini, 1975]. Radioactive drugs containing ^{82}Br, ^{123}I, and ^{131}I isotopes have been prepared [Liu, 1985] by the halogen exchange technique using cown ether catalysts.

1.1.3. With Oxide Anions

The displacement of alkyl halides with hydroxide ion under liquid-liquid phase transfer catalysis is not particularly straightforward, although high conversion of 1-bromooctane to *n*-octanol has been realized using ammonium salts with four *n*-butyl or larger alkyl groups [Herriott, 1972], there are intrinsic problems in such reactions. Ether formation and elimination are serious side reactions, and the hydrated hydroxide ion is quite a difficult to transfer into the organic layer. In order to maintain an effective level of Q^+OH^- it requires high concentration of metal hydroxide in the aqueous phase which is detrimental to the quaternary cations at even moderate temperatures, thus limiting operating temperatures in the 50°-60°C range.

The formation of 1-adamantanol in 92% yield from 1-bromoadamantane [Slobodin, 1988] is an extraordinary case, contributed by steric effect.

Betaine salts $R_3N^+CH_2COO^-$ seem to be 10- to 50-fold more active than ordinary quaternary alkylammonium salts in promoting the hydroxide displacement reaction [Starks, unpublished]. Involvement of carboxylate intermediates is likely.

An amazingly efficient interconversion of mesitoic acid and its esters can be performed under PTC conditions. The esters are formed in essentially a solid-liquid phase transfer conditions [Loupy, 1986]. Mesitoic esters (even *t*-butyl mesitoate) are saponified on treatment with KOH-crown ether complexes in benzene or toluene [Pederson, 1967] or KOH-cryptands [Dietrich, 1973], by way of acyl-oxygen bond cleavage. (-)-Menthol was recovered from the menthyl ester. (For the more conventional transesterification between phenol and isopropenyl acetate under solid-liquid phase transfer conditions, see [Barry, 1988].)

<u>*n*-Octyl Mesitoate</u>. [Loupy, 1986]

Mesitoic acid is added to finely powdered KOH (containing 15% w/w water) and a small amount of Aliquat 336. The resulting mixture is shaken at room temperature for 10 min, while it is treated with 1-bromooctane at midpoint. The reaction is completed by heating at 85°C for 2 h. Yield of the ester is 88%.

While the alkylation of secondary alcohols with chloromethyl phenyl sulfide under standard conditions gives only 10-30% yield of the ethers, along with large quantities of bis(phenylthio)methane, the PTC route is superior for the preparation [Rawal, 1993].

OH, Ph — $PhSCH_2Cl$; 50% NaOH, hexanes, Bu_4NBr → O–SPh, Ph — 55-60%

For the synthesis of unsymmetrical ethers under PTC it is better to pair larger alcohols with smaller alkyl halides. Optimal conditions consist of excess alkyl chloride (as solvent if possible), more than fivefold excess of 50% NaOH over the alcohol, and 3-5 mole % of n-Bu_4N^+ HSO_4^- as catalyst at 25°-70°C [Freedman, 1975].

Optically pure oxetanes are similarly made by cyclization of 1,2,2-trisubstituted 1,3-propanediol monomesylates [X. Hu, 1995]. The corresponding reaction of the tosylates is slow.

Enediol ethers are readily formed by reaction of benzoin with alkyl tosylates under PTC conditions [Merz, 1977]. It gives unsaturated crown ethers in better yields than conventional method.

Ph OH Ph O + TsO–O–O–OTs → (n-$Bu_4N^+Br^-$, 50% NaOH) → 19.5% + 14%

Replacement of the chlorine atom of 2- and 4-chloropyridine with an alkoxy group (e.g., benzyloxy) is accomplished by heating the substrate with a suspension of powdered KOH and K_2CO_3 in dry toluene containing the alcohol and a little tris(3,6-dioxaheptyl)amine [Ballesteros, 1987].

2-chloropyridine + $PhCH_2OH$ → (KOH, K_2CO_3, 120°C, 4 h, PTC) → 2-OCH_2Ph-pyridine 84%

Ordinarily aliphatic alcohols cannot be methylated with dimethyl sulfate, but the reaction proceeds readily in the presence of a tetra-n-butylammonium salt [Merz, 1973a].

O-Alkylation of phenols is easily conducted by their addition to a two-phase system of aqueous NaOH, dichloromethane solution of the alkylating agent and $Bn(nBu)_3N^+ Cl^-$ [McKillop, 1974]. The extremely high chemoselectivity which is not complicated by *C*-alkylation is noteworthy. Under similar conditions catechols are converted in good yields (76-86%) to methylenedioxyarenes using dibromomethane as alkylating agent [Bashall, 1975].

<u>Methylenedioxyarenes</u>. [Bashall, 1975]

To a vigorously stirred refluxing mixture of dibromomethane (0.15 mol) and Adogen 464 (1 mmol) in water (20 mL) is added a solition of a catechol (0.1 mol) and NaOH (0.25 mol) in water (50 mL) during 2 h, while oxygen is being excluded from the system. After the addition the mixture is heated for a further 2 h before workup.

A recent significant employment of PTC alkylation is in a convergent synthesis of lycoramine [Parker, 1992]. Formation of the *cis*-2-aryloxy-4-siloxycyclohexanone was the result of a kinetic process.

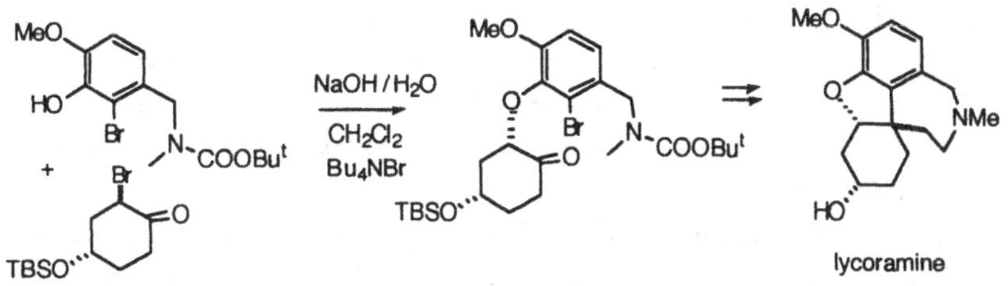

The PTC alkylation of 2,5-dimethylphenol with 1-bromo-3-chloropropane [Reinholz, 1990] illustrates the importance of catalyst structure. In the presence of accessible catalysts (e.g., crown ether complexes, $RNMe_3X$), which form relatively tight ion pairs with anions such as OH^- and which do not activate anion appreciatively, the major product is the aryl 3-chloropropyl ether. On the other hand, bulky ammonium salts which belong to the anion activating group also gives considerable amounts of the 3-bromopropyl ether, owing to activation of Br^- to effect an exchange of the chlorine in the initiate product.

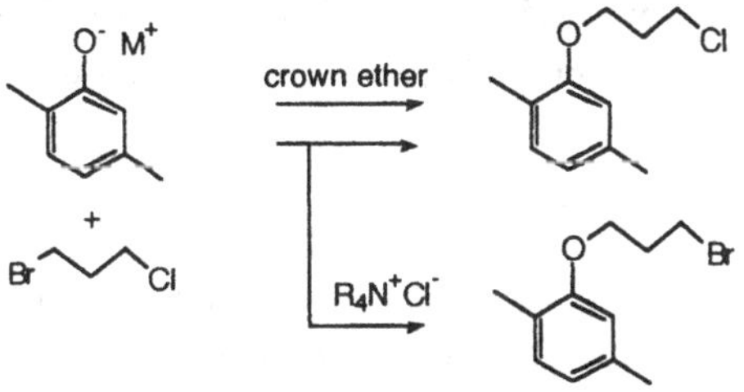

An unusual procedure for *O*-alkylation of 2-naphthol consists of mixing sodium 2-naphthoxide and the alkylating agent in a molten quaternary salt (best results in molten *n*-Bu_4NBr) [Badri, 1992]. The reaction rates and selectivity are better than in DMF. Work up is by ether extraction of the cooled mixture.

Formation of ester from $RCOONa/H_2O$ and $R'CH_2X$ in a two-phase system was observed when an amine was added [BASF, 1912]. However, more than fifty years lapsed before the catalytic role of an ammonium salt generated in situ was determined [Hennis, 1967]. One alkyl group of the ammonium ion must be quite long, and benzylammonium salts are not useful.

Anion exchange macroreticular resins (e.g., Amberlite IRA-904) in which the anion is replaced by a carboxylate undergo alkylation (best with alkyl iodides). Esters are released in 50-90% yield [Cainelli, 1975].

Various other PTC techniques are effective for *O*-alkylation of the carboxylate ion. A correlation of crown ether/cryptand structure with rate of benzyl acetate formation from benzyl chloride and potassium acetate in acetonitrile at room temperature has been made [Knöchel, 1975]. The reaction of an acyl chloride with an alkyl halide in the presence of an alkali metal bicarbonate and a phase transfer catalyst (Bu_4NBr or PEG-400) in acetonitrile at 80°C to give an ester [Y. Hu, 1992] actually proceeds via the metal carboxylate.

Methyl esters can be converted to other alkyl esters by sequential saponification and *O*-alkylation [Doecke, 1991; O'Donnell, 1991]. Amino acid methyl esters generally undergo transesterification without racemization, except the highly sensitive compounds such as the phenylglycine derivative.

$$\text{RCOOMe} \xrightarrow[\text{Bu}_4\text{NBr, r.t.}]{\text{NaOH, CH}_2\text{Cl}_2/\text{H}_2\text{O}} \text{RCOONa} \xrightarrow[\text{Bu}_4\text{NBr, r.t.}]{\text{R'Br, CH}_2\text{Cl}_2/\text{H}_2\text{O}} \text{RCOOR'}$$

In the solid-phase peptide synthesis, immobilization of the first amino acid to the Merrifield resin is quantitative when carried out by a 18-crown-6 promoted reaction with the Boc-amino acid potassium salt in DMF [Roeske, 1976]. Microwave as energy source shortens the time for a solid-liquid phase transfer reaction [Bram, 1990].

n-Octyl Acetate. [Bram, 1990]

In an open flask is placed potassium acetate, 1-bromooctane and 10 mol% Aliquat 336. The mixture is irradiated in a commercial microwave oven (600 W; final temp. 187°C) for 1-2 min. The yield of the ester is 98% by GC.

The displacement of α-halocarbonyl compounds, including 3-chloro-2,4-pentanedione [Cativiela, 1995], is readily achieved. A 2-phenyl-2-butenolide can be obtained on continued refluxing a mixture of potassium phenylacetate, an α-bromocarbonyl compound in acetonitrile in the presence of 18-crown-6 [Dehm, 1975]. Occurrence of an intramolecular Claisen condensation from the ester is reconcilable with various factors.

$$R^3C(=O)C(R')(R)Br + PhCH_2COOK \xrightarrow{\text{18-c-6}} R^3C(=O)C(R')(R)OC(=O)CH_2Ph \xrightarrow{\Delta} \text{butenolide}$$

A solution to the technical difficulties in permethylation of sugars is to subject the peracetyl derivatives to bromomethane in benzene-water in the presence of *n*-

Bu_4N^+ HSO_4^- [Di Cesare, 1980]. Sequential saponification and methylation are catalyzed by the phase transfer agent.

Methyl Tetra-*O*-methyl-D-glucopyranosides. [Di Cesare, 1980]

Penta-*O*-acetyl-α-D-glucopyranoside (1.17 g, 3 mmol) and tetrabutylammonium hydrogen sulfate (0.102 g) are added to benzene (15 mL) which contains a suspension of powdered NaOH (5 g). The vessel is blanketed under an atmosphere of bromomethane and stirred at room temperature for 4 h. The reaction mixture is filtered over Celite which is washed with ether (20 mL). The combined organic solutions are washed with water (2x20 mL), dried over $MgSO_4$, concentrated under reduced pressure, and chromatographed on silica gel to afford methyl tetra-*O*-methyl-α-D-glucopyranoside (292 mg, 39%) and methyl tetra-*O*-methyl-β-D-glucopyranoside (421 mg, 56%).

A modified condition for the Koenigs-Knorr reaction in the presence of a crown ether has been reported [Knöchel, 1974b].

AcO, AcO, AcO, AcO, Br (glycosyl bromide) + ROH —$AgNO_3$ / crown ether→ AcO, AcO, AcO, AcO, OR (glycoside) + AgBr + HNO_3

A unusual report claimed the successful reaction of 2-propenyl bromide with carboxylate ions [Klan, 1992].

Acetylation of alcohols (and anilines) with acetyl chloride is rapid using powdered sodium hydroxide and *n*-Bu_4N^+ HSO_4^- in an organic solvent [Illi, 1979]. Under such conditions it is possible to selectively acetylate the phenolic hydroxyl group of estradiol. Better results for *t*-butoxycarbonylation of alcohols and phenols with di-*t*-butyl carbonate are obtained when phase transfer technique is employed [Houlihan, 1985]. Of course all these *O*-acylation processes do not proceed by direct displacement.

The radical anion superoxide is available commercially in K and Na salts. The use of these salts in organic synthesis has been limited by their poor solubility in the common solvents. With the advent of crown ethers the problem is largely solved, although polar organic solvents are still required. The dialkyl peroxide product from reaction with an alkyl bromide is the result of two S_N2 steps intervened by reduction of the alkylperoxy radical intermediate with another superoxide radical anion [R.A. Johnson, 1975].

When DMSO is used as solvent or cosolvent the superoxide substitution reaction leads directly to the inverted alcohol. Approximately four equivalents of KO_2 are needed for completing the reaction [Corey, 1975].

4x KO_2
4.5x 18-c-6
DMSO-DMF-DME
(1:1:1)
0°C, 20 min

AcO, OMs, C_5H_{11} → AcO, OH, C_5H_{11} 75%

4x KO_2
5x 18-c-6
DMSO
25°C, 3 min

MsO, OMs, Ph → O-O, Ph 35%

A linear polycarbonate is produced when potassium carbonate is mixed with α,α'-dibromo-*p*-xylene in the presence of a crown ether [Soga, 1977].

1.1.4. With Other Nucleophiles

Studies on the phase transfer reaction of amines, amides and imides are extensive. *N*-Alkylation of amines, particularly the heteroaromatics, have been studied [Dou, 1976]. The reaction of α-amino acids with 2,3-dibromo-2-methylpropanoyl chloride in the presence of an ion-exchange resin (Duolite-A109) and 30% NaOH to afford β-lactams [Okawara, 1981] involves *N*-acylation and intramolecular *N*-alkylation.

Br, Br, COCl + H_2N, R, COOH —resin, 30% NaOH→ Br, O, N, R, COOH 45-65%

Introduction of a *N*-dialkylaminoethyl group to the ala^2 residue of an antitumor cyclic hexapeptide, without racemization, is readily performed by the PTC technique [Itokawa, 1993].

It is definitely more convenient to apply the solid-liquid PTC method to the Gabriel synthesis [Landini, 1976]. With solid potassium phthalimide and a chiral secondary bromide the reaction was demonstrated to be an S_N2 process. A method for α-amino acid synthesis involves alkylation of trifluoroacetamide with α-bromocarboxylic esters under the solid-liquid PTC conditions [Landini, 1991b]. In connection to indole alkaloid synthesis this technique saved the effort [Angle, 1993] while the allylic chloride failed to react with sodium trifluoroacetamide.

NHBoc Cl $\xrightarrow{CF_3CONH_2}$ K_2CO_3, $n\text{-}Bu_4N^+Br^-$ NHBoc $NHCOCF_3$ 69% ⇒ N Boc

<u>*N*-Octylphthalimide</u>. [Landini, 1976]

A stirred mixture of 1-bromooctane (3.86 g, 20 mmol) potassium phthalimide (4.63 g, 25 mmol) and hexadecyltributylphosphonium bromide (1.01 g, 2 mmol) in toluene (10 mL) is refluxed until all the bromo compound disappears (GC monitoring). After 2 h (98% conversion) the reaction is terminated by cooling, the solid is filtered and washed with ether (30 mL). The combined solutions are eluted through silica (10 g) and the eluent is washed with 10% NaOH, H_2O, and dried over Na_2SO_4. Evaporation of the solvent affords the product (4.66 g, 90%); m.p. 47°-49°C.

The imino nitrogen of sulfoximines is not easy to alkylate by conventional methods. Here the solid-liquid PTC technique offers an excellent solution [C.R. Johnson, 1993].

N–S(O)(=NH)–Me $\xrightarrow{KH/DME}$ RBr, $Bu_4N^+Br^-$ N–S(O)(=NR)–Me

Phosphinamides are also readily *N*-alkylated. The phosphinamide alkylation [Zwierzak, 1977] can be controlled to proceed to the monoalkylated or dialkylated stage by temperature and quantities of the alkylating agent. The products are hydrolyzed to give the corresponding amines on treatment with hydrogen chloride.

$Ph_2P(O)NH_2$ — $n\text{-}Bu_4N^+ HSO_4^-$, 50% NaOH

1x RX / PhH Δ 1.5 h → $Ph_2P(O)NHR$ (45-91%) → RNH_2

4x RX / PhMe Δ 4 h → $Ph_2P(O)NR_2$ → R_2NH

Since alkyl azides are relatively easy to prepare by conventional displacement method, the PTC protocol [Reeves, 1976; Hassner, 1986] has a somewhat diminished value. A more interesting report is that the quaternary ammonium chloride catalyzed reaction of poly(vinyl chloride) in moderately nonpolar solvents such as THF [Takeishi, 1973].

Acyl azides are similarly prepared [Brandström, 1974]. At higher temperature they undergo Curtius rearrangement to provide the isocyanates. The latter compounds are also obtainable from reaction of alkyl halides and sodium cyanate at temperatures <100°C [Zenner, 1972].

$$RCOCl + Q^+N_3^- \longrightarrow RCON_3 + Q^+Cl^- \xrightarrow{\Delta,\ -N_2} R\text{-}N\text{=}C\text{=}O$$

Primary nitroalkanes are the major products when alkyl bromides and an alkali metal nitrite react under phase transfer conditions [Kimura, 1976, Zubrick, 1975]. Yields are slightly lower than the $NaNO_2$/DMSO method. It is remarkable that the reaction of 2,4-dinitrochlorobenzene with dry and wet sodium nitrite leads to 1,2,4-trinitrobenzene and 2,4-dinitrophenol, respectively [Bhati, 1991].

<u>Reaction of 2,4-Dinitrochlorobenzene with Potassium Nitrite</u>. [Bhati, 1991]

2,4-Dinitrochlorobenzene (2.02 g, 0.01 mol), KNO_2 (1.70 g, 0.02 mol) and benzyltriethylammonium chloride (0.227 g, 0.001 mol) in toluene (25 mL) are stirred magnetically at 70°C for 6 h. The toluene layer is separated, washed with water (3x10 mL), dried over Na_2SO_4, filtered, and evaporated to leave 1,2,4-trinitrobenzene which is recrystallized from methanol (1.51 g, 71%).

2,4-Dinitrochlorobenzene (2.02 g, 0.01 mol), KNO_2 (1.70 g, 0.02 mol) and benzyltriethylammonium chloride (0.227 g, 0.001 mol) in toluene (25 mL) is stirred with $NaHCO_3$ (3.3 g, 0.04 mol) in water (5 mL) at 70°C for 5 h. Workup in the same manner gives 2,4-dinitrophenol (1.63 g, 89%).

1,2,4-trinitrobenzene $\xleftarrow[\text{dry } NaNO_2]{BnNEt_3^+Cl^-}$ 2,4-dinitrochlorobenzene $\xrightarrow[\text{wet } NaNO_2]{BnNEt_3^+Cl^-}$ 2,4-dinitrophenol

Thiols and symmetrical dialkyl sulfides are generated, often in yields approaching quantitative, when sodium hydrogen sulfide and sodium sulfide, respectively, are alkylated [Jursic, 1989; Landini, 1974b]. Sulfides are also easily synthesized from alkyl- and arylthiols using NaOH as base in the aqueous phase. The high nucleophilicity of thiolate anions enables the use of secondary halides as alkylating agents [Herriott, 1975].

$$R\text{-}X + NaSH(aq) \xrightarrow{PTC} R\text{-}SH + NaX$$

$$2R\text{-}X + Na_2S(aq) \xrightarrow{PTC} R\text{-}S\text{-}R + 2NaX$$

Activated chloroarenes undergo reaction with sodium sulfide under PTC conditions. Of interest are the results from liquid-liquid and solid-liquid reactions of *p*-chloronitrobenzene [Pradhan, 1992]; only nitro group reduction was observed in the former but bis(4-nitrophenyl) sulfide was produced when Na_2S and the catalyst were present as solid.

Thioacetals are formed from *gem*-dihalides by the PTC displacement reaction. Interestingly, activated 1,3-dithianes thus prepared undergo allylation with transposition [Lissel, 1982], in contrast to reactions with NaH as the base.

SH SH + $Cl_2CHCOOEt$ $\xrightarrow[\text{Aliquat, PhMe}]{K_2CO_3}$ S S COOEt 63%

Br; K_2CO_3, Aliquat/PhMe

S S COOEt + S S COOEt (95 : 5) 68%

Dialkyl disulfides are now available from the reaction of alkyl chlorides or bromides with sulfur in the presence of sodium hydroxide and polyethylene glycol-400 in benzene at 65°C [Wang, 1995].

For reaction with ambident anions the choice of catalyst and conditions may be critical for product formation. Thus the displacement of an alkyl halide with KSCN is effectively catalyzed by sulfonium salts with bulky and hydrophobic groups [Kondo, 1988]. With benzyl chloride the affords BnSCN exclusively by liquid-liquid PTC, but solid PTC at 180°C can give rise to variable amounts of BnNCS also, the ratio of the two isomers being dependent on the structure of the catalyst [Dehmlov, 1990].

$$PhCH_2SCN \xleftarrow[QCl]{(liq.-liq.)} PhCH_2Cl + KNCS \xrightarrow[QCl]{(solid-liq.)} PhCH_2NCS + PhCH_2SCN$$

benzyl thiocyanate; benzyl isothiocyanate

Preparatively viable displacement of alkyl halides with sodium bisulfite [Lantzsch, 1977] is surprising because a divalent anion needs to be transferred. Dichloromethane has been converted to $ClCH_2SO_3Na$ in 72% yield during 6 h at 90°C under the influence of Et_4NCl.

Aqueous hydrochloric acid is a mediocre reagent for the conversion of *n*-alkanols to 1-chloroalkanes. Improved rate and yield of the reaction are observed in the presence of a quaternary phosphonium salt [Landini, 1974a] or ammonium salt [Jursic, 1988]. Alternatively, the transformation is more expediently achieved in a column of silica gel containing a quaternary phosphonium salt [Tundo, 1987]. The method is particularly useful for the conversion of water soluble alcohols.

1-Chlorooctane from Octanol. [Jursic, 1988]

Octanol (13 g, 0.1 mol) hexadecyltrimethylammonium bromide (1 g, 0.0025 mol) and 37% HCl (205 mL, 2.5 mol) are heated at reflux overnight. After cooling the organic layer is separated and the aqueous solution is extracted with petroleum ether (3x100 mL). The organic solutions are combined, washed with 10% Na_2CO_3 solution (50 mL), dried over Na_2SO_4 and evaporated. The oily residue is taken up with ether, passed through silica gel to remove some of the ammonium salt, and distilled to afford 1-chlorooctane (13 g, 87%).

$$RCH_2OH + HCl \xrightarrow{Q^+Cl^-} RCH_2Cl$$

A triphase reaction among an organic halide, carbon monoxide, and sodium hydroxide is catalyzed by a combination of n-Bu_4NI and $(Ph_3P)_4Pd$ to give the sodium alkanoate [Cassar, 1976]. Benzyl, aryl, heterocyclic, and vinyl halides are reactive.

$$PhCH_2Cl + CO + NaOH \xrightarrow[(Ph_3P)_4Pd]{n\text{-}Bu_4N^+I^-} PhCH_2COONa \quad 83\%$$

The reductive carbonylation of *gem*-dibromocyclopropanes with a synthesis gas and a Ni-Co system in the presence of a phase transfer catalyst [Grushin, 1991] is an interesting process.

Ph Br Ph Br $\xrightarrow[\text{2 KCN, KOH; PhMe, PEG-400; CO-}H_2\text{ (3:1)}]{Ni(CN)_2 \cdot 4H_2O,\ CoCl_2 \cdot 6H_2O}$ Ph Ph COOH 47%

Br Br [Co] → Br $Co^{(III)}$ L_n [Ni] → Br $Ni(CO)_2CN$ CO → Br $Ni(CO)_2CN$ O

↓ OH^-

COOH H_2O ← $Ni(CO)_2CN$ O^- OH^- ← $Ni(CO)_2CN$ COOH

1.2. ALKYLATIONS

Terminal alkynes are acidic enough to be alkylated by alkyl iodides in the system of KOH/18-crown-6 in refluxing benzene [Lissel, 1985]. Only catalytic amounts of the crown ether are needed. Alkylation with carbonyl compounds in a two-phase system is practical only in the presence of a PTC [Dehmlow, 1982b].

OH + Me_2CO $\xrightarrow[H_2O/\text{ PhMe}]{\text{NaOH}}$ OH OH

$n\text{-}Bu_4N^+Br^-$	100%
no catalyst	7%

The alkylation of carbonyl compounds by the PTC technique is effectively applied to α,α-disubstituted aldehydes and activated ketones. Allylation and benzylation of such aldehydes [Dietl, 1973] appear to be synthetically useful, even though the yield are moderate most of the time (30-75% for isobutyraldehyde). Arylacetones can be alkylated under various PTC conditions [Jarrouse, 1951; Cinquini, 1974; Makosza, 1975a]. However, extensive dialkylation occurs with more active alkylating agent, while the dialkylation cannot be avoided in all cases of cyclic ketones (i.e., β-tetralone, etc.), regardless of electrophile. A series of geminal disulfoxides, sulfinylphosphonates, diphosphonates have been found to exhibit excellent catalytic activities for the ketone alkylation [Mikolajczk, 1975].

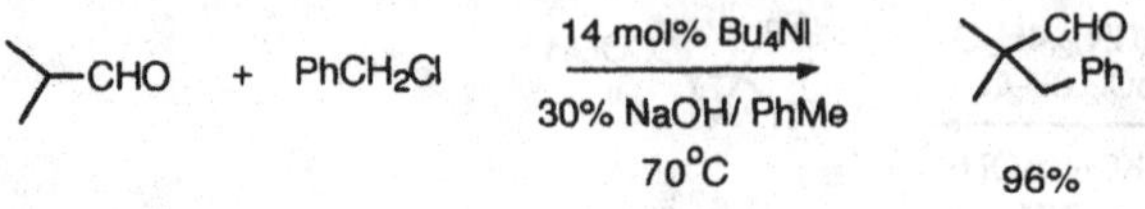

Monoalkylation or dialkylation of benzyl methyl ketone in the absence of solvent is subjected to control by regulating the ratio of the base and alkylating agent [Aranda, 1992].

Sequential alkylation of cyclopentanone with 1,4-dibromo-2-pentene by the PTC technique led to a spirocyclic product in 40-55% yield [Näf, 1978]. The product is a key intermediate for jasmonoid synthesis.

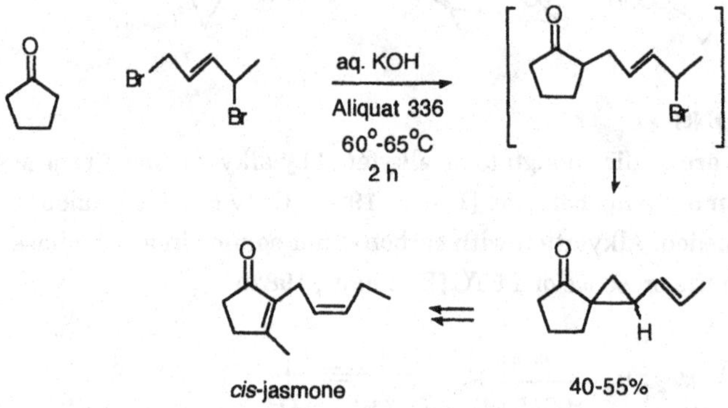

The ambident behavior of enolates toward alkylating agents is influenced by the onium ion structure. With accessible catalysts, the more exposed nitrogen atom such as that in $RNMe_3^+$ binds more tightly to the oxygen atom of the enolate, enabling higher percentage of *C*-alkylation.

A suspension of methyl sodioacetoacetate in organic solvent undergoes *C*-benzylation (>90%) with the help of a quaternary ammonium chloride [Durst, 1974]. The result is in contrast to the preferred *O*-benzylation in dipolar aprotic solvents. Note that in benzoylation the nature of the electrophile dominates [Jones, 1977].

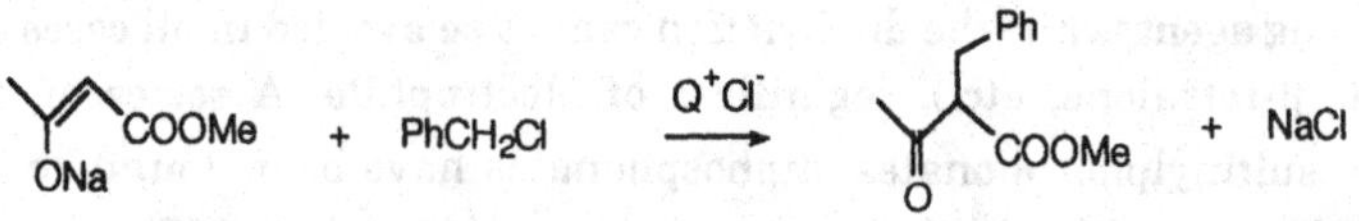

Aqueous sodium hydroxide rapidly saponifies esters therefore one must resort to using *t*-butyl esters in those β-ketoesters and malonic esters to be alkylated

[Jonczyk, 1973]. (Note a method for deliberate removal of one ester group from a malonic ester is by treatment with KOH/18-crown-6 and heat [Hunter, 1977]) On the other hand, methyl and ethyl esters can be used in the ion pair extraction procedure [Brandström, 1972]. With the latter method and the presence of a chiral catalyst it has been possible to obtain optically active allylated products [Fiaud, 1975]. The optical yields are poor.

An expedient procedure [Heiszman, 1987] for alkylation of malonic esters espouses the use of catalytic Q^+Cl^-, K_2CO_3 as base, and 5-10% water just enough to promote the reaction. Further improvement by using a combination of the onium salt and a crown ether (or polyethylene glycol) is indicated [Szabo, 1987].

<u>Diethyl cyclopropane-1,1-dicarboxylate.</u> [Heiszman, 1987]

Diethyl malonate (32.0 g, 0.2 mol) and 1,2-dichloroethane (40 g, 0.4 mol) are mixed with dry potassium carbonate (69.0 g, 0.5 mol), tetra-*n*-butylammonium bromide (2.0 g, 6 mmol) in benzene (100 mL) which contains water (1mL). The stirred mixture is refluxed for ca. 20 h, cooled, and filtered. The filtrate together with benzene washings (for the solid) is evaporated, and the residue distilled to give the product (31.6 g, 85%); b.p. 115°C/22 torr.

Benzylation of α-sulfeinyl esters followed by in situ β-elimination constitutes a useful method for the synthesis of cinnamic esters. This process is adaptable to solid-liquid phase transfer technique [Xu, 1987].

Ar–CH₂Cl + PhSO–CH₂–COOMe → (K_2CO_3; Aliquat 336; DMF 25-> 60°C) → Ar–CH=CH–COOMe

When arylacetic esters are complexed by the tricarbonylchromium group, the enhanced acidity of the benzylic proton enables an extremely rapid alkylation [des Abbayes, 1977b] such that even a methyl ester can be largely preserved.

(ArCH₂COOMe)Cr(CO)₃ + RX → ($C_{16}H_{33}NMe_3^+Br^-$; aq. NaOH) → (ArCH(R)COOMe)Cr(CO)₃ + NaX

A practical methylation of arylacetic acids which leads to naproxen and ibuprofen consists of three PTC reactions [Canicio, 1985]. Esterification followed by *C*-methylation is terminated by saponification. The base is solid KOH suspended in dichloromethane, 3.2-3.3 equivalents of dimethyl sulfate is present.

$$\text{MeO-naphthyl-CH}_2\text{COOH} + (MeO)_2SO_2\ (3.2\ x) \xrightarrow[\text{KOH (8.2 x)},\ CH_2Cl_2\ \Delta]{BnNEt_3^+Cl^-\ (31\ mol\%)} \text{naproxen (84\%)}$$

$$\text{isobutylphenyl-CH}_2\text{COOH} + (MeO)_2SO_2\ (3.3\ x) \xrightarrow[\text{KOH (9.6 x)},\ CH_2Cl_2\ \Delta]{BnNEt_3^+Cl^-\ (31\ mol\%)} \text{ibuprofen (74\%)}$$

Amino acid synthesis from glycine is facilitated by protection of the functional groups as imine and ester or amide as such derivatives undergo *C*-alkylation readily. Phase transfer technique is advantageous for the transformation [O'Donnell, 1978] which is also adaptable to asymmetric synthesis [O'Donnell, 1994; Oppolzer, 1994] (For analogous alkylation, see [Eddine, 1995]). The condensation of the benzylidene derivative of glycine ester with an aromatic aldehyde is efficient in the presence of potassium carbonate and benzyltriethylammonium chloride in acetonitrile [Wu, 1986]. Thus β-hydroxy α-amino acids are readily obtained from such oxazolidine products by acid hydrolysis.

(Scheme: sultam glycinamide, N=C(SMe)$_2$ → LiOH.H$_2$O, CH$_2$Cl$_2$, RX, Bu$_4$NHSO$_4$,))) → alkylated product ⇒ HO$_2$C–CH(R)–NH$_2$)

Alkylation of a Chiral N^{α}-Bis(methylthio)methyleneglycinamide. [Oppolzer, 1994]

lithium hydroxide hydrate (50 equiv.) is added to the sultam-derived *N*-protected glycinamide (1.0 equiv.), Bu_4NHSO_4 (1.1 mol. equiv.) and an alkyl halide (1.2 mol. equiv.) in CH_2Cl_2/H_2O (30:1, 10 mL/mmol) at -10°C, and the mixture is immediately sonicated for 2-8 min. The workup procedure involves filtration, evaporation of the filtrate, trituration of the residue with ether (50 mL/mmol) to remove Bu_4NX, and washing the solution (with water and brine), drying, and concentration. The product is recrystallized.

Activated nitriles, particularly arylacetonitriles, have been subjected to extensive PTC alkylation studies using NaOH as base. Under such conditions a secondary or tertiary alkyl ester group present in the alkylating agent can survive [Makosza, 1969a]. Arylation by *p*-chloronitrobenzene and 4-nitrobenzophenone has been realized [Makosza, 1974b]. A remarkable vinylation with acetylenes was observed [Makosza, 1966b] in the presence of DMSO.

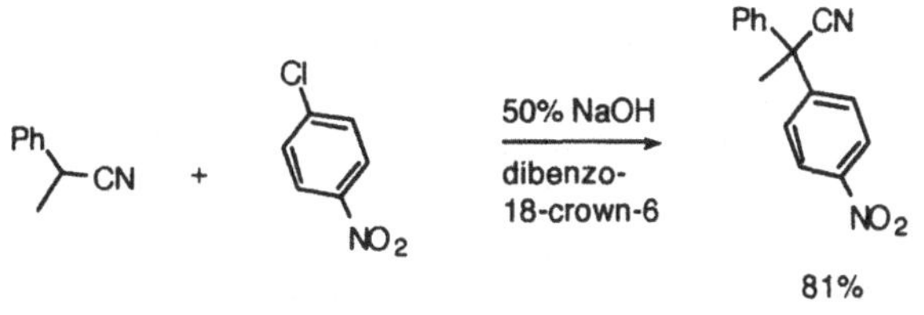

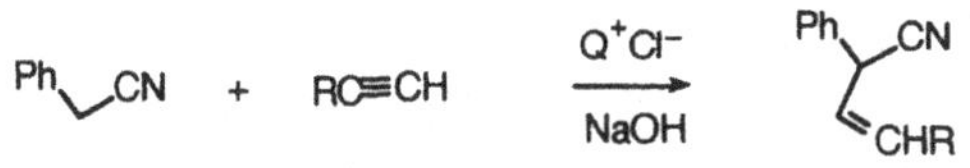

An excellent methylation of phenylacetonitrile with dimethyl carbonate is by passing a gaseous mixture (dimethyl carbonate in fourfold excess) over a catalytic bed of corundum spheres coated with 5 wt% K_2CO_3 and 5 wt% polyethylene glycol 6000 at 180°C [Tundo, 1989]. The selectivity is 99% at 98% conversion, only trace amounts of the dimethylated product are detectable.

Alkynylation of α-substituted phenylacetonitriles occurs with β,β-dichlorostyrene as alkylating agent under PTC [Jonczyk, 1993].

The diethylaminodithiocarbamate of thioglycolenitrile has been developed into a formyl anion synthon or carbonyl 1,1-dianion synthon by virtue of PTC alkylation [Masuyama, 1976]. Thus, reaction with 1,4-dibromobutane followed by acid hydrolysis leads to cyclopentanone.

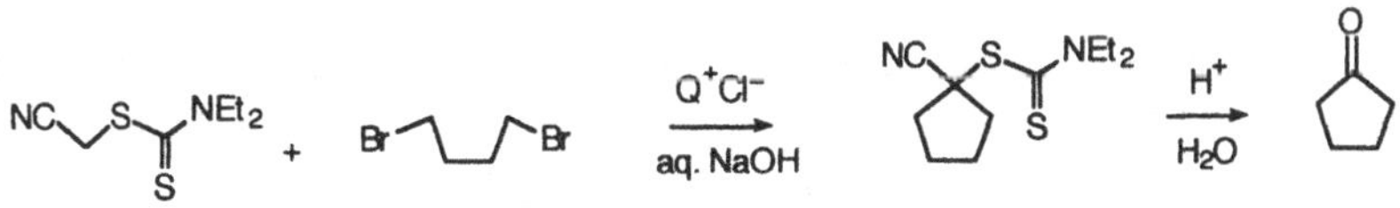

The ion pair extraction procedure allows the use of alkyl iodides because stoichiometric quantity of the quaternary ammonium salt is added [Brandström, 1972]. (For alkylation of β-ketosulfoxides by this protocol, see [Samuelsson, 1971]).

When aromatic aldehydes and cyanomethanephosphonates react under such conditions the products are the unsaturated nitriles (Emmons-Wadsworth reaction)

[D'Incan, 1977]. Actually the direct condensation of carbonyl compounds with excess acetonitrile is the most cost effective method [Gokel, 1976].

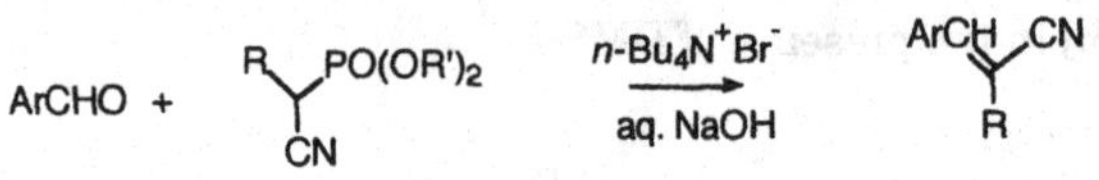

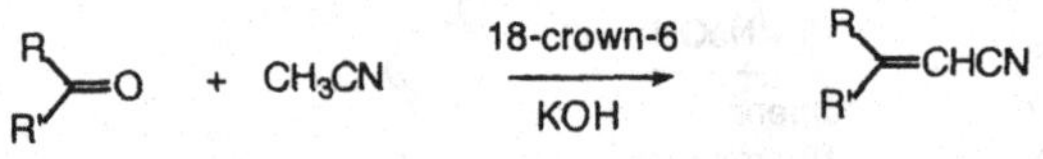

n-$Bu_4N^+Br^-$, aq. NaOH

18-crown-6, KOH

The Darzens reaction is mechanistically akin to the above condensation; only that after the C-C bond formation an intramolecular displacement occurs instead of dehydration. Accordingly the PTC technique is well suited for the Darzens reaction [Jonczyk, 1972; Makosza, 1974]. An interesting variant pertains to the formation of an epoxide by a tandem Michael addition-Darzens reaction [McIntosh, 1977].

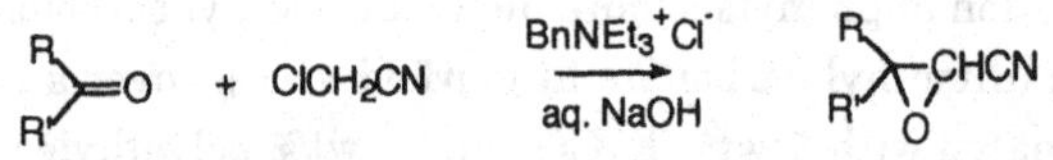

$BnNEt_3^+Cl^-$, aq. NaOH

R	Ph	Me	Ph	-$(CH_2)_4$-	-$(CH_2)_4CH(Me)$-	Ph
R'	H	Me	Me			Ph
Yield (%)	75	60	80	65	78	55

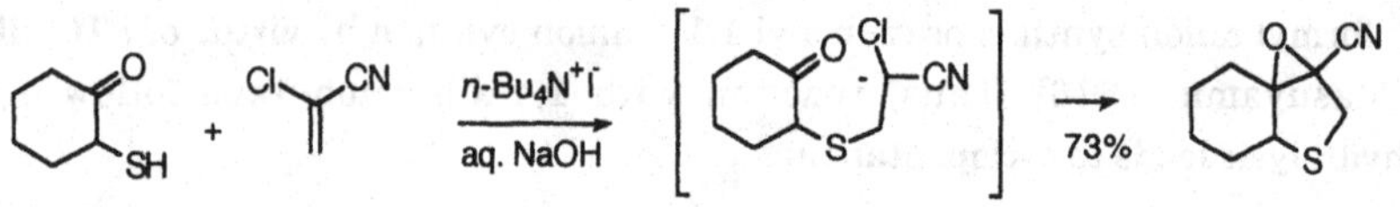

n-$Bu_4N^+I^-$, aq. NaOH; 73%

The less well known analogy, that is cyclopropanation, can also be effected [Jonczyk, 1976]. However, the electronic characteristics of the reaction partner play an important role. Thus phenyl vinyl sulfone gives open-chain Michael adducts ony.

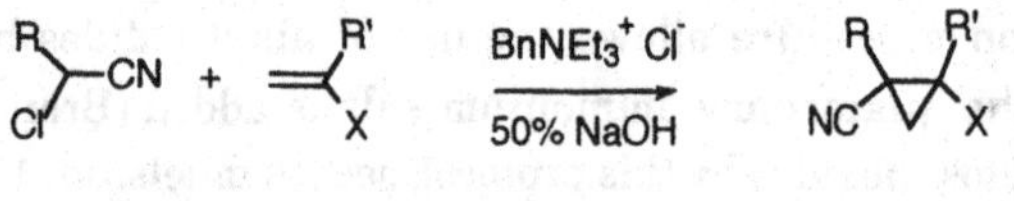

$BnNEt_3^+Cl^-$, 50% NaOH

cis + trans

X = CN, SO_2Ph, $COOBu^t$

Other compounds that have been submitted to PTC alkylation include tosylmethyl isocyanide [Van Leusen, 1975], cyclopentadiene [Makosza, 1968], indene [Makosza, 1966a], fluorene [Makosza, 1967]. The indene products are 3-substituted, due to isomerization.

1.3. CONDENSATIONS

Symmetrical 1,4-diaryl-1,4-butadiynes are obtained when β,β-dibromostyrenes are treated with carbon monoxide in the presence of a Pd(0) catalyst under PTC conditions [Galamb, 1983]. The function of CO is unknown.

$$2\ ArCH{=}CBr_2 \xrightarrow[\text{BnNEt}_3\text{Cl, CO, Pd(diphos)}_2]{5N\ \text{NaOH / PhH}} Ar{-}C{\equiv}C{-}C{\equiv}C{-}Ar \quad 54\text{-}74\%$$

Like conventional alkylation reactions, the Wittig and related reactions suffer from the stringent requirement of anhydrous and hence expensive solvents and reagents. Also like the results discussed in the above section, relief by the application of PTC method has been partially fulfilled. For the Wittig reaction, the use of aqueous NaOH as base is successful for aldehydes [Märkl, 1973; Hünig, 1974; Tagaki, 1974].

$$RCHO + Ph_3\overset{+}{P}\text{-}CH_2R'\ Cl^- \xrightarrow[CH_2Cl_2]{NaOH} RCH{=}CHR'$$

The condensation involving α-phosphonocarbanions and aldehydes may be effected without adding a phase transfer catalyst [Mikolajczyk, 1976], indicating an autocatalytic activity of the phosphonates.

Formation of epoxides from carbonyl compounds and sulfonium ylides under the PTC conditions ($Me_3S^+I^-$, $n\text{-}Bu_4N^+I^-$ / CH_2Cl_2-NaOH{aq.}) has been confirmed [Merz, 1973b], as the quaternary ammonium salt is an essential component. Excellent yields of epoxides are produced from aldehydes, and the reaction delivers moderate enantiomeric excess when chiral catalysts are used [Hiyama, 1975]. Ketones react very sluggishly, perhaps because iodide ion is present. The use of sulfonium chloride and benzyltriethylammonium chloride appears to remedy the situation [Rosenberger, 1980].

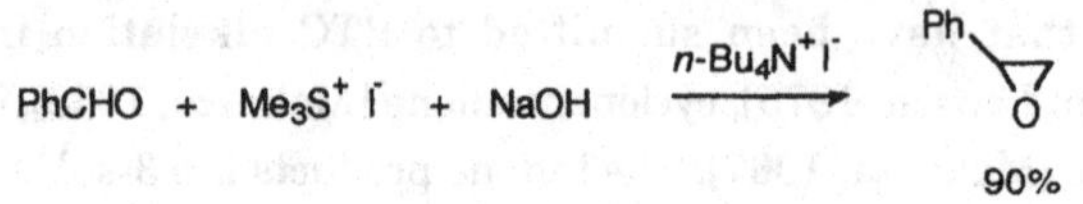

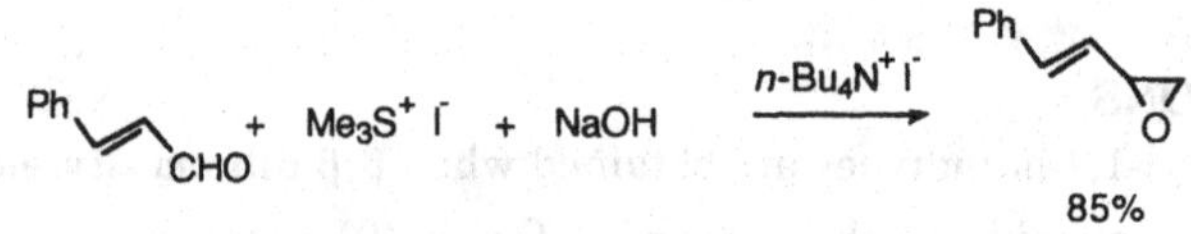

PhCHO + $Me_3S^+ I^-$ + NaOH $\xrightarrow{n\text{-}Bu_4N^+I^-}$ Ph-oxirane 90%

Ph-CH=CH-CHO + $Me_3S^+ I^-$ + NaOH $\xrightarrow{n\text{-}Bu_4N^+I^-}$ Ph-CH=CH-oxirane 85%

2-Phenyloxirane. [Merz, 1973]

To a solution of benzaldehyde (10.6 g, 0.1 mol) and n-Bu_4NI (0.5 g, 1.35 mmol) in CH_2Cl_2 (100 mL) is underlayed 50% NaOH. After addition of $Me_3S^+I^-$ (20.4 g, 0.1 mol) the mixture is vigorously stirred and heated at 50°C for 48 h, and poured onto ice. The organic layer is separated, washed with water, and dried. Distillation gives the oxirane product (11 g, 92%).

The behavior of the dimethyloxosulfonium methylide give inferior results, and it effects cyclopropanation of conjugated ketones in the same manner as the homogeneous reaction.

Aldol condensation such as that shown below [Dehmlow, 1981b] is very useful for the construction of polyunsaturated carboxylic acid derivatives.

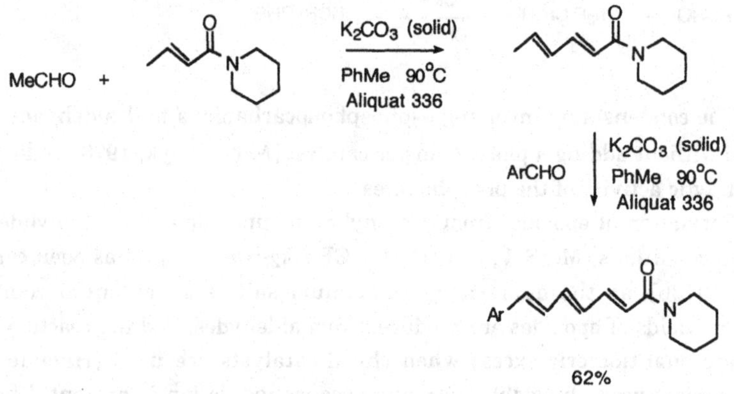

MeCHO + crotonyl piperidide $\xrightarrow[\text{PhMe } 90^\circ C \text{, Aliquat 336}]{K_2CO_3 \text{ (solid)}}$ sorbyl piperidide

ArCHO, K_2CO_3 (solid), PhMe 90°C, Aliquat 336 → Ar-(CH=CH)$_3$-CO-piperidide 62%

It is possible to achieve enantioselective aldol condensation [Ando, 1993] and Robinson annulation [Bhattacharya, 1986] by using chiral catalysts such as the quaternized quinine alkaloids. Formation of tight ion-pairs (hydrogen bonded) of the enolates are indicated [Reetz, 1993].

Aldol Condensation of 2-Methyl-α-tetralone. [Ando, 1993]

A column of Amberlyst-A26 (Cl^- form, 2.5 g, 10 mmol) is transformed into the OH^- form by passing through a 1N NaOH solution and washed to neutral pH with water and rinsed with methanol. The column is treated with *N*-benzylcinchonium bromide (464 mg, 1.0 mmol) in methanol (20 mL). Another wash with methanol the elutes are combined and neutralized with 1N hydrofluoric acid Evaporation of the solvent leaves a residue which is dehydrated by coevaporated three times with 1:1 benzene-acetonitrile. Further drying over P_2O_5 at 40°C overnight gives the quaternary ammonium fluoride (428 mg).

The catalyst (49 mg, 0.12 mmol) in THF (8 mL) at -78°C under argon is successively added benzaldehyde (0.11 mL, 1.1 mmol) and a solution of 1-trimethoxysilyl-2-methyl-3,4-dihydronaphthalene (233 mg, 1.0 mmol) in THF (2 mL). After 6 h, water (2 mL) is added and the mixture is warmed to room temperature and concentrated. The residue is taken up in methanol (10 mL) and treated with 1N HCl (0.3 mL) for 2 h. Workup followed by chromatography gives the aldol products (198 mg, 74%, Erythro : threo = 7:3) [70% ee for the erythro aldol and 20% ee for the threo aldol].

The preparation of a γ-lactone in 66% yield from an α-bromoaldehyde, potassium ethyl malonate [Sakai, 1980] greatly simplifies the synthesis of avenaciolide. The lactone formation under phase transfer catalysis involves esterification (*O*-alkylation), intramolecular Claisen condensation, Michael addition and decarboxylation.

C_8H_{17} CHO Br + COOK COOEt $\xrightarrow[PhH / H_2O]{Bu_4N^+Br^-}$ C_8H_{17} COOEt COOEt 66% ⇒ avenaciolide

1.4. ELIMINATIONS AND ADDITIONS

1.4.1. β-Eliminations

Dehydrohalogenation by the PTC procedure is of relatively little synthetic value. However, it may be mentioned that 18-crown-6 is a promoter for the conversion of *cis*-β-bromo-*p*-styrene to the acetylene by KF [Naso, 1974]. The effect is particularly prominent in acetonitrile.

Generally, the ion-exchange equilibrium hinders β-dehydrohalogenation under PTC conditions. As elimination proceeds the concentration of halide ion increases, rendering the $[OH^-]$ negligible in the organic phase and the reaction stops. Ion pair extraction conditions are highly successful but the requirement of equimolar Q^+OH^- is uneconomical. A better method [Makosza, 1991] involves the addition of a weak organic acid YH (e.g., CF_3CH_2OH, Ph_3COH, 2-methylindole) to the system to form a lipophilic anion (strong base, weak nucleophile) and enter the organic layer as Q^+Y^-. which is an effective reagent. The YH product suffers deprotonation at the phase boundary and the reagent is replenished.

It has been observed that dehydrobromination is strongly promoted by the formation of a third liquid phase which is rich in the catalyst [Mason, 1991].

A useful method for the synthesis of unsymmetrical 1,2-bis(perfluoroalkyl)-ethynes is to dehydroiodinate the iodoalkenes with sodium hydroxide in the presence of a quaternary ammonium salt [Sanchez, 1993].

$R^f(I)C{=}CHR'^f \xrightarrow[Bu_4NHSO_4]{NaOH} R^f{-}C{\equiv}C{-}R'^f$

vic-Dibromides are debrominated by sodium thiosulfate when catalytic amount of a phosphonium bromide and sodium iodide are added [Landini, 1975]. The results are comparable to the conventional method using NaI in acetone.

Double dehydrohalogenation of *vic*-dihalides by use of powdered potassium hydroxide is effective with various phase transfer catalysts [Dehmlow, 1981a]. The efficiency is lower with the dichlorides than the dibromides.

The reaction of 2-chloromethyl-1,1-dichlorocyclopropane with α-substituted phenylacetonitriles leads to 2,3-disubstituted 1-methylenecyclopropanes [Jonczyk, 1994] through a series of dehydrochlorination. With phenylacetonitrile itself only displacement of the primary chloride is observed.

NaOH(s), DMSO, $[BnNEt_3]^+Cl^-$, PhCHRCN

1.4.2. α-Eliminations

Trihalomethanes undergo α-elimination on exposure to a strong base (e.g., *t*-BuOK) to give dihalocarbenes which can be intercepted with alkenes. The α-elimination is a stepwise process via the trihalomethide ions. In the presence of relatively good nucleophiles the carbene intermediates are rapidly destroyed and dihalocyclopropane formation is suppressed. The discovery of the dramatic increase in the yield of dichlorocarbene adduct of cyclohexene (0.5% -> 60-70%) on addition of a quaternary ammonium chloride [Starks, 1968, 1971; Makosza, 1969b] was indeed exciting. Of particular interest is the fact that accessible catalysts ($RNMe_3Br$, Et_4NBr, etc.) are among the best, yet they are very poor catalysts for displacement reactions. Tetramethylammonium chloride is apparently the most active catalyst to promote formation of the monoadduct from a diene [Dehmlow, 1982a].

$$CHCl_3 + NaOH + Q^+X^- \longrightarrow Q^+ CCl_3^- + NaX + H_2O$$

$$Q^+ CCl_3^- \rightleftharpoons Q^+ Cl^- + :CCl_2 \xrightarrow{\text{alkene}} \text{gem-dichlorocyclopropane}$$

1-Phenyl-2,2-dichlorocyclopropane. [Makosza, 1969b]

A mixture of styrene (10.4 g, 0.1 mol), chloroform (12 g, 0.1 mol), benzyltriethylammonium chloride (0.4 g) and 50% NaOH (20 mL) is vigorously stirred at 40°C for 4 h. Upon dilution with water, separation into layers and distillation, 1-phenyl-2,2-dichlorocyclopropane (15 g, 80%) is obtained.

Under PTC conditions, efficient dichlorocyclopropanation of alkenes is due to rapid transfer of the Cl_3C^- anion to the organic phase so that once the carbene is surrounded by a great number of alkene molecules once it is generated, there is much less opportunity for the carbene to encounter the destructive hydroxide ion. From the quaternary ammonium ion structure one might surmise a close ion pairing with the Cl_3C^- species.

Trialkylamines are also active in promoting the reaction [Isagawa, 1974], and in fact they are the preferred catalysts for $:CBr_2$ addition to alkenes [Makosza, 1975b]. (Addition of small amount of ethanol [Makosza, 1973] or pinacol [Dehmlow, 1988] to the reaction media also gives better results in the $:CBr_2$ addition.) The transferred reagent may be $R_3N^+CHX_2$ CCl_3^- which are formed by reaction of trialkylammonium dihalomethylide ($R_3N^+CX_2^-$) with the trihalomethane.

For synthetic application to substrates which are sensitive to aqueous NaOH a suspension of sodium trichloroacetate may be used as the dichlorocarbene source [Dehmlow, 1976]. Dichlorocarbene is also generated from ethylene oxide, chloroform in the presence of Q^+Cl^- [Nerdel, 1965]. The Cl_3C^- ion is formed by proton abstraction of chloroform by the 2-chloroethoxide ion. Various epoxides are converted to dichlorocyclopropanes by reaction with dichlorocarbene [Tabushi, 1976].

$$Q^+ Cl^- + \text{(ethylene oxide)} \rightarrow Q^+ \,^-OCH_2CH_2Cl$$

$$\downarrow CHCl_3$$

$$Q^+ \,^-CCl_3 + HOCH_2CH_2Cl$$

$$\downarrow$$

$$Q^+ Cl^- + :CCl_2$$

Dihalocarenes are electrophilic, therefore they show discrimination toward alkenes. Electron-deficient alkenes such as α,β-unsaturated sulfones often react with X_3C^- instead of the carbene [Shostakovskii, 1974].

Dichlorocarbene adds to the C=N bond of imines. Some adducts suffer ring opening [Makosza, 1974a]. Cycloaddition involving the C=C double bond of pyrrole and indole is followed by ring expansion in situ [Kwon, 1976]. Strained dihalocyclopropanes (e.g., derived from norbornene and norbornadiene) behave similarly [Kraus, 1972; Dehmlow, 1972].

Bromochlorocarbene is generated from a mixture of dibromomethane, trichloromethylbenzene, and potassium hydroxide in the presence of tetrabutylammonium hydrosulfate [Balcerzak, 1991]. The probable intermediates are dibromomethide ion and chlorodibromomethane which are formed by deprotonation and chlorine abstraction, respectively.

exo-3,4-Dibromobicyclo[3.2.1]oct-2-ene. [Kraus, 1972]

Norbornene (5.0 g), bromoform (135 g), benzyltriethylammonium chloride (0.1 g) and 50% NaOH (110 g) are mixed and stirred vigorously at room temperature for 91 h. After dilution with water and filtration the mixture is washed several times with water, dried over $CaCl_2$, evaporated, and distilled to afford the dibromo compound (6.9 g, 48%); b.p. 100°C/0.5 torr.

Ar, :CCl2, Ar, N, Ar, Cl, Cl, Cl, N, Ar, Ar, Cl, OH⁻, Cl, H, N, Ar, Ar, O

BnNEt3+ Cl⁻, aq. NaOH, PhH, N, H, Cl, Cl, N, H, Cl, N, 68%

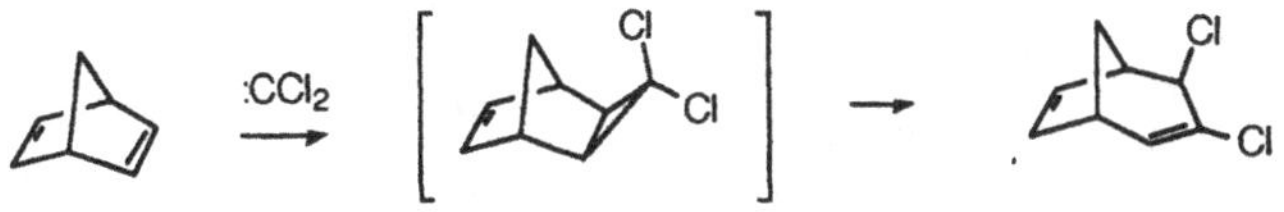

P,P'-Diaryl-3-dichloromethylenediphosphiranes are prepared [Yoshifuji, 1991] by a cycloaddition-rearrangement process.

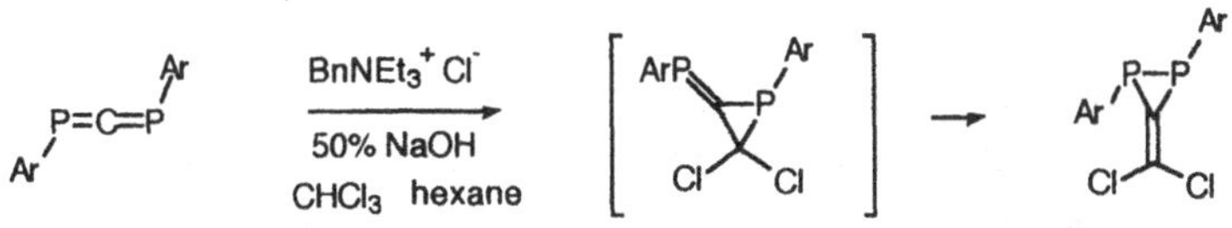

Certain cationic transition metal complexes serve a dual catalytic role in homogeneous and phase transfer catalysis. For example, after effecting the semi-hydrogenation of phenylacetylene, dichlorobis(1,10-phenanthroline)iridium(III) chloride is capable of promoting dichloropropanation [Goldberg, 1986].

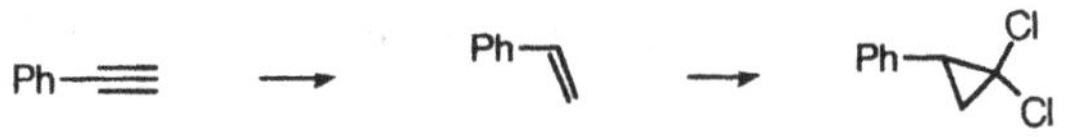

1,1-Dichloro-2-phenylcyclopropane. [Goldberg, 1986]

A 1% solution of phenylacetylene in chloroform is hydrogenation in the presence of ichlorobis(1,10-phenanthroline)iridium(III) chloride at 60°C under 10 atm hydrogen for 2 h. The mixture (containing styrene and ca. 10% ethylbenzene) is stirred with 50% NaOH at 40°C for 12 h to give the dichlorocyclopropane in 80% yield (85% styrene conversion).

Dichlorocarbene insertion into C-H bond is of little synthetic value, except for the preparation of 1-dichloromethyladamantane [Tabushi, 1970] and dichloromethyl ketones from acetals of aldehydes [Steinbeck, 1978]. The formation of α-chloromethylenation of *N*-methyl-δ-valerolactam [Marcos, 1986] is a formal C-H insertion, but more likely it proceeds by alkylation of the lactam with dichlorocarbene.

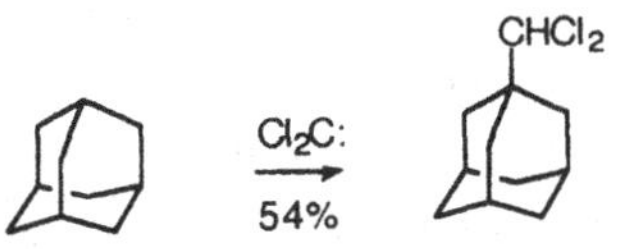

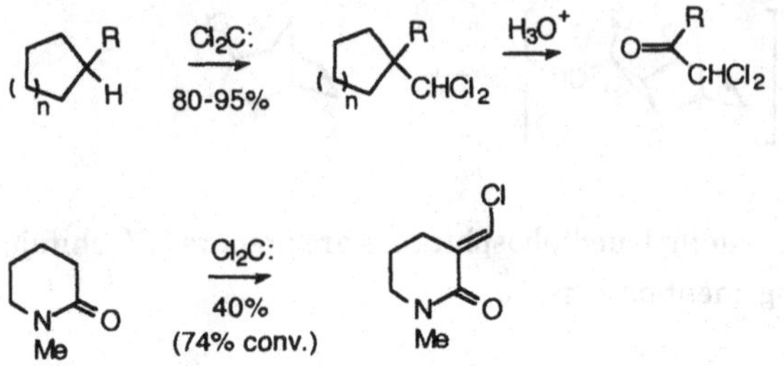

$\xrightarrow[80\text{-}95\%]{Cl_2C:}$ $CHCl_2$ $\xrightarrow{H_3O^+}$ $O{=}C(R)CHCl_2$

$\xrightarrow[40\%\ (74\%\ \text{conv.})]{Cl_2C:}$

1-Dichloromethyladamantane. [Tabushi, 1970]

To a vigorously stirred mixture of adamantane (13.6 g), 50% NaOH (200 mL), benzene (20 mL) and benzyltrimethylammonium chloride (0.4 g) at 40°C was dropwise added chloroform (80 mL) during 6 h. After a further stirring of 0.5 h workup gives the product (b.p. 114°-115°C/8 torr) in 54% yield (91% based on consumed adamantane).

The one-step synthesis of isonitriles [Weber, 1972a, b] from primary amines constitutes a superior method to the standard procedure via the formamides. Secondary amines afford formamides under the same conditions [Graefe, 1974]. An economical method for the generation of diazomethane from hydrazine is based on PTC [Sepp, 1974].

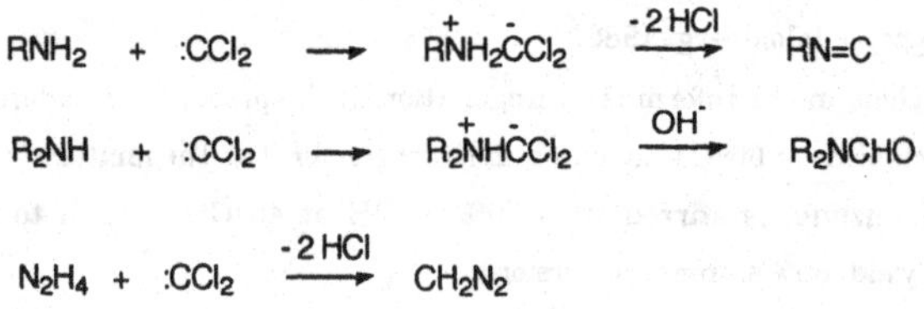

$$RNH_2 + :CCl_2 \longrightarrow R\overset{+}{N}H_2\overset{-}{C}Cl_2 \xrightarrow{-2\,HCl} RN{=}C$$

$$R_2NH + :CCl_2 \longrightarrow R_2\overset{+}{N}H\overset{-}{C}Cl_2 \xrightarrow{OH^-} R_2NCHO$$

$$N_2H_4 + :CCl_2 \xrightarrow{-2\,HCl} CH_2N_2$$

Diazomethane. [Sepp, 1974]

A Vigreux column (with takeoff sidearm connected to a condenser and topped by an addition funnel) is mounted on a two-neck flask which is then charged with KOH pellets (80 g, 1.4 mol), water (20 mL), chloroform (48 g, 0.4 mol), ether (200 mL), 18-crown-6 (0.2 g, 0.8 mmol), and finally 85% hydrazine hydrate (11.76 g, 0.2 mol). Upon sealing the second neck with a rubber stopper the reaction mixture is vigorously stirred and heated. Foaming starts and yellow distillate appears. After collection of about 100 mL distillate in the ice-cooled 500 mL Erlenmeyer receiver (originally containing 50 mL ether), ether is replenished from the addition funnel at the rate approximately equal to the distillation, which is stopped when the distillate becomes colorless. A 48% yield based on hydrazine is obtained. (Small amount of hydrazine can be removed if dichloromethane is used as solvent, and the distillate is washed with water {10 mL per 100 mL distillate}. The yield is decreased by about 10%. Also a lower yield of diazomethane is observed when NaOH/n-Bu_4NOH are used instead of KOH/crown ether.

Dehydration of primary amides, aldoximes, and elimination of hydrogen sulfide from thioamides have been observed [Saraie, 1973]. These processes may involve interaction of the carbene with the oxygen or sulfur atom of the substrates.

$$\begin{array}{ccccc} & & RCH{=}NOH & & \\ & & \downarrow :CCl_2 & & \\ RCONH_2 & \xrightarrow{:CCl_2} & RCN & \xleftarrow{:CCl_2} & RCSNH_2 \end{array}$$

Of limited value is the conversion of alcohols to chlorides by dichlorocarbene generated in the PTC system [Sasaki, 1973].

Dimethylvinylidenecarbene is generated from 3-chloro-3-methylbutyne by 1,3-dehydrochlorination under anhydrous conditions. The PTC method [Julia, 1984; Patrick, 1974; Sasaki, 1974] is better not only in terms of convenience but also in yields.

$$(CH_3)_2C(Cl)C{\equiv}CH \xrightarrow{-HCl} [(CH_3)_2C{=}C{=}C:] \xrightarrow{RCH=CHR} (CH_3)_2C{=}C{=}C\langle\begin{matrix}CHR\\ |\\ CHR\end{matrix}$$

The generation of N-carbethoxynitrene by α-elimination and trapping with a triorganoborane is best conducted in a two-phase system with a quaternary ammonium chloride [Akimoto, 1981].

1.4.3. Addition Reactions

Homologation of halides to form carboxylic acids involving carbon monoxide is usually considered as carbonylation reaction. As carbon monoxide is concerned the two bond-forming steps to its carbon atom constitute an α-addition. The use of PTC to effect such a reaction in the presence of a metal carbonyl (e.g., $Co_2(CO)_8$) is most fruitful, as the halides are protected from solvolytic destruction, active metal carbonyl anions are readily transferred to the organic phase and activated by a bulky quaternary ammonium ion in the displacement of the halides. The decomposition of the intermediates by hydroxide ion after acyl migration is facilitated by the presence of the transferred Q^+OH^- [des Abbayes, 1983, 1985].

$$RCH_2Br + Q^+\,Co(CO)_4^- \longrightarrow RCH_2Co(CO)_4 + Q^+\,Br^-$$

organic: $RCH_2COOH \longleftarrow RCH_2CO\text{-}Co(CO)_4$

aqueous: $RCH_2COO^-\,Na^+ \quad Q^+\,OH^- + NaBr \rightleftarrows NaOH + Q^+\,Br^-$

With the use of iron pentacarbonyl as catalyst benzylic halides can be largely converted to dibenzyl ketones [des Abbayes, 1988] by manipulating the conditions to maximize the concentration of the tetracrbonylacyliron anion in the organic solution while decreasing the transfer rate of OH^-.

Phenol undergoes carbonylation to give diphenyl carbonate when subjected to treatment with carbon monoxide and oxygen in the presence of a phase transfer catalyst, $PdBr_2$, $Mn(acac)_2$, molecular sieves and NaOH [Hallgren, 1981].

$$PhOH + CO + 0.5\,O_2 \xrightarrow[\text{Bu}_4\text{NBr, NaOH, mol. sieves}]{PdBr_2\,/Mn(acac)_2} PhO\text{-}C(=O)\text{-}OPh + H_2O$$

It is interesting that a phase transfer catalyst changes the course for ammonium polysulfide to react with α,β-unsaturated ketones [Krein, 1993].

Ketone arylhydrazones are converted to α-cyanoketones on treatment with sodium cyanide in a biphasic system in the presence of acetic acid and air [Chiba, 1991]. The three stages involved in this transformation are nucleophilic addition, oxidation, and extrusion of nitrogen.

$$R^1R^2C{=}NNHCOR^3 \xrightarrow[\text{Oc}_3\text{NMe Cl}^-\ (\text{HOAc})]{NaCN} R^1R^2C(NC)\text{-}NHNHCOR^3 \xrightarrow[\text{Oc}_3\text{NMe Cl}^-]{OH^-\,/O_2} R^1R^2C(NC)\text{-}N{=}NCOR^3 \xrightarrow{-N_2} R^1R^2C(NC)\text{-}COR^3$$

A Michael addition of active methylene compounds to an unsaturated nitrosugar in benzene requires catalysis by a quaternary phosphonium salt [Sakakibara, 1975]. The reaction is highly stereoselective, producing the thermodynamically less stable *manno*-isomers. For accomplishing more routine Michael additions the solid PTC method warrants consideration [Bram, 1985].

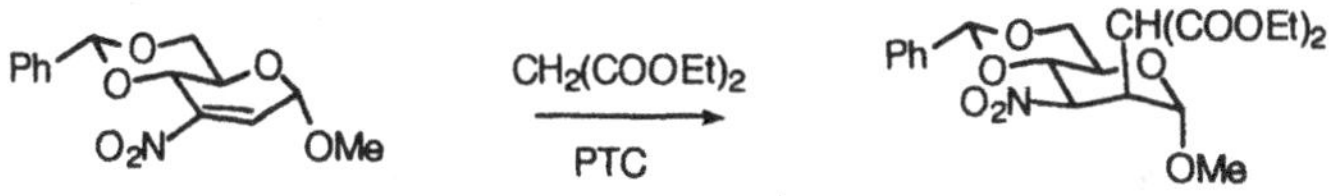

3-Substituted glutamic acid derivatives have been synthesized by a Michael addition. The solid-liquid PTC process is *anti*-selective [Alvarez-Ibarra, 1995].

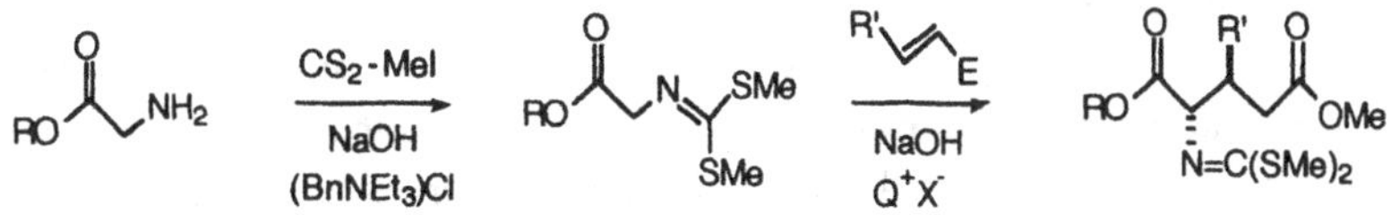

o-Allylarenes containing moderately electron-withdrawing substitutents form adducts with phenylacetonitriles under PTC conditions (e.g., NaOH/PhCl, $Bu_4N^+Br^-$) via the isomerized propenylarenes [Lasek, 1993].

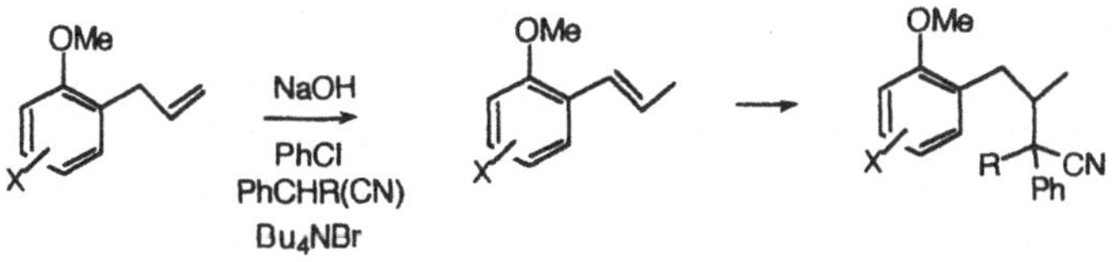

The reaction of an arylmercury chloride with methyl vinyl ketone which is Pd(II)-catalyzed [Cacchi, 1985] is a formal Michael addition. This and other Heck reactions benefit from phase transfer catalysis [Jeffery, 1984; Hoffmann, 1989], particularly in dealing with thermally labile substrates. The phase transfer catalyst facilitates the regeneration of Pd(0) species in a Heck reaction [Jeffery, 1994].

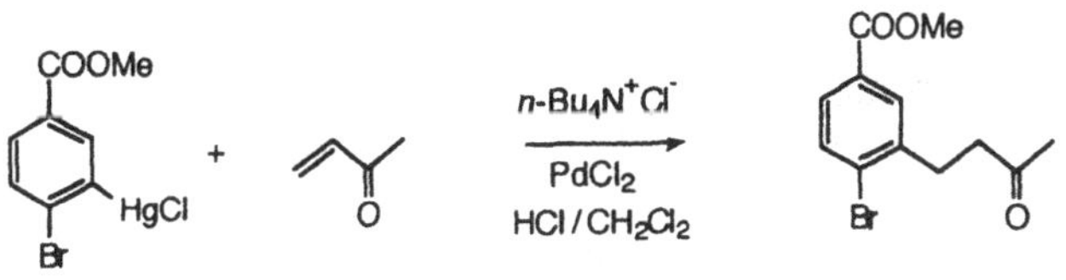

An interesting Heck-type reaction has been achieved using an arenesulfonyl chloride instead of an aryl halide under PTC conditions [Miura, 1989].

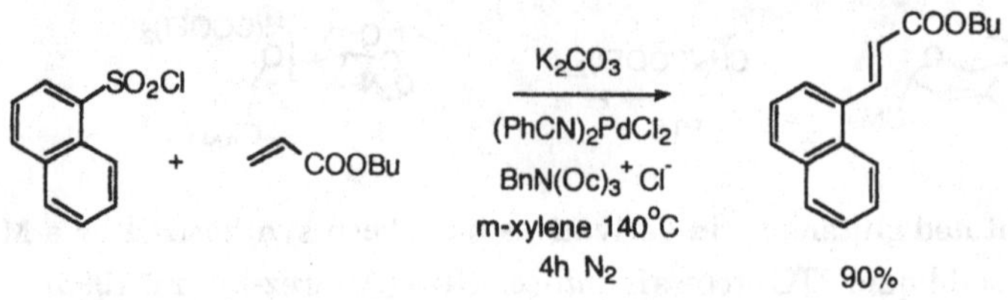
SO2Cl
+
COOBu
K2CO3
(PhCN)2PdCl2
BnN(Oc)3+ Cl-
m-xylene 140°C
4h N2
COOBu
90%

1.5. OXIDATIONS

The assistance from PTC to solving problems in oxidation of organic molecules is even more profound than in other areas. An often encountered difficulty concerning the availability of a compatible solvent is obviated. Strong oxidants which are based on anions with a high valent transition metal core can be transferred from water into inert organic solvents to perform oxidation of substrates. For permanganate anion, all sorts of phase transfer agents are suitable to achieve the objective.

1.5.1. Oxidation of Alkenes, Alkynes, and Arenes

Permanganate oxidation of alkenes by the phase transfer technique is extremely versatile because different sets of products are obtainable by varying conditions. By regulating the pH of the aqueous oxidant which is added to the organic solution containing the catalyst and the alkene, one type of product predominates: *cis*-1,2-diol (basic), α-ketol (neutral), aldehydes/carboxylic acids (acidic), as shown in the case of dicyclopentadiene [Ogino, 1980]. Complex mixture of products usually result if the pH is not carefully controlled. In the solid-liquid PTC system (solid $KMnO_4$/catalyst $N\{CH_2CH_2OCH_2CH_2OCH_3\}_3$) the workup procedure is an important factor: *cis*-1,2-diol (dry workup or basic water), aldehydes (acidic water).

It should be noted that chemoselective cleavage of double bonds are possible [Rathore, 1986].

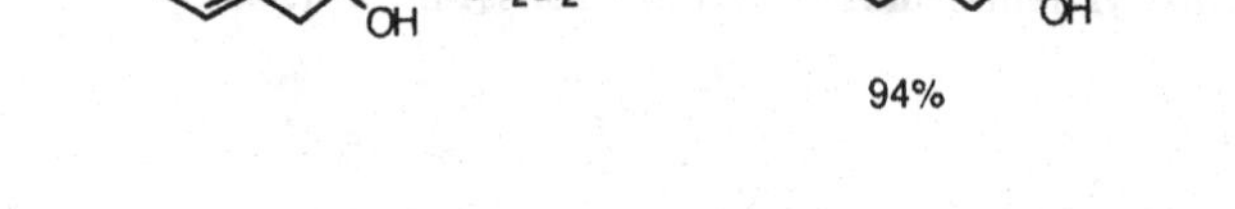
Ph
Ph
Q+ MnO4-
CH2Cl2
Ph
Ph
O
+
O
83%
71%

Ph
OH
Q+ MnO4-
CH2Cl2
O
Ph
OH
94%

Crown ethers also render $KMnO_4$ soluble in benzene, therefore many nonpolar compounds are readily oxidized [Sam, 1972]: *(E)*-stilbene -> benzoic acid (100%), α-pinene -> pinonic acid (90%), *p*-xylene -> toluic acid (100%).

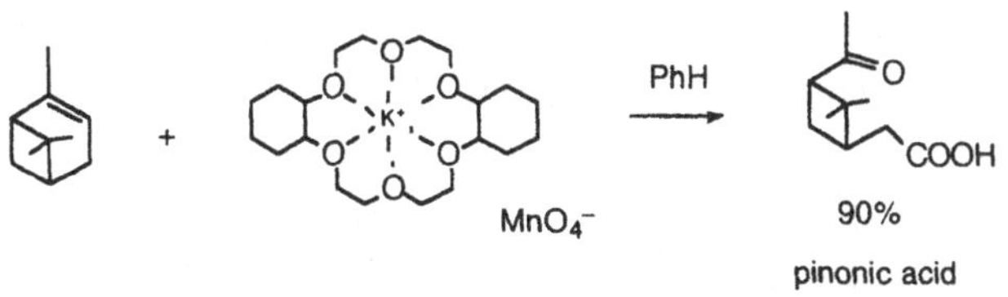

PhH
MnO_4^-
COOH
90%
pinonic acid

Alkynes afford carboxylic acids [D.G. Lee, 1978] and α-diketones [Gannon, 1987], respectively, when subjected to $KMnO_4$ oxidation under liquid-liquid and dry solid-liquidconditions.

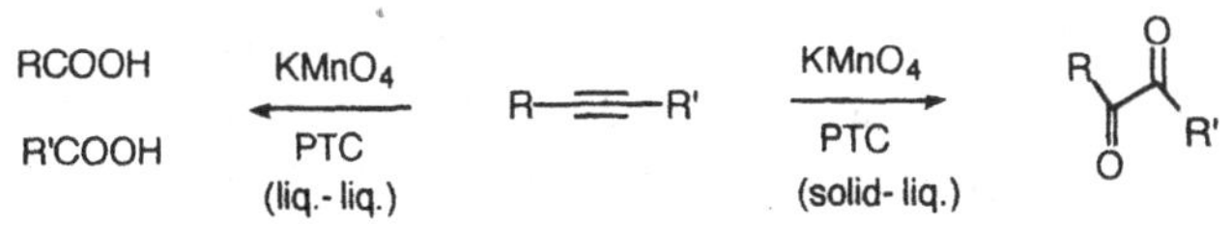

RCOOH
R'COOH
$KMnO_4$
PTC
(liq.- liq.)
R—≡—R'
$KMnO_4$
PTC
(solid- liq.)

Industrial epoxidation of alkenes with hydrogen peroxide, cocatalyzed by a phase transfer agent and metalate ion such as tungstate and molybdate [Payne, 1959] is improved by the addition of phosphoric acid [Venturello, 1981]. For long-chain 1-alkenes the optimal pH is about 1.6. Epoxidation of allyl methacrylate gives glycidyl methacrylate with 94% selectivity at 74% conversion [Fort, 1992].

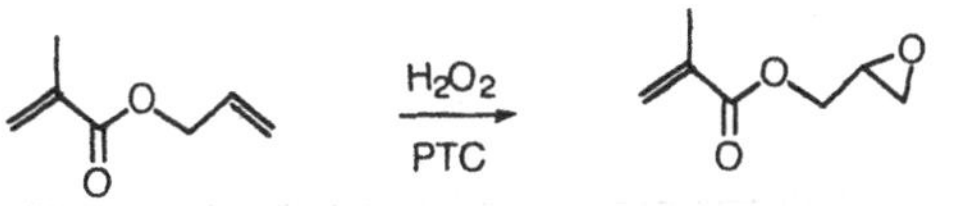

H_2O_2
PTC

Terminal alkynes afford α-ketoaldehydes with excellent selectivity on oxidation with acidic hydrogen peroxide in the presence of Q^+Cl^- and sodium molybdate [Ballistreri, 1988].

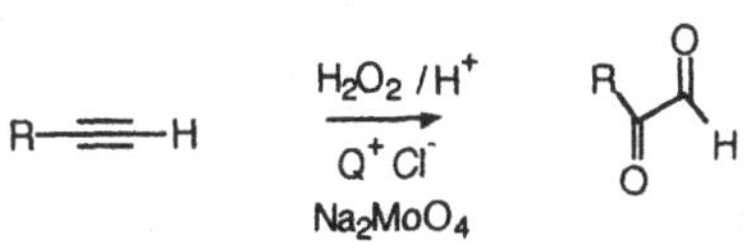

R—≡—H
H_2O_2/H^+
Q^+Cl^-
Na_2MoO_4

Osmium tetroxide is probably the best reagent for the conversion of alkenes to *cis*-1,2-diols owing to its high selectivity and mildness. Its serious shortcomings in terms of cost and toxicity have been addressed. In situ recycle of Os(VIII) species by many innocuous and inexpensive cooxidants is quite successful. However, when periodic acid or sodium periodate is used the reaction for nonpolar substrates is very slow. Phase transfer technique is definitely a remedy for this situation; furthermore, the two phase system allows for electrochemical regeneration of HIO_4 so that the process becomes very economical [M.A. Johnson, 1972].

(cat.) K_2OsO_4 / $NaIO_4$ PTC → [OsO_2] → CHO CHO

The stronger oxidant RuO_4 is also too expensive to be used stoichiometrically. The same tactic can be applied to oxidation of alkenes to carboxylic acids, and in this case aqueous sodium hypochlorite (+ n-Bu_4NBr) is an excellent agent to recondition the ruthenium [Foglia, 1977]. The PTC/$RuCl_3$ cocatalyst system is also effective for promoting oxidative cleavage of alkenes with hydrogen peroxide [Barak, 1987].

The base-catalyzed epoxidation of electron-deficient alkenes with hydrogen peroxide or t-butyl hydroperoxide can take advantage of chiral phase transfer agents (e.g., N-benzylquinine chloride) to produce optically active epoxides [Helder, 1976].

Ph, Ph → ROOH / base, Q^{*+} Cl^- → Ph, O, Ph

<u>Chalcone Oxide</u>. [Helder, 1976]

Chalcone (20.4 g) in toluene (125 mL) is treated with NaOH (7.5 g), 30% hydrogen peroxide (90 mL) and cinchonine benzochloride (0.75g) at room temperature. After vigorous stirring for 24 h, a levorotatory product (21.8 g, 99%) is obtained.

The phase transfer technique obviates the problem of nuclear chlorination during epoxidation of dimethyl o-propargyloxybenzylidenemalonate with sodium hypochlorite [Bernaus, 1991]. The substrate and the product are less exposed to the reagent under the PTC conditions.

The Wacker oxidation is an industrially important process which is limited to lower alkenes due to solubility and hence rate problems. Using β-cyclodextrin as an inverse PTC to deliver sparingly soluble alkenes to the aqueous phase enhances the rates dramatically [Zahalka, 1986]. This method furnishes very clean products.

O_2
$PdCl_2/CuCl$
β-CD/H_2O

As indicated above, permanganate functions as a benzylic oxidant. A similar capability is shown by biphasic ruthenium(III)-H_2O_2 and RuO_4-NaOCl systems [Sasson, 1986] in combination with a phase transfer agent. For rapid oxidation of fluorene, xanthene, acridane, and analogues it only requires air (or oxygen) supply to a benzene solution containing a PTC and a base (aq. NaOH or solid KOH/18-crown-6) [Alneri, 1977].

O_2
PhH / MOH
PTC

X = O, NR, void

Some polycyclic arenes undergo nuclear oxidation. For example, an excellent yield of the 9,10-oxide has been obtained from phenanthrene using PTC hypochlorite reaction [Fonouni, 1983]. On subjecting to the Q^+Cl^--tungstate-H_3PO_4-H_2O_2 combination, such compounds suffer ring cleavage to furnish the diacids [Saito, 1986].

n-Bu_4N^+ $HSO4^-$
aq. NaOCl / CH_2Cl_2
ph 8.5, 40°C

96%

1.5.2. Oxidation of Heterofunctional Compounds

The oxidation of alcohols to carbonyl products under PTC conditions can be achieved by a number of reagents. Controlled oxidation of primary alcohols to the aldehyde stage is often desirable; for benzylic alcohols the system consisting of solid $KMnO_4$ and $N\{CH_2CH_2OCH_2CH_2OCH_3\}_3$ [Okimoto, 1977] and the liquid-liquid reaction involving Q^+Cl^- and aqueous hypochlorite [G.A. Lee, 1976] or $Q^+HSO_4^-$ and aqueous dichromic acid [Landini, 1979] give satisfactory results. Secondary alcohols are of course oxidized to ketones.

An unusual oxidation of benzyl alcohol to benzaldehyde at room temperature involves stirring with iodobenzene in the presence of sodium bicarbonate and catalytic amounts of palladium(II) acetate and tetrabutylammonium chloride [Choudary, 1985].

Chromic oxide oxidation of secondary alcohol in the presence of R_4NX has been reported [Gelbard, 1980]. *N*-Bromoquinuclidine in combination with pyridine can oxidize secondary alcohols at room temperature. The two-phase system is better than dichloromethane alone [Blair, 1992]. Apparently quinuclidine functions as a phase transfer catalyst.

$$RR'CH\text{-}OH \xrightarrow[\text{pyridine, } H_2O/CH_2Cl_2]{N\text{-bromoquinuclidinium } Br^-} RR'C{=}O$$

For the oxidation of aromatic aldehydes with aqueous $KMnO_4$ a small amount of [2-(dodecylmethyl-*t*-butylsiloxy)ethyl]trimethylammonium nitrate is recommended as catalyst [Jaeger, 1988]. This salt is stable in water at pH 3-11 and it can be decomposed by fluoride ion in aqueous or non aqueous media. Consequently emulsions which are often formed and interfere with workup when conventional tetra-alkylammonium salts are used are avoided.

Excellent yield of benzoic acid is obtained by treatment of benzoin with a combination of potassium superoxide and 18-crown-6 in benzene at room temperature [San Filippo, 1976].

Primary amines of the RCH_2NH_2 type undergo remarkably facile transformation to the nitriles by a PTC hypochlorite, whereas secondary amines give ketones [G.A. Lee, 1976].

$$RCH_2NH_2 + NaOCl \xrightarrow{Q^+Cl^-} RCN$$

$$RR'CH\text{-}NH_2 + NaOCl \xrightarrow{Q^+Cl^-} RR'C{=}O$$

Deoximation with a mixture of chlorotrimethylsilane, $NaNO_x$ (x=2,3) and a phase transfer agent [J.G. Lee, 1990] probably involves transphasic delivery of NO_x^- to to react with the chlorosilane, generating the silyl nitrite or nitrate and thence the purported reagent NO_xCl.

$$R(R')C{=}NOH + ClSiMe_3 + NaNO_x \xrightarrow{Q^+Cl^-} R(R')C{=}O$$

> 90%

Improvement of the nitrile hydration to primary amides with alkaline hydrogen peroxide by phase transfer catalysts is definite [Cacchi, 1980].

$$RCN \xrightarrow[PTC]{H_2O_2/NaOH} RCONH_2$$

Asymmetric oxygenation of ketones in the presence of chiral phase transfer catalysts derived from cinchonine has good results [Masui, 1988].

O_2/NaOH, PhMe, $Q^{*+}Br^-$

73% (55% ee)

Br^- CF_3

O_2 attacks from frontside of enolate

<u>Chiral α-Hydroxyketones</u>. [[Masui, 1988].

Addition of the ketone (1 mmol) in toluene (10 mL) containing triethyl phosphite (0.2 mmol) and the cinchonium salt (5 mol%) to 50% NaOH (5 mL) is followed by bublling of an oxygen stream at room temperature. At complete consumption of the ketone the mixture is quenched with water (10 mL) and extracted with benzene. The organic solutions are washed with 10% HCl, H_2O, and brine. Drying, evaporation and chromatography over silica gel furnishes the product.

1.6. REDUCTIONS

Lithium aluminum hydride cannot be exposed to water therefore PTC must involve solid-liquid technique. 15-Crown-5 seems to be a suitable sequester for the lithium ion thereby solubilizing the reducing agent in aprotic solvents. Indeed various functional groups which are affected by lithium aluminum hydride can be reduced in this manner [Gevorgyan, 1985].

Borohydride reduction under phase transfer conditions can discriminate aldehydes from ketones [Rao, 1986]. For rate acceleration in ketone reduction one may use of a quaternary ammonium salt bearing a β-hydroxyl group [Colonna, 1978]. The method is applicable to reduction of acid chlorides to alcohols [Costello, 1985].

CHO + (acetophenone) —$NaBH_4/H_2O$, PhH, $n\text{-}Bu_4N^+Br^-$→ (alcohol) 95% + (alcohol) 7%

Acid chloride attaching to insoluble polymers can be converted to the aldehydes in 70-96% yields by treatment with iron pentacarbonyl in the presence of Bu_4NBr and NaOH, followed by neutralization with acetic acid [Ungurenasu, 1989].

Reductive dehalogenation with sodium borohydride is usually very sluggish. However, octadecane is obtained from the alkyl bromide in 80% yield when the reaction is conducted in hot toluene in the presence of a polyethylene-bound crown ether and tri-*n*-butyltin chloride as catalysts [Bergbreiter, 1987].

Organoazides do not undergo smooth reduction with borohydride, but there is great improvement if the PTC technique is employed [Rolla, 1982]. This reaction can be combined with azide displacement to enable a one-pot preparation of amines from alkyl halides or mesylates.

n-Octylamine. [Rolla, 1982]

The stirred mixture of 1-bromoctane (19.3 g, 0.1 mol), sodium azide (16.2 g, 0.25 mol), hexadecyltributylphosphonium bromide (5.1 g, 0.01 mol) in water (50 mL) is heated at 80°C for 8 h. Thereafter the aqueous phase is removed, replaced with toluene (15.5 mL), and treated with a solution of sodium borohydride (11.7 g, 0.3 mol) in water (30 mL) during 0.5 h. Heating at 80°C is resumed and maintained for 16 h to complete the reduction. The amine is extracted from the cooled organic solution (which retains the catalyst) with 10% hydrochloric acid and then recovered (11.4 g, 88%).

It is possible that the reduction of aromatic diazonium salts in the presence of 18-crown-6 [Korzenowski, 1977] is facilitated by complexation of the cation to loosen the C-N bond.

MeO–C₆H₄–N_2^+ BF_4^- —Cu_2O, H_3PO_2, 18-c-6, $CHCl_3$→ MeO–C₆H₅ 88%

Nitroarenes are reduced to anilines by the combination of a PTC and $Fe_3(CO)_{12}$-NaOH [des Abbayes, 1977a], $CpV(CO)_4$-NaOH [Falicki, 1988] or the bimetallic catalyst (1,5-hexadiene$RhCl)_2$/ $Co_2(CO)_8$-CO-NaOH [Hashem, 1984]. The actual reducing agents in these reactions may be the carbonylmetal hydride anions. The vanadium hydride is also useful for dehalogenation.

$$ArNO_2 \xrightarrow[H_2O\ /\ org.\quad Q^+Cl^-]{L_mM(CO)_n\quad NaOH} ArNH_2$$

The treatment of allylic nitro compounds with carbon disulfide in the presence of a solid base and a quaternary ammonium salt results in the aldoximes which on further reaction under liquid-liquid phase transfer conditions give conjugated nitriles [Albanese, 1990]. The overall process comprises a reduction followed by dehydration.

K_2CO_3, Bu_4NBr, CS_2, $CH_2Cl_2(H_2O)$; 15% NaOH — [CH=NOH] → CN, 63%

1-Cyano-3,4-dihydronaphthalene. [Albanese, 1990]

A mixture of 1-nitromethyl-3,4-dihydronaphthalene (1.89 g, 10 mmol), tetra-*n*-butylammonium bromide (0.32 g, 1 mmol), K_2CO_3 (0.69 g, 5 mmol) in dichloromethane (100 mL, containing 36μL H_2O) is stirred at room temperature for 15 min, treated with carbon disulfide (0.9 mL, 15 mmol). After 8 h, more carbon disulfide (4 mL, 67 mmol) is added followed by 15% NaOH (10 mL, 44 mmol) in a dropwise fashion over 1.5 h. Stirring continues for 2 h before workup which affords the nitrile (0.98 g, 63%); b.p. 88-89°C/0.3 torr.

Aliquat 336 effectively transfers $RhCl_3$ from aqueous solution in the form of $[(C_8H_{17})_3NMe]^+ [RhCl_4]^-$ to organic solvents (preferably 1,2-dichloroethane) to act as hydrogenation catalyst [Blum, 1983]. Selective reduction of unsaturated carbonyl compounds has been demonstrated. Note that tertiary phosphines are poisons.

$$PhCH{=}CHCOPh + H_2 \xrightarrow[H_2O\quad ClCH_2CH_2Cl]{Q^+[RhCl_4]^-} PhCH_2CH_2COPh$$

Exclusive reduction of the double bond of conjugated ketones with sodium dithionite ($Na_2S_2O_4$-$NaHCO_3$/Aliquat) has also been observed [Louis-Andre, 1985].

An excellent hydrogenation system which is active for arenes and some heteroaromatic compounds operates in two-phase containing the chloro(1,5-hexadiene)rhodium dimer and a Q^+X^- [Januszkiewicz, 1983]. Functional groups such as ketone, ester, and amides are not affected.

Monographs and Reviews

Dehmlow, E.V.; Dehmlow, S. (1993) *Phase Transfer Catalysis*, 3rd. Ed., VCH: Weinheim.

Dockx, J. (1973) *Synthesis* 441.

Jones, R.A. (1976) *Aldrichimica Acta* **9**/3: 35.

Keller, W.E. (1979) *Compendium of Phase-transfer Reactions and Related Synthetic Methods*, Fluka.

Makosza, M. (1980) *Survey Prog. Chem.* **9:** 1.

Starks, C.M.; Liotta, C.L.; Halpern, M. (1994) *Phase Transfer Catalysis*, Chapman & Hall: New York.

Weber, W.P.; Gokel, G.W. (1977) *Phase Transfer Catalysis in Organic Synthesis*, Springer-Verlag: New York.

References

Akimoto, I.; Suzuki, A. (1981) *Synth. Commun.* **11:** 475.

Albanese, D.; Landini, D.; Penso, M.; Pozzi, G. (1990) *Synth. Commun.* **20:** 965.

Alneri, E.; Bottaccio, G.; Carletti, V. (1977) *Tetrahedron Lett.* 2117.

Alvarez-Ibarra, C.; Csakÿ, A.G.; Maroto, M.; Quiroga, M.L. (1995) *J. Org. Chem.* **60:** 6700.

Ando, A.; Miura, T.; Tatematsu, T.; Shioiri, T. (1993) *Tetrahedron Lett.* **34:** 1507.

Angle, S.R.; Fevig, J.M.; Knight, S.D.; Marquis, R.W.; Overman, L.E. (1993) *J..Am. Chem. Soc.* **115:** 3966.

Aranda, A.; Diaz, A.; Diez-Barra, E.; de la Hoz, A.; Moreno, A.; Sanchez-Verdu, P. (1992) *J. Chem. Soc. Perkin Trans. 1* 2427.

Badri, M.; Brunet, J.-J.; Perron, R. (1992) *Tetrahedron Lett.* **33:** 4435.

Balcerzak, P.; Jonczyk, A. (1991) *Synthesis* 857.

Ballesteros, P.; Claramunt, R.M.; Elguero, J. (1987) *Tetrahedron* **43:** 2557.

Ballistreri, F.P.; Failla, S.; Tomaselli,.G.A. (1988) *J. Org. Chem.* **53:** 830.

Barak, G.; Sasson, Y. (1987) *Chem. Commun.* 1266.

Barry, J.; Bram, G.; Petit, A. (1988) *Tetrahedron Lett.* **29:** 4567.

BASF (1912) *Ger. Pat.* 268621.

Bashall, A.P.; Collins, J.F. (1975) *Tetrahedron Lett.* 3489.

Bergbreiter, D.E.; Blanton, J.R. (1987) *J. Org. Chem.* **52:** 472.

Bernaus, C.; Font, J.; de March, P. (1991) *Tetrahedron* **47:** 7713.

Bhalerao, U.T.; Mathur, S.N.; Rao, S.N. (1992) *Synth. Commun.* **22:** 1645.

Bhati, N.; Kabra, A.; Narang, C.K.; Mathur, N.K. (1991) *J. Org. Chem.* **56:** 4967.

Bhattacharya, A.; Dolling, U.-H.; Grabowski, E.J.J.; Karady, S.; Ryan, K.M.; Weinstock, L.M. (1986) *Angew. Chem. Int. Ed. Engl.* **25:** 476.

Bianchi, T.A.; Cate, L.A. (1977) *J. Org. Chem.* **42:** 2031.

Blair, L.K.; Hobbs, S.; Bagnoli, N.; Husband, L.; Badika, N. (1992) *J. Org. Chem.* **57:** 1600.

Blum, J.; Amer, I.; Zoran, A.; Sasson, Y. (1983) *Tetrahedron Lett.* **24:** 4139.

Bram, G.; Sansoulet, J.; Galons, H.; Bensaid, Y.; Combet-Farnoux, C.; Miocque, M. (1985) *Tetrahedron Lett.* **26:** 4601.

Bram, G.; Loupy, A.; Pigeon, P. (1988) *Synth. Commun.* **18:** 1661.

Bram, G.; Loupy, A.; Majdoub, M. (1990) *Synth. Commun.* **20:** 125.

Brändström, A.; Gustavii, K. (1969) *Acta Chem. Scand.* **23:** 1215.

Brändström, A.; Junggren, U. (1972) *Tetrahedron Lett.* 473.

Brändström, A.; Lamm, B.; Palmertz, I. (1974) *Acta Chem. Scand. Ser. B* **28:** 699.

Cacchi, C.; Misiti, D.; LaTorre, F. (1980) *Synthesis* 243.

Cacchi, S.; Palmieri, G. (1985) *J. Organomet. Chem.* **282:** C3.

Cainelli, G.; Manescalchi, F. (1975) *Synthesis* 723.

Cainelli, G.; Manescalchi, F. (1976) *Synthesis* 472.

Canicio, J.A.; Ginebreda, A.; Canela, R. (1985) *Anal. Quim. Ser. C* **81:** 181.

Cassar, L.; Foa, M.; Gardano, A. (1976) *J. Organomet. Chem.* **121:** C55.

Cativela, C.; Serrano, J.L.; Zurbano, M.M. (1995) *J. Org. Chem.* **60:** 3074.

Chiba, T.; Okimoto, M. (1991) *J. Org. Chem.* **56:** 6163.

Childs, M.E.; Weber, W.P. (1976) *J. Org. Chem.* **41:** 3486.

Choudary, B.M.; Reddy, N.P.; Kantam, M.L.; Jamil, Z. (1985) *Tetrahedron Lett.* **26:** 6257.

Cinquini, M.; Montanari, F.; Tundo, P. (1974) *Chem. Commun.* 878.

Cinquini, M.; Montanari, F.; Tundo, P. (1975) *Chem. Commun.* 393.

Coates, H.; Barker, R.L.; Guest, R.; Kent, A. (1973) *Brit. Pat.* 1336883.

Colonna, S.; Fornasier, R. (1978) *J. Chem. Soc. Perkin Trans. 1* 371.

Corey, E.J.; Nicolaou, K.C.; Shibasaki, M.; Machida, Y.; Shiner, C.S. (1975) *Tetrahedron Lett.* 3183.

Costello, A.T.; Milner, D.J. (1985) *Brit. Pat. Appl.* 2155464.

Dehm, D.; Padwa, A. (1975) *J. Org. Chem.* **40:** 3139.

Dehmlow, E.V. (1972) *Tetrahedron* **28:** 175.
Dehmlow, E.V. (1976) *Tetrahedron Lett.* 91.
Dehmlow, E.V.; Lissel, M. (1981a) *Tetrahedron* **37:** 1653.
Dehmlow, E.V.; Shamout, A.R. (1981b) *J. Chem. Res. (Mini.)* 1178.
Dehmlow, E.V.; Prashad, M. (1982a) *J. Chem. Res. (Synop.)* 354.
Dehmlow, E.V.; Shamout, A.R. (1982b) *Liebigs Ann. Chem.* 1750.
Dehmlow, E.V.; Raths, H.-C.; Soufi, J. (1988) *J. Chem. Res. (Synop.)* 334.
Dehmlow, E.V.; Knufinke, V. (1989) *J. Chem. Res. (Synop.)* 224.
Dehmlow, E.V.; Torossian, G.O. (1990) *Z. Naturforsch. B: Chem. Sci.* **45:** 1091.
Dermeik, S.; Sasson, Y. (1985) *J. Org. Chem.* **50:** 879.
des Abbayes, H.; Alper, H. (1977a) *J.Am. Chem. Soc.* **99:** 98.
des Abbayes, H.; Boudeville, M.-A. (1977b) *J. Org. Chem.* **42:** 4104.
des Abbayes, H.; Buloup, A.; Tanguy, G. (1983) *Organometallics* **2:** 1730.
des Abbayes, H. (1985) *Israel J. Chem.* **26:** 249.
des Abbayes, H.; Clement, J.-C.; Laurent, P.; Tanguy, G.; Thilmont, N. (1988) *Organometallics* **7:** 2293.
Di Cesare, P.; Duchaussoy, P.; Gross, B. (1980) *Synthesis* 953.
Dietl, H.; Brannock, K.C. (1973) *Tetrahedron Lett.* 1273.
Dietrich, B.; Lehn, J.-M. (1973) *Tetrahedron Lett.* 1225.
D'Incan, E. (1977) *Tetrahedron* **33:** 951.
Doecke, C.W.; Staszak, M.A.; Luke, W.D. (1991) *Synthesis* 985.
Dou, H.J.-M.; Metzger, J. (1976) *Bull. Soc. Chim. Fr.* 1861.
Durst, H.D.; Liebeskind, L. (1974) *J. Org. Chem.* **39:** 3271.
Eddine, J.J.; Cherquaoui, M. (1995) *Tetrahedron: Asymmetry* **6:** 1225.
Falicki, S.; Alper, H. (1988) *Organometallics* **7:** 2548.
Fiaud, J.-C. (1975) *Tetrahedron Lett.* 3495.
Foglia, T.A.; Barr, P.A.; Malloy, A.J. (1977) *J. Am. Oil Chem. Soc.* **54:** 858A.
Fonouni, H.E.; Krishnan, S.; Kuhn, D.G.; Hamilton, G.A. (1983) *J. Am. Chem. Soc.* **105:** 7672.
Fort, Y.; Olszewski-Ortar, A.; Caubere, P. (1992) *Tetrahedron* **48:** 5099.
Freedman, H.H.; Dubois, R.A. (1975) *Tetrahedron Lett.* 3251.
Galamb, V.; Gopal, M.; Alper, H. (1983) *Organometallics* **2:** 801.
Gannon, S.M.; Krause, J.G. (1987) *Synthesis* 915.
Gelbard, G.; Brunelet, T.; Jouitteau, C. (1980) *Tetrahedron Lett.* **21:** 4653.
Gevorgyan, V.; Lukevics, E. (1985) *Chem. Commun.* 1234.
Gobbi, A.; Landini, D.; Maia, A.; Delogu, G.; Podda, G. (1994) *J. Org. Chem.* **59:** 5059.
Gokel, G.W.; DiBiase, S.A.; Lipisko, B.A (1976) *Tetrahedron Lett.* 3495.
Goldberg, Y.; Alper, H. (1993) *J. Org. Chem.* **58:** 3072.

Goldberg, Y.Sh.; Iovel, I.G.; Shymanska, M.V. (1986) *Chem. Commun.* 286.
Graefe, J.; Froehlick, I.; Muehlstaedt, M. (1974) *Z. Chem.* **14:** 434.
Grushin, V.V.; Alper, H. (1991) *Tetrahedron Lett.* **32:** 3349.
Hallgren, J.E.; Lucas, G.M. (1981) *J. Organomet. Chem.* **212:** 135.
Hampl, F. (1990) *Czech Pat.* CS 265360.
Hashem, K.E.; Petrignani, J.-F.; Alper, H. (1984) *J. Mol. Catal.* **26:** 285.
Hassner, A.; Stern, M. (1986) *Angew. Chem. Int. Ed. Engl.* **25:** 478.
Heiszman, J.; Bitter, I.; Harsanyi, K.; Töke, L. (1987) *Synthesis* 738.
Helder, R.; Hummelen, J.C.; Laane, R.W.P.M.; Wiering, J.S.; Wynberg, H. (1976) *Tetrahedron Lett.* 1831.
Hennis, H.E.; Easterly, Jr, J.P.; Collins, L.R.; Thompson, L.R. (1967) *Ind. Eng. Chem. Prod. Res. Dev.* **6:** 193.
Herriott, A.W.; Picker, D. (1972) *Tetrahedron Lett.* 4521.
Herriott, A.W.; Picker, D. (1975) *Synthesis* 447.
Hiyama, T.; Mishima, T.; Sawada, H.; Nozaki, H. (1975) *J. Am. Chem. Soc.* **97:** 1626.
Hoffmann, H.M.R.; Schmidt, B.; Wolff, S. (1989) *Tetrahedron* **45:** 6113.
Horvath, I.T.; Rabai, J. (1994) *Science* **266:** 72.
Houlihan, F.; Bouchard, F.; Frechet, J.M.J.; Willson, C.G. (1985) *Can. J. Chem.* **63:** 153.
Hu, X.; Kellogg, R.M. (1995) *Synthesis* 533.
Hu, Y.; Pa, W.; Cui, W.; Wang, J. (1992) *Synth. Commun.* **22:** 2763.
Hünig, S.; Stemmler, I. (1974) *Tetrahedron Lett.* 3151.
Hunter, D.H.; Perry, R.A. (1977) *Synthesis* 37.
Ichikawa, J.; Kobayashi, H.; Sonoda, T. (1988) *J. Org. Synth. Chem.* **46:** 943.
Illi, V.O. (1979) *Tetrahedron Lett.* 2431.
Isagawa, K.; Kimura, Y.; Kwon, S. (1974) *J. Org. Chem.* **39:** 3171.
Itokawa, H.; Suzuki, J.; Hitotsuyanagi, Y.; Kondo, K.; Takeya, K. (1993) *Chem. Lett.* 695.
Iwamoto, H.; Sonoda, T.; Kobayashi, H. (1983a) *Tetrahedron Lett.* **24:** 4703.
Iwamoto, H.; Yoshimura, M.; Sonoda, T.; Kobayashi, H. (1983b) *Bull. Chem. Soc. Jpn.* **56:** 796.
Jaeger, D.A.; Ward, M.D.; Dutta, A.K. (1988) *J. Org. Chem.* **53:** 1577.
Januszkiewicz, K.R.; Alper, H. (1983) *Organometallics* **2:** 1055.
Jarrouse, J. (1951) *C.R. Hebd. Seances Acad. Sci. Ser. C* **232:** 1424.
Jeffery, T. (1984) *Chem. Commun.* 1287.
Jeffery, T.; Galland, J.-C. (1994) *Tetrahedron Lett.* **35:** 4103.
Johnson, C.R.; Lavergne, O.M. (1993) *J. Org. Chem.* **58:** 1922.
Johnson, M.A.; Washecheck, P.H.; Yang, K.; Starks, C.M. (1972) *U.S. Patent*

3650918.
Johnson, R.A.; Nidy, E.G. (1975) *J. Org. Chem.* **40:** 1680.
Jonczyk, A.; Fedorynski, M.; Makosza, M. (1972) *Tetrahedron Lett.* 2395.
Jonczyk, A.; Ludwikow, M.; Makosza, M. (1973) *Rocz. Chem.* **47:** 89.
Jonczyk, A.; Makosza, M. (1976) *Synthesis* 387.
Jonczyk, A.; Kulinski, T. (1993) *Synth. Commun.* **23:** 1801.
Jonczyk, A.; Kmiotek-Skarzynska, I.; Zdrojewski, T. (1994) *J. Chem. Soc. Perkin Trans. 1* 1605.
Jones, R.A.; Nokked, S.; Singh, S. (1977) *Synthesis* 195.
Julia, S.; Michelot, D.; Linstrumelle, G. (1974) *C.R. Hebd. Seances Acad. Sci. Ser. C* **278:** 1523.
Jursic, B. (1988) *Synthesis* 868.
Jursic, B. (1989) *J. Chem. Res. Synop.* 104.
Kimura, C.; Murai, K.; Ishikawa, Y.; Kashiwara, K. (1976) *Sekiyu Gakkaishi* **19:** 49. *Chem. Abstr.* (1977) **87:** 133781.
Klan, P.; Benovsky, P. (1992) *Monatsh. Chem.* **123:** 469.
Knöchel, A.; Rudolph, G.; Thiem, J. (1974) *Tetrahedron Lett.* 551.
Knöchel, A.; Oehler, J.; Rudolph, G. (1975) *Tetrahedron Lett.* 3167.
Koenig, K.E.; Weber, W.P. (1974) *Tetrahedron Lett.* 2275.
Kondo, S.; Takeda, Y.; Tsuda, K. (1988) *Synthesis* 403.
Kondo, S.; Takeda, Y.; Tsuda, K. (1989) *Synthesis* 862.
Kondo, S.; Shibata, A.; Kunisada, H.; Yuki, Y. (1992) *Bull. Chem. Soc. Jpn.* **65:** 2555.
Korzenowski, S.H.; Blum, L. (1977) *J. Org. Chem.* **42:** 1469.
Kraus, W.; Klein, G.; Sadlo, H.; Rothenwohrer, W. (1972) *Synthesis* 485.
Krein, E.B.; Aizenshtat, Z. (1993) *J. Org. Chem.* **58:** 6103.
Kwon, S.; Nishimura, Y.; Ikeda, M.; Tamura, Y. (1976) *Synthesis* 249.
Landini, D.; Montanari, F.; Rolla, F. (1974a) *Synthesis* 428.
Landini, D.; Rolla, F. (1974b) *Synthesis* 565.
Landini, D.; Quici, S.; Rolla, F. (1975) *Synthesis* 397.
Landini, D.; Rolla, F. (1976) *Synthesis* 389.
Landini, D.; Montanari, F.; Rolla, F. (1979) *Synthesis* 134.
Landini, D.; Maia, A. (1991a) *Tetrahedron* **47:** 1285.
Landini, D.; Penso, M. (1991b) *J. Org. Chem.* **56:** 420.
Landini, D.; Albanese, D.; Mottadelli, S.; Penso, M. (1992) *J. Chem. Soc. Perkin Trans. 1* 2309.
Lantzsch, R.; Marhold, A.; Lehment, K.-F. (1977) *German Patent* 2545644.
Lasek, W.; Makosza, M. (1993) *Synthesis* 780.
Lee, D.G.; Chang, V.S. (1978) *Synthesis* 462.

Lee, G.A.; Freedman, H.H. (1976) *Tetrahedron Lett.* 1641.

Lee, J.G.; Kwak, K.H.; Hwang, J.P. (1990) *Tetrahedron Lett.* **31:** 6677.

Li, Z.-H.; Mori, A.; Kato, N.; Takeshita, H. (1991) *Bull. Chem. Soc. Jpn.* **64:** 2778.

Lin, C.L.; Pinnavaia, T.J. (1991) *Chem. Mater.* **3:** 213.

Liotta, C.L.; Cook, F.L.; Bowers, C.W. (1974a) *J. Org. Chem.* **39:** 3416.

Liotta, C.L.; Harris, H.P. (1974b) *J. Am. Chem. Soc.* **96:** 2250.

Lissel, M. (1982) *Liebigs Ann. Chem.* 1589.

Lissel, M. (1985) *Tetrahedron Lett.* **26:** 1843.

Liu, B.L.; Jin, Y.T.; Liu, Z.H.; Luo, C.; Kojima, M.; Maeda, M. (1985) *Int. J. Appl. Radiat. Isot.* **36:** 561.

Louis-Andre, O.; gelbard, G. (1985) *Tetrahedron Lett.* **26:** 831.

Loupy, A.; Pedoussaut, M.; Sansoulet, J. (1986) *J. Org. Chem.* **51:** 740.

McIntosh, J.M.; Khalil, H. (1977) *J. Org. Chem.* **42:** 2123.

McKillop, A.; Fiaud, J.-C.; Hug, R.P. (1974) *Tetrahedron* **30:** 1379.

Märkl, G.; Merz, A. (1973) *Synthesis* 295.

Makosza, M.; Serafin, B. (1965) *Rocz. Chem.* **39:** 1223; and later papers.

Makosza, M. (1966a) *Tetrahedron Lett.* 4621.

Makosza, M. (1966b) *Tetrahedron Lett.* 5489.

Makosza, M. (1967) *Bull. Acad. Pol. Sci. Ser. Sci. Chim.* **15:** 165.

Makosza, M. (1968) *Polish Patent* 55571.

Makosza, M. (1969a) *Rocz. Chem.* **43:** 79.

Makosza, M.; Wawrzyniewicz, M. (1969b) *Tetrahedron Lett.* 4659.

Makosza, M.; Fedorynski, M. (1973) *Synthesis* 305.

Makosza, M.; Kacprowicz, A. (1974a) *Rocz. Chem.* **48:** 2129.

Makosza, M.; Ludwikow, M. (1974b) *Angew. Chem. Int. Ed. Engl.* **13:** 655.

Makosza, M. (1975a) *Pure Appl. Chem.* **43:** 439.

Makosza, M.; Kacprowicz, A.; Fedorynski, M. (1975b) *Tetrahedron Lett.* 2119.

Makosza, M.; Lasek, W. (1991) *Tetrahedron* **47:** 2843.

Marcos, M.; Castro, J.L.; Castedo, L.; Riguera, R. (1986) *Tetrahedron* **42:** 649.

Mason, D.; Magdassi, S.; Sasson, Y. (1991) *J. Org. Chem.* **56:** 7229.

Masui, M.; Ando, A.; Shioiri, T. (1988) *Tetrahedron Lett.* **29:** 2835.

Masuyama, Y.; Ueno, Y.; Okawara, M. (1976) *Tetrahedron Lett.* 2967.

Merz, A. (1973a) *Angew. Chem. Int. Ed. Engl.* **12:** 846.

Merz, A.; Märkl, G. (1973b) *Angew. Chem. Int. Ed. Engl.* **12:** 845.

Merz, A. (1977) *Angew. Chem.* **89:** 484.

Mikolajczyk, M.; Grzejszczak, S.; Zatorski, A. (1975) *Tetrahedron Lett.* 3757.

Mikolajczyk, M.; Grzejszczak, S.; Midura, W.; Zatorski, A. (1976) *Synthesis* 396.

Miura, M.; Hashimoto, H.; Itoh, K.; Nomura, M. (1989) *Tetrahedron Lett.* **30:** 975.

Näf, F.; Decorzant, R. (1978) *Helv. Chim. Acta* **61:** 2524.

Nanba, H.; Takahashi, N.; Abe, K.; Saito, M. (1988) *Jpn. Kokai Tokkyo Koho* JP 63/196547, 63/196549.

Naso, F.; Rozini, L. (1974) *J. Chem. Soc. Perkin Trans. 1* 340.

Nerdel, F.; Buddrus, J. (1965) *Tetrahedron Lett.* 3585.

O'Donnell, M.J; Boniece, J.M.; Earp, S.E. (1978) *Tetrahedron Lett.* 2641.

O'Donnell, M.J; Cook, G.K.; Rusterholz, D.B. (1991) *Synthesis* 989.

Ogino, T. (1980) *Tetrahedron Lett.* **21:** 177.

Okawara, T.; Noguchi, Y.; Matsuda, T.; Furukawa, M. (1981) *Chem. Lett.* 185.

Okimoto, T.; Swern, D. (1977) *J. Am. Oil Chem. Soc.* **54:** 862A.

Oppolzer, W.; Moretti, R.; Zhou, C. (1994) *Helv. Chim. Acta* **77:** 2363.

Parker, K.A.; Kim, H.-J. (1992) *J. Org. Chem.* **57:** 752.

Patrick, T.B. (1974) *Tetrahedron Lett.* 1407.

Payne, G.B.; Williams, P.H. (1959) *J. Org. Chem.* **24:** 54.

Pedersen, C.J. (1967) *J. Am. Chem. Soc.* **89:** 7017.

Rao, C.; Someswara, D.; Deshmukh, A.A.; Patel, B.J. (1986) *Ind. J. Chem.* **25B:** 626.

Rathore, R.; Chandrasekaran, S. (1986) *J. Chem. Res. (S)* 458.

Rawal, V.H.; Singh, S.P.; Dufour, C.; Michoud, C. (1993) *J. Org. Chem.* **58:** 7718.

Reetz, M.T.; Hütte, S.; Goddard, R. (1993) *J. Am. Chem. Soc.* **115:** 9339.

Reeves, W.P.; Bahr, M.L. (1976) *Synthesis* 823.

Regen, S.L. (1982a) *Nouv. J. Chem.* **6:** 629.

Regen, S.L.; Kimura, Y. (1982b) *J. Am. Chem. Soc.* **104:** 2064.

Reinholz, E.; Becker, A.; Hagenbruch, B.; Schäfer, S.; Schmitt, A. (1990) *Synthesis* 1069.

Roeske, R.W.; Gessellchen, P.D. (1976) *Tetrahedron Lett.* 3369.

Rolla, F. (1982) *J. Org. Chem.* **47:** 4327.

Rosenberger, M.; Jackson, W.; Saucy, G. (1980) *Helv. Chim. Acta* **63:** 1665.

Saito, Y.; Araki, S.; Sugita, Y.; Kurata, N. (1986) *Eur. Pat. Appl.* EP 193368.

Sakai, T.; Horikawa, H.; Takeda, A. (1980) *J. Org. Chem.* **45:** 2039.

Sakakibara, T.; Sudoh, R. (1975) *J. Org. Chem.* **40:** 2823.

Sam, D.J.; Simmons, H.E. (1972) *J. Am. Chem. Soc.* **94:** 4024.

Samuelsson, B.; Lamm, B. (1971) *Acta Chem. Scand.* **25:** 1555.

San Filippo, J.; Chern, C.I. (1976) *J. Org. Chem.* **41:** 1079.

Sanchez, V.; Greiner, J. (1993) *Tetrahedron Lett.* **34:** 2931.

Saraie, T.; Ishiguro, T.; Kawashima, K.; Morita, K. (1973) *Tetrahedron Lett.* 2121.

Sasaki, T.; Eguchi, S.; Kiriyama, T.; Sakito, Y. (1973) *J. Org. Chem.* **38:** 1648.

Sasaki, T.; Eguchi, S.; Ogawa, T. (1974) *J. Org. Chem.* **39:** 1927.

Sasson, Y.; Zappi, G.D.; Neumann, R. (1986) *J. Org. Chem.* **51:** 2880.

Sepp, D.T.; Scherer, K.V.; Weber, W.P. (1974) *Tetrahedron Lett.* 2983.
Shostakovskii, S.M.; Nikol'skii, N.S.; Makosha, M.; Artst, Ya.V. (1974) *Tezisy Dokl. Nauchn. Sess. Khim. Technol. Org. Soedin. Sery Sernistykh Neftei 13th.* 144. *Chem. Abstr.* (1976) **85:** 176884.
Simchen, G.; Kobler, H. (1975) *Synthesis* 605.
Slobodin, Ya.M. (1988) *Zh. Org. Khim.* **24:** 2621. *Chem. Abstr.* **110:** 231162.
Smiley, R.A.; Arnold, C. (1960) *J. Org. Chem.* **25:** 257.
Soga, K.; Hosoda, S.; Ikeda, S. (1977) *J. Polym. Sci. Polym. Lett. Ed.* **15:** 611.
Soula, G. (1985) *J. Org. Chem.* **50:** 3717.
Starks, C.M.; Napier, D.R. (1968) *Australian Pat.* 439286; *Netherlands Pat.* 6804687.
Starks, C.M. (1971) *J. Am. Chem. Soc.* **93:** 195.
Starks, C.M. (unpublished) results quoted in Starks, C.M.; Liotta, C.L.; Halpern, M. (1994) *Phase Transfer Catalysis*, Chapman & Hall: New York; p.152.
Steinbeck, K. (1978) *Tetrahedron Lett.* 1103.
Szabo, G.T.; Aranyosi, K.; Csiba, M.; Töke, L. (1987) *Synthesis* 565.
Tabushi, I.; Yoshida, Z.; Takahashi, N. (1970) *J. Am. Chem. Soc.* **92:** 6670.
Tabushi, I.; Kuroda, Y.; Yoshida, Z. (1976) *Tetrahedron* **32:** 997.
Tagaki, W.; Inoue, I.; Yano, Y.; Okonogi, T. (1974) *Tetrahedron Lett.* 2587.
Takeishi, M.; Kawashima, R.; Okawara, M. (1973) *Makromol. Chem.* **167:** 26.
Tan, S.N.; Dryfe, R.A.; Girault, H.H. (1994) *Helv. Chim. Acta* **77:** 231.
Tundo, P.; Venturello, P.; Angeletti, E. (1987) *J. Chem. Soc. Perkin Trans. 1* 2157.
Tundo, P.; Trotta, F.; Moraglio, G. (1989) *J. Chem. Soc. Perkin Trans. 1* 1070.
Ungurenasu, C.; Cotzur, C. (1989) *Polymer Bull.* **22:** 151.
Van Leusen, A.M.; Bouma, R.J.; Possel, O. (1975) *Tetrahedron Lett.* 3487.
Venturello, C.; Alneri, E.; Lana, G. (1981) *Ger. Pat.* 3027349.
Weber, W.P.; Gokel, G.W. (1972a) *Tetrahedron Lett.* 1637.
Weber, W.P.; Gokel, G.W.; Ugi, I.K. (1972b) *Angew. Chem. Int. Ed. Engl.* **11:** 530.
Wu, S.; Zhou, C.; Jiang, Y. (1986) *Synth. Commun.* **16:** 1479.
Xu, C.-y.; Liu, G.-j.; Zhang, Z. (1987) *Synth. Commun.* **17:** 1839.
Yamamura, K.; Murahashi, S.-I. (1977) *Tetrahedron Lett.* 4429.
Yoshifuji, M.; Toyota, K.; Yoshimura, H.; Hirotsu, K.; Okamoto, A. (1991) *Chem. Commun.* 124.
Zahalka, H.A.; Januszkiewicz, K.; Alper, H. (1986) *J. Mol. Catal.* **35:** 249.
Zenner, K.F.; Appel, H.G. (1972) *Ger. Pat.* 2126296.
Zubrick, J.W.; Dunbar, B.I.; Durst, H.D. (1975) *Tetrahedron Lett.* 71.
Zwierzak, A.; Podstawczynska, I. (1977) *Angew. Chem. Int. Ed. Engl.* **16:** 702.

2 REACTIONS WITH SOLID-SUPPORTED REAGENTS AND CATALYSTS

The development of solid phase polypeptide synthesis [Merrifield, 1963] has had a tremendous impact on synthetic methodology in general, although it was recognized early on that the method is most likely adaptable with equal efficiency to the preparation of substances containing repeating subunits. However, the pioneering work provided a stimulus to chemists in the functionalization of polymers and utilization of such as reagents and catalysts.

There are several practical advantages in attaching a reagent or catalyst to a solid support. It simplifies workup as often filtration is sufficient to separate the spent reagent/catalyst from the product; it is usually possible to recycle the spent reagent/catalyst; reaction can be forced to completion by using an excess of reagent can be readily removed; insoluble reagents/catalysts are nonvolatile therefore they are odorless and of little danger to cause toxication. The reactivity of the reagent may also be enhanced due to its dispersion over high surface areas. Sometimes the change of a support completely alters the reaction course [Onaka, 1989].

The term *supported* refers to either ionic or covalent bonding of reactive species to the solid which may be organic or inorganic. It is almost equivalent to the term *immobilized* commonly found in a biochemical context, although an immobilized species may only be physically trapped. Besides covalent bonding methods for the preparation of supported reagents are: ion exchange, deposition, precipitation or coprecipitation, impregnation, adsorption from solution, and mixing/grinding. For reagents on inorganic suports post-treatment such as thermal activation to control water content is commonly implemented.

Directly related to the topic of synthesis on solid supports is combinatorial chemistry [Lowe, 1995; Terrett, 1995; Fruchtel, 1996; Hermkens, 1996], the purpose of which is to generate in one sequence of operations a huge number of molecular specimens with the same skeleton but bearing different substituents at critical sites, so that defined characteristics such as pharmacological properties of each member of the library can be evaluated. The enormous implications of the technique in drug development is self-evident.

A small band of organic chemists have devoted themselves to the study of reactions in the solid state [M.D. Cohen, 1973; Toda, 1993]. While many fascinating aspects have been discovered, a broad generality and synthetic applications are yet to be established. Consequently the area will not be discussed in this book.

It should also be mentioned that reactions carried out with soluble polymer-bound catalysts have received some attention. Presently, development in that area is well behind although its utility for synthesis of peptides, oligonucleotides, oligosaccharides, has been demonstrated [Gravert, 1997].

2.1. POLYMER-BASED SUPPORTS

Reagents and catalysts can be incorporated in both linear and cross-linked polymers. A required functional group is either introduced in the preformed support or present in the monomer which is to be used in copolymerization. The preference for these routes depends on the availability of unmodified polymers and the ease of copolymerization. Generally the copolymerization method allows for better control of functional group distribution.

2.1.1. Preparation of Functionalized Polymers by Copolymerization

Polymers derived from 4-vinylimidazole, 4-vinylpyridine, and *N*-acryloylmorpholine are themselves useful, whereas maleic anhydride and maleimide readily undergo copolymerization with styrene to afford supported materials suitable for synthetic applications. In addition, a number of substituted styrenes have been made to serve as functionalized monomers.

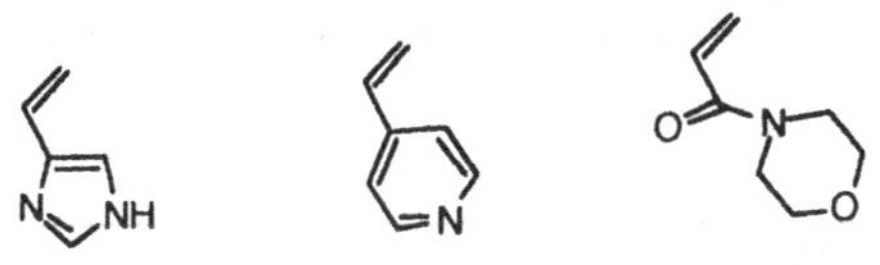

N NH N O N O

When mild reactions are needed in the modification, styrene monomers can be used. Otherwise the sensitive double bond should be protected, e.g., in the form of β-bromoethyl or acetyl precursors, until modification is complete. The former category includes the use of *p*-vinylphenylmagnesium chloride and analogous Grignard reagents to prepare phosphines [Rabinowitz, 1961], silanes [Senear, 1990], stannanes [Leebrick, 1958], alcohols and carboxylic acid [Leebrick, 1958; Letsinger, 1964]. Of course a styrene precursor tolerate harsh conditions to permit Friedel-Crafts reactions, sulfonation, nitration, and others to introduce other functionalities. The chloromethyl derivative is particularly versatile, but direct chloromethylation alkylbenzenes generates a mixture of isomers. Pure *p*-chloromethylstyrene is accessible by much longer reaction sequences starting from either *p*-dibromobenzene or *p*-methylacetophenone [Arshady, 1976, 1978]. Acrylic/methacrylic monomers are readily polymerized, but the ester and amide groups in them tend to limit their role as supports.

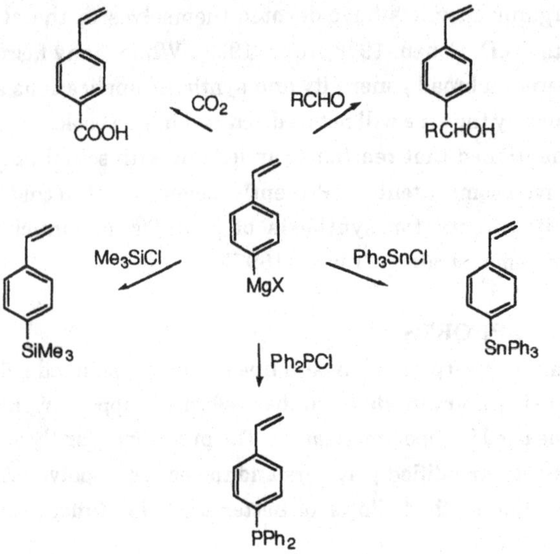

2.1.2. Chemical Modification of Polymers

This alternative takes advantage of the availability of many *bare* supports and it skips the polymerization step which might require some investigation for a new monomer. However, the approach is not without its shortcomings: reactivity difference due to conformational effects may complicate the application of standard protocols which work well in the functionalization of the corresponding monomers. Low conversions as a result of phase separation and/or undesirable aggregation of the functionality owing to local concentration of reagents under inhomogeneous conditions are not uncommon. When the initial product bears a residue (e.g., chloromethyl and chlorosulfonyl groups) it could lead to cross-linking.

For linear polystyrene-type supports most of the common functionalization steps starts from lithiation of the aromatic ring using *n*-BuLi-TMEDA [Chalk, 1968]. Alternatively, halogen-lithium exchange of nuclear halogenated polystyrenes accomplish the same purpose. The lithiated polymers can then be used to react with various reagents.

Poly(vinyl alcohol) is also a valuable support into which functional groups are readily introduced [Wiley, 1971].

Crosslinked polystyrene resins have been subjected to functionalization. Generally, it is more difficult to control accurately the loading, characterize the structural changes and reproduce reactions from run to run, and there is often problems of reagent penetration into the crosslinked networks. Two popular functionalization methods involve lithiation [Farrall, 1976; Grubbs, 1976] and electrophilic substitutions, especially chloromethylation. Direct lithiation works reasonably well when low loading is required and it occurs mainly in the meta position. On the other hand, halogen-lithium exchange is notoriously difficult to reproduce. Noteworthy is the phosphines can be made by displacement of an aryl bromide with lithium diphenylphosphide [Heitz, 1972].

2.1.3. Reactions with Polymer-supported Reagents

Many reagents, particularly those for achieving displacement reactions, are conveniently prepared from ion-exchange resins. From the strongly basic resins which are crosslinked polystyrenes bearing quaternary (e.g., trimethylbenzylammonium) ammonium groups the desired anion can be introduced by displacing a more weakly bound anion or by treating the hydroxide form of the resin with the required acid. The affinity order according to anions is: $I^- > PhO^- > Br^- > CN^- > NO_2^- > Cl^- > IO_3^- > AcO^- > OH^- > F^-$. Common cation-exchange resins which carry either a benzenesulfonic acid or carboxylic acid group may be treated similarly to incorporate the cation. The strongly acidic sulfonic acid resins show a relative trend of cation association in the order of: $Ni^{2+} > Cu^{2+} > Co^{2+} > Mn^{2+} > Ag^+ > Na^+ > H^+$; whereas the weak carboxylic acid resins prefer $H^+ >> Ag^+ > Na^+$. If is often possible to conduct reactions in aprotic solvents such as toluene or dichloromethane if a macroporous ion-exchange resin is used. It is actually a phase transfer technique.

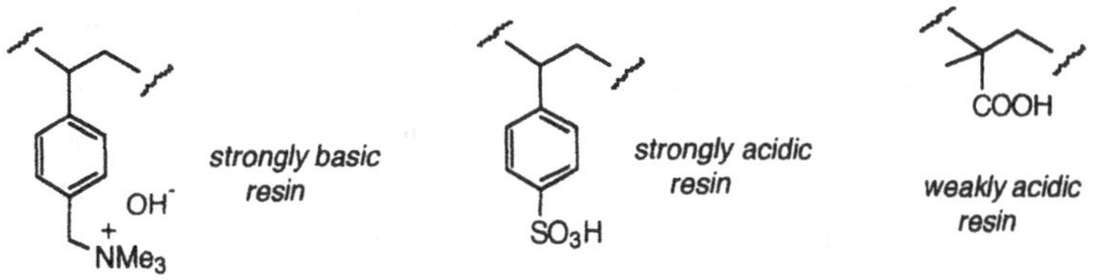

OH^- $\overset{+}{N}Me_3$
strongly basic resin

SO_3H
strongly acidic resin

COOH
weakly acidic resin

2.1.3.1. Substitutions

Alkyl fluorides are obtained by treatment of sulfonates and halides with fluoride form of resin, usually in large excess [Cainelli, 1976b]. Secondary alkyl halides are prone to elimination therefore the displacement is better carried out with the sulfonates. Preparation of sulfonyl fluorides from the corresponding chlorides is relatively straightforward, acetonitrile being the preferred solvent for the reaction [Borders, 1972].

Halides are converted to phenylacetonitriles in reasonable to excellent yields in either hot ethanol [Gordon, 1963] or benzene/toluene. Alkyl thiocyanates (isothiocyanates) and nitroalkanes [Gelbard, 1977] have also been prepared in a similar way.

The combination of carbon tetrachloride and a polymer-supported phosphine effectively transforms alcohols into alkyl chlorides and acids to acid chlorides [Hodge, 1975; Regen, 1975]. It is interesting that the rates are generally faster than the the corresponding reactions using monomeric triphenylphosphine. Reactant formation might have involved two juxtaposed phosphines, a situation favored when they are fixed on a polymer backbone.

$$ROH + Ⓟ\!-\!PPh_2 + CCl_4 \longrightarrow R\text{-}Cl + CHCl_3 + Ⓟ\!-\!P(=O)Ph_2$$

Homologation of an iodoalkane by way of double displacement using the malodorous thioanisole as pivot has obvious disadvantages which can be overcome in a polymeric version [Crosby, 1977]. The reagent is automatically regenerated in the final step.

Ⓟ—SMe —(*n*-BuLi, DABCO; RI)→ Ⓟ—SCH$_2$R —(MeI / NaI, DMF)→ Ⓟ—SMe + RCH$_2$I

The Koenigs-Knorr glycoside formation involves displacement at the anomeric site. The silver salt of a maleic acid/1,4-(bisvinyloxy)butane copolymer is a promoter [Eschenfelder, 1975] with the advantage of simplified workup of the products.

O-Alkylation of polymer-bound carboxylate [Cainelli, 1975] and aryloxide [Gelbard, 1977] anions by column or batch technique is possible, but the synthetic utility for such products is low. A more interesting report pertains to *C*-prenylation of acylphloroglucinols [Collins, 1973], albeit in moderate to low yields.

Resin hydroxides can be used as base in the alkylation of stabilzed enolates such as those derived from malonic esters, cyanoacetates and cyanamides [Shimo, 1963] and β-keto esters [Gelbard, 1977]. *C*-Alkylation predominates under such conditions.

By virtue of strong hydrogen bonding ability of the anion, polymer-supported fluoride is an excellent mediator of *C*-alkylation and sulfenylation of β-diketones, *O*- and *S*-alkylation of phenols and thiophenols, respectively [Miller, 1978]. Yields range from 50-80%.

Conversion of an alkyl halide to the homologous aldehyde can be accomplished by reaction with sodium tetracarbonylhydridoferrate. This process is further improved by replacing sodium with a polymer-bound ammonium ion [Cainelli, 1978] such that the inconvenience in product isolation is removed.

$HFe(CO)_4^-$ + R-X ⟶ R-CHO

2.1.3.2. Group Transfer Reactions

Polymer-bound iodobenzene dihalide reagents are able to transfer the halogen to alkenes. The dichloro derivative react normally [Hallensleben, 1972], difluoroiodoaryl polymers convert a series of styrene derivatives to *gem*-difluorides in excellent yields [Zupan, 1975].

Polypyridinium hydrobromide perbromide reacts with alkenes and ketones smoothly [Frechet, 1977]. Remarkably, *(Z)*-stilbene affords the *dl*-dibromide in 98% yield which contrasts with the nonstereoselective results of conventional bromination (with bromine). (Note that polymeric analogues of *N*-bromosuccinimide are not selective in benzylic bromination. They often introduce more than one bromine atom into the substrates.)

Ph Ph + (P)–C₅H₄NH⁺ Br_3^- → Ph Br / Br Ph + (P)–C₅H₄NH⁺ Br^-

Sulfonyl azide reagents bound to a polystyrene polymer backbone are readily prepared (via chlorosulfonylation and displacement). Not only can such a reagent smoothly deliver a diazo group to active methylene compounds [Roush, 1974], its insensitivity to shock (in contrast to tosyl azide) is an asset.

O R / O R + (P)–SO_2N_3 → (Et_3N) O R / =N_2 / O R + (P)–SO_2N_3

Bis(triphenylphosphine)carbonylrhodium(I) chloride is to expensive to be a practical mediator for ketone synthesis which involves the treatment with an organolithium species and then an acid chloride. Using polymer-bound phosphine to ligate the rhodium atom [Pittman, 1977] permits the recovery and reuse of the most expensive component.

(P)–P(Ph)₂–Rh(Cl)(CO)–P(Ph)₂–(P) → (RLi, -78°C) → (P)–P(Ph)₂–Rh(R)(CO)–P(Ph)₂–(P) → (R'COCl, -78°C) → (P)–P(Ph)₂–Rh(R)(COR')(Cl)(CO)–P(Ph)₂–(P) → RCOR'

Difficulty in Grignard reagent formation can be overcome by using the complex [Mg(anthracene)(thf)3] to deliver the metal. Anthracene contamination can be avoided when it is incorporated into a polymer through a silicon linker at C-9 [Harvey, 1988].

2.1.3.3. Dehydration and Condensations

The polymeric phosphine-carbon tetrachloride is a simple system useful for the dehydration of aldoximes and primary amides to give nitriles [C.R. Harrison, 1977]. The same reagent effects the condensation of carboxylic acids and amines to afford simple amides [Hodge, 1975] and dipeptides [Appel, 1976]. Addition of 1-hydroxybenztriazole before the reaction suppresses racemization of the amino acid residues.

The polymeric version of the Mukaiyama method for peptide synthesis employs a supported phosphine [Horiki, 1976]. Reported yields of dipeptides are in the 67-88% range.

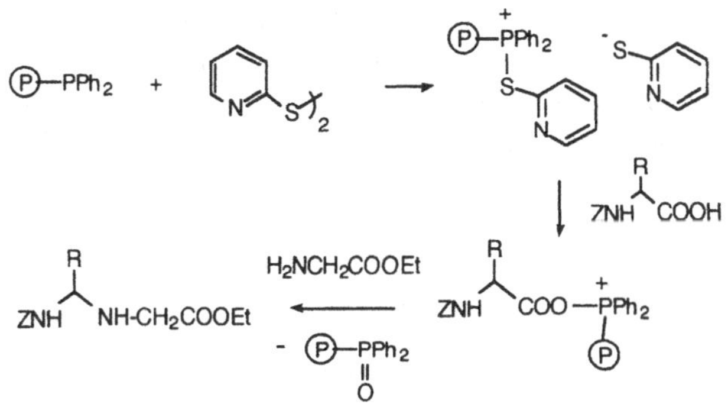

Mixed anhydrides derived from crosslinked polystyrenes functionalized with sulfonic acid, carboxylic acid, and carbonic acid groups are potential reagents for acyl transfer. The second class of such compounds include activated aryl esters (e.g., *o*-nitroaryl esters [Fridkin, 1966]). Copolymers containing *N*-hydroxysuccinimide units have also been esterified for the same purpose. Most of these reagents were prepared also in connection with peptide synthesis in which the growing peptide is not anchored to the solid support.

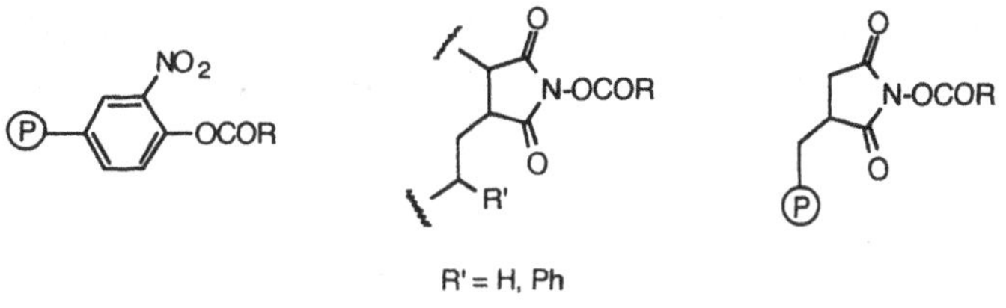

R' = H, Ph

The *o*-nitroaryl esters seem to be among the best as successful application to a synthesis of bradykinin [Fridkin, 1968] and cyclic peptides, the latter by intramolecular transacylation.

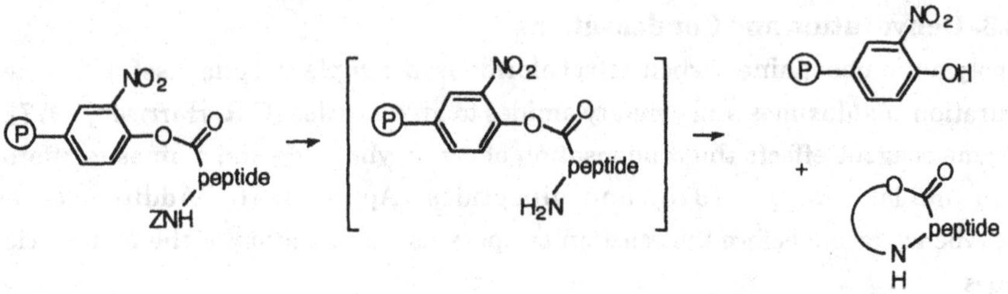

That two or more functionally incompatible polymers ("wolf and lamb") can be mixed or suspended in a solvent without destruction is due to their mutual inaccessibility. This characteristic can be exploited in reactions such as condensation of acetophenone with a polymeric *o*-nitrophenyl benzoate, using polymeric trityllithium as base [B.J. Cohen, 1981]. The acylating agent is immune to direct attack by the strong base, but deprotonates acetophenone in the solvent (THF, DME) which is free to approach the electrophilic polymer.

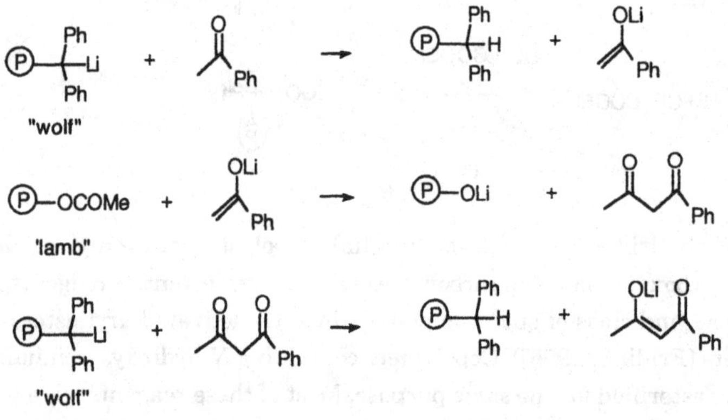

Activation of phosphate for the preparation of diesters usually employs a hindered arenesulfonyl chloride. As removal of the sulfonic acid byproduct may be problematic, incorporation of the chlorosulfonyl group to a polymer backbone clearly obviates the issue [Rubinstein, 1972].

NBS; - HBr → polymerize; $ClSO_3H$ → SO_2Cl

RO-P(=O)(O⁻)-OH + R'OH —(P)–SO_2Cl → (P)–SO_3H→ RO-P(=O)(O⁻)-OR'

A method for the stereocontrolled construction of a polyketide chain on a polymer support has been developed [Reggelin, 1996]. Essentially it involves cycles of aldol condensation of an oxidized Wang resin (aldehyde form) with an enol boronate derived from a chiral *N*-acyloxazolidone (Evans's auxiliary), hydroxyl protection, and conversion of the amide group to the aldehyde.

$OBBu_2$ ... Ph → OH O ... Ph

↓ iPr_3SiOTf, lutidine; MeNH(OMe).HCl, Me_3Al

iPr_3SiO ... H ← iBu_2AlH — iPr_3SiO ... N-OMe

The Dieckmann cyclization of unsymmetrical diesters leads to a mixture of products which may be difficult to separate. One solution to the technical problem arising from nonregioselective cyclization is to link one of the carboxyl group to a polymer [Crowley, 1970]. Although the regioselectivity may not improve, retention of one of cyclic β-keto ester isomers on the polymer support facilitates the separation.

$OCPh_3$ ... R —base→ $COOCPh_3$ ... R + ... R

In the synthesis of one of the first member of the rotaxane family (originally named "hooplane") the use of statistical method led to dismal yield and caused enormous problems in its isolation. While inherent characteristics of the approach prevent the yield to increase, attaching the "hoop" to a polymer does permit many cycles of "rod" insertion [I.T. Harrison, 1967]. Thus capping of 1,10-decanediol with trityl groups while in the inserted state holds it indirectly onto the polymer, which is readily freed from the uncaptured diether by washing with solvent. As the reaction sequence is repeated more hooplanes are formed, and after sufficient number of cycles are performed the hooplane is released by hydrolysis.

2.1.3.4. Reaction with Ylides

One major defect of the popular Wittig reaction is the generation of triphenylphosphine oxide which often requires careful chromatography for its removal. Thus it is not surprising that much attention has been devoted to the polymeric version [Castells, 1979; Heitz, 1973; McKinley, 1972] in which both reagent and side products are easily separated. Because of the polymer sweeling properties of DMSO it is the solvent of choice. Comparable yields and stereselectivity to the conventional Wittig reaction are obtained.

Isophthaldehyde and terephthaldehyde give monoalkene products [Castells, 1979] even the ylide is in excess.

<u>Polystryryldiphenylphosphine</u>. [Relles, 1974]

A suspension of 2% crosslinked polystyrene (16-25 mesh, 83.68 g, 0.8 mol of repeating units) in nitrobenzene (500 mL) containing boron trifluoride (56 g, 0.8 mol) is stirred at 25°C in the dark while bromine (256 g, 1.6 mol) is added dropwise during 0.5 h. Vigorous evolution of HBr is observed during the addition, and if necessary external cooling is applied. The mixture is stirred for an additional 18 h and filtered. The beads are washed with once each with CH_2Cl_2-MeOH (9:1, 3:1, 2:3, 3:1, 9:1) and CH_2Cl_2, ecuated at 100°C to give the brominated resin (147.7 g, 100%).

With initial external cooling, the brominated resin (36.92 g, 0.2 mol) is suspended in THF (900 mL), stirred under nitrogen for 1 h, and treated with a solution of chlorodiphenylphosphine (88 g, 0.4 mol) in THF (300 mL), and lithium metal (6.4 g, 0.93 mol) in small pieces. After stirring at room temperature for 18 h the excess metal is removed, the reaction mixture is refluxed for 4.5 h. The beads are filtered on addition of some methanol, washed with CH_2Cl_2-MeOH (2:3, 3:1, 9:1) and CH_2Cl_2, ecuated at 100°C to afford the polymer bound phosphine (53.55 g).

Preparation of 2-Phenyl-2-hexene. [Heitz, 1972]

A stirred mixture of 2% crosslinked polystyryldiphenylphosphine (30 g, 7.3% P = 70.5 mmol), *n*-bromobutane (11.6 g, 84.1 mmol) in THF (80 mL) is refluxed for 24 h. The resin is filtered, washed with THF, and dried in vauo at 40°C (39 g, + 65.7 mmol). To a suspension of this polymeric phosphonium salt (19 g) in dioxane (40 mL) is added *n*-butyllithium (20%, 13 mL) at 20°C, followed by acetophenone (30 mL in dioxane 10 mL) after 0.5 h. The mixture is maintained at 60°C for 8 h, the resin beads are transferred to a Soxhlet extractor (1:1 THF/H_2O). Aqueous layer is separated and again extracted. Product is isolated from the organic solution after evaporation and distillation (3.85 g, 97%).

2-Substituted indoles are readily formed by a reaction sequence of: *P*-alkylation of the resin-linked phosphine with *o*-nitrobenzyl bromide, reduction of the nitro group, acylation, and intramolecular Wittig reaction [Hughes, 1996]. The exceptionally mild conditions for the Emmons-Wadsworth condensation is especially suitable for preparation of unsaturated amides which are bound to resins [Johnson, 1995].

Ph P Ph → Ph +P Ph Br⁻ NO2 ⇄ Ph +P Ph Br⁻ NH O R

t-BuOK ↓ PhMe / DMF

Ph P=O Ph + NH R

P N H O PO(OEt)2 + RCHO LiBr, Et3N / MeCN → P N H O R

PEG-PAL resin

Crosslinked polystyrenes in which the aromatic ring carries a methylthio group can be converted to the sulfonium salts by reaction with FSO_3Me. Ylide generation with aqueous NaOH and reaction with aromatic aldehydes and ketones in dichloromethane (triphase system) afford epoxides [Farrall, 1979]. The spent polymer reagent can be recycled.

Ⓟ–SMe → Ⓟ–$\overset{+}{S}Me_2$ FSO_3^-

ArCHO / NaOH

Ar-epoxide

2.1.3.5. Oxidations

Treatment of crosslinked poly(4-vinylpyridine) with chromium trioxide in hydrochloric acid leads to an analogue of PCC. The polymeric reagent is usually used in large excess in a solvent such as cyclohexane and benzene [Frechet, 1978]. When the chloride ion of the more common anion-exchange resin is replaced by chromate ion another oxidizing agent is obtained [Cainelli, 1976a]. Both reagents are effective in oxidation of primary and secondary alcohols to give carbonyl products in high yields, and the latter is also capable of converting benzylic and allylic halides in refluxing benzene to aldehydes or ketones [Cardillo, 1976].

The *N*-chloro derivative of nylon-66 is readily accessible [Schuttenberg, 1973]. Its very attractive property as an oxidant is the solubility in benzene while the spent reagent precipitates. Note that aliphatic primary alcohols give ester products, probably via acetals of the aldehydes.

RCH_2OH —(N-Cl amide)→ $RCHO$ —($2\ RCH_2OH$)→ $RCH(OCH_2R)_2$ → $RCOOCH_2R$

A serious problem associated with the Pfitzner-Moffatt oxidation is the persistent contamination of the substituted urea byproducts. A polymer-supported carbodiimide offers a cleancut solution. Such a reagent can be prepared from chloromethylated crosslinked polystyrene [Weinshenker, 1972a] and used at ambient temperature in a cosolvent system containing benzene [Weinshenker, 1972b].

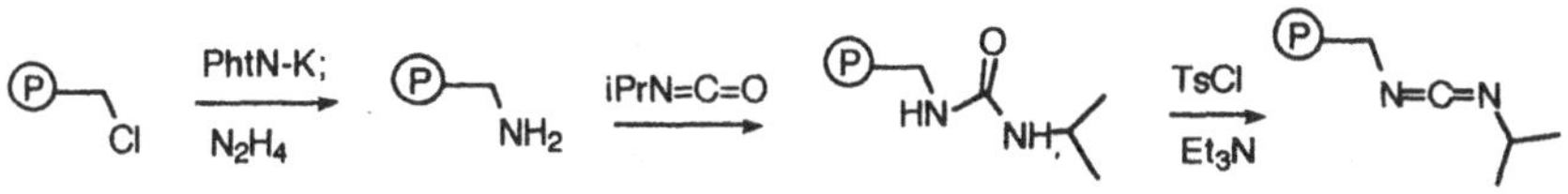

N-Polystyrylmethyl-*N'*-isopropylcarbodiimide. [Weinshenker, 1972a]

Chloromethylated 2% crosslinked-polystyrene (11.5% Cl, 3.1 mmol Cl/g) is treated with potassium phthalimide (2.5 equiv. per Cl) and DMF at 100°C for 5 h, then with hydrazine (1 mL/g polymer) in ethanol at reflux. After 6 h the amine resin is filtered, reacted with isopropyl isocyanate (2 equiv.) in THF at 25°C for 15 h. The resulting urea is dehydrated with tosyl chloride (2 equiv.) and triethylamine in dichloromethane (15 h reflux). The polymeric carbodiimide thus obtained has a maximum activity of 2.42 mmol/g.

Chlorination of the crosslinked polystyrene units containing a methylthio substituent furnish an oxidizing agent [Crosby, 1975] equivalent to that derived from thioanisole.

The polymeric periodate reagent prepared from ion-exchange resin has been employed to oxidize catechols and quinols, cleavage *vic*-diols in nonpolar solvents. Oxidation of sulfides to sulfoxides proceeds in reasonable rates only in methanol or water.

2-Methylnaphthalene has been converted to 2-naphthaldehyde in 62% yield by a polymer bearing a diaryl selenoxide group [Michels, 1976]. The reagent is akin to SeO_2 in its reactivity.

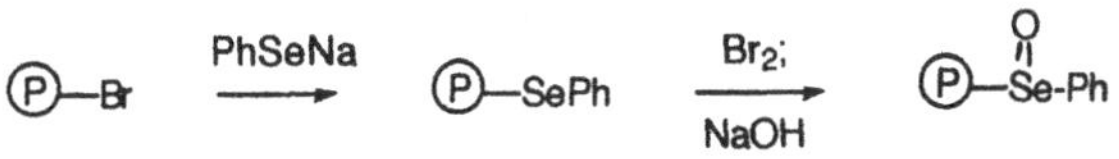

Epoxidation of alkenes with polymeric versions [Frechet, 1975; C.R. Harrison, 1976a] of perbenzoic acid seems to work better with less crosslinking. These species can be prepared from the carboxylic acid resins and concentrated hydrogen peroxide in the presence of a sulfonic acid, from the acid chloride resin by reaction with Na_2O_2-H_2O_2, or from an aldehyde resin by treatment with ozone. The best solvent seems to be THF and not CH_2Cl_2 or $CHCl_3$. With respect to the alkene structure 1-alkenes are hardly touched. Aliphatic peroxyacid analogues are not suitable because they are explosive in the dry state.

One good use of the polymeric perbenzoic acid is in the conversion of penicillins and deacetoxycephalosporins to the sulfoxides [C.R. Harrsion, 1976b].

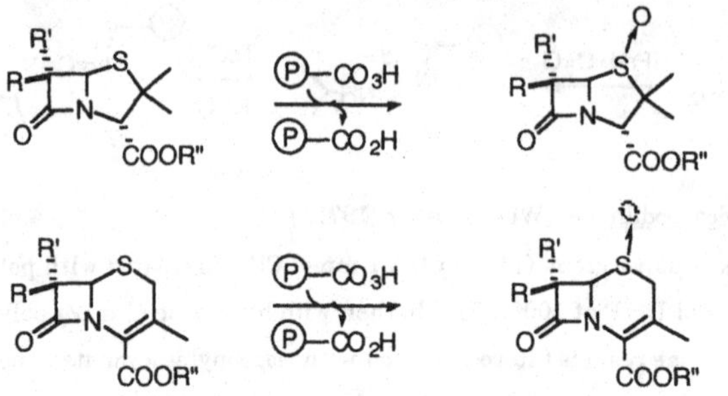

<u>Cyclooctene Oxide</u>. [C.R. Harrison, 1976a]

To a stirred mixture of the carboxy-resin (5g, Biobeads SX1) and methanesulfonic acid (15 mL) is added dropwise hydrogen peroxide (85%, 10 mL) in methanesulfonic acid (10 mL), while maintaining the temperature at 25-30°C. When the addition is complete, stirring continues for 16 h at 20°C. The filtered resin, after washing successively with THF and CH_2Cl_2, and drying under vacuum, has an activity of 3.9 mmol active oxygen /g.

A mixture of cyclooctene (55 mg, 0.5 mmol) and the peracid resin (256 mg) in THF (2 mL) is stirred at 40°C for 4 h. The cooled and filtered resin is washed with THF which is combined with the filtrate. Analysis by GC indicates the presence of cyclooctene oxide (95% yield) and unchanged cyclooctene (2%).

After undergoing diacetoxylation the *p*-iodostyrene polymer can be employed to oxidize aniline into azobenzene [Hallensleben, 1972].

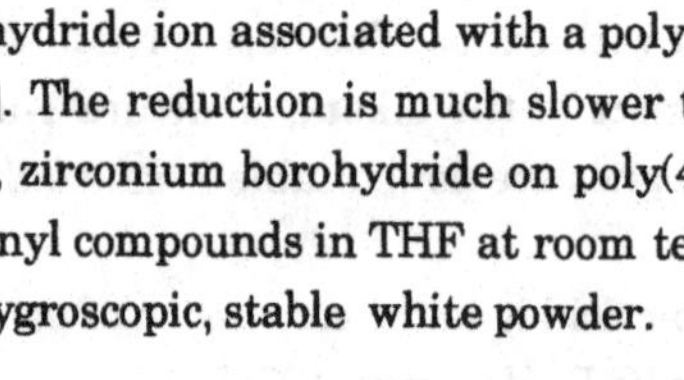

2.1.3.6. Reductions

Borohydride ion associated with a polymer support is a mild reducing agent [Gibson, 1977]. The reduction is much slower than the conventional method. On the other hand, zirconium borohydride on poly(4-vinylpyridine) is very effective in reducing carbonyl compounds in THF at room temperature [Tamami, 1994]. The reagent is a nonhygroscopic, stable white powder.

$N\text{--}Zr(BH_4)_4$

Sodium cyanoborohydride supported on Amberlyst 26 is indefinitely stable yet retains all the reducing potency [Hutchins, 1978]. Another great advantage of this supported reagent is that the solvent is free from contamination by cyanide ion released during the reaction.

Polymer-supported alkoxyaluminum hydrides can be prepared by reacting crosslinked polymers containing hydroxyl groups with excess lithium aluminum hydride in ether.

Among this group of reagents the most extensively studied is the macroporous crosslinked polystyryltin dihydride. It is odorless and nontoxic, capable of reducing aldehydes and ketones in fair to good yields [Weinshenker, 1975]. A 6:1 mixture of the hydroxy aldehyde and diol obtained from reduction of terephthaldehyde is a desirable result, the selectivity might indicate separation of the reductants and restricted mobility of the substrate in the polymer network.

Li $\xrightarrow[BuSnCl_3]{MgBr_2;}$ Bu, Sn-Cl, Cl $\xrightarrow{LiAlH_4}$ Bu, Sn-H, H

As expected, the possibilities of replacing monomeric tin hydrides with polymer-bound forms to achieve other reactions have been studied. Free radical formation from organobromine compounds and in situ trapping by activated alkenes have been demonstrated [Bokelmann, 1992]. The tin hydride can be regenerated by treatment with sodium borohydride.

(P) Bu, Sn-H, Bu + R-Br + CN ⟶ R CN + (P) Bu, Sn-Br, Bu

The borane complexes of poly(4-vinylpyridine) and of cross-linked polymers containing pyridine moieties reduce aldehydes and ketones in benzene [Hallensleben, 1974].

Columns packed with titanium(III) or chromium(II) forms of ion-exchange resin have been used to reduced quinones, nitroarenes, azoarenes, benzils and benzoins [Dalibor, 1958; Inczedy, 1961].

A crosslinked polymer containing a 9,10-dihydroxyanthracene unit can be used to release the carboxylic acid from 2-acyloxymethylanthraquinone through a redox process [Kemp, 1977].

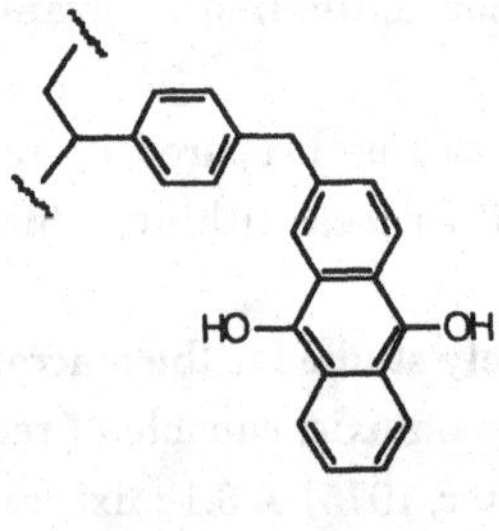

HO— —OH

2.1.4. Polymeric Protecting Groups

In the original Merrifield solid-phase synthesis of polypeptides the polymer support served as both an anchor as well as a protecting group. Due to physical properties of polymers that once a reaction occurs with a polyfunctional molecule at one site the other functional group(s) of the same nature is largely excluded, the difficulties in selective reaction of symmetrically disubstituted compounds can be overcome, at least in principle, by prior protection of one of the reactive groups with a suitable polymeric reagent[Leznoff, 1978]. Thus the technique also represents an alternative method of high dilution. In integrating the reactant to the polymer a large excess can be used to further ensure the 1:1 purchase. Even if the reactant is expensive there is hardly any sacrifice because the unreacted material can be recovered by filtration (either it is used in a neat form or solution).

Monoacylation and monotritylation of symmetrical diols with polymer-bound acid chlorides [Leznoff, 1972] and trityl chlorides [Frechet, 1976] have been successfully demonstrated. A synthetic potential of this tactic is fulfilled in the preparation of several insect pheromones [Leznoff, 1977].

The protection of 1,2-diols and 1,3-diols as cyclic boronates less less than ideal because such derivatives are quite sensitive to moisture. The polymeric version of phenylboronic acid forms cyclic boronates with diols without any difficulties [Seymour, 1976]. Such derivatives possess enhanced hydrolytic stability due to the relative inaccessibilty of the electrophilic boron atom to water molecules.

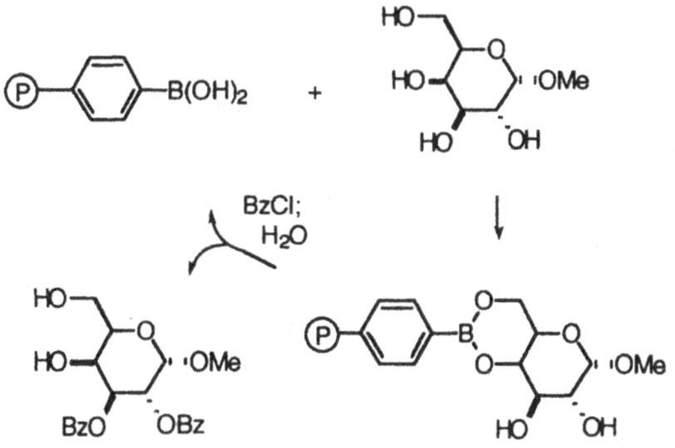

For modification of symmetrical dialdehydes, the first step is acetalization with a polymeric *vic*-diol prepared from glycerol acetonide [Leznoff, 1973]. The final product is disengaged from the polymer by acid cleavage whereby the reagent is also recovered. The synthesis of *p*-formylstilbene by a Wittig reaction as shown below is representative.

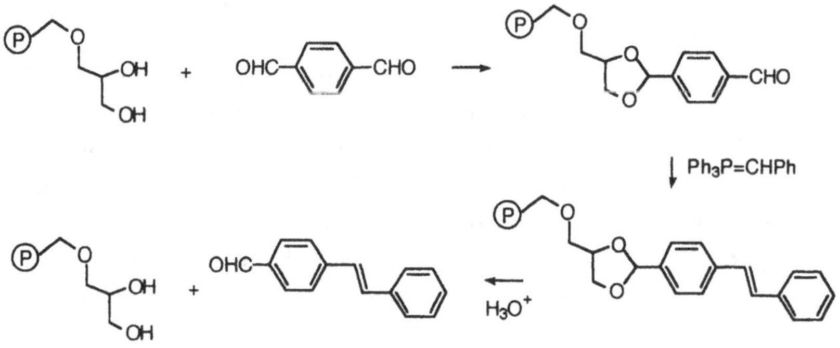

p-Formylstilbene. [Leznoff, 1973]

Sodium (2 g) is dissolved in 2,2-dimethyl-4-hydroxymethyl-1,3-dioxolane (60 mL) and the solution is treated with chloromethylated 2% crosslinked polystyrene beads (1.7 mmol Cl/g resin) at room temperature overnight and at 80°C for 24 h. The collected solid is washed with dioxane, water, ethanol-water, ethanol, and dry ether. After stirring with a 1:1 mixture of dioxane and 1N HCl (60 mL) at room temperature for 48 h, the finished resin is filtered, washed with water, acetone, ethanol, and dry ether.

A suspension of the polymer (4.25 g) in dioxane (60 mL) is stirred with terephthalaldhyde (2 g), *m*-benzenedisulfonic acid (0.1 g) and anhydrous sodium sulfate for 48 h at room temperature, filtered, washed twice with aqueous pyridine (1:1), water, methanol, and dry ether.
Benzylidenetriphenylphosphorane, formed by reaction of the bromide salt (4 g) with NaOMe (2.5 g) in DMF (60 mL) during 5 min, is treated with the polymer at room temperature for 24 h. The mixture is filtered, washed with methanol, ether, DMF, DMF-H_2O, water, and suspended in a 1:1 mixture of dioxane and dilute hydrochloric acid. After 48 h the product is isolated by filtration and extraction of the filtrate with ether. Purification by preparative TLC gives p-formylstilbene in 76% yield.

2.1.5. Reactions with Polymer-supported Catalysts

Ion exchange resins bearing a strong or weak acid function and those with an ammonium ion associated with the hydroxide ion are readily accessible, therefore they are useful catalysts for relevant reactions. The most important resins are the chemically modified gel-type polystyrene beads of nominal crosslink ratio ranging from ~2% to ~10% and similarly modified macroporous polystyrenes. The acid resin can be stored as H^+ or Na^+ form, whereas the strong basic resins should be generated from the Cl^- form prior to use owing to the tendency of the hydroxide ion to react with carbon dioxide in the air.

It should be mentioned that Nafion-H^+ (perfluoroalkanesulfonic acids) resins possess superior physical and catalytic properties. One major drawback of the Nafion is the high cost (although the resin is in principle completely recoverable).

Practically all types of organic reactions promoted by homogeneous acid or base have been reinvestigated in the presence of the related ion-exchange resin [Polyanskii, 1970]. These include acetalization, elimination, hydration, hydrolysis, etherification, esterification, cyclization, condensation, alkylation, isomerization and polymerization.

Superacids are prepared by admixture of strong Lewis acids and protonic acids. A combination of sulfonic acid resin and boron trifluoride constitutes an analogous system [Kelly, 1958] which can be used to alkylate olefins with isoparaffins. When a perfluorocopolymer containing the sulfonic acid group is impregnated with aluminum halides the enhanced acidity renders its catalytic service in the disproportionation of benzene and diethylbenzene [Olah, 1977]. Other related reactions are also catalyzed by the solid superacid systems.

2.2. INORGANIC SUPPORTS

A number of inorganic solids are suitable supports for reagents. The nature of these supports determine whether reactions occur on the surface or within the pore structures. Silica, alumina, Celite, graphite, charcoal powder, molecular sieves and several kinds of clay are the most common support materials. The use of zeolites as support has not been explored extensively probably because they themselves (esp. custom designed zeolites) are excellent catalysts for many reactions.

2.2.1. Acid- and Base-catalyzed Reactions

Silica gel absorbs and retains water such that it surface is sufficiently acidic to promote many organic reactions. Thus silica gel not only catalyzes the Diels-Alder reaction between benzoquinone and methyl pentadienoate at room temperature to give 1,4-naphthoquinone, it also promotes the removal of the ester group [Hudlicky, 1974]. Quantitative isomerization of an angular *o*-quinone to a linear *p*-quinone [Kato, 1977] has been reported. While the intramolecular ene reaction of an unsaturated aldehyde leading to the formation of a seven membered ring proceeds in the presence of tin(IV) chloride, a simpler protocol is by silica gel chromatography [Marshall, 1970].

The acid clays, particularly montmorillonite which is an aluminum hydrosilicate mineral, have a broader scope of utility as selective acid catalysts. Their effectiveness in various types of Friedel-Crafts reactions has been demonstrated.

The above depicts some organic processes which occur on or in solid supports themselves, and such are outside the scope of this book. In this chapter we address mainly reactions involving reagents and catalysts which are bound to such supports.

Dehydration of allylic, tertiary and sterically strained secondary alcohols is readily performed at room temperature with iron(III) chloride on silica gel [Keinan, 1978]. Regioselective monodehydration of steroidal polyols has been demonstrated.

β-Keto esters (methyl esters being the best) undergo dealkoxycarbonylation on heating with a suspension of alumina (containing 1.5% H_2O) in dioxane [Greene, 1976]. The reaction of α,α-disubstituted esters is sluggish.

In the presence of water, deacetalization by silica gel without affecting sensitive functions has been observed [Huet, 1977]. For example, 3-methylene-2-heptanone is produced in 95% yield. Quantitative dethioacetalization is possible when wet silica is first treated with sulfuryl chloride [Hojo, 1976b]. This system oxidizes sulfides to sulfoxides [Hojo, 1976a].

$$RR'C(SR'')_2 \xrightarrow{SiO_2\text{-}SO_2Cl_2} RR'C{=}O$$

Ketones can be regenerated from dithioacetals [M. Balogh, 1984] and various nitrogenous derivatives (*N,N*-dimethylhydrazones, phenylhydrazones, tosylhydrazones, and semicarbazones) [Laszlo, 1984b, 1985b] with the K-10 montmorillonite clay impregnated with iron(III) nitrate in dichloromethane.

The deep blue crystals of lamellar $C_{24}^{+}HSO_4^{-}\cdot 2H_2SO_4$ which are formed on electrolysis of 98% sulfuric acid with a graphite anode function as an excellent esterification catalyst [Bertin, 1974]. Primary and secondary alcohols (but not phenol) are acylated with various aliphatic acids at room temperature in cyclohexane. Sparingly soluble acids such as tartaric acid enter reaction gradually. Esterification of benzoic acid and cinnamic acid requires higher temperatures (70°C).

$$\text{RCOOH} + \text{R'OH} \xrightarrow[(-H_2O)]{C_{24}^{+}HSO_4^{-}.2H_2SO_4} \text{RCOOR'}$$

Many substituted cyclohexadienones undergo rearrangement/aromatization on treatment with K-10 montmorillonite clay impregnated with iron(III) nitrate [Chalais, 1986].

K-10 (Fe^{3+})
25°C
OH
R
87-98%
K-10 (Fe^{3+})
25°C
R = benzyl, allyl, crotyl, propargyl

K-10 (Fe^{3+})
25°C
OH
75-95%

Alkylation of phenylacetonitrile and ethyl phenylacetate in moderate yields has been reported using graphite-intercalated C_8K [Savoia, 1977]. Comparing with traditional methods there is no obvious advantage of this technique. Dispersions of sodium on charcoal and on graphite or alumina have high surfaces which can effect monoalkylation of ketones in hexane [Hart, 1977]. Interestingly, alkylation of 2-methylcyclohexanone affords >90% the 2,2-isomer.

Tetraalkylammonium fluorides no longer show hygroscopicity upon deposition on silica gel. However, potassium fluoride on alumina is perhaps even more versatile. The supported materials behave in the same fashion as the naked fluoride ion in catalyzing aldol condensation, Darzens reaction, Henry reaction, Michael addition, *C*-alkylation of β-diketones, and related processes [Clark, 1978]. Michael addition can also be effected in the presence of potassium *t*-butoxide deposite on the clay xonotlite in THF [Laszlo, 1985].

An excellent method for carrying out the Nef reaction of nitroalkanes is to treat them with NaOMe-silica gel at room temperature [Keinan, 1977; Hogg, 1978]. The environment which provides both basic and acidic sites is ideal for nitronate anion formation as well as the subsequent protonation as testified by the excellent yields of the carbonyl products. The reagent is prepared by evaporating a mixture of silica gel and methanolic NaOH and heating the resulting mass at 400°C for several hours.

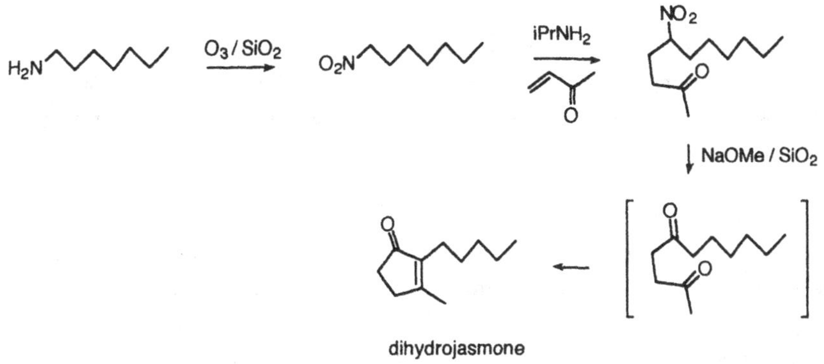

dihydrojasmone

Homoallyl ether formation by the reaction of an allylsilane with a carbonyl compound or its acetal is catalyzed by aluminum(III) salts deposited on montmorillonite K-10 [Kawai, 1986].

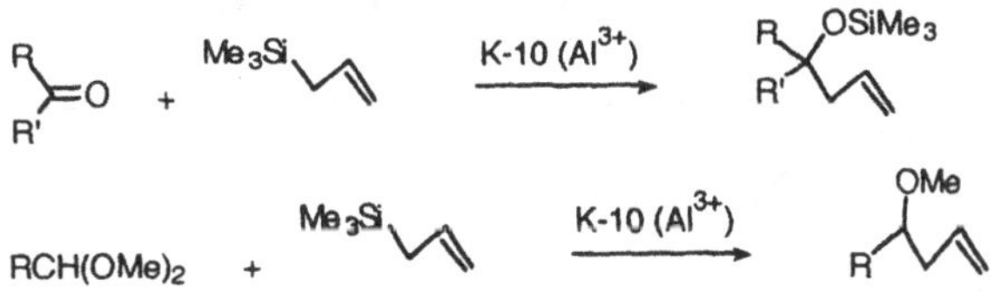

Efficient Michael addition between β-diketones and enones in the presence of a clay supported nickel halide and iron(III) chloride [Laszlo, 1990] manifests the cooperative catalysis.

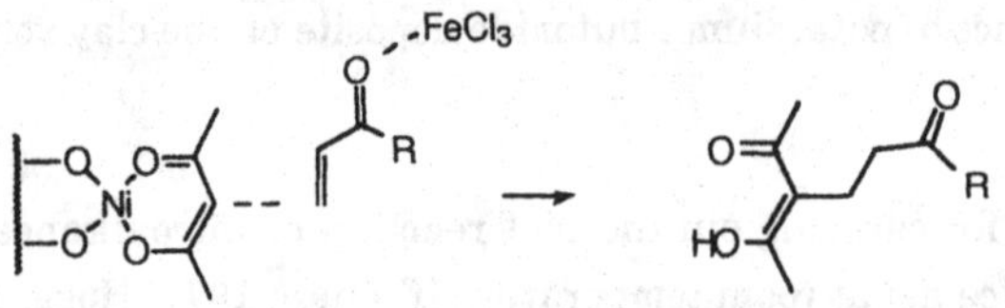

Condensation of Acetylacetone with Benzalacetone. [Laszlo, 1990]

Anhydrous nickel chloride in acetonitrile (1 g/mL) is stirred with moderate heating while K-10 montmorillonite or kaolinite in the same weight of Ni metal (0.34 g) is added. After evaporation of solvent in vacuo the residue is placed in a 280°C oven overnight to dry. It is stored in oven at 150°C.

Acetylacetone (20 mmol) and benzalacetone (4 mmol) in dioxane (2 mL) is stirred under nitrogen while the nickel salt impregnated K-10 clay (30 mg) and iron(III) chloeide (20 mg) are added all at once. The reaction is allowed to proceed at room temperature and monitored by GC. Virtually pure product is isolated by filtration through a short length of silica gel.

2.2.2. Substitutions

Through a series of studies, commercial Woelm alumina (W-200 neutral) impregnated with several equivalents of alcohols, amines, acetic acid, and benzeneselenol has been found to effect opening of epoxides at room temperature [Posner, 1977c]. Most other functional groups tolerate the mild conditions of this regioselective substitution. One inconvenience of this technique is the apparent need to handle the reagent preparation in a glove box.

Of significance is the observation that cyclopentadiene monoxide gives mixtures of products on reaction with *n*-BuOH-alumina and with HOAc-alumina but the impregnated *n*-butylamine attacks only at the allylic position [Posner, 1977d].

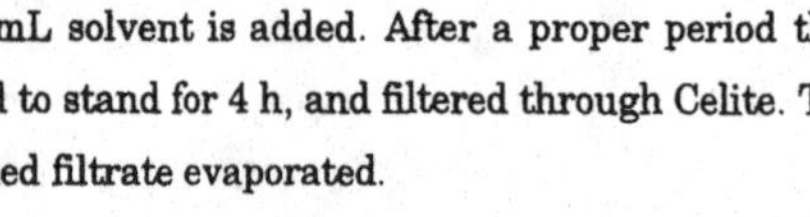

Epoxide Opening. [Posner, 1977c]

Alumina (Woelm W-200 grade) is slurried with sufficient amount of a solvent (usually ether) and treated with reagent RXH (4% wt of alumina). After 5 min. the epoxide (1 mmol per 7.5 g alumina) in 2-3 mL solvent is added. After a proper period the slurry is poured into MeOH (50-100 mL), allowed to stand for 4 h, and filtered through Celite. The solid is washed with more MeOH and the combined filtrate evaporated.

An impressive improvement of stereoselectivity has been witnessed in the displacement of an allylic acetate with diethylamine [Trost, 1978]. With tetrakis-(triphenylphosphine)palladium supported on a phosphinated silica gel as catalyst the reaction giving rise to a 2:1 product mixture of *cis*- and *trans*-isomers returns the *cis*-amino ester only.

COOMe, OAc + $EtNH_2$ → COOMe, NEt_2 + COOMe, NEt_2

$(Ph_3P)_4Pd$ (yield 85%)	67	:	33
$(Ph_3P)_4Pd$ / SiO_2 (yield 72%)	100	:	0

2.2.3. Halogenation

A definite role is played by silica gel in promoting chlorination of arenes with sulfuryl chloride [Hojo, 1975] as hardly any reaction occurs in its absence. When nuclear chlorination is impossible in the case of fully methylated arenes reaction site is changed to the benzylic position(s). The degree of chlorination can be controlled by varying the substrate: SO_2Cl_2 ratio.

SO_2Cl_2/SiO_2, CCl_4 → Cl 56%

SO_2Cl_2/SiO_2, CCl_4

1x SO_2Cl_2	96%		
2x SO_2Cl_2		87%	
4x SO_2Cl_2			91%

The lamellar structure of graphite allows intercalation of many substances. The antimony pentachloride intercalate is an interesting reagent for the introduction of a β-chlorine atom to bromoalkanes [Luche, 1974]. For example, a 81% yield of *trans*-1,2-bromochlorocyclohexane has been obtained.

SbCl$_5$ / graphite, CCl$_4$ — 81%

Bromine absorbed by 5A molecular sieves adds to styrene but not cyclohexene [Risbood, 1978]. It has been rationalized by the penetrability of the substrates into the cavities of the molecular sieves where bromine resides.

2.2.4. Cycloadditions

The Diels-Alder reaction is affected by Lewis acids. Improved rates and stereoselectivity have been observed. In the presence of a Fe(III) salt impregnated clay (K-10 montmorillonite) the Diels-Alder reaction [Laszlo, 1984a]. Some reactions involving furan as the diene, which proceed efficiently only under high pressures, are promoted by the supported catalyst.

CHO + K-10 (Fe^{3+}), CH_2Cl_2, - 24°C — CHO, 60%

R^2, O, R^5 + O, R — K-10 (Fe^{3+}), CH_2Cl_2, - 43°C — R^2, O, R, O, R^5, 24-45%

2.2.5. Oxidations

2.2.5.1. Hydrocarbons

The tertiary C-H group of saturated hydrocarbons undergoes oxygen insertion with ozone absorbed in silica [Cohen, 1975]. A convenient procedure consists of saturating the cooled (-78°C) silica gel in which the substrate is already absorbed with ozone, then warming up slowly to room temperature and eluting the product.

A methylene group adjacent to a cyclopropane ring can be converted to ketone [Proksch, 1976]. The same reagent destroys aromatic rings unless they contain strong electron-withdrawing substituents [Klein, 1975]. Thus phenylcyclohexane and tetralin give cyclohexanecarboxylic acid and adipic acid in 90% and 50% yields, repectively, both at 95% conversion. Interestingly, addition of about 25% water to the silica increases the yield of the latter reaction to 75%.

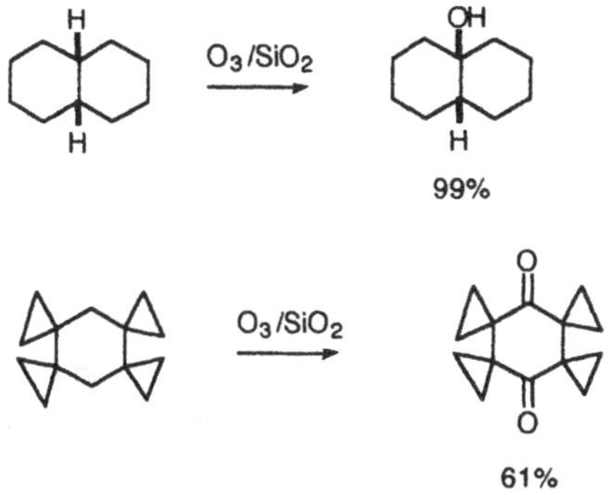

O_3/SiO_2 → 99%

O_3/SiO_2 → 61%

Clay-supported thallium(III) nitrate has been used in the preparation of certain α-formylpyrroles from the methylpyrroles [Jackson, 1984].

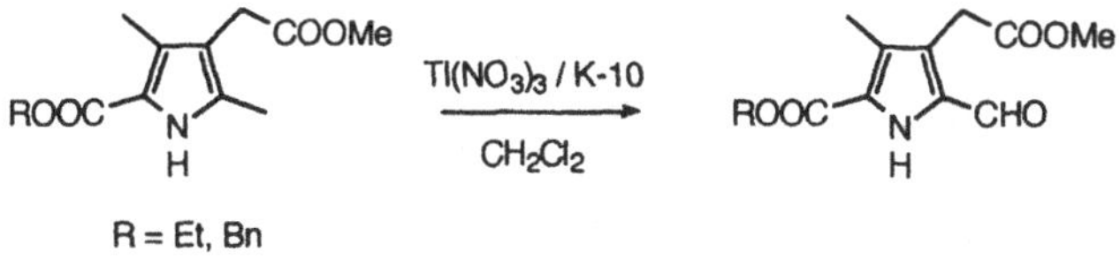

$Tl(NO_3)_3$ / K-10, CH_2Cl_2

R = Et, Bn

Epoxidation of alkenes with hydrogen peroxide has been carried out with a catalyst prepared from vanadyl chloride and montmorillonite [Choudary, 1990b]. Its activity is comparable to $VO(acac)_2$. Furthermore, using an analogous catalyst it is possible to achieve asymmetric epoxidation of primary allylic alcohols.

2.2.5.2. Oxygenated Compounds

An excellent variant of the Oppenauer oxidation consists of replacing the aluminum alkoxide-ketone system with alumina-supported aldehyde [Posner, 1976, 1977a,b]. Alcohols containing sulfide and selenide can be selectively oxidized; highly hindered as well as strained alcohols (e.g., cyclobutanol) undergo oxidation smoothly. Since there is a great difference in the oxidation rates secondary alcohols are converted to ketones preferentially.

OH ... OH —PhCHO / Al_2O_3→ O ... OH 65%

—Cl_3CCHO / Al_2O_3→ 68%

Graphite nitrate ($C_{24}^+NO_3^-\cdot HNO_3$), obtained from nitric acid and graphite, is a mild oxidant for oxidation of alcohols to aldehydes and ketones at room temperature [Kagan, 1976]. Cyclohexane is the solvent to use.

The deposition of Collins reagent [$(pyridine)_2CrO_3$] on Celite for the oxidation of allylic alcohols enables nonaqueous workup [Andersen, 1973], therefore acid-sensitive compounds survive better.

—$CrO_3.2$ py / Celite→ CHO

The solid prepared from chromic acid (H_2CrO_4) and silica gel is a useful oxidants for alcohols [Santaniello, 1978], giving product yields in the range of 68-90%. However, the reagent cannot be kept for a long period. Chromyl chloride oxidation of alcohols seems to be more selective when adsorbed on silica-alumina [Filippo, 1977].

For alcohol oxidation with potassium permanganate in benzene the reagent is best supported on Linde 13X molecular sieves [Regen, 1977]. However, primary aliphatic alcohols only furnish low yields of the aldehydes.

Vanadia-pillared montmorillonite with small interlamellar spacings (5.7A) can be used to effect selective oxidation of benzylic alcohols with hydrogen peroxide [Choudary, 1990a]. Thus *p*-substituted benzyl alcohols are readily converted to the acids (and esters), whereas the *o*-isomers remain unchanged.

X— —OH —V / montmorillonite, H_2O_2→ X— —COOR

R = H, CH_2Ar

A brown-black powder produced by admixing activated carbon and aqueous potassium permanganate is believed to be MnO_2-C [Carpino, 1970]. Its ability to oxidize benzylic and allylic alcohols is similar to that of the commonly used active manganese dioxide.

Silver carbonate-on-Celite, also known as Fetizon reagent, smoothly oxidizes alcohols under essentially neutral conditions [V. Balogh, 1971]. The reaction is carried out in a refluxing solution of a substrate with an excess of the reagent. Commonly used solvents are benzene, toluene and dichloromethane. Product isolation usually involves filtration and evaporation.

Silver carbonate-on-Celite. [V. Balogh, 1971]

Celite is washed with methanol containing 10% conc. hydrochloric acid and then distilled water until neutral. After drying at 120°C the purified Celite is added to a mechanically stirred solution of silver nitrate (34 g, 200 mmol) in distilled water (200 mL), followed by aq. $Na_2CO_3 \cdot 10H_2O$ (30 g in 300 mL H_2O) more slowly. After an additional stirring of 10 min, the precipitate is collected and dried in a rotary evaporator. The material contains ca. 1 mmol Ag_2CO_3 per 0.57 g which should be stored in the dark. Residue water can be removed by azeotropic distillation with benzene under a Dean-Stark trap.

(E)-3-Methyl-2-hexenal.

To the reagent (8g) which has been dehydrated and suspended in benzene (100 mL) is added *(E)*-3-methyl-2-hexenol (800 mg) in benzene (10 mL). The mixture is refluxed for 1 h, cooled, and evaporated. Distillation affords the aldehyde in 88% yield.

A secondary alcohol is oxidized faster than a primary alcohol, and such selectivity is valuable. In contrast to many oxidants, many 1,2-diols give α-ketols because cleavage is much slower [Fetizon, 1969a](Cf. the $NaIO_4$-SiO_2 system which cleaves 1,2-diols [Gupta, 1981]). More interestingly, 1,5-diene-3,4-diols and 1-aryl-3-alkene-1,2-diols are found to behave according to their relative configurations [Thuan, 1978]. *erythro*-Diols are converted to the α-ketols but the *threo*-diols undergo C-C bond cleavage.

When the hydroxyl groups of an alkanediol are primary-primary or primary-tertiary and properly separated the oxidation leads to a lactol which suffers a more rapid conversion to the lactone [Fetizon, 1969b]. α-Methylene-γ-butyrolactone is obtained selectively from the enediol [Fetizon, 1975a].

OH OH $\xrightarrow{Ag_2CO_3\,/\,Celite}$ O O

Related to lactone formation which indicates a more rapid oxidation of the lactols, certain aldehydes can be converted to methyl esters in the presence of methanol [Morgenlie, 1973].

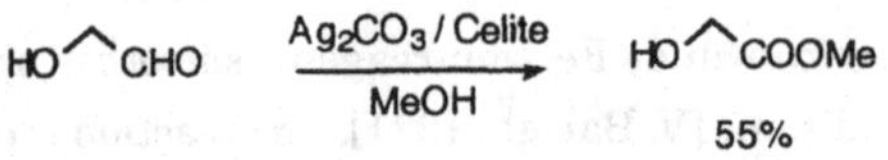

HO⌃CHO —(Ag_2CO_3 / Celite, MeOH)→ HO⌃COOMe 55%

Two oxidation products of 2-(1-hydroxyethyl)piperidine have been identified as the α-aminoenone and the iminoketone [Büchi, 1971]. The iminoketone is the primary product which isomerizes under the reaction conditions.

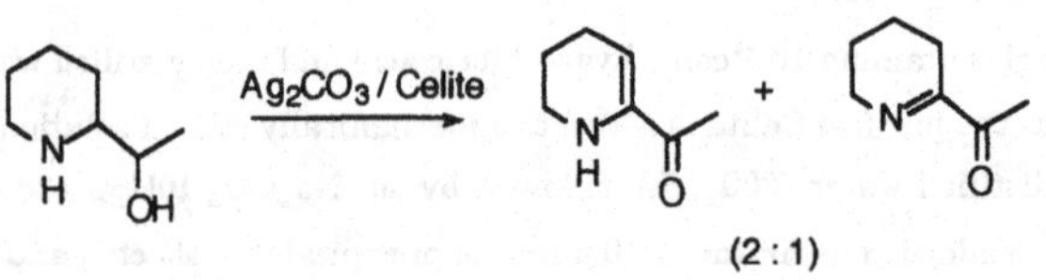

Ag_2CO_3 / Celite (2 : 1)

Note that tertiary propagylic alcohols and cyanohydrins undergo degradation with loss of the ethynyl chain [Lenz, 1972] and the cyano group, respectively. Coordination of the triple bond and the nitrile function to Ag(I) ion triggers the fragmentation. The behavior of halohydrins is also dominated by silver ion-assisted processes, giving either 1,2-rearrangment products or epoxides [Fetizon, 1973].

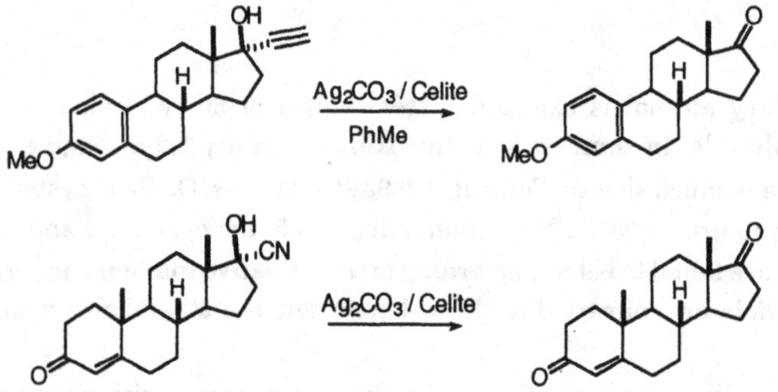

Ag_2CO_3 / Celite, PhMe

Ag_2CO_3 / Celite

Catechols and hydroquinols are converted to the corresponding quinones and *o*-aminophenol to the phenoxazone by this reagent [V. Balogh, 1971]. Interestingly, some simple phenols undergo oxidative dimerization to afford diphenoquinones and stilbenequinones. *p*-Cresol gives Pummerer's ketone in 63% yield [Anderson, 1977].

Ag_2CO_3 / Celite

Ag_2CO_3 / Celite

R = Me, iPr, tBu

Ag_2CO_3 / Celite

Pummerer's Ketone

Silica gel supported metallic nitrates, particularly copper and zinc, effect oxidative cleavage of ethers to provide aldehydes and/or ketones [Nishiguchi, 1990]. Silica gel is essential and the active oxidant is nitrogen dioxide.

Enamines are degraded to the norketones in good yields (75-91%) by $KMnO_4/Al_2O_3$ [Harris, 1997].

Supported thallium(III) nitrate on montmorillonite, prepared from a solution in trimethyl orthoformate containing methanol, is an even more selective and convenient reagent for inducing oxidative rearrangement of acetophenones and alkenes [Taylor, 1976]. Some inert substrates or those giving rise to complex mixtures on treatment with methanolic TTN react normally.

$Tl(NO_3)_3$ / K-10, $HC(OMe)_3$, 5-10 min → Ar–CH(R)–COOMe, 84-98%

$Tl(NO_3)_3$ / K-10, $HC(OMe)_3$, < 1 min → cyclopentyl–$CH(OMe)_2$, 85%

<u>Oxidative Rearrangement of Alkenes and Enolizable Ketones</u>. [Taylor, 1976]

To a stirred mixture of trimethyl orthoformate (125 mL) and methanol (100 mL) is added thallium(III) nitrate trihydrate (49 g, 0.11 mol). Upon dissolution of the salt Montmorillonite K-10 (110 g) is introduced in one portion through a powder funnel. The funnel is rinsed with methanol (25 mL) and the mixture is stirred for 5 min more. The flask is then attached to a rotary evaporator to remove the excess solvent (bath temp. 60-70°C) to give a free flowing, pale yellow powder (170 g). This reagent (1.1 equiv.) is added to a solution of the substrate (1 equiv.) in an inert solvent (CH_2Cl_2, CCl_4, dioxane, heptane, or toluene) and stirred at room temperature until a starch-iodide test is negative. The spent reagent is filtered, filtrate is washed with aq. $NaHCO_3$, H_2O, dried, evaporated. Distillation or recrystallization affords the product.

2.2.5.3. Nitrogenous compounds

Simple aliphatic amines are subject to rapid oxidation with ozone-silica to afford nitroalkanes [Keinan, 1977]. Only very small amounts of ketones or carboxylic acids accompany the major products. However, the method cannot be extended to anilines with equal effectiveness.

Nitroalkanes from Amines. [Keinan, 1977]

A stream of ozone (3% in oxygen) is passed through silica gel (ca. 30 g, Merck Kieselgel 60, 70-230 mesh, predried at 450°C for 24 h) which has been well mixed with the amine (0.1-0.2 wt%) and cooled to -78°C. Thereafter the mixture is warmed to room temperature and eluted with theproper solvent.

The efficient dehydrogenation of aldoximes to nitrile oxides and disubstituted hydroxylamines to nitrones by Ag_2CO_3-Celite [Fetizon, 1975b] facilitates the preparation of isoxazoles or isoxazolines and isoxazolidines. On the other hand it is possible to synthesize nitroso compounds rapidly from the monosubstituted hydroxylamines [Maasen, 1971].

ArCH=NOH —(Ag_2CO_3 / Celite)→ [$Ar\text{-}\overset{+}{C}{=}N\text{-}O^-$] —(ArCH=NOH)→ 3,5-diaryl-1,2,4-oxadiazole (Ar, N, N, O, Ar)

Ph⌒N(OH)⌒Ph —(Ag_2CO_3 / Celite)→ Ph⌒$\overset{+}{N}$(O$^-$)=⌒Ph

Dehydrogenation occurs when many nitrogenous compounds are exposed to the Fetizon reagent. Thus hydrazones give diazoalkanes rapidly which lead to azines. Camphor hydrazone loses nitrogen completely, resulting in a mixture of tricyclene and bornylene (ratio 30:1) [Fetizon, 1975b]. The MnO_2-carbon mixture mentioned above also oxidizes hydrazines to azo or diazoalkanes [Carpino, 1970].

$Ar_2C{=}N\text{-}NH_2$ —(Ag_2CO_3 / Celite, short time)→ $Ar_2C{=}N_2$ —(Ag_2CO_3 / Celite, long time)→ $Ar_2C{=}N\text{-}N{=}CAr_2$ + $Ar_2C{=}O$

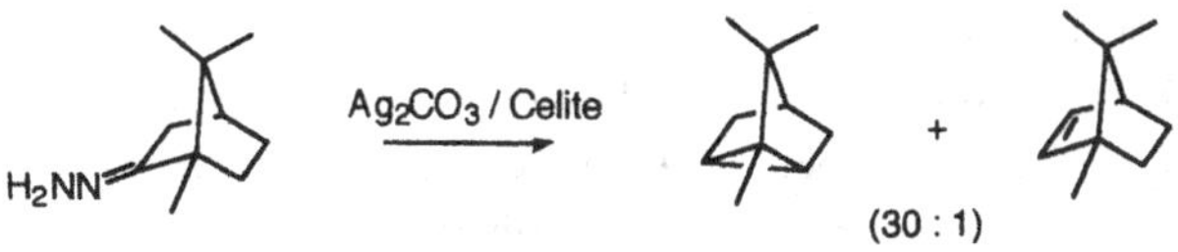

Ag_2CO_3 / Celite

(30 : 1)

2.2.5.4. Miscellaneous Substrates

Nonaromatic organoselenoxides are thermally unstable and they undergo elimination readily. Thus this property is exploitable for synthesis where introduction of unsaturatrion is required. Alkyl phenyl selenides can be oxidized with a great variety of oxidants but their treatment with *t*-butyl hydroperoxide in the presence of basic alumina at 55°C constitutes a superior protocol [Labar, 1978a,b].

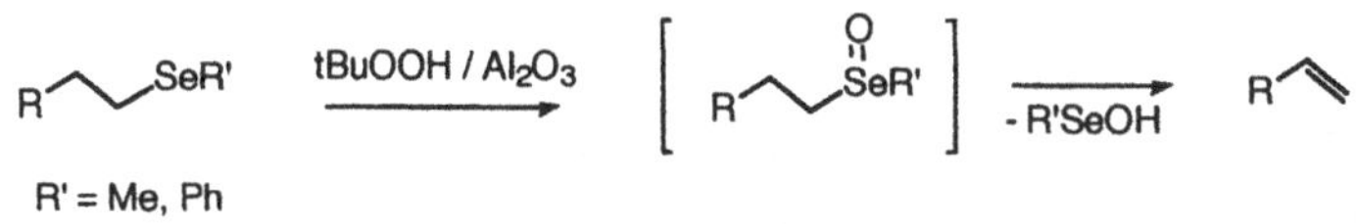

tBuOOH / Al_2O_3 ; - R'SeOH

R' = Me, Ph

For oxidation of sulfides to sulfoxides one might employ alumina-supported thallium(III) nitrate [Liu, 1978] or sodium periodate [Liu, 1978], and magnesium monoperoxyphthalate on silica gel [Ali, 1997]..

2.2.6. Reductions

Sodium borohydride deposited on alumina (after evaporation over P_2O_5 of a slurry with concentrated aqueous $NaBH_4$) has the advantage of reducing base-sensitive carbonyl compounds because a nonpolar solvent such as benzene can be used [Hodosan, 1969]. For example, the reduction of 3-acetoxyandrosta-3,5-dien-17-one is achieved without affecting the dienol acetate.

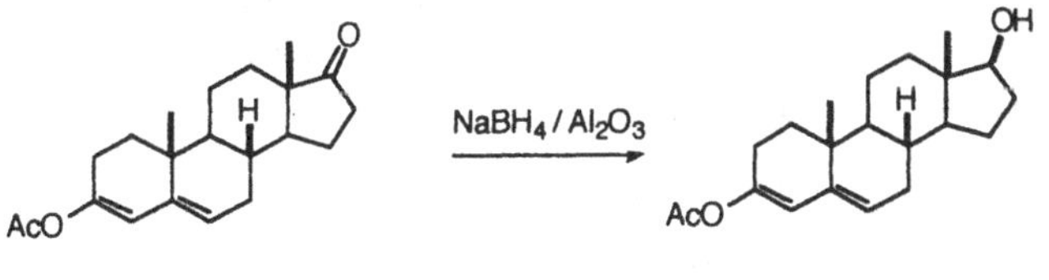

$NaBH_4$ / Al_2O_3

Silica gel supported zinc borohydride is a chemoselective reducing agent for conjugated carbonyl compounds [Ranu, 1991], giving allylic alcohols as products. While the reduction normally proceeds in THF at -5°~-10°C for many substrates, cyclopentenone and cyclohexenone are best treated with the reagent at lower temperatures without solvent.

Silica gel apparently increases the reducing power of tributyltin hydride [Fung, 1978], as under such conditions not only carbonyl compounds containing strong electron-withdrawing groups are affected. Aldehydes are reduced in preference to ketones, while esters, nitriles, nitro compounds, sulfoxides, and halides survive the treatment.

The use of isopropanol adsorbed on activated alumina (Woelm W-200) to reduce carbonyl compounds [Posner, 1977e] can be considered as an improvement of the Meerwein-Pondorff-Verley method. Excellent discrimination of carbonyl groups is observed. Most importantly, this procedure enables the synthesis of primary alcohols specifically deuterated at the α-position by using Me_2CDOH.

CHO H iPrOH / Al_2O_3 OH H O 61%

Reduction of Aldehydes. [Posner, 1977e]

Alumina (Woelm W-200) is dehydrated in a quartz vessel at 400°C/0.6 torr for 24 h. In a N_2-filled glove bag a portion (~5 g) of the alumina is transferred into an oven-dried flask containing a stirring bar which is then stoppered. The flask is charged with an inert solvent (~ 5 mL) ($CHCl_3$, CCl_4, Et_2O, hexane), isopropanol (~ 500 mg), and after 0.5h, the aldehyde (~ 1 mmol) in a solvent (~ 1 mL). The mixture is stirred at room temperature for 2 h, quenched with methanol (25 mL) and filtered. From the combined filtrate and washings the product is isolated.

Ketone reduction with the graphite-potassium intercalate C_8K [Lalancette, 1972] seems to proceed in a manner akin to electrochemical reduction, via adsorption of the ketone on the surface of the reagent. In the enone cases saturation of the conjugated double bond is the major course.

O C_8K O 75%

Monographs and Reviews

Akelah, A. (1981) *Synthesis* 413.

Balogh, M.; Laszlo, P. (1993) *Organic Chemistry Using Clays*, Springer: Berlin.
Clark, J.H. (1994) *Catalysis of Organic Reactions by Supported Inorganic Reagents*, VCH: New York.
Ford, W.T. (Ed.) (1986) *Polymeric Reagents and Catalysts*, ACS: Washinton, D.C.
Frechet, J.M.J. (1981) *Tetrahedron Lett.* **37:** 663.
Hodge, P.; Sherrington, D.C. (Eds.) (1980) *Polymer-supported Reactions in Organic Synthesis*, Wiley: Chichester.
Kraus, M.A.; Patchornik, A. (1980) *Macromol. Rev.* **15:** 55.
McKillop, A.; Young, D.W. (1979) *Synthesis* 401, 481.
Manecke, G.; Storck, W. (1978) *Angew. Chem. Int. Ed. Engl.* **17:** 657.
Mathur, N.K.; Narang, C.K.; Williams, R.E. (1980) *Polymers as Aids in Organic Chemistry*, Academic Press: New York.
Posner, G.H. (1978) *Angew. Chem. Int. Ed. Engl.* **17:** 487.
Smith, K. (Ed.) (1992) *Solid Supports and Catalystss in Organic Synthesis*, Prentice-Hall: New York.

References

Ali, M.H.; Stevens, W.C. (1997) *Synlett* 764.
Anderson, R.A.; Dalgleish, D.T.; Nonhebel, D.C.; Pauson, P.L. (1977) *J. Chem. Res. (Synop.)* 12.
Andersen, N.H.; Uh, H.-S. (1973) *Synth. Commun.* **3:** 115; *Tetrahedron Lett.* 2079.
Appel, R.; Strüver, W.; Willms, L. (1976) *Tetrahedron Lett.* 905.
Arshady, R.; Kenner, G.W.; Ledwith, A. (1976) *Makromol. Chem.* **177:** 2911.
Arshady, R.; Ledwith, A. (1978) *Makromol. Chem.* **179:** 819.
Balogh, M.; Cornelis, A.; Laszlo, P. (1984) *Tetrahedron Lett.* **25:** 3313.
Balogh, V.; Fetizon, M.; Golfier, M. (1971) *J. Org. Chem.* **36:** 1339.
Bertin, J.; Kagan, H.B.; Luche, J.-L.; Setton, R. (1974) *J. Am. Chem. Soc.* **96:** 8113.
Bokelmann, C.; Neumann, W.P.; Peterseim, M. (1992) *J. Chem. Soc. Perkin Trans. 1* 3165.
Borders, C.L.; MacDonell, D.L.; Chambers, J.L. (1972) *J. Org. Chem.* **37:** 3549.
Büchi, G.; Wüest, H. (1971) *J. Org. Chem.* **36:** 609.
Chalk, A.J. (1968) *J. Polymer Sci., Polymer Lett.* 649.
Cainelli, G.; Manescalchi, F. (1975) *Synthesis* 723.
Cainelli, G.; Cardillo, G.; Orena, M; Sandri, S. (1976a) *J. Am. Chem. Soc.* **98:** 6737.
Cainelli, G.; Manescalchi, F. (1976b) *Synthesis* 472.
Cainelli, G.; Manescalchi, F.; Ronchi, A.U. (1978) *J. Org. Chem.* **43:** 1598.
Cardillo, G.; Orena, M; Sandri, S. (1976) *Tetrahedron Lett.* 3985.
Carpino, L.A. (1970) *J. Org. Chem.* **35:** 3971.

Castells, J.; Font, J.; Virgili, A. (1979) *J. Chem. Soc. Perkin Trans. 1* 1.
Chalais, S.; Laszlo, P.; Mathy, A. (1986) *Tetrahedron Lett.* **27:** 2627.
Choudary, B.M.; Valli, V.L.K. (1990a) *Chem. Commun.* 1115.
Choudary, B.M.; Valli, V.L.K.; Prasad, A.D. (1990b) *Chem. Commun.* 721.
Clark, J.H. (1978) *Chem. Commun.* 789.
Cohen, B.J.; Kraus, M.A.; Patchornik, A. (1981) *J. Am. Chem. Soc.* **103:** 7620.
Cohen, M.D.; Green, B.S. (1973) *Chem. Brit.* **9:** 490.
Cohen, Z.; Keinan, E.; Mazur, Y.; Varkony, T.H. (1975) *J. Org. Chem.* **40:** 2141.
Collins, M.; Laws, D.R.J. (1973) *J. Chem. Soc. Perkin Trans. 1* 2013.
Crosby, G.A.; Weinshenker, N.M.; Uh, H.-S. (1975) *J. Am. Chem. Soc.* **97:** 2232.
Crosby, G.A.; Kato, M. (1977) *J. Am. Chem. Soc.* **99:** 278.
Crowley, J.I.; Rapoport, H. (1970) *J. Am. Chem. Soc.* **92:** 6363.
Dalibor, H. (1958) *Chem. Ber.* **91:** 1955.
Eschenfelder, V.; Brossmer, R.; Wachter, M. (1975) *Angew. Chem. Int. Ed. Engl.* **14:** 715.
Farrall, M.J.; Frechet, J.M.J. (1976) *J. Org. Chem.* **41:** 3877.
Farrall, M.J.; Durst, T.; Frechet, J.M.J. (1979) *Tetrahedron Lett.* 203.
Fetizon, M.; Golfier, M.; Louis, J.-M. (1969a) *Chem. Commun.* 1102.
Fetizon, M.; Golfier, M.; Louis, J.-M. (1969b) *Chem. Commun.* 1118.
Fetizon, M.; Golfier, M.; Louis, J.-M. (1973) *Tetrahedron Lett.* 1931.
Fetizon, M.; Golfier, M.; Louis, J.-M. (1975a) *Tetrahedron* **31:** 171.
Fetizon, M.; Golfier, M.; Milcent, R.; Papadakis, I. (1975b) *Tetrahedron* **31:** 165.
Filippo, J.S.; Chern, C.-I. (1977) *J. Org. Chem.* **42:** 2182.
Frechet, J.M.J.; Haque, K.E. (1975) *Macromolecules* **8:** 130.
Frechet, J.M.J.; Nuyens, L.J. (1976) *Can. J. Chem.* **54:** 926.
Frechet, J.M.J.; Farrall, M.J.; Nuyens, L.J. (1977) *J. Macromol. Sci., Chem. A* **11:** 507.
Frechet, J.M.J.; Warnock, J.; Farrall, M.J. (1978) *J. Org. Chem.* **43:** 2618.
Fridkin, M.; Patchornik, A.; Katchalski, E. (1966) *J. Am. Chem. Soc.* **88:** 3164.
Fridkin, M.; Patchornik, A.; Katchalski, E. (1968) *J. Am. Chem. Soc.* **90:** 2953.
Fruchtel, J.S.; Jung, G. (1996) *Angew. Chem. Int. Ed. Engl.* **35:** 17.
Fung, N.Y.M.; de Mayo, P.; Schauble, J.H.; Weedon, A.C. (1978) *J. Org. Chem.* **43:** 3977.
Gelbard, G.; Colonna, S. (1977) *Synthesis* 113.
Gibson, H.W.; Bailey, F.C. (1977) *Chem. Commun.* 815.
Gordon, M.; DePamphilis, M.L.; Griffin, C.E. (1963) *J. Org. Chem.* **28:** 698.
Gravert, D.J.; Janda, K.D. (1997) *Chem. Rev.* **97:** 489.
Greene, A.E.; Cruz, A.; Crabbe, P. (1976) *Tetrahedron Lett.* 2707.
Grubbs, R.H.; Su, S.C.H. (1976) *J. Organomet. Chem.* **122:** 151.

Gupta, D.N.; Hodge, P.; Davies, J.E. (1981) *J. Chem. Soc. Perkin Trans. 1* 2970.
Hallensleben, M.L. (1972) *Angew. Makromol. Sci.* **27:** 223.
Hallensleben, M.L. (1974) *J. Polymer Sci., Polymer Symp.* **47:** 1.
Harris, C.E.; Chrisman, W.; Bickford, S.A.; Lee, L.Y.; Torreblanca, A.E.; Singaram, B. (1997) *Tetrahedron Lett.* **38:** 981.
Harrison, C.R.; Hodge, P. (1976a) *J. Chem. Soc. Perkin Trans. 1* 605.
Harrison, C.R.; Hodge, P. (1976b) *J. Chem. Soc. Perkin Trans. 1* 2252.
Harrison, C.R.; Hodge, P.; Rogers, W.J. (1977) *Synthesis* 41.
Harrison, I.T.; Harrison, S. (1967) *J. Am. Chem. Soc.* **89:** 5723.
Hart, H.; Chen, B.-l.; Peng, C.-t. (1977) *Tetrahedron Lett.* 3121.
Harvey, S.; Raston, C.L. (1988) *Chem. Commun.* 652.
Heitz, W.; Michels, R. (1972) *Angew. Chem. Int. Ed. Engl.* **11:** 298.
Heitz, W.; Michels, R. (1973) *Liebigs Ann. Chem.* 227.
Hermkens, P.H.H.; Ottenheijm, H.C.J.; Rees, D. (1996) *Tetrahedron* **52:** 4527.
Hodge, P.; Richardson, G. (1975) *Chem. Commun.* 622.
Hodosan, F.; Serban, N. (1969) *Rev. Roumaine Chim.* **14:** 121.
Hogg, J.L.; Goodwin, T.E.; Nave, D.W. (1978) *Org. Prep. Proc. Int.* **10:** 9.
Hojo, M.; Masuda, R. (1975) *Synth. Commun.* **5:** 169.
Hojo, M.; Masuda, R. (1976a) *Tetrahedron Lett.* 613.
Hojo, M.; Masuda, R. (1976b) *Synthesis* 678.
Horiki, K. (1976) *Tetrahedron Lett.* 4103.
Hudlicky, T. (1974) *J. Org. Chem.* **39:** 3461.
Huet, F.; Pellet, M.; Conia, J.M. (1977) *Tetrahedron Lett.* 3505.
Hughes, I. (1996) *WOP* 9600378.
Hutchins, R.O.; Natale, N.R.; Taffer, I.M. (1978) *Chem. Commun.* 1088.
Inczedy, J. (1961) *Acta Chim. Acad. Hung.* **27:** 185.
Jackson, A.H.; Rao, K.R.N.; Ooi, S.N.; Adelakun, E.; (1984) *Tetrahedron Lett.* **25:** 6049.
Johnson, C.R.; Zhang, B. (1995) *Tetrahedron Lett.* **36:** 9253.
Kagan, H.B. (1976) *Chemtech* **6:** 510.
Kato, T.; Katagiri, N.; Nakano, J.; Kawamura, H. (1977) *Chem. Commun.* 645.
Kawai, M.; Onaka, M.; Izumi, Y. (1986) *Chem. Lett.* 381.
Keinan, E.; Mazur, Y. (1977) *J. Org. Chem.* **42:** 844.
Keinan, E.; Mazur, Y. (1978) *J. Org. Chem.* **43:** 1020.
Kelly, J.T. (1958) *U.S. Pat.* 2843642.
Kemp, D.S.; Reczek, J. (1977) *Tetrahedron Lett.* 1031.
Klein, H.; Steinmetz, A. (1975) *Tetrahedron Lett.* 4249.
Labar, D.; Dumont, W.; Hevesi, L.; Krief, A. (1978a) *Tetrahedron Lett.* 1145.

Labar, D.; Hevesi, L.; Dumont, W.; Krief, A. (1978b) *Tetrahedron Lett.* 1141.
Laszlo, P.; Lucchetti, J. (1984a) *Tetrahedron Lett.* **25:** 2147, 4387.
Laszlo, P.; Polla, E. (1984b) *Tetrahedron Lett.* **25:** 3309.
Laszlo, P.; Pennetreau, P. (1985a) *Tetrahedron Lett.* **26:** 2645.
Laszlo, P.; Polla, E. (1985b) *Synthesis* 439.
Laszlo, P.; Montaufier, M.T.; Randriamahefa, S.L. (1990) *Tetrahedron Lett.* **31:** 4867.
Leebrick, J.R.; Ramsden, H.E. (1958) *J. Org. Chem.* **23:** 935.
Lenz, G.R. (1972) *Chem. Commun.* 468.
Lalancette, J.-M.; Rollin, G.; Dumas, P. (1972) *Can. J. Chem.* **50:** 3058.
Letsinger, RL.; Kornet, M.J.; Mahadevan, V.; Jerina, D.M. (1964) *J. Am. Chem. Soc.* **86:** 5163.
Leznoff, C.C.; Wong, J.Y. (1972) *Can. J. Chem.* **50:** 2892.
Leznoff, C.C.; Wong, J.Y. (1973) *Can. J. Chem.* **51:** 3756.
Leznoff, C.C.; Fyles, T.M.; Weatherston, J. (1977) *Can. J. Chem.* **55:** 1143.
Leznoff, C.C. (1978) *Acc. Chem. Res.* **11:** 327.
Liu, K.-T. (1977) *J. Chin. Chem. Soc.* **24:** 217.
Liu, K.-T.; Tong, Y.-C. (1978) *J. Org. Chem.* **43:** 2717.
Lowe, G. (1995) *Chem. Soc. Rev.* **24:** 309.
Luche, J.-L.; Bertin, J.; Kagan, H.B. (1974) *Tetrahedron Lett.* 759.
McKinley, S.V.; Rakshys, J.W. (1972) *Chem. Commun.* 134.
Maasen, J.A.; De Boer, T.J. (1971) *Recl. Trav. Chim. Pays-Bas* **90:** 373.
Marshall, J.A.; Andersen, N.H.; Johnson, P.C. (1970) *J. Org. Chem.* **35:** 186.
Merrifield, R.B. (1963) *J. Am. Chem. Soc.* **85:** 2149.
Michels, R.; Kato, M.; Heitz, W. (1976) *Makromol. Chem.* **177:** 2311.
Miller, J.M.; So, K.-H.; Clark, J.H. (1978) *Chem. Commun.* 466.
Morgenlie, S. (1973) *Acta Chem. Scand.* **27:** 3009.
Nishiguchi, T.; Bougauchi, M. (1990) *J. Org. Chem.* **55:** 5606.
Olah, G.A.; Kaspi, J.; Bukala, J. (1977) *J. Org. Chem.* **42:** 4187.
Onaka, M.; Sugita, K.; Izumi, Y. (1989) *J. Org. Chem.* **54:** 1116.
Pittman, C.U.; Hanes, R.M. (1977) *J. Org. Chem.* **42:** 1194.
Polyanskii, N.G. (1970) *Russ. Chem. Rev.* **39:** 244.
Posner, G.H.; Perfetti, R.B.; Runquist, A.W. (1976) *Tetrahedron Lett.* 3499.
Posner, G.H.; Chapdelaine, M.J. (1977a) *Tetrahedron Lett.* 3227.
Posner, G.H.; Chapdelaine, M.J. (1977b) *Synthesis* 555.
Posner, G.H.; Rogers, D.Z. (1977c) *J. Am. Chem. Soc.* **99:** 8208.
Posner, G.H.; Rogers, D.Z. (1977d) *J. Am. Chem. Soc.* **99:** 8214.
Posner, G.H.; Runquist, A.W.; Chapdelaine, M.J. (1977e) *J. Org. Chem.* **42:** 1202.
Proksch, E.; de Meijere, A. (1976) *Tetrahedron Lett.* 4851.

Rabinowitz, R.; Marcus, R. (1961) *J. Org. Chem.* **25:** 4157.
Ranu, B.C.; Das, A.R. (1991) *J. Org. Chem.* **56:** 4796.
Regen, S.L.; Lee, D.P. (1975) *J. Org. Chem.* **40:** 1669.
Regen, S.L.; Koteel, C. (1977) *J. Am. Chem. Soc.* **99:** 3837.
Reggelin, M.; Brenig, V. (1996) *Tetrahedron Lett.* **37:** 6851.
Relles, H.M.; Schluenz, R.W. (1974) *J. Am. Chem. Soc.* **96:** 6469.
Risbood, P.A.; Ruthven, D.M. (1978) *J. Am. Chem. Soc.* **100:** 4919.
Roush, W.R.; Feitler, D.; Rebek, J. (1974) *Tetrahedron Lett.* 1391.
Rubinstein, M.; Patchornik, A. (1972) *Tetrahedron Lett.* 2881.
Santaniello, E.; Ponti, F.; Manzocchi, A. (1978) *Synthesis* 534.
Savoia, D.; Trombini, C.; Umani-Ronchi, A. (1977) *Tetrahedron Lett.* 653.
Schuttenberg, H.; Klumpp, G.; Kaczmar, U.; Turner, S.R.; Schulz, R.C. (1973) *J. Macromol. Sci., Chem. A* **7:** 1085.
Senear, A.E.; Wirth, J.; Neville, R.G. (1960) *J. Org. Chem.* **25:** 807.
Seymour, E.; Frechet, J.M.J. (1976) *Tetrahedron Lett.* 1149.
Shimo, K.; Wakamatsu, S. (1963) *J. Org. Chem.* **28:** 504.
Tamami, B.; Goudarzian, N. (1994) *Chem. Commun.* 1079.
Taylor, E.C.; Chiang, C.-S.; McKillop, A.; White, J.F. (1976) *J. Am. Chem. Soc.* **98:** 6750.
Terrett, N.K.; Gardner, M.; Gordon, D.W.; Kobylecki, R.J.; Steele, J. (1995) *Tetrahedron* **51:** 8135.
Thuan, S.L.T.; Maitte, P. (1978) *Tetrahedron* **34:** 1469.
Toda, F. (1993) *Synlett* 303.
Trost, B.M.; Keinan, E. (1978) *J. Am. Chem. Soc.* **100:** 7779.
Weinshenker, N.M.; Shen, C.-M. (1972a) *Tetrahedron Lett.* 3281.
Weinshenker, N.M.; Shen, C.-M. (1972b) *Tetrahedron Lett.* 3285.
Weinshenker, N.M.; Crosby, G.A.; Wong, J.Y. (1975) *J. Org. Chem.* **40:** 1966.
Wiley, R.H.; Kim, K.S.; Rao, S.P. (1971) *J. Polymer Sci., A1* **9:** 805.
Zupan, M.; Pollak, A. (1975) *Chem. Commun.* 715.

3 SONOCHEMISTRY

The application of ultrasound is well known for medical diagnostics, teeth cleaning, and echo location (SONAR) to detect underwater objects, but its widespread use in synthetic chemistry has a short history. Only in the 1980s was the potential of ultrasound for promotion of various organic reactions was recognized.

Ultrasound generally designates electromagnetic radiation in the range of 20 to 10,000 KHz. The energy is insufficient to cause chemical reactions but when ultrasound travels through media it creates a series of compressions and rarefactions, and the rarefaction of solvents can lead to cavities. Bubbles thus formed would grow by attracting surrounding vapor until the next compression starts. Eventually the sudden and violent collapse (implosion) of the bubbles during the compression phase after a few acoustic cycles causes an enormous rise in local pressures and temperatures. Thus it is evident that the solvent is the energy transmitter in sonochemical reactions; however, there is a significant difference between the thermal processes promoted by ultrasound and the conventional method. The beneficial effects of acceleration by sonochemical radiation are frequently dramatic on account of the relatively short time required for completing the reaction such that decomposition of thermally labile products is greatly reduced. Of significance is the fact that ultrasound influences both homogeneous and heterogeneous reactions.

The frequency of ultrasound has surprisingly little influence on reactions within the range in which cavitation occurs. On the other hand, the power input to the transducer intimately affects reaction rates because cavitation of the volume of liquid is dependent on the energy supplied to the system.

3.1. ISOMERIZATIONS

The effect of ultrasound on the isomerization of alkenes has been studied in very few cases. The conversion of maleic acid to fumaric acid in the presence of bromine or alkyl bromide [Elpiner, 1965] may involve sonic generation of bromine atom which adds to the double bond of the acid. The carbon radical is free to undergo C-C bond rotation before expulsion of bromine.

Br_2 —))), H_2O→ 2 Br·

HOOC–CH=CH–COOH (cis) —Br·→ HOOC–CH(Br)–CH·–COOH —))), -Br·→ HOOC–CH=CH–COOH (trans)

Iron pentacarbonyl responds differently to heat, light and ultrasound. Sonic irradiation splits off two molecules of carbon monoxide resulting in the coordinatively unsaturated $Fe(CO)_3$. The latter species is an avid scavenger of ligands such as alkenes, therefore double bond isomerization can be initiated, e.g., 1-pentene to 2-pentenes [Suslick, 1981]. Only catalytic amounts of $Fe(CO)_5$ is required, but the isomerization proceeds with a rate increase of 10^5 fold.

$Fe(CO)_5$ —)))→ $Fe(CO)_3$ —2 L→ $Fe(CO)_3L_2$

L = alkene,...

1-pentene —))), $Fe(CO)_5$ (cat.)→ 2-pentenes

Internal triple bonds migrate to the terminal position by treatment with potassium 3-aminopropylamide, when the intervening positions of the carbon chain is unblocked. The reaction is thermodynamically driven. The required base is conveniently prepared from the diamine and the metal in the presence of ferric nitrate by sonication [Kimmel, 1984].

$H_2N(CH_2)_3NH_2$ —))), K / $Fe(NO)_3$, THF, 90°, 15 min→ $KNH(CH_2)_3NH_2$

alkynol (internal triple bond)–OH —$KNH(CH_2)_3NH_2$, 0°, 30 min→ terminal alkynol–OH (97%)

3.2. HYDROLYSIS

Ester hydrolysis can be conducted without external heating when irradiated with ultrasound [Moon, 1979]. Comparing the results of saponification of methyl 2,4-dimethylbenzoate in water definite benefit is shown. The rate enhancement has been attributed to secondary effects of cavitation such as emuldification. The hydrolysis of *p*-nitrophenyl esters has also been studied [Kristol, 1981].

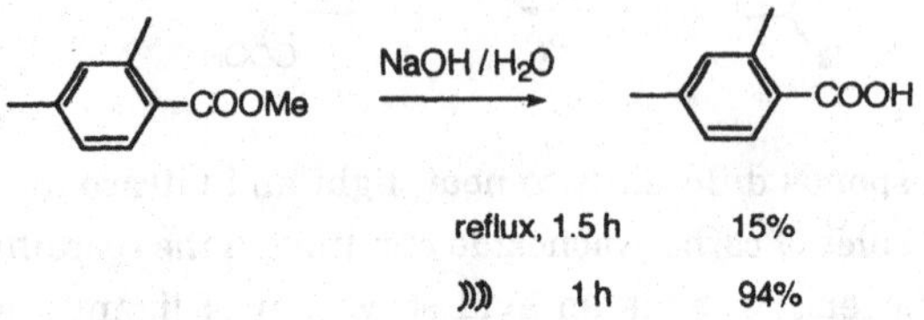

-COOMe → NaOH / H_2O → -COOH

reflux, 1.5 h 15%

))) 1 h 94%

The yields of carboxylic acids from nitriles are greatly increased under sonication [Elguero, 1984]. However, good results are achieved at elevated temperatures.

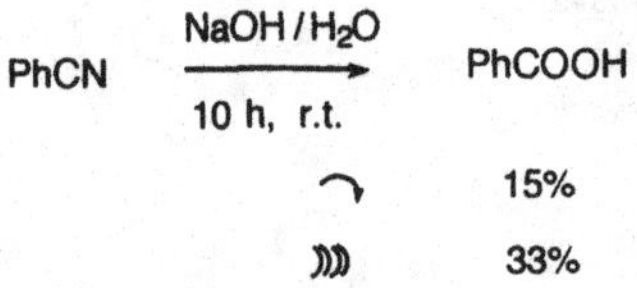

PhCN → NaOH / H_2O, 10 h, r.t. → PhCOOH

15%

))) 33%

Carbonyl group regeneration from acetals and enol ethers by aqueous acid is quite easy. In pursuing synthesis of a milbemycin-D fragment [Crimmins, 1987] the double hydrolysis was conducted with the aid of sonication. The sonic role is unclear.

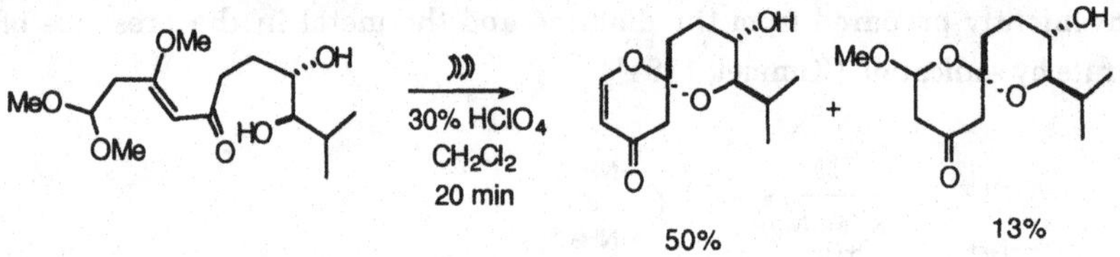

))) 30% $HClO_4$, CH_2Cl_2, 20 min

50% 13%

3.3. OXIDATIONS

Aerial oxidation of aldehydes including sugars have been subjected to simultaneous sonication. More conventional oxidations carried out in an ultrasound bath include the preparation of hindered nitroxyl radicals [Kaliska, 1987] which failed with stirring.

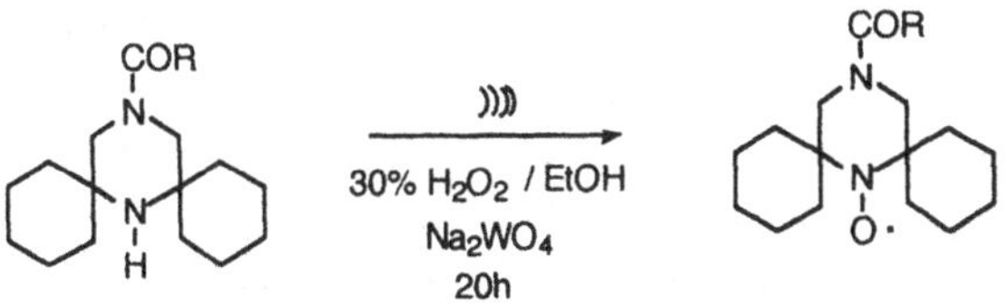

COR

N

N

H

)))

30% H_2O_2 / EtOH

Na_2WO_4

20h

COR

N

N

O·

Oxidation of alcohols with a $KMnO_4$ suspension in benzene or hexane becomes possible with irradiation of ultrasound [Yamawaki, 1983]. (Cf. reaction in the presence of a crown ether, i.e., under phase transfer conditions).

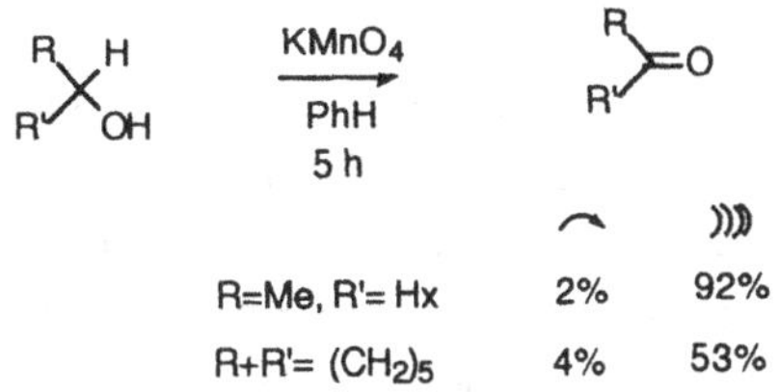

R H / R' OH → ($KMnO_4$, PhH, 5 h) → R, R' C=O

	⌒	)))
R=Me, R'= Hx	2%	92%
R+R'= $(CH_2)_5$	4%	53%

Crystalline manganese dioxide of low reactivity for oxidation of allylic and benzylic alcohols can be activated by ultrasonic treatment [Kimura, 1988]. The use of a higher boiling solvent increases the yield of the product [Zikra, unpubl.].

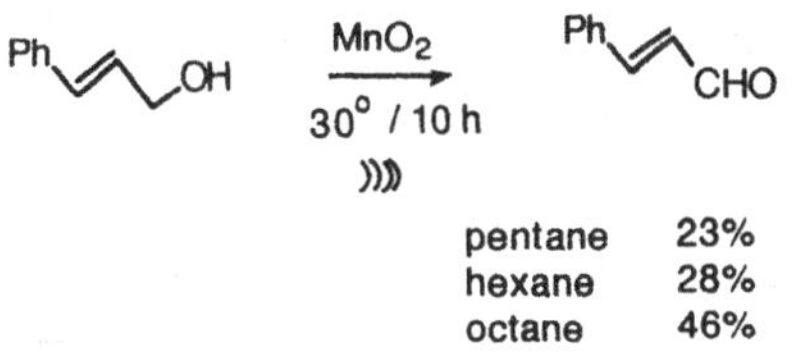

Ph-CH=CH-CH₂OH → (MnO_2, 30° / 10 h,)))) → Ph-CH=CH-CHO

pentane	23%
hexane	28%
octane	46%

The reaction of primary alcohols with 60% nitric acid can take different courses, depending whether ultrasound is used [C. Einhorn, 1990]. Oxidation to the carboxylic acids occurs with ultrasonic assistance, otherwise formation of the nitrate ester is observed.

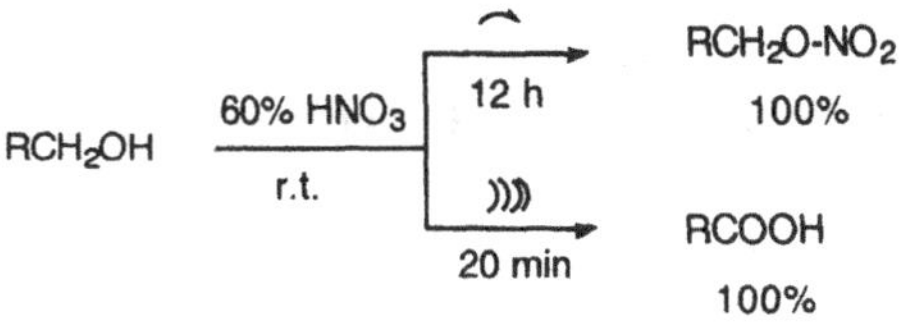

RCH_2OH → (60% HNO_3, r.t.)

⌒, 12 h → $RCH_2O\text{-}NO_2$ 100%

))), 20 min → RCOOH 100%

Oxygenation at a benzylic position by chromyl chloride in high yield can be accomplished with ultrasound [Luzzio, 1993]. A very interesting observation pertains to autoxidation of *p*-nitrotoluene [Newmann, 1985], giving the carboxylic acid. On the other hand, with the same conditions (O_2, KOH, polyethylene glycol) but omitting sonication only the substituted styrene and bibenzyl were obtained.

Ph_2CX_2 Ph_3CY

CrO_2Cl_2 ox.)))
X = H->O
Y = H->OH

99% 80% 99%

Indanone. [Luzzio, 1993]

To a cooled (0°), sonicated (high intensity) solution of freshly prepared chromyl chloride (2.53 mmol) in dichloromethane (50 mL) is introduced indane (0.1 g, 0.85 mmol) in dichloromethane (10 mL). The mixture is then allowed to warm to room temperature while sonication continues. At this point, the addition of zinc dust (0.41 g, 6.3 mmol) and the sonication level is reduced to a low level. After 0.5 h, the mixture is cooled, quenched with ice, and resubjected to maximum intensity ultrasound. Indanone is isolated in 80% yield (cf. 45% in the absence of ultrasound irradiation) after extractive workup, proper washing and drying, and flash chromatography.

NO_2 KOH PEG O_2))) O_2N–C6H4–COOK

O_2N–C6H4–CH=CH–C6H4–NO_2 + O_2N–C6H4–CH2CH2–C6H4–NO_2

3.4. REDUCTION, REDUCTIVE ELIMINATION AND COUPLING

As many organic reduction reactions are carried out with metals, often under heterogeneous conditions, the value of sonication is amply shown.

Deamination of 6-aminopenicillanic acid has been achieved [Brennan, 1985] via the bromide which is obtainable by diazotization in the presence of NaBr. Treatment with zinc (after esterification) and hydrolysis of the organozinc species terminate the process.

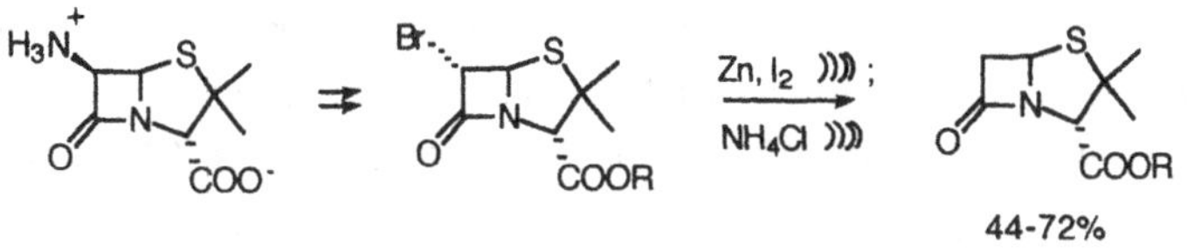

Much shorter time is needed and higher yields of reduction products are routinely obtained when aryl halides are treated with $LiAlH_4$ in DME (vs. stirring in THF) [Han, 1982a] or nickel chloride-NaI in aqueous HMPA [Yamashita, 1985]. The feasibility of electrochemical decontamination of polychlorinated chlorinated biphenyls (PCB's) at stirred mercury pool electrodes has been evaluated, and sonication increases the rate by 2- to 3-fold [Connors, 1984].

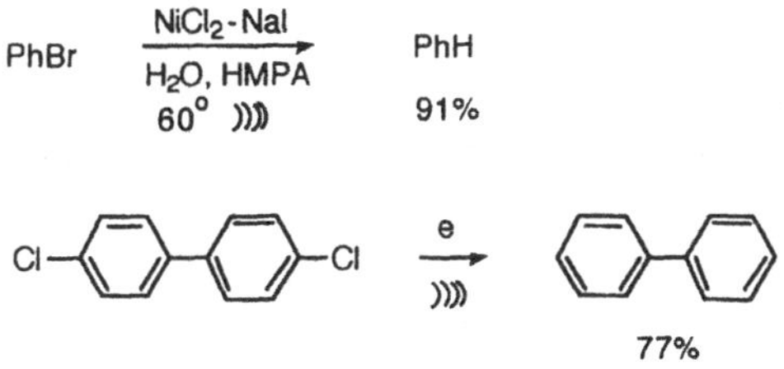

Ultrasonically dispersed mercury in acetic acid reacts with α,α'-dibromocycloalkanones (ring size: 5~12) to give the parent ketones and α-acetoxy ketones in varying amounts [Fry, 1979].

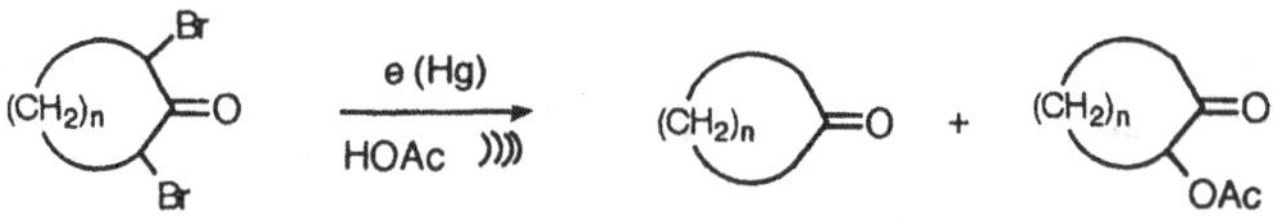

Reductive elimination of bromo-epoxides [Sarandeses, 1992] is a key step in the conversion of ionones to damascones. Subjecting the heterogeneous mixture of zinc, copper(I) iodide and the substrate to sonic treatment ensures good results.

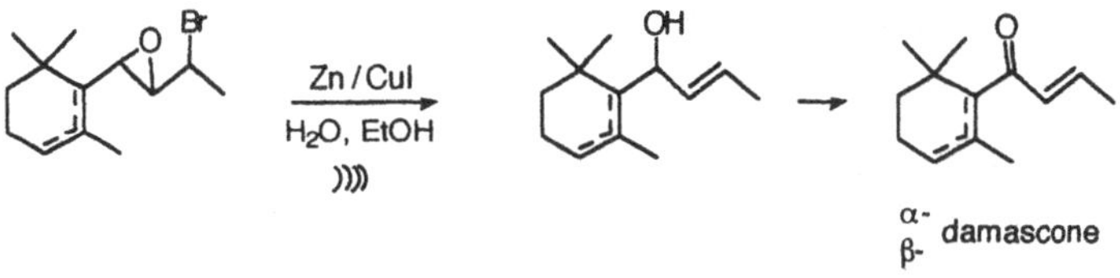

α,α'-Dibromo-*o*-xylene is debrominated by zinc, leading to *o*-quinodimethane [Boudjouk, 1986c]. Similar sonication with lithium effects only single bromine-lithium exchange. *o*-Quinodimethane can be trapped in situ with dienophiles [Chew, 1986].

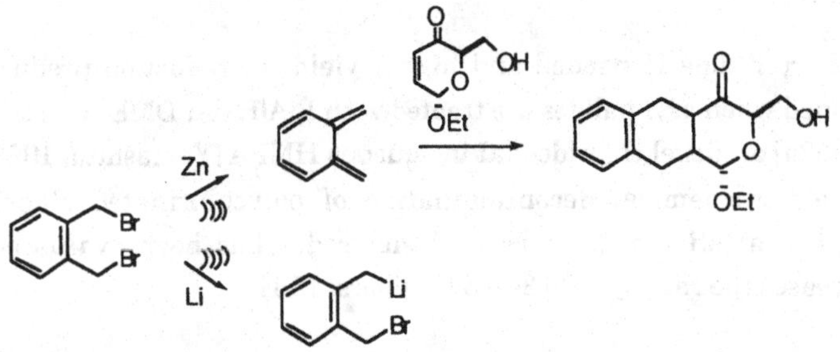

π-Allylpalladium complexes are formed from allylic bromides by sonication with palladium black [Inoue, 1984]. Many synthetic uses of these species are known [Tsuji, 1995]..

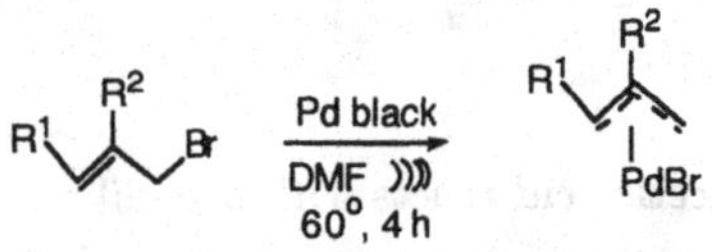

A convenient method for the generation of cyclopropylcarbenoids is by reacting *gem*-dihalocyclopropanes with Li, Na, or Mg in THF with sonication at ambient temperature [L. Xu, 1985]. These carbenoids undergo ring opening to give allenes or intramolecular C-H insertion after losing metal halide.

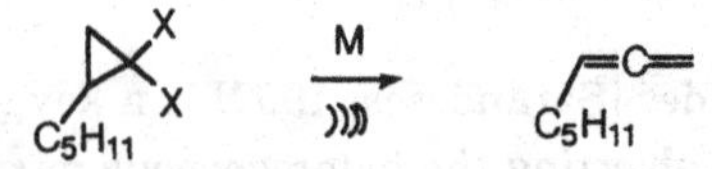

X	M	time	yield
X= Br	M= Li	10 min	81%
	M= Mg	10 min	87%
X= Cl	M= Na	5 min	68%

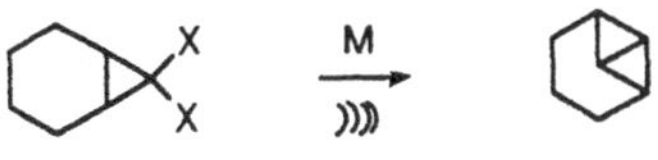

X= Br	M= Li	15 min	46%
	M= Mg	15 min	44%
X= Cl	M= Na	5 min	30%

Dehalogenation of *gem*-Dihalocyclopropanes. [L. Xu, 1985]

A flask containing suspension of the metal (Li, Mg, Na; 50 mmol) in dry THF (20 mL) is added the *gem*-dihalide (20 mmol) and then placed in a ultrasonic cleaning bath which is maintained at 20° After irradiation for 5-15 min, excess metal is removed by filtration, the mixture is treated with water (25 mL) and filtered. The filtrate is extracted with pentane, dried, and distilled afford the dehalogenated product.

Dehalogenation by a tin hydride coupled with aeration furnishes hydroperoxides [Nakamura, 1995].

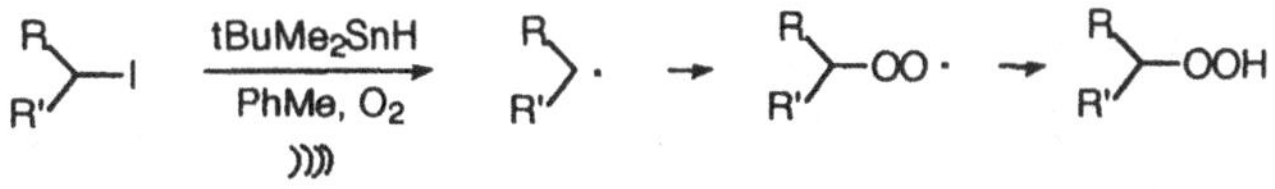

Tetrahalomethanes are homolyzed on exposure to ultrasound. The trihalomethyl radical has been used to advantage to propagate the chain reaction sequence of Barton's decarboxylation [Dauben, 1990].

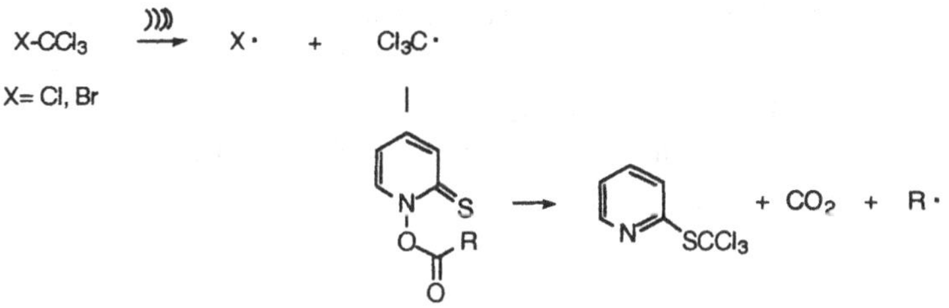

Organometal hydrides are readily formed by reaction of the corresponding halides (and alkoxides) with lithium aluminum hydride in a hydrocarbon solvent [Lukevics, 1984].

$$R_3M\text{-}Cl \xrightarrow[25\text{-}40^{\circ},\ 2.5\text{-}4.5\ h]{LiAlH_4\ /\ RH} R_3M\text{-}H \quad 80\text{-}95\%$$

M = Si, Ge, Sn

Aromatic nitro compounds undergo reduction to anilines by the hydrazine-iron system in the presence of activated charcoal [Han, 1985]. Aliphatic nitro compounds do not respond in the same way.

$$ArNO_2 \xrightarrow[C]{Fe\text{-}N_2H_4} ArNH_2$$

The Birch reduction benefits from sonication. The method can be used to prepare 3,6-dihydro-3,6-disilyl-1,4-cyclohexadienes [Barrett, 1984, 1988].

Li / Me_3SiCl, THF, r.t. 16 h — 55%

The ultrasonic palladium-catalyzed hydrogenation, using formic acid in ethanol as hydrogen source [Boudjouk, 1983] is very efficient for double and triple bonds, it also cleaves phenylcyclopropane to afford phenylpropane in quantitative yield.

A catalytic hydrogenation system can be generated by sonochemical reduction of nickel salts with zinc powder [Petrier, 1987a]. The activated zinc also reduces water in situ to provide hydrogen. This system shows selectivity for hydrogenation of conjugated double bonds, which can be further enhanced by adjusting reactions conditions (particularly pH) [Petrier, 1987b].

Zn - $NiCl_2$ (9:1), EtOH / H_2O, r.t. — 3 h: 96%; 1.5 h, pH 8: 95%

The method for reduction of conjugated ketones, amides (lactams), and nitriles, using activated magnesium in methanol [Brettle, 1981] is particularly valuable for a synthesis of indolactam-V [de Laszlo, 1986]. The indole nucleus is preserved.

HN COOMe → Mg / MeOH))) → HN COOMe

100%

Ketone reduction with alkali metal suspensions in tetrahydrofuran appears to involve electron transfer [Huffman, 1987], as camphor gives predominantly the *endo*-alcohol (Li, 73%, Na, 68%; but K 42%). Benzoquinone and camphorquinone are easily reduced with zinc and the hydroquinone may be trapped as the bistrimethylsilyl ethers [Boudjouk, 1986a].

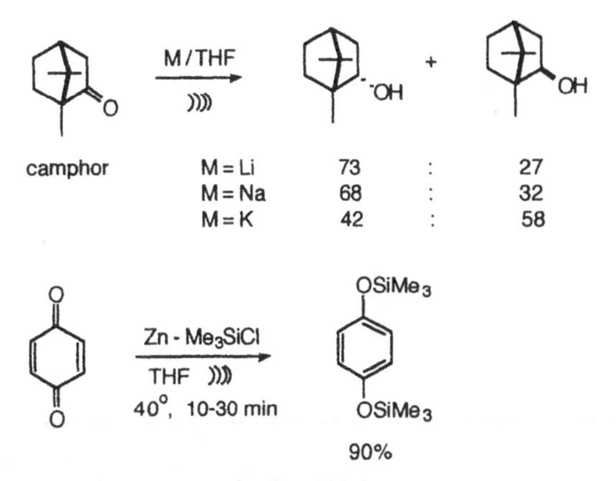

Reduction of Diketones. [Boudjouk, 1986a]

A flask charged with activated zinc powder (2.0 g, 30 mmol), chlorotrimethylsilane (2.1 g, 19 mmol) and ether (25 mL) is placed in an ultrasound cleaning bath (150W, 55 kHz) which is filled with water and maintained at 40°. After sonication for 1 min, the diketone (7.5 mmol) is added, and the progress of reaction is monitored by nmr. At the conclusion of the reaction the mixture is filtered, washed, evaporated, and chromatographed on a column.

Ultrasound promotion of amide reduction with lithium aluminum hydride has been observed [Eremeev, 1985].

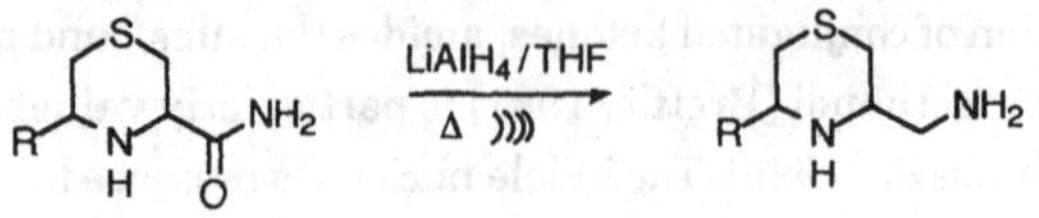

Pinacol formation by the reductive coupling route is rapid when ultrasound is provided to the system [So, 1988]. With acetone as solvent, enones undergo cross coupling [Delair, 1989]. The McMurry reaction exhibits rate enhancement by sonication [Nayak, 1991]. Solvent effects include the change of *E:Z* ratio of the alkene products. In DME the reaction stops at the pinacol stage.

1-(1-Hydroxy-1-methyl)ethyl-2-cyclohexen-1-ol. [Delair, 1989]

A mixture of zinc dust (600 mg, 9.2 mmol), $CuCl_2.2H_2O$ (250 mg, 1.47 mmol) in acetone (4 mL) and water (1 mL) is sonicated for 2-3 min. 2-Cyclohexenone (2.2 mmol) and ethyleneglycol monomethyl ether (300 mg) are added to initiate the reaction. After 1 5 h, the stopped reaction is treated with saturated aq. NH_4Cl, extracted with ethyl acetate. The organic extracts are processed by drying, evaporation and silica gel chromatography to provide the diol in 87% yield.

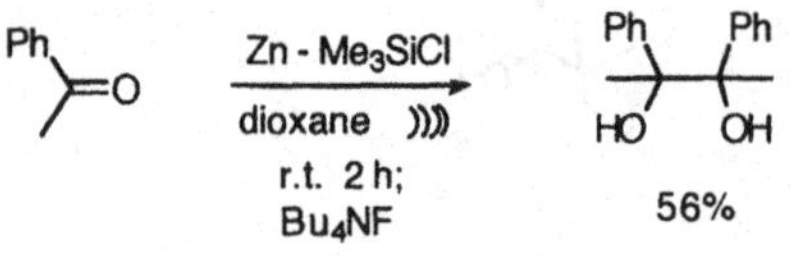

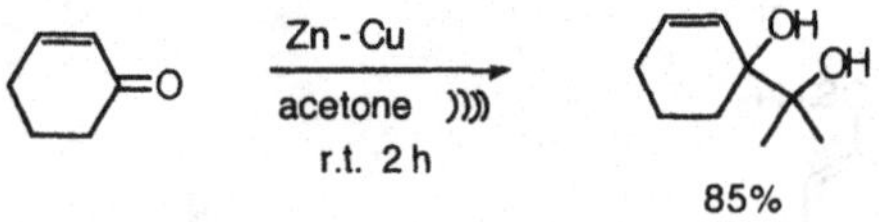

Activated C-X bonds are subject to reductive cleavage. Often such a process is aided by ultrasound. Thus the removal of the benzyl group from a phenyl ether [Townsend, 1981], the transpositional insertion of a carbonyl group to alkenyl-oxiranes which has great potential for lactone synthesis [Horton, 1984] with reagent derived from sonolysis of $Fe_2(CO)_9$, and desulfurization with Raney nickel [Schreiber, 1988] are exemplary.

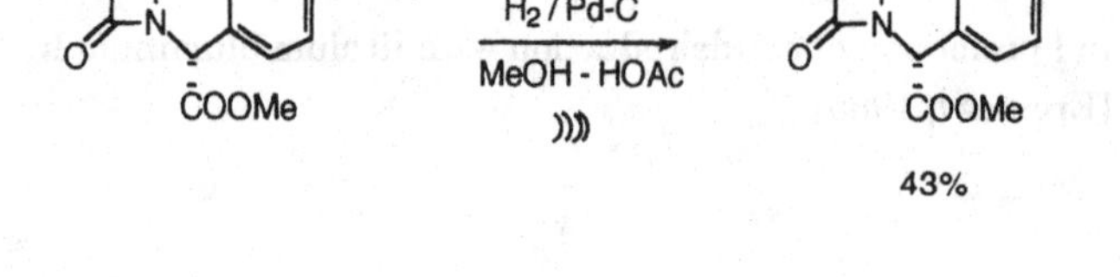

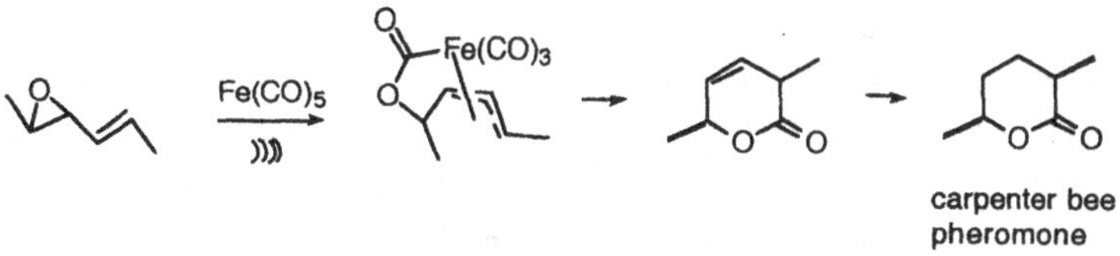

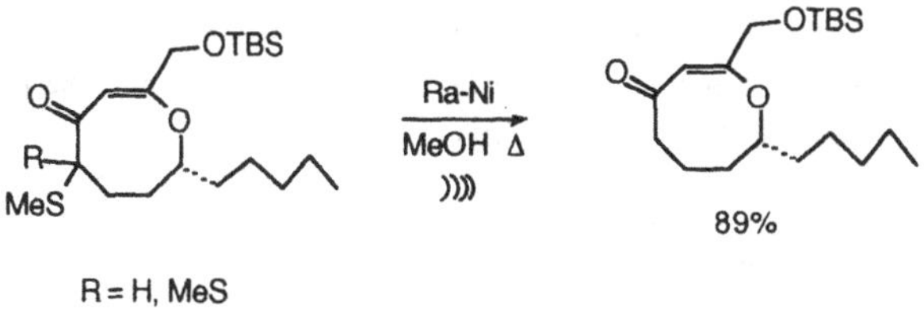

Mechanistic switching as a result of ultrasonic effects is shown in the reaction of *o*-allylbenzamides with lithium [Luche, 1990]. With irradiation the radical pathway for cyclization predominates, leading to 2-methylindanone, whereas under ordinary conditions an ionic mechanism prevails. As expected, sonication is useful for assisting reductive cleavage of the N-N bond of hydrazines [Alexakis, 1991].

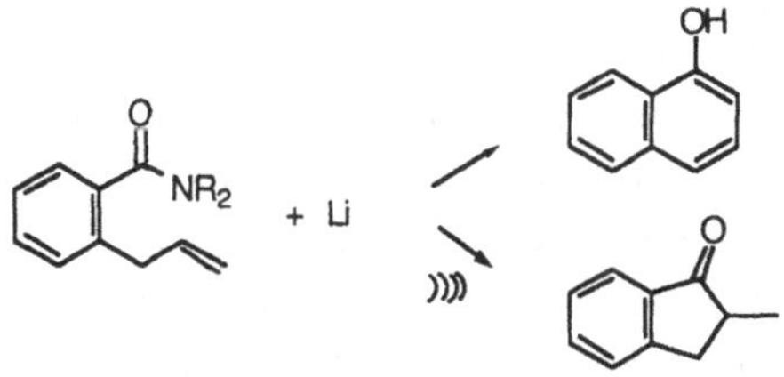

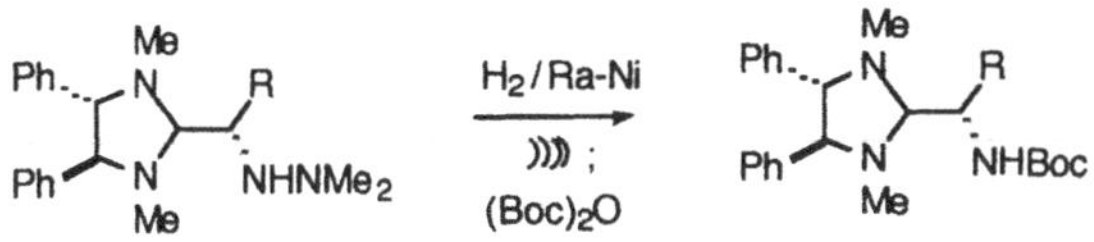

Preparation of alkyldiarylphosphines can be accomplished by cleavage of triarylphosphines with lithium and alkylation of the resulting phosphide anions. The reduction step is greatly accelerated by ultrasound [Chou, 1985a]. On lowering the temperature the reaction becomes cleaner [Chou, 1985b]. Cyclic sulfones also suffer regioselective ring cleavage [Chou, 1985c]. Desulfonation with superior results in the area of spiroacetals has been made [Ley, 1986b], the conventional method is unpredictable and requires substantial amounts of reagent.

Alkylphosphines from Phenylphosphines. [Chou, 1985a]

A suspension of chopped Li metal (11.5 mmol) in dry THF (5 mL) together with the tertiary phenylphosphine (1.15 mmol in 1 mL THF) is sonicated <30°. After consumption of the phosphine (ca. 35 min) the excess metal is removed and the dark red mixture is treated with freshly distilled *t*-butyl chloride (97 mg, 1.05 mmol) to destroy phenyllithium. The resulting solution is cooled to 0° and an alkyl halide (1.05 mmol in 1 mL THF) is introduced. The pale yellow mixture is stirred at room temperature for 1 h to complete the reaction. The salt is dissolved in water and the product is obtained by extraction with benzene and vacuum distillation.

R(Ph)P–Ph → [Li/THF,))), 0° - r.t.] → R(Ph)P–Li → [tBuCl; MeI] → R(Ph)P–Me 71-84%

R = Ph, $Ph_2PCH_2CH_2$-

→ [K / PhMe,))) ; MeI] → 82% + 8%

Reductive coupling of organic halides with alkali metals is much slower than the formation of organometallic species under ultrasonic conditions. Such reactions require long times to complete [Han, 1981] and the products are more efficiently prepared by treatment of the triflates with a zinc-nickel system in which low-valent nickel compounds are the active coupling reagents [Yamashita, 1986].

–OTf → [Zn - $NiCl_2$ Ph_3P, NaI / DMF,))) 60°] → 85%

For coupling of aryl halides the classical Ullmann reaction specifies copper as the promoter. The high temperature is obviated on application of ultrasound [Lindley, 1987] whereby more available metal surfaces in a clean state are exposed for the reaction. However, solvent is an important factor.

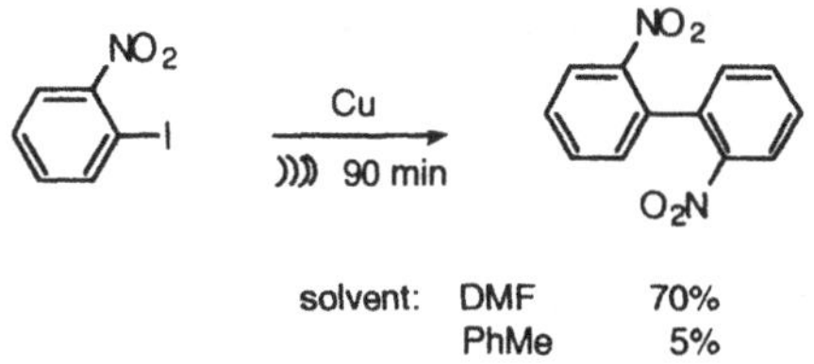

Cu,))) 90 min

solvent: DMF 70%
PhMe 5%

Chlorosilanes and chlorostannanes give disilanes and distannanes on treatment with lithium wire in tetrahydrofuran [Boudjouk, 1981]. Without ultrasonic assistance all these reactions fail. Dichlorodimesitylsilane provides tetramesityldisilene in an excellent yield [Boudjouk, 1982].

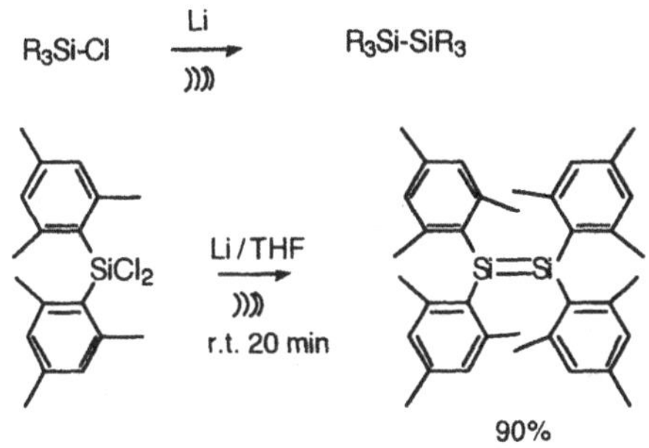

$R_3Si\text{-}Cl \xrightarrow[)))]{Li} R_3Si\text{-}SiR_3$

$SiCl_2$ Li / THF,))), r.t. 20 min, Si=Si

90%

Alkenyl and allylic halides can couple with perfluoroalkyl bromides and iodides by treatment with zinc in the presence of a palladium catalyst [Kitazume, 1985]. γ-Selectivity is observed for the allylic halides.

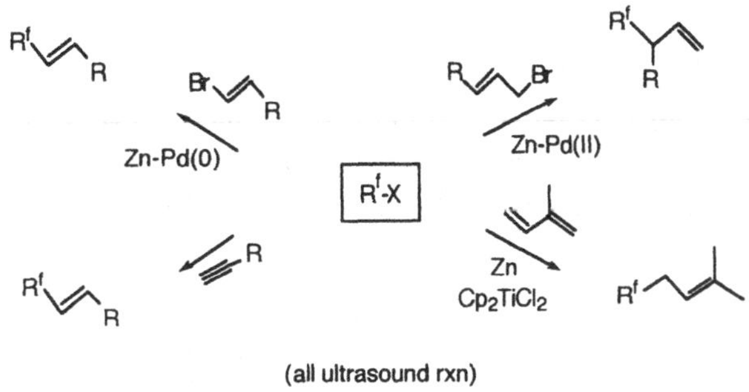

Zn-Pd(0) Zn-Pd(II) R^f-X Zn, Cp_2TiCl_2

(all ultrasound rxn)

3.5. ELIMINATIONS AND ADDITIONS

Very few elimination reactions have been subjected to ultrasonic modification. However, an example of dehydrobromination with potassium hydroxide promoted by ultrasound and phase transfer catalyst in the absence of solvent [Diez-Barra, 1992] is very appealing.

KOH
Bu_4NBr
75°)))

Hydration of γ-alkynyl ketones is directed by the existing carbonyl group. The role of ultrasound in the palladium-catalyzed reaction [Imi, 1987] has not been clarified. It is interesting that most products are 1,4-diketones, however, 2-(alk-2-ynyl)cyclopentanone furnishes the 1,5-diketone.

$PdCl_2(MeCN)_2$
MeCN - H_2O
)))

$PdCl_2(MeCN)_2$
MeCN - H_2O
)))

5-Oxoprostaglandin-E_1 Methyl Ester. [Imi, 1987]
5,6-Didehydro-PGE_2 methyl ester (50 mg, 0.137 mmol) and $PdCl_2(MeCN)_2$ (1 mg, 3 mol%) and MeCN-H_2O (25:1, 0.5 mL) is irradiated with ultrasound at room temperature to give the oxo compound (40.6 mg, 77%).

Hydration of double bonds by the oxymercuration-demercuration protocol can be improved [Dumas, 1986] and even further in the sense that instead of mercury(II) acetate or trifluoroacetate a combination of mercury(II) oxide and any carboxylic acid under sonication may be used [Einhorn, 1989]. This modified method also has the advantages of much shorter reaction time and better yields of the alcohol products.

THF - H_2O

r.t.

OH

$Hg(OAc)_2$	⌒	48%
HgO / tBuCOOH	)))	80%
HgO /C_7F_{15}COOH	)))	80%

The remarkable enhancement of the rate of hydroboration by ultrasound [Brown, 1985] enables rapid access to organoboranes and ensures their preservation. Similarly, hydrosilylation of alkenes and alkynes at a much lower temperature [Han, 1983] is feasible.

Hydroboration of Alkenes. [Brown, 1985]

A 100 mL-flask fitted with a reflux condenser and connecting tube is flushed with nitrogen, charged with an alkene (10 mmol) and a borane reagent, and irradiation with ultrasound at 25°. The reaction is followed by ^{11}B nmr, and at its completion the resulting organoborane is processed according to the need (e.g., oxidation to give the alcohol, etc.).

BH

THF

B

65°, 5 h	⌒	99%
r.t., 1 h	)))	

R_3Si-H

)))

H

SiR_3

R_3 = Et_3, $(EtO)_3$, Cl_3, Cl_2Me

Side-chain nitration of styrene and cyclic analogues to give conjugated nitro compounds using nitryl iodide involves an addition-elimination sequel. The reagent can be made from iodine and potassium nitrite in the presence of 18-crown-6 which is improved by ultrasound [Ghosh, 1996].

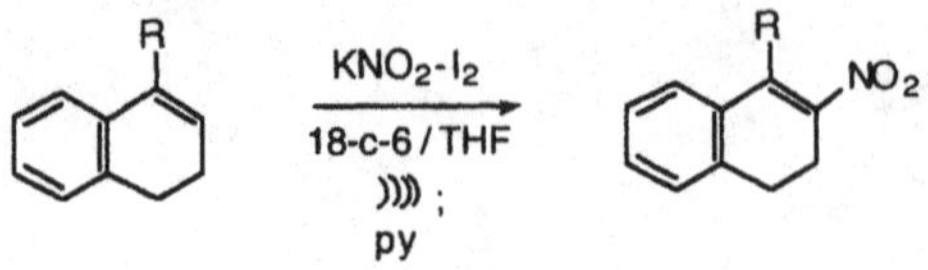

Formation of β-azidoalkenoic esters by Michael addition of the azide ion to alkynoic esters in a biphasic system is complicated by side reactions due to decomposition and self-condensation of the products. Improvement by the ultrasound technique has been noted [Priebe, 1987].

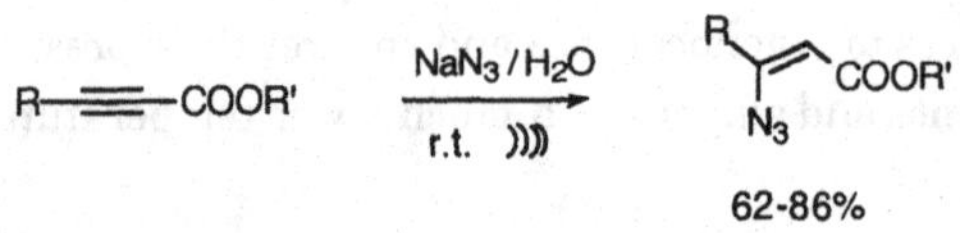

62-86%

Allylzinc bromides are readily formed under sonication and they add to terminal alkynes to give 1,4-dienes [Knochel, 1984]. It is possible to prepare alkenes with a perfluoroalkyl substituent by reaction of bimetallic perfluoroalkyl derivatives with 1-alkynes [Kitazume, 1982]. The reagents are formed by ultrasonic reactions of perfluoroalkyl iodides with zinc and then copper(I) iodide.

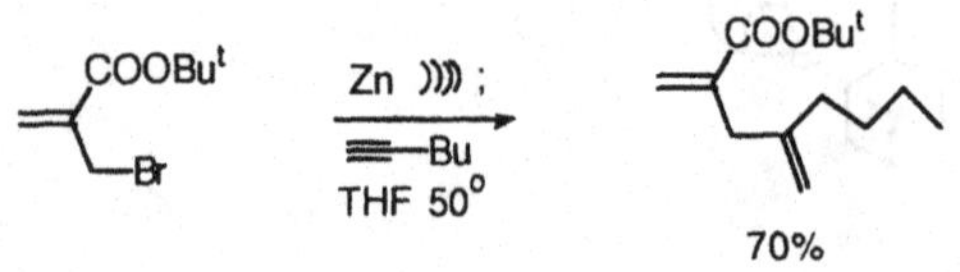

70%

$R^fI \xrightarrow[\text{CuI / THF)))}]{\text{Zn / THF))) ;}} (R^f)_2CuZnI \xrightarrow[H_3O^+]{\equiv\!-R\,;} R^fCH{=}CHR$

3,3,3-Trfluoro-1-phenylpropene. [Kitazume, 1985]

In a flask equipped with a dry ice-acetone reflux condenser is charged with Zn dust (1.30 g, 0.02 mmol), CuI (0.72 g), CF_3Br (3.0 g, 20 mmol), phenylethyne (1.02 g, 10 mmol) and THF (30 mL). The flask is placed in an ultrasonic bath to irradiate for 2 h, and the mixture is poured into 2% hydrochloric acid. Workup by extraction with ether furnishes the alkene in 55% yield.

Hydroperfluoroalkylation (1,4-addition) of isoprene by the zinc reagent is catalyzed by titanocene.

R^fX + [isoprene] $\xrightarrow[Cp_2TiCl_2]{Zn\ )))}$ R^f–[prenyl]

X = Br, I

A very interesting method for cyclopentenone synthesis involves conjugate addition of the ester homoenolate, which is derived by ring cleavage of a 1-alkoxy-1-trimethylsiloxycyclopropane with zinc chloride, to 2-alkynoic esters [Crimmins, 1990]. The addition conducted with ultrasound assistance leads to allenic intermediates.

1-OEt-1-$OSiMe_3$-cyclopropane $\xrightarrow{ZnCl_2}$ $Zn(CH_2CH_2COOEt)_2$ — COOMe–C≡C–CH(OR)R' + $CuBr.SMe_2$, Me_3SiCl / HMPA $\xrightarrow{)))}$ 2-COOEt-3-(CH(OR)R')-cyclopent-2-enone

Hydrostannylation of triple bond with a tin hydride does not require an initiator if ultrasound is applied to the system [Nakamura, 1989]. Generation of carbon radical by halogen abstraction is similarly facilitated.

Bu–C≡CH + $HSnPh_3$ $\xrightarrow[)))]{7^\circ}$ Bu–CH=CH–$SnPh_3$ (95%)

A method for acquiring active magnesium powder involves anthracene as a transfer agent [Bonnemann, 1983]. On sonication the (anthracene)Mg·3THF adduct forms rapidly and then releases finely powdered magnesium. The active metal (Mg*) usually reacts very readily with an organohalogen compound to give Grignard reagents without difficulty, including those from allylic halides [Oppolzer, 1984]. The troublesome reductive dimerization of such halides is suppressed.

anthracene (THF) $\underset{-\ Mg^*}{\overset{+\ Mg\ )))}{\rightleftharpoons}}$ (anthracene)Mg · 2THF

This development is significant to the realization of intramolecular magnesio-ene reaction which has since found many applications, e.g., in the synthesis of chokol-A [Oppolzer, 1986a], 6-protoilludene [Oppolzer, 1986b] and others.

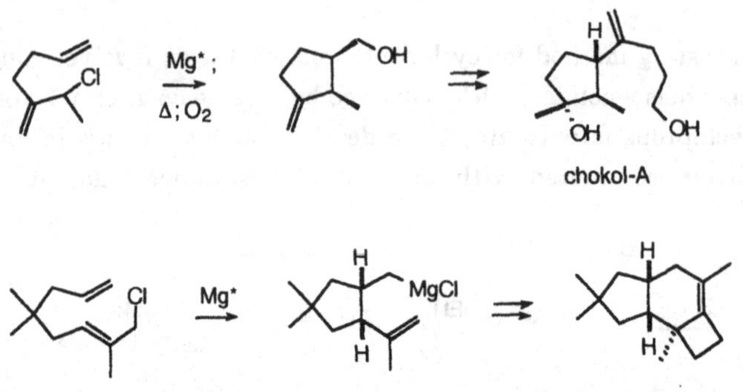

An operational inconvenience associated with most main group organometallic reagents is due to their intolerance of moisture and hydroxylic materials. Accordingly, reactions unaffected by water are always welcome [Li, 1993, 1997; Lubineau, 1994] as the costs in dehydration of solvents are also saved. The first report on the preparation of Grignard reagents and organolithiums and organoaluminums in wet diethyl ether was published more than 40 years ago [Renaud, 1950]. The most dramatic effect is that the decrease in the initiation period which is likely the result of removal of water molecules from the metal surface upon sonication. The presence of water in the bulk solvent does not have any influence on this step.

Organic halides including *n*-propyl, *n*-pentyl, and phenyl halides yield the corresponding organolithiums even in wet tetrahydrofuran when the reaction with lithium is assisted by ultrasound. Secondary and tertiary alkyl halides are less reactive but it is still possible to generate those reagents.

The preparation of an alcohol from a mixture of organic halide, carbonyl compound and magnesium was discovered by Barbier, but due to several drawbacks this reaction was subsequently eclipsed by the Grignard protocol of preforming the organomagnesium reagent. The current renaissance of the Barbier method owes much to the use of lithium [Luche, 1980] instead of magnesium, and above all the assistance by ultrasound. Under such conditions various side reactions (e.g., enolization, reduction) are minimized. Ultrasonic generation of reagents from vinylic, allylic and benzylic halides is not complicated by Wurtz coupling [Burkow, 1987]. Allyl diethyl phosphonates have been shown to be as useful as the corresponding halides [Araki, 1988].

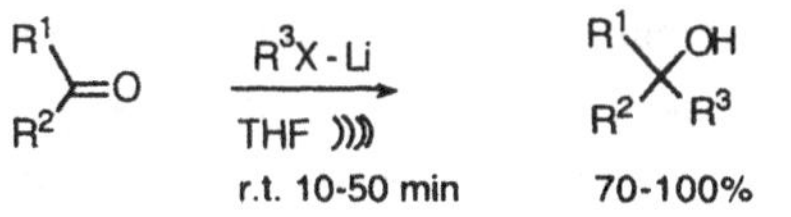

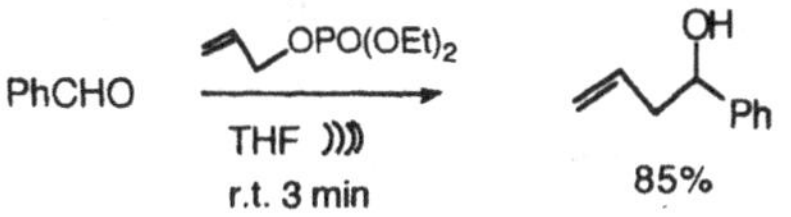

It is not surprising that the convenient method has attracted attention from chemists engaging in the synthesis of complex molecules: trichodiene [Snowden, 1984], sesquicarene [Uyehara, 1985], and pentalenic acid [Ihara, 1987]. Both intermolecular and intramolecular versions are represented.

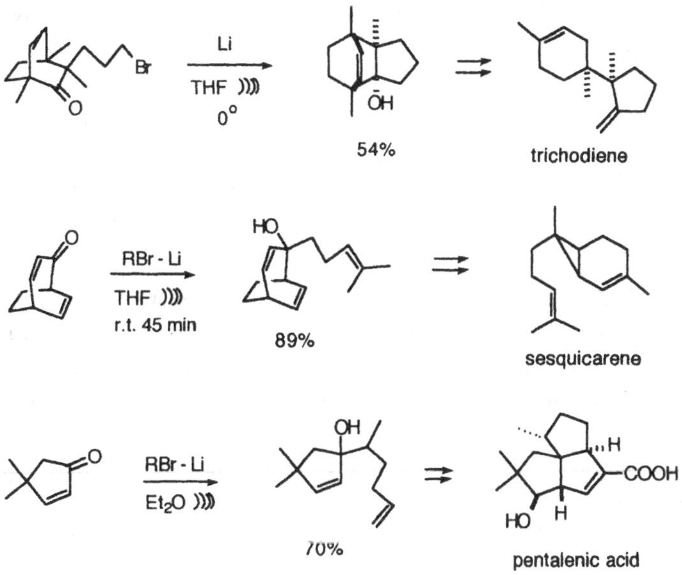

A mechanistic study [de Souza-Barbosa, 1988] reveals a temperature effect which can be explained by the dependence of bubble generation on solvent viscosity and vapor pressure. Low vapor pressure favors more energetic cavitation collapses but higher viscosity suppresses bubbles. There is a temperature range for optimal reaction rate for a particular solvent.

Formation of α-chloromethyllithium from bromochloromethane and lithium and its in situ reaction with carbonyl compounds [Einhorn, 1988] indicates the advantage of ultrasound to permit the use of cheaper reagents. The products are epoxides which result from addition followed by ring closure. To avoid destruction of the epoxides it seems necessary to employ a probe instead of a bath as the ultrasound source so that the rate of the reagent formation stays far above that of the cyclization of the *O*-lithiochlorohydrin intermediates.

Li - $ClCH_2Br$, THF -20°, 20 min,))) ;
$R^1R^2C=O$ → epoxide + chlorohydrin (R^1, R^2, OH, Cl)

4-Bromo-2-sulfolene forms an organometallic reagent with magnesium amalgam which reacts with carbonyl compounds at $C_{(2)}$ [Tso, 1987]. However, a regioretentive reaction occurs by using a zinc-silver couple.

Zn-Ag, THF))), 78-92% ; Mg- $HgCl_2$, THF))), 30%

Perfluoroalkyl carbinols are readily obtained from reaction of perfluoroalkyl halides (iodides and bromides) with carbonyl compounds in the presence of zinc under sonication in DMF [Kitazume, 1981, 1985]. For ketone substrates the addition of titanocene dichloride to generate a catalyst is needed. An aryl ring ligated tricarbonylchromium group is not affected by ultrasound, at least during the period for completing reaction of the araldehyde with perfluoroalkylzinc reagents [Solladie-Cavallo, 1984].

CHO, $Cr(CO)_3$ + R^fI —Zn / DMF, r.t. 0.5-2.5 h,)))→ OH, R^f, $Cr(CO)_3$; 85%

In media containing a large amount of water, the 1,2-addition of allylzinc species (tin can be used as well as zinc) to carbonyl compounds is greatly accelerated by ultrasound [Petrier, 1985b, 1985c], and the reaction is discriminative between aldehyde and ketone groups. The method is particularly valuable for hydrophilic compounds [Einhorn, 1987b].

Allylation of Carbonyl Compounds. [Einhorn, 1987b]

To a mixture of the carbonyl compound, an allylic bromide, and tin (1.2-2 equiv.) in 50% aq. THF (2 mL) is added saturated aq. NH_4Cl (1-4 mL) and sonicated in a bath (50 kHz) at 15-18° until all the metal disappears and a thick suspension results (ca. 30-60 min). Dilution with THF, filtration through Celite, and workup of the filtrate by ether extraction affords the homoallylic alcohol.

CHO + Br; Sn, THF-H_2O (1:5), r.t. 0.5 h,))); OH; 98%

Allylic alcohols are converted to nucleophiles by sonication with a combination of palladium and tin(II) chlorides in a nonpolar solvent. Interestingly, the reaction with carbonyl compounds in situ [Masuyama, 1992] affords homoallylic alcohols with regiochemistry opposite to that observed in polar solvents.

OH + PhCHO; $SnCl_2$, $PdCl_2(PhCN)_2$, Et_2O))); OH, Ph (major)

Although the Reformatsky reaction has undergone many modifications [Rathke, 1975] with respect to solvent, additive, and activation procedure the most expedient improve-ment seems to be the application of ultrasound [Han, 1982b]. Thus by conventional procedure the reaction with benzaldehyde is complete in 12 h, giving the β-hydroxy ester in 61% yield, whereas the ultrasonic reaction is over in 5 min and the yield is increased to 98%. Also by replacing the carbonyl substrates with aldimines, it provides a convenient method for synthesizing monocyclic β-lactams [Bose, 1984; Oguni, 1986].

N,4-Diaryl-β-lactams. [Bose, 1984]

A flask containing the Schiff base, ethyl bromoacetate, activated zinc (HNO_3 washed) and a crystal of iodine is placed in a ultrasonic cleaning bath to carry out the reaction which is monitored by thin layer chromatography. The β-lactam product is obtained in the 70-95% range, comparing to 25-50% yield in refluxing toluene without sonication.

$R^1R^2C{=}O$ + $BrCH_2COOEt$ —Zn (I_2) /dioxane, r.t., 5-30 min,))) → $R^1R^2C(OH)CH_2COOEt$ 97-98%

ArCH=NAr' + $BrCH_2COOEt$ —Zn (I_2) /dioxane, r.t.))) → β-lactam (Ar, Ar') 70-95%

By changing the electrophilic component of the Reformatsky reaction to a *N*-acyloxazolidinone or thione, the method can be applied to the synthesis of β-keto-esters [Kashima, 1993]. A palladium-catalyzed synthesis of γ-oxo α-amino acid derivatives [Jackson, 1989] is a similar process.

RCO–N(oxazolidin-2-one/thione) + BrCH(R')COOEt —Zn / THF,))) ; reflux → RCOCH(R')COOEt

X = O 40-86%
X = S 23-97%

RCOCl + $IZnCH_2CH(NHBoc)COOBn$ —$(PhP_3)_2PdCl_2$,))) → $RCOCH_2CH(NHBoc)COOBn$

The formation of *N-t*-butylbenzamide by reaction of *N-t*-butyl isocyanate with phenylmetals [Einhorn, 1986a] indicates that magnesium is the best among several metals tested, in terms of yields (Mg, 91%; Na, 78%, Li, 51%). Note that in many other reactions lithium seems to be superior.

*t*BuN=C=O + PhBr —M / THF,))) → *t*BuNHCOPh

The Bouveault reaction is similarly enhanced by ultrasound [Petrier, 1982]. Aldehydes are obtained in one operation (reaction through an addition-elimination sequence), but a curious observation is that when certain β-amino formamides served as the formyl donor the reaction required 500 kHz acoustic wave [Einhorn, 1986b]. By this method *o*-substituted benzaldehydes including phthalaldehyde are accessible from bromobenzene.

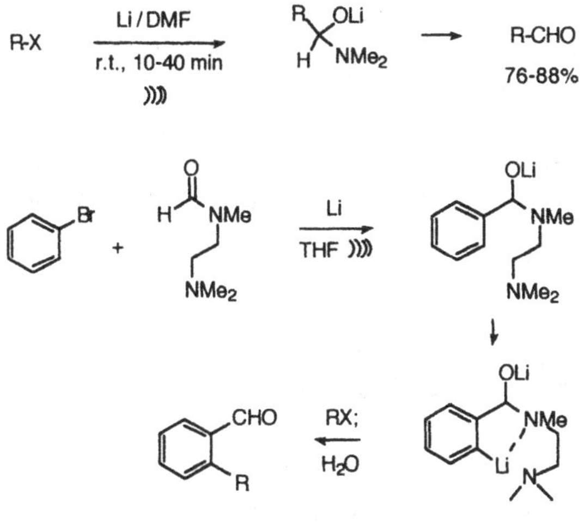

Homologous Aldehydes from Halides. [Petrier, 1982]

A mixture of an organic halide (1 mmol), dry DMF (1 mmol), lithium sand in oil (50 mg, 2.1 mmol) in dry THF (4 mL) is sonicated in a bath (96W, 40 kHz). The reaction lasts about 15 min at 10-20°. Workup is by addition of ether, washing with dilute hydrochloric acid, brine, and drying, evaporation. The crude product is uncontaminated with the starting material.

The method of ketone synthesis based on electrophilic lithium carboxylates [Danhui, 1994] can be extended to 2-furanyl ketones [Aurell, 1995] by taking advantage of the rapid deprotonation of furan. There is limitation due to low solubility of some carboxylate salts (e.g., acrylate, crotonate) in tetrahydrofuran.

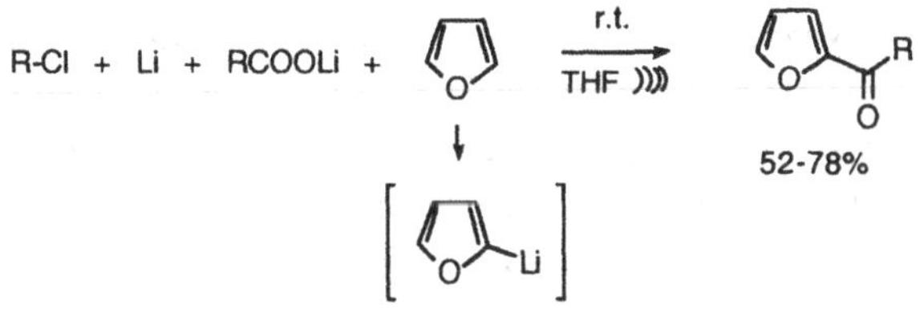

Organocuprate reagents typically add to conjugate carbonyl compounds in the Michael fashion. In situ generation [Luche, 1982] of these valuable reagents in THF with the aid of ultrasound is definitely expedient. When the solvent is wet, a slight change of reaction conditions can accomplish the same goal. Thus after a black suspension (Zn-Cu couple) is formed by ultrasonic irradiation of a zinc and copper(I) iodide mixture, the two organic reactants are added. The reaction is sustained for a period of time. Such a method is more convenient and economical than that involving organocuprates. A procedure for reaction in a solvent containing substantial amount of water [Petrier, 1986] is interesting.

An evaluation of the reaction parameters leads to the conclusion that halide reactivity is tert.>sec.>>prim. and I>Br (Cl being inert). Copper(I) iodide and copper(II) chloride are the most satisfactory activators. The preferred solvent system is aqueous ethanol in terms of best yields.

Zn-Cu / $I(CH_2)_3Cl$
py - H_2O (1:4))))
2 h
Cl
85%

Based on the efficient hydrozirconation of 1-alkenes under sonication and metal exchange with copper, a new procedure for conjugate addition of an alkyl group to enones has been developed [Wipf, 1991]. Only catalytic quantities of $CuBr.SMe_2$ (or other Cu(I) and Cu(II) salts) are required.

Bu
$Cp_2Zr(H)Cl$
r.t. 20 min
)))
Bu
$ZrCp_2$
Cl
$CuBr.SMe_2$

Diorganozinc compounds may be prepared in situ by metal exchange, after rapid formation of organolithium reagents under ultrasonic irradiation [Petrier, 1985a]. However, it should be noted that diarylzinc compounds are more easily accessible [Luche, 1983], whereas a more energetic source (e.g., probe generator vs. cleaning bath) is needed to form dialkylzincs [Petrier, 1984]. An experiment indicated the formation of mixed organozincs by the introduction of two halides [Petrier, 1985a].

Other than noting 1,4-reduction as a side reaction when the organozinc reagent contains a β-hydrogen, the synthesis of β-cuparenone [Greene, 1984] and a penitrem-D precursor [Smith, 1995] demonstrate the usefulness of such organometallics (with a nickel catalyst) to effect transformations that organocopper reagents may fail.

Br
Li - $ZnBr_2$
Et_2O)))
r.t., 30 min
Zn
2
$Ni(acac)_2$
67%
β-cuparenone

Lithium triorganozincates ($R_3ZnLi{\cdot}2LiCl$) also add to enones in the Michael fashion [Isobe, 1977]. Preparation of the reactive reagents is simplified by sonicating a mixture of RBr, $ZnBr_2$, Li wire in ether.

An accelerating effect of ultrasound on the conjugate addition of nitroalkanes to α,β-unsaturated esters under solid-liquid phase transfer conditions has been witnessed [Jouglet, 1991].

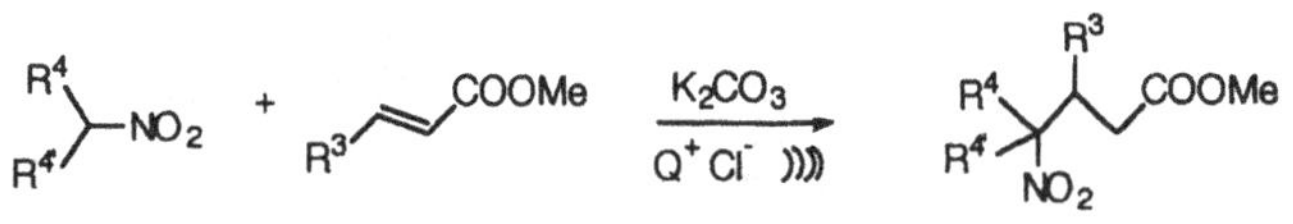

Condensation of aldehydes with acetonitrile under the influence of potassium superoxide is favored by sonication to generate radical anion pair [Shibata, 1987]. Aromatic aldehydes give cinnamonitriles which tend undergo conjugate addition.

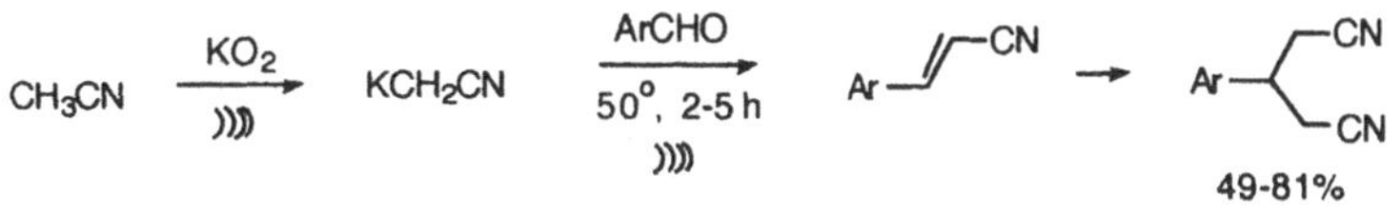

3.6. CONDENSATIONS

Acetalization of sugars under sonication [Akiya, 1950; Einhorn, 1986c] is more efficient. In other words, it requires lower temperature, shorter time and gives higher yields. Besides acetonides and cyclohexanonides, benzylidene derivatives can also be obtained [Chittenden, 1988].

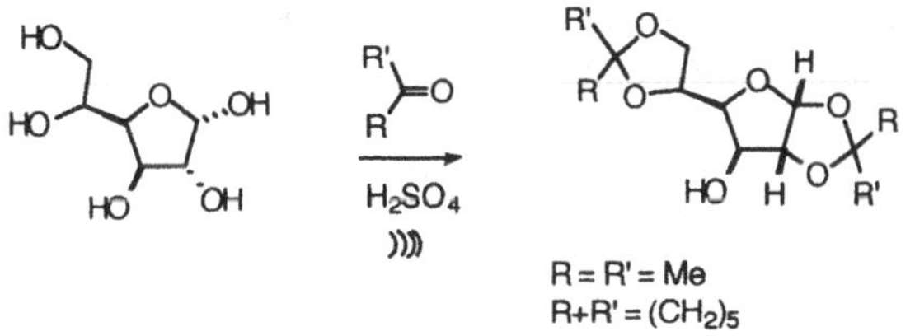

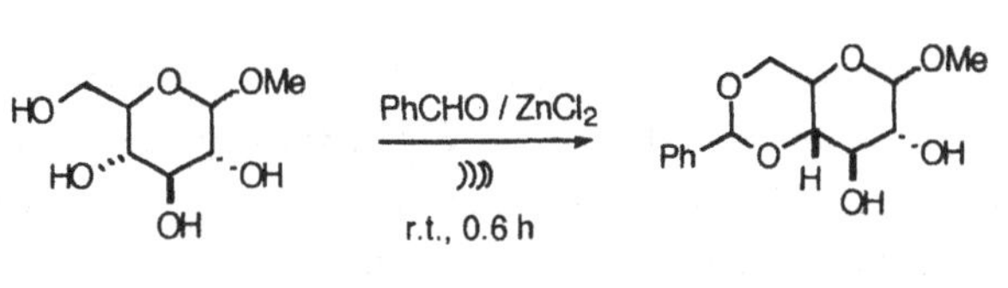

While thioamides are usually made from amides by treatment with phosphorus pentasulfide or Lawesson's reagent, problems can arise in certain cases, for example the requirement of a large excess of the reagent. It might be possible to accelerate a sluggish reaction by ultrasound [Raucher, 1981] and thereby removing the attendant complications (complex mixture of products), such as in the case of a tricyclic lactam [Heathcock, 1992].

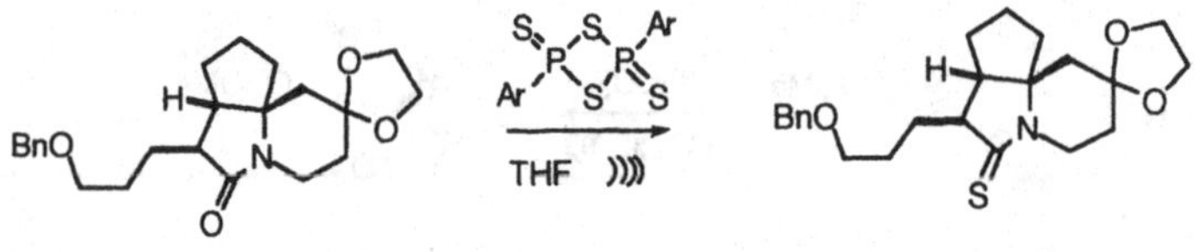

The Wittig reaction is a highly valued synthetic method. When using a two-pronged Wittig reaction to access polycyclic arenes from *o*-quinones, ultrasound can have positive effects of rate enhancement and yield augmentation [Yang, 1992].

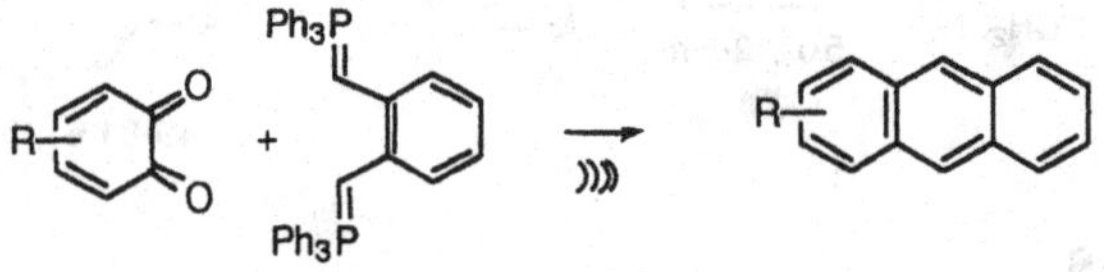

It has been argued that a sonochemical version of the Emmons-Wadsworth reaction is promoted by hydroxyl radical generated from water [Sinisterra, 1987].

An obvious deficiency of the Wittig reaction is related to the large triphenylphosphine moiety in the reagents and the removal of triphenylphosphine oxide can be problematical. In certain cases transylidation can become serious [Fadel, 1993] and the problem may be overcome by ultrasound.

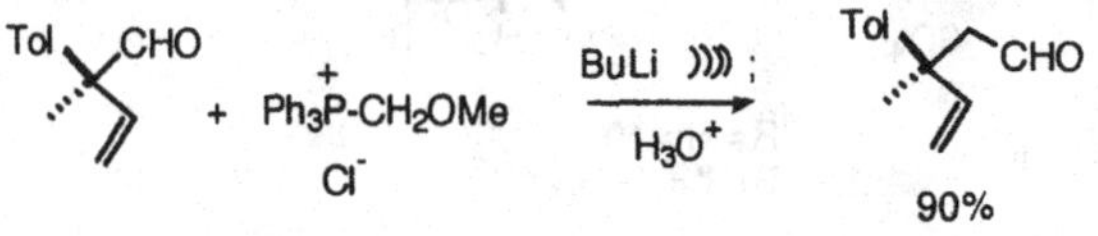

Although many alternative methods for methylenation has been devised, its achievement by simple sonicating a mixture of the carbonyl compound with zinc and diiodomethane [Yamashita, 1984] is most rewarding, although only aldehydes seem to be reactive enough to furnish reasonable yields of the alkenes. Note the conditions are essentially those of the Simmons-Smith reaction.

$$\text{R-CHO} \xrightarrow[\text{)))}]{\text{Zn - CH}_2\text{I}_2} \text{R-CH=CH}_2$$

Styrene from Benzaldehyde. [Yamashita, 1984]

To a 3-necked flask containing dried zinc powder (1.57 g, 24 mmol) are added benzaldehyde (0.53 g, 5 mmol) diiodomethane (2.68 g, 10 mmol) and durene (as internal standard for GLC determination). Ultrasonic irradiation (cleaning bath, 53W, 41 kHz) under nitrogen for 20 min at room temperature afford a 70% yield of styrene.

The reaction of metal phenolates with aldehydes to give *o*-hydroxyalkylphenols is an aromatic version of aldol condensation. Very intriguing results concerning *syn-/anti*-selectivity by varying the metal from magnesium to titanium have been obtained [Casiraghi, 1987]. These reactions are low yielding without ultrasound assistance.

ML_n = MgBr (in CH_2Cl_2) 91% de

ML_n = $Ti(OPr^i)_3$ (in PhMe) 90% de

Improvement is reported for the Strecker synthesis of α-aminonitriles from carbonyl compounds both in homogeneous solution [Menendez, 1986] or on solid alumina support [Hanafusa, 1987], as side reactions are suppressed under ultrasonic irradiation.

$$\xrightarrow[\text{HOAc}]{\text{RNH}_2\text{ - KCN}}$$

R	Conditions	Time	Yield
R = H	stirring	12 d	62%
	)))	25 h	100%
R = Bu	stirring	12 d	60%
	)))	25 h	87.5%

The condensation of salicylaldehydes and β-nitrostyrenes in the presence of alumina as support and base also proceeds more readily with ultrasound [Varma,

1985]. Dihydropyrimidinethione formation from a chalcone and thiourea benefits from ultrasound [Toma, 1987] in comparison with the thermal process.

Al_2O_3, 30-35°, 2-4 h,))))

EtONa-EtOH, 50°, 3.5-11 h,)))) — 58-79%

A colloidal potassium in toluene or xylene which is effective to promote Dieckmann condensation at room temperature can be prepared in a few minutes by ultrasonic irradiation [Luche, 1984]. Rather unfortunately, the scope of the reaction is limited to the formation of 5- and 6-membered ring β-ketoesters.

K / PhMe, r.t. 5 min,)))) — 83%

Ultrasonic activation of the reagents simplifies the acyloin condensation (Rühlmann version) [Fadel, 1990] so that sodium dispersion is not needed and chlorotrimethylsilane in lower purity is useful.

<u>1,2-Bistrimethylsiloxycyclobutene</u>. [Fadel, 1990]

To the suspended Na cubes (3.32 g, 144 mmol) kept under argon in ether or THF (30 mL) at 0° is added chlorotrimethylsilane (19.5 g, 180 mmol) and diethyl succinate (5.22 g, 30 mmol) in ether or THF (10 mL). The flask containing the mixture is immersed in a 0-5° cleaning bath and irradiated with ultrasound for ca. 2 h. The mixture is filtered through Celite, evaporated and distilled to give the product (5.65 g, 82%).

COOEt / COOEt + Me_3SiCl — Na/THF, 0-5°)))) — $OSiMe_3$ / $OSiMe_3$, 82%

3.7. CYCLOADDITIONS

The Simmons-Smith cyclopropanation [Simmons, 1973] requires activation of zinc metal, commonly in the form of Zn-Cu and Zn-Ag couples, to form an organometallic adduct as reagent. However, chemical activation is no longer necessary if the reaction is carried out sonically assisgted in refluxing DME under nitrogen, and traces of iodine are removed from diiodomethane [Repic, 1982]. Further advantages of the new protocol include shortened induction period, easy control, reproducible and higher yields. Interestingly, acyclic β,γ-unsaturated ketones give furans [Repic, 1984; Jie, 1987].

Zn / CH_2I_2, DME → H

12%

))) 67%

Of economic significance is possibility of using dibromomethane in the cyclopropanantion [Friedrich, 1985], although the zinc must be activated and reaction time is longer and yields are lower. Using dimethyl maleate to probe the reaction course it was found that cobalt chloride but not nickel chloride catalyzes a stereoselective process [X. Xu, 1988].

MeOOC COOMe + Zn + CH_2Br_2

$CoCl_2$,))) → MeOOC COOMe

$NiCl_2$,))) → MeOOC COOMe + MeOOC COOMe

A most practical phase transfer reaction is related to the formation of *gem*-dichlorocyclopropanes. The reaction can be accomplished using ultrasound instead of the phase transfer catalyst, while also employing solid sodium hydroxide [Regen, 1982]. However, additional stirring seems necessary. The high density of chloroform would hinder the operation, thus it may call for a high power acoustic source.

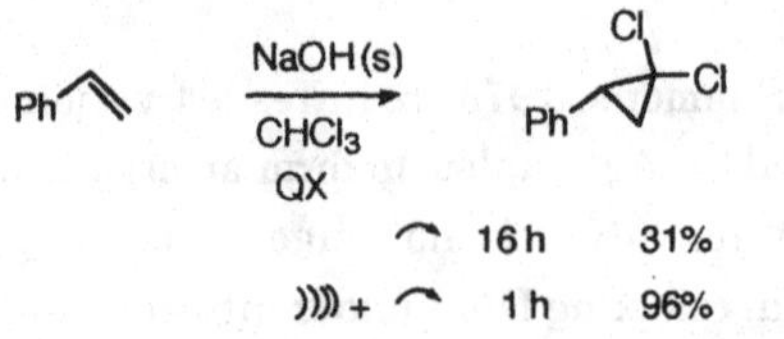

Benefits such as shortened reaction time are evident for the generation of dichloroketene from trichloroacetyl chloride and zinc [Mehta, 1985], the purpose of which is to effect [2+2]cycloaddition to alkenes in situ.

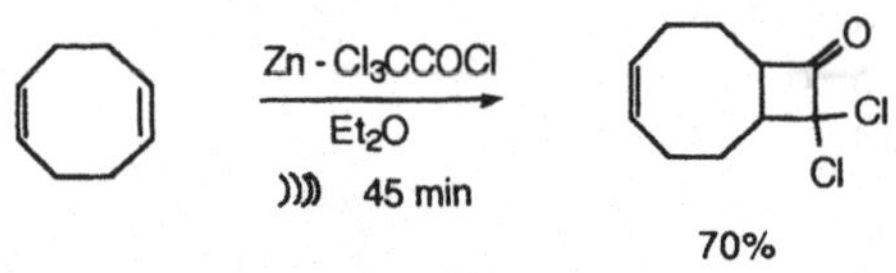

<u>2,2-Dichlorocyclobutanones</u>. [Mehta, 1985]

Under nitrogen a flask is partially submerged in an ultrasound cleaning bath and charged with an alkene (5 mmol), zinc dust (10 mmol) and dry ether (50 mL). A solution of trichloroacetyl chloride (7.5 mmol) in ether (25 mL) is added during 30 min while the bath temperature is maintained between 15° to 20°. Reaction which requires about 60 min to complete is quenched with wet ether, filtered through Celite and washed with water, saturated sodium bicarbonate solution, and brine. After drying (Na_2SO_4) the solvent is evaporated and the product is purified by silica gel chromatography.

In a previous section the reductive debromination is mentioned. In a nonhydroxylic solvent the reaction leads to 2-oxyallyl zwitterions which can be trapped with cyclic dienes (furan, thiophene,...) to furnish bicyclo[3.2.1]oct-6-en-3-ones [Joshi, 1986].

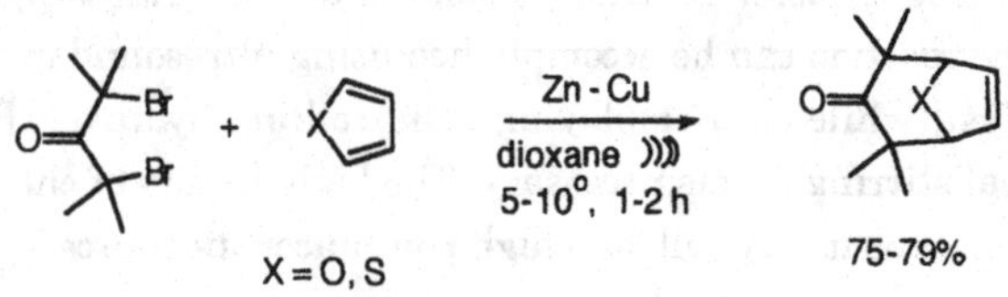

The Pauson-Khand reaction is a useful method for the construction of cyclopentenones. It has also been studied with respect to the effects of ultrasound [Billington, 1988; Bladon, 1988].

Ph — Co$_2$(CO)$_6$ PhMe — Ph

110°, 36 h 42%
))) 3 h 49%

For 1,3-dipolar cycloaddition of nitrones to alkenes, ultrasound significantly decreases reaction times in comparison with conventional stirring [Borthakur, 1988].

PhMe, 100°

Ar = 4-ClC$_6$H$_4$
24 h 75%
))) 1 h 75%

Since cavitation produces local heating as well as pressure, ultrasound is expected to ameliorate the efficiency of Diels-Alder reactions. This expectation has indeed been borne out by an investigation [Lee, 1989] which indicates a very similar effect to high pressure (11 kbar), in terms of increased yields and regioselectivity. The reaction was carried out without solvent.

D.A; DDQ

PhH Δ 2h	30%	15%
PhH 11 kbar 2h	58%	9%
)))	59%	7%

By conventional method the condensation of 1-dimethylamino-1-azabuta-1,3-dienes with electron-deficient dienophiles is difficult. However, ultrasound facilitates the reaction [Villacampa, 1994] and a very expedient route to pyridines is available.

air,)))

60%

3.8. SUBSTITUTIONS

3.8.1. On Carbon

Formation of spirocyclic systems involving twofold alkylation of cyclic ketones with 1,n-dibromoalkanes is greatly facilitated by ultrasound [Fujita, 1986]. The reduction of reaction from 80°C to 40°C is particularly important when cyclopentanone is the substrate because the ketone undergoes self-condensation more rapidly than alkylation at the higher temperature.

tBuOK, PhH 40°,)))

Perfluoroalkylation of chiral enamines proceeds in moderate yields and enantiomer excess [Kitazume, 1985]. The reagents are composed of perfluoroalkyl iodide, zinc, and titanocene dichloride, while the reaction mechanism remains to be clarified.

$Zn-R^fX$))), Cp_2TiCl_2 ; H_3O^+

Ultrasonic irradiation increases the yields and reduces reaction time in the alkylation of Reissert compound under phase transfer catalysis [Ezquerra, 1984].

RX-NaOH, Q^+X^-)))

X = Cl, Br

R = $ArCH_2$, $EtOOCCH_2$

Ethoxythiocarbonylation of ketone enolates in acceptable yields only with ultrasonic irradiation [Palominos, 1986].

NaH,))) ; $(EtOC(S))_2S$,))) ; H_3O^+

2-Ethoxythiocarbonylcyclohexanone. [Palominos, 1986]

Under nitrogen a mixture of cyclohexanone (15 mmol) and NaH (15 mmol) in THF (20 mL) is irradiated in an ultrasound cleaning bath filled with water (temp. 40°) for 1.5 h. *O,O*-Diethyl trithiodicarbonate (7 mmol in 15 mL THF) is added via a syringe and a color change is noted at the end of about 0.5 h. The reaction mixture is poured into cold 5% HCl and extracted with dichloromethane. After drying and evaporation 2-ethoxythiocarbonylcyclohexanone is isolated in 93% yield.

Nitriles can be obtained (in 44-92% yield) from allylic, benzylic and primary alkyl halides when they subjected to ultrasound together with KCN, aluminum oxide in toluene which contains water [Ando, 1984]. Comparing with the phase transfer technique the ultrasound method seems superior as it requires lower temperatures and somewhat better yields. Aroyl chlorides undergo displacement by reaction with powdered KCN in acetonitrile, when assisted with ultrasound [Ando, 1983]. Solvent selection is important to achieve good results.

$$PhCH_2Br \xrightarrow[PhMe\ 50^\circ]{KCN - Al_2O_3} PhCH_2CN\ (67\%)$$

Nitrile Preparation. [Ando, 1984]

A heterogeneous mixture of KCN (16 mmol), aluminum oxide (1.6 g) and water (9.6 mmol) in a solvent (toluene, hexane, etc.) is irradiated with ultrasound at 50° for 1-3 h. An alkyl halide (3.2 mmol) is added and the reaction is allowed to proceed until completion.

It should be noted the above process is a truely sonochemical, involving free radical intermediates. Ionic reactions are not affected by ultrasound in the strict sense. Thus without ultrasonic irradiation, substituted diphenylmethanes are formed because Friedel-Crafts alkylation is more favorable.

Improvements of Friedel-Crafts reaction by ultrasound can be judged by the excellent two-fold reaction on 1,3,6,8-tetramethylnaphthalene [Boudjouk, 1986b].

COCl

$AlCl_3$ hexane

2.5 h)))

O

75%

Ligand displacement on metal centers by π-systems can be considered as a substitution reaction. The formation of the tricarbonyl iron complexes of conjugated dienes [Ley, 1986a] and of a pentamethylcyclopentadienyliron chloride [Roger, 1987] are examples. It is remarkable that in the latter preparation the ultrasonic reaction is complete in 10 min, whereas conventional procedure requires 7 days.

AcO — $Fe_2(CO)_9$, PhH))), 25° → AcO — $Fe(CO)_3$ 100%

K / PhMe-THF, - 10°, 1 h))) → 50%

A unique method for the synthesis of a dinuclear iron complex from anthracene [Begley, 1987] depends on ultrasound.

+ $Fe_2(CO)_9$ — hexane, -20°, 20 h))) → $Fe(CO)_3$ / $Fe(CO)_3$

3.8.2. On Heteroatoms

Ether formation from both aliphatic alcohols [Walkup, 1987] and phenols [Mason, 1988] in high yields is readily achieved by sonication. Its combination with phase transfer catalysts enables the facile preparation of 2-aryloxy-2-cyclohexenones from 2,3-epoxycyclohexanone [Jung, 1995]. Oxidation of the products affords 2-aryloxyphenols.

HO, OH, COOMe — BnBr - Ag_2O, THF 10°))) → BnO, OH, COOMe 60-70%

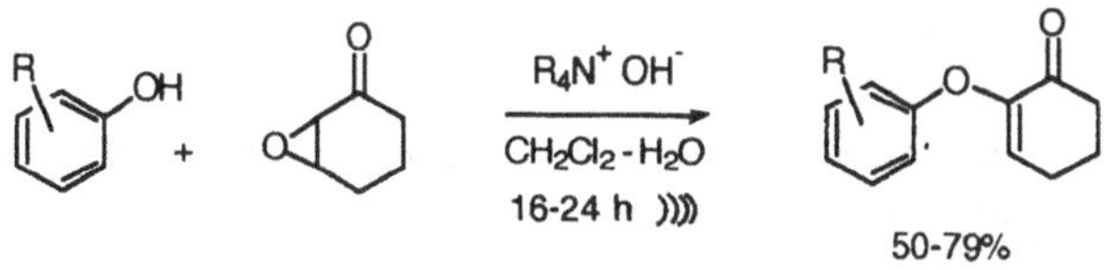

$R_4N^+ OH^-$, CH_2Cl_2 - H_2O, 16-24 h)))

50-79%

N-Alkylation of indole, carbazole and diphenylamine under phase transfer conditions is affected positively by ultrasound [Davidson, 1983]. *N*-Benzylglycines have been prepared in good yields from Cu(gly)$_2$ and benzyl chlorides when the benzene ring does not carry electron-rich substituents [Reddy, 1987]. Very mild conditions are required for alkylation of *N,N*-diazacoronands using aqueous KOH as base [Jurczak, 1988].

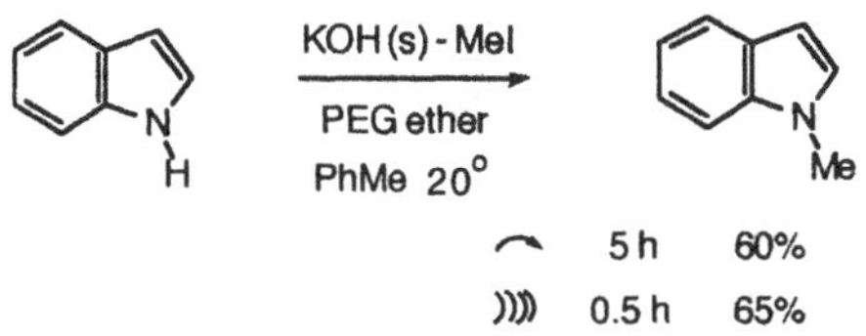

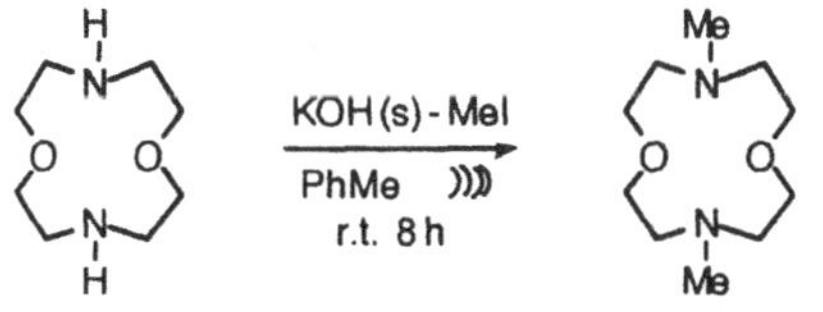

KOH(s) - MeI, PEG ether, PhMe 20°

⌒	5 h	60%
)))	0.5 h	65%

KOH(s) - MeI, PhMe))), r.t. 8 h

92%

Access to allylic and propargylic azides from the corresponding halides by a procedure involving ultrasound irradiation of a two-phase system (aqueous and organic) appears to be quite safe [Priebe, 1984]. Unfortunately, unactivated halides are recalcitrant.

<u>Azides from Halides</u>. [Priebe, 1984]

An organic halide (allylic or propargylic, 40 mmol) is added to 30% aq. solution of sodium azide (160 mmol) in an Erlenmeyer flask equipped with a reflux condenser and immersed in an ultrasonic cleaning bath. During the reaction the bath temperature is allowed to reach 65°. The product is either distilled into a cold trap under reduced pressure (for low boiling azides) or extracted into ether.

$$R\text{-}X + NaN_3 \xrightarrow[)))]{H_2O} R\text{-}N_3$$

S-Alkylation to form alkylthiocyanates [Reeves, 1983] are advantageously conducted under the influence of ultrasonic waves. The ultrasound promotes reductive cleavage of diphenyl diselenide with sodium furnishes a superior nucleophilic reagent [Ley, 1986c] due to its colloidal state, with which substitutions are more efficient.

SNa + PrCl —H_2O, 40°→ SPr

60 min <10%

))) 40 min 100%

PhSe-SePh —Na/THF, r.t.)))→ PhSeNa —OTs→ SePh 93%

Pyrolytic decomposition of aryldiazonium tetrafluoroborate gives fluoroarenes (Schiemann reaction). The high temperature is avoidable by conducting the reaction with ultrasound irradiation, in the presence of triethylamine trishydrofluoride [Müller, 1986].

$-N_2^+$ BF_4^- —$Et_3N{\cdot}3HF$, Freon 113, 40°)))→ –F 92-95%

3.9. VARIOUS PREPARATIONS

Alkyllithium reagents (*n*-, *t*-butyl..) are readily prepared which can then be used in situ to deprotonate carbon acids such as furan, 1,3-dithane, and carboxylic acids (forming dianions). Similarly the very popular nonnucleophilic base LDA (lithium diisopropylamide) can be rapidly prepared (15 min, 91% yield) by sonicating a mixture of 1-chlorobutane, diisopropylamine, and lithium in THF [Einhorn, 1987a], although only demonstrated in <10 mmol quantities. In situ preparation of *s*-BuLi and use in Wittig reaction and *o*-lithiation of anisole were reported.

N-H —Li - BuCl, r.t. 20 min)))→ N-Li 92%

Sodium methylsulfinylmethylide obtained under sonication [Sjoberg, 1966] exhibits good stability up to two months. This modified preparation protocol also has other advantages associated with ultrasound.

$$CH_3S(O)CH_3 \xrightarrow[50^\circ \;\; 1\ h \;\;)))]{NaH} CH_3S(O)CH_2^- \; Na^+$$

Active powders of Mg can be obtained by sonication in the presence of 1,2-dichloroethane in THF to give $MgCl_2$ which is then reduced with lithium. Such species are comparable to the Rieke magnesium [Rieke, 1977] in their use for the formation of Grignard reagents.

Certain primary Grignard reagents are obtainable by hydromagnesiation of 1-alkenes in the presence of titanium or zirconium catalysts.

Trialkylaluminums can be prepared by reaction of aluminum powder with Grignard reagents [Renaud, 1950] in situ, when promoted by ultrasound. Direct treatment of alkyl halides with Al under sonication leads to $R_3Al_2X_3$. Thus it makes boranes available from alkyl borates [Liou, 1985] in another manner.

Organoborane formation mediated by magnesium is also accelerated by ultrasound [Brown, 1986]. The procedure is very useful for the access to those which cannot be obtained by the hydroboration method.

$$3\ RX + Mg \xrightarrow[Et_2O\ )))\ \ 15\text{-}30\ min]{BF_3.OEt_2} R_3B$$

Monographs and Reviews

Abdulla, R.F. (1988) *Aldrichimica Acta* **21:** 31.

Einhorn, C.; Einhorn, J.; Luche, J.-L. (1989) *Synthesis* 787.

Ley, S.V.; Low, C.M.R. (1989) *Ultrasound in Synthesis*, Springer Verlag: Berlin.

References

Akiya, S.; Okui, S. (1950) *J. Pharm. Soc. Jpn.* **71:** 182.

Alexakis, A.; Lensen, N.; Mangeney, P. (1991) *Synlett* 625.

Ando, T.; Kawate, T.; Yamawake, J.; Hanafusa, T. (1983) *Synthesis* 637.

Ando, T.; Kawate, T.; Ichihara, J.; Hanafusa, T. (1983) *Chem. Lett.* 725.

Araki, S.; Butsugan, Y. (1988) *Chem. Lett.* 457.

Aurell, M.J.; Einhorn, C.; Einhorn, J.; Luche, J.-L. (1995) *J. Org. Chem.* **60:** 8.

Barrett, A.G.M.; Dauzonne, D.; O'Neill, I.A.; Renaud, A. (1984) *J. Org. Chem.* **49:** 4409.

Barrett, A.G.M.; O'Neill, I.A. (1988) *J. Org. Chem.* **53:** 1815.

Begley, M.J.; Puntambekar, S.G.; Wright, A.H. (1987) *Chem. Commun.* 1251.

Billington, D.C.; Helps, I.M.; Pauson, P.L.; Thomson, W.; Willison, D. (1988) *J. Organomet. Chem.* **354:** 233.

Bladon, P.; Pauson, P.L.; Brunner, H.; Eder, R. (1988) *J. Organomet. Chem.* **355:** 449.

Bonnemann, H.; Bogdanovic, B.; Brinkmann, R.; He, D.W.; Spliethoff, B. (1983) *Angew. Chem. Int. Ed. Engl.* **22:** 728.

Borthakur, D.R.; Sandhur, J.S. (1988) *Chem. Commun.* 1444.

Bose, A.K.; Gupta, K.; Manhas, M.S. (1984) *Chem. Commun.* 86.

Boudjouk, P.; Han, B.H. (1981) *Tetrahedron Lett.* **22:** 3813.

Boudjouk, P.; Han, B.H.; Anderson, K.R. (1982) *J. Am. Chem. Soc.* **104:** 4992.

Boudjouk, P.; Han, B.H. (1983) *J. Catal.* **79:** 489.

Boudjouk, P.; So, J. (1986a) *Synth. Commun.* **16:** 775.

Boudjouk, P.; Ohrbom, P.; Woell, W.H. (1986b) *Synth. Commun.* **16:** 401.

Boudjouk, P.; Sooriyakumaran, R.; Han, B.H. (1986c) *J. Org. Chem.* **51:** 2818.

Brennan, J.; Hussain, F.H.S. (1985) *Synthesis* 749.

Brettle, R.; Shibib, S.M. (1981) *J. Chem. Soc. Perkin Trans. I* 2912.

Brown, H.C.; Racherla, U.S. (1985) *Tetrahedron Lett.* **26:** 2187.

Brown, H.C.; Racherla, U.S. (1986) *J. Org. Chem.* **51:** 427.

Burkow, I.C.; Sydnes, L.K.; Ubeda, D.C.N. (1987) *Acta Chem. Scand. Ser. B* **41:** 235.

Casiraghi, G.; Cornia, M.; Casnati, G.; Fava, G.G.; Belicchi, M.F.; Zetta, L. (1987) *Chem. Commun.* 794.

Chew, S.; Ferrier, R.J. (1986) *Chem. Commun.* 911.

Chittenden, G. (1988) *Rec.Trav. Chim. Pays Bas* **107:** 607.

Chou, T.-s.; Yuan, J.-J.; Tsao, C.-.H. (1985a) *J. Chem. Res. (S)* 18.

Chou, T.-s.; Tsao, C.-.H.; Hung, S.C. (1985b) *J. Org. Chem.* **50:** 4329.

Chou, T.-s.; You, M.L. (1985c) *Tetrahedron Lett.* **26:** 4495.

Connors, T.F.; Rushing, J.F. (1984) *Chemosphere* **13:** 415.

Crimmins, M.T.; Hollis, W.G.; Bankaitis-Davis, D.M. (1987) *Tetrahedron Lett.* **28:** 3651.

Crimmins, M.T.; Nantermet, P.G. (1990) *J. Org. Chem.* **55:** 4235.

Danhui, Y.; Einhorn, J.; Einhorn, C.; Aurell, M.J.; Luche, J.-L. (1994) *Chem. Commun.* 1815.

Dauben, W.G.; Bridon, D.P.; Kowalczyk, B.A. (1990) *J. Org. Chem.* **55:** 376.

Davidson, R.S.; Patel, A.M.; Safder, A.; Thornthwaite, D. (1983) *Tetrahedron Lett.* **24:** 5907.

Delair, P.; Luche, J.-L. (1989) *Chem. Commun.* 398.

de Laszlo, S.E.; Ley, S.V.; Porter, R.A. (1986) *Chem. Commun.* 344.

de Souza-Barbosa, J.C.; Petrier, C.; Luche, J.-L. (1988) *J. Org. Chem.* **53:** 1212.

Diez-Barra, E.; de la Hoz, A.; Diaz-Ortiz, A.; Prieto, P. (1992) *Synlett* 893.

Dumas, F.; d'Angelo, J. (1986) *Tetrahedron Lett.* **27:** 3725.

Einhorn, C.; Einhorn, J.; Dickens, M.J.; Luche, J.-L. (1990) *Tetrahedron Lett.* **31:** 4129.

Einhorn, J.; Luche, J.-L. (1986a) *Tetrahedron Lett.* **27:** 501.

Einhorn, J.; Luche, J.-L. (1986b) *Tetrahedron Lett.* **27:** 1791, 1793.

Einhorn, J.; Luche, J.-L. (1986c) *Carbohydr. Res.* **155:** 258.

Einhorn, J.; Luche, J.-L. (1987a) *J. Org. Chem.* **52:** 4124.

Einhorn, J.; Luche, J.-L. (1987b) *J. Organometal. Chem.* **322:** 177.

Einhorn, J.; Allavena, C.; Luche, J.-L. (1988) *Chem. Commun.* 333.

Einhorn, J.; Einhorn, C.; Luche, J.-L. (1989) *J. Org. Chem.* **54:** 4479.

Elguero, J.; Goyer, P.; Lissavetzky, J.; Valdeomillos, A.M. (1984) *C.R. Acad. Sci. Paris* **298:** 877.

Elpiner, I.E.; Sokolskaya, A.V.; Margulis, M.A. (1965) *Nature* **208:** 945.

Eremeev, A.V.; Nurdinov, R.; Polyak, F.D. (1985) *Zh. Org. Khim.* **21:** 2239.

Ezquerra, J.; Alvarez-Builla, J. (1984) *Chem. Commun.* 54.

Fadel, A.; Canet, J.-L.; Salaün, J. (1990) *Synlett* 89.

Fadel, A.; Canet, J.-L.; Salaün, J. (1993) *Tetrahedron: Asymmetry* **4:** 27.

Fry, A.J.; Donaldson, W.A.; Ginsburg, G.S. (1979) *J. Org. Chem.* **44:** 349.

Fujita, T.; Watanabe, S.; Sakamoto, M.; Hashimoto, H. (1986) *Chem. Ind. (London)* 427.

Friedrich, E.C.; Domek, J.M.; Pong, R.Y. (1985) *J. Org. Chem.* **50:** 4640.

Ghosh, D.; Nichols, D.E. (1996) *Synthesis* 195.

Greene, A.E.; Lansard, J.P.; Luche, J.-L.; Petrier, C. (1984) *J. Org. Chem.* **49:** 931.

Han, B.H.; Boudjouk, P. (1981) *Tetrahedron Lett.* **22:** 2757.

Han, B.H.; Boudjouk, P. (1982a) *Tetrahedron Lett.* **23:** 1643.

Han, B.H.; Boudjouk, P. (1982b) *J. Org. Chem.* **47:** 5030.

Han, B.H.; Boudjouk, P. (1983) *Organometallics* **2:** 769.

Han, B.H.; Shin, D.H.; Cho, S.Y. (1985) *Bull. Kor. Chem. Soc.* **6:** 320.

Hanafusa, T.; Ichihara, J.; Ashida, T. (1987) *Chem. Lett.* 687.

Heathcock, C.H.; Davidsen, S.K.; Mills, S.G.; Sanner, M.A. (1992) *J. Org. Chem.* **57:** 2531.

Horton, A.M.; Hollinshead, D.M.; Ley, S.V. (1984) *Tetrahedron* **40:** 1737.

Huffman, J.W.; Liao, W.; Wallace, R.H. (1987) *Tetrahedron Lett.* **28:** 3315.

Ihara, M.; Katogi, M.; Fukumoto, K.; Kametani, T. (1987) *Chem. Commun.* 721.
Imi, K.; Imai, K.; Utimoto, K. (1987) *Tetrahedron Lett.* **28:** 3127.
Inoue,Y.; Yamashita, J.; Hashimoto, H. (1984) *Synthesis* 244.
Isobe, M.; Kondo, S.; Nagasawa, M.; Goto, T. (1977) *Chem. Lett.* 679.
Jackson, R.F.W.; James, K.; Wythes, M.J.; Wood, A. (1989) *Chem. Commun.* 644.
Jie, M.S.F.L.K.; Lam, W.L.K. (1987) *Chem. Commun.* 1460.
Joshi, N.N.; Hoffmann, H.M.R. (1986) *Tetrahedron Lett.* **27:** 687.
Jouglet, B.; Blanco, L.; Rousseau, G. (1991) *Synlett* 907.
Jung, M.E.; Starkey, L.S. (1995) *Tetrahedron Lett.* **36:** 7363.
Jurczak, J.; Ostaszewski, R. (1988) *Tetrahedron Lett.* **29:** 959.
Kaliska, V.; Toma, S.; Leski, J. (1987) *Coll. Czech. Chem. Commun.* **52:** 2266.
Kashima, C.; Huang, X.C.; Harada, Y.; Hosomi, A. (1993) *J. Org. Chem.* **58:** 793.
Kimmel, T.; Becker, D. (1984) *J. Org. Chem.* **49:** 2494.
Kimura, T.; Fujita, M.; Ando, T. (1988) *Chem. Lett.* 137.
Kitazume, T.; Ishikawa, N. (1981) *Chem. Lett.* 1679.
Kitazume, T.; Ishikawa, N. (1982) *Chem. Lett.* 1453.
Kitazume, T.; Ishikawa, N. (1985) *J. Am. Chem. Soc.* **107:** 5186.
Knochel, P.; Normant, J.F. (1984) *Tetrahedron Lett.* **25:** 1475.
Kristol, D.S.; Klotz, H.; Parker, R.C. (1981) *Tetrahedron Lett.* **22:** 907.
Kuchin, A.V.; Nurushev, R.A.; Tolstikov, G.A. (1983) *Zh. Obshch. Khim.* **53:** 2519.
Lee, J.; Snyder, J.K. (1989) *J. Am. Chem. Soc.* **111:** 1522.
Ley, S.V.; Low, C.M.R.; White, A.D. (1986a) *J. Organomet. Chem.* **302:** C13.
Ley, S.V.; Lygo, B.; Sternfeld, F.; Wonnacott, A. (1986b) *Tetrahedron* **42:** 4333.
Ley, S.V.; O'Neil, I.A.; Low, C.M.R. (1986c) *Tetrahedron* **42:** 5363.
Li, C.-J. (1993) *Chem. Rev.* **93:** 2023.
Li, C.-J.; Chan, T.-H. (1997) *Organic Reactions in Aqueous Media*, Wiley: New York.
Lindley, J.; Mason, T.J.; Lorimer, J.P. (1987) *Ultrasonics* **25:** 45.
Liou, K.F.; Yang, P.H.; Lin, Y.T. (1985) *J. Organomet. Chem.* **294:** 145.
Lubineau, A.; Auge, J.; Queneau, Y. (1994) *Synthesis* 741.
Luche, J.-L.; Damiano, J.C. (1980) *J. Am. Chem. Soc.* **102:** 7926.
Luche, J.-L.; Petrier, C.; Gemol, A.L.; Zikra, N. (1982) *J. Org. Chem.* **47:** 3805.
Luche, J.-L.; Petrier, C.; Lansard, J.P.; Greene, A.E. (1983) *J. Org. Chem.* **48:** 3837.
Luche, J.-L.; Petrier, C.; Dupuy, C. (1984) *Tetrahedron Lett.* **25:** 753.
Luche, J.-L.; Einhorn, C.; Einhorn, J.; Sinisterra-Gago, J.V. (1990) *Tetrahedron Lett.* **31:** 4125.
Lukevics, E.; Gevorgyan, V.N.; Goldberg, Y.S. (1984) *Tetrahedron Lett.* **25:** 1415.
Luzzio, F.A.; Moore, W.J. (1993) *J. Org. Chem.* **58:** 512.
Mason, T.J.; Lorimer, J.P.; Turner, A.T.; Harris, A.R. (1988) *J. Chem. Res. (S)* 80.

Masuyama, Y.; Hayakawa, A.; Kurusu, Y. (1992) *Chem. Commun.* 1102.
Mehta, G.; Rao, H.S.P. (1985) *Synth. Commun.* **15:** 991.
Menendez, J.C.; Trigo, G.G.; Söllhuber, M.M. (1986) *Tetrahedron Lett.* **27:** 3285.
Moon, S.; Duchin, L.; Cooney, J.V. (1979) *Tetrahedron Lett.* 3917.
Müller, A.; Roth, U.; Siegert, S.; Miethchen, R. (1986) *Z. Chemie* **26:** 169.
Nakamura, E.; Machii, D.; Inubushi, T. (1989) *J. Am. Chem. Soc.* **111:** 6849.
Nakamura, E.; Sato, K.; Imanishi, Y. (1995) *Synlett* 525.
Nayak, S.K.; Banerji, A. (1991) *J. Org. Chem.* **56:** 1940.
Newmann, R.; Sasson, Y. (1985) *Chem. Commun.* 616.
Oguni, N.; Tomago, T.; Nagata, N. (1986) *Chem. Express* **1:** 495.
Oppolzer, W.; Schneider, P. (1984) *Tetrahedron Lett.* **25:** 3305.
Oppolzer, W.; Cunningham, A.F. (1986a) *Tetrahedron Lett.* **27:** 5467.
Oppolzer, W.; Nakao, A. (1986b) *Tetrahedron Lett.* **27:** 5471.
Palominos, M.A.; Rodriguez, R.; Vega, J.C. (1986) *Chem. Lett.* 1251.
Petrier, C.; Gemal, A.L.; Luche, J.-L. (1982) *Tetrahedron Lett.* **23:** 3361.
Petrier, C.; Luche, J.-L.; Dupuy, C. (1984) *Tetrahedron Lett.* **25:** 3463.
Petrier, C.; de Souza Barbosa, J.C.; Dupuy, C.; Luche, J.-L. (1985a) *J. Org. Chem.* **50:** 5761.
Petrier, C.; Einhorn, J.; Luche, J.-L. (1985b) *Tetrahedron Lett.* **26:** 1449.
Petrier, C.; Luche, J.-L. (1985c) *J. Org. Chem.* **50:** 910.
Petrier, C.; Luche, J.-L. (1986) *Tetrahedron Lett.* **27:** 3149.
Petrier, C.; Dupuy, C.; Luche, J.-L. (1987a) *Tetrahedron Lett.* **28:** 2347.
Petrier, C.; Luche, J.-L. (1987b) *Tetrahedron Lett.* **28:** 2351.
Priebe, H. (1984) *Acta Chem. Scand.* **38B:** 895.
Rathke, M.W. (1975) *Org. React.* **22:** 423.
Raucher, S.; Klein, P. (1981) *J. Org. Chem.* **46:** 3558.
Reddy, G.S.; Smith, G.G. (1987) *Inorg. Chim. Acta* **133:** 1.
Regen, S.L.; Singh, A. (1982) *J. Org. Chem.* **47:** 1587.
Reeves, P.; McClusky, J.V. (1983) *Tetrahedron Lett.* **24:** 1585.
Renaud, P. (1950) *Bull. Soc. Chim. Fr. (Ser. S)* 1044.
Repic, O.; Vogt, S. (1982) *Tetrahedron Lett.* **23:** 2729.
Repic, O.; Lee, P.G.; Giger, N. (1984) *Org. Prep. Proc. Int.* **16:** 25.
Rieke, R.D. (1977) *Acc. Chem. Res.* **10:** 301.
Roger, C.; Marseille, P.; Salus, C.; Hamon, J.R.; Lapinte, C. (1987) *J. Organomet. Chem.* **336:** C13.
Sarandeses, L.A.; Luche, J.-L. (1992) *J. Org. Chem.* **57:** 2757.
Schreiber, S.L.; Kelly, S.E.; Porco, J.A.; Sammakia, T.; Suh, E.M. (1988) *J. Am. Chem. Soc.* **110:** 6210.

Shibata, K.; Kondo, H.; Urano, K.; Matsui, M. (1987) *Chem. Express* **2:** 117, 169.
Simmons, H.E.; Cairns, T.L.; Vladuchick, S.A.; Hoiness, C.M. (1973) *Org. React.* **20:** 1.
Sinisterra, J,V.; Fuentes, A.; Marinas, J.M. (1987) *J. Org. Chem.* **52:** 3875.
Sjoberg, K. (1966) *Tetrahedron Lett.* 6383.
Smith, A.B., III; Nolen, E.G.; Shirai, R.; Blase, F.R.; Ohta, M.; Chida, N.; Hartz, R.A.; Fitch, D.M.; Clark, W.M.; Sprengeler, P.A. (1995) *J. Org. Chem.* **60:** 7837.
Snowden, R.L.; Sonnay, P. (1984) *J. Org. Chem.* **49:** 1465.
So, J.; Park, M.; Boudjouk, P. (1988) *J. Org. Chem.* **53:** 5871.
Solladie-Cavallo, A.; Farkhani, D.; Fritz, S.; Lazrak, T.; Suffert, J. (1984) *Tetrahedron Lett.* **25:** 4117.
Suslick, K.S.; Schubert, P.F.; Goodale, J.W. (1981) *J. Am. Chem. Soc.* **103:** 7342.
Toma, S.; Putala, M.; Salisova, M. (1987) *Coll. Czech. Chem. Commun.* **52:** 395.
Townsend, C.A.; Nguyen, L.T. (1981) *J. Am. Chem. Soc.* **103:** 4582.
Tso, H.-H.; Chou, T.-s.; Hung, S.C. (1987) *Chem. Commun.* 1552.
Tsuji, J. (1995) *Palladium Reagents and Catalysts*, Wiley: Chichester.
Uyehara, T.; Yamada, J.; Ogata, K.; Kato, T. (1985) *Bull. Chem. Soc. Jpn.* **58:** 211.
Varma, R.S.; Kabalka, G.W. (1985) *Heterocycles* **23:** 139.
Villacampa, M.; Perez, J.M.; Avendano, C.; Menendez, J.C. (1994) *Tetrahedron* **50:** 10047.
Walkup, R.D.; Cunningham, R.T. (1987) *Tetrahedron Lett.* **28:** 4019.
Wipf, P.; Smitroich, J.H. (1991) *J. Org. Chem.* **56:** 6494.
Xu, L.; Tao, F.; Yu, T. (1985) *Tetrahedron Lett.* **26:** 4231.
Xu, X.; Li, Z.; Na, Y.; Liu, G. (1987) *Yingyong Huaxue* **4:** 73.
Yamashita, J.; Inoue, Y.; Kondo, T.; Hashimoto, H. (1984) *Bull. Chem. Soc. Jpn.* **57:** 2335.
Yamashita, J.; Inoue, Y.; Kondo, T.; Hashimoto, H. (1985) *Bull. Chem. Soc. Jpn.* **58:** 2709.
Yamashita, J.; Inoue, Y.; Kondo, T.; Hashimoto, H. (1986) *Chem. Lett.* 407.
Yamawaki, J.; Sumi, S.; Ando, T.; Hanafusa, T. (1983) *Chem. Lett.* 379.
Yang, C.-X.; Yang, D.T.C.; Harvey, R.G. (1992) *Synlett* 799.
Zikra, N.; Luche, J.L. (unpubl.) quoted in [Einhorn, 1989].

4 HIGH PRESSURE REACTIONS

The beneficial effects of high pressure in promoting chemical reactions have been witnessed for quite a long time. Thus Moissan's attempts at conversion of graphite to diamond under pressure is quite well known to the scientific community. Actually polyethene was accidentally discovered in 1933 in an ICI laboratory while E.W. Fawcett and R.O. Gibson embarked on a study of the behavior of various chemicals under high pressure.

Generally, a chemical reaction is speeded up by raising temperature or by increasing the reactant concentrations. When considered in terms of activation volume $\Delta V^{\ddagger}$ of a reaction, which is the difference in partial molal volume between the transition state and the initial state, the dependence of rate to pressure is evident.

Table 1 shows pressure required for doubling the reaction rate at a specified activation volume [Neumann, 1972]. Accordingly, a change in pressure provides a means to control reaction rates. For a typical organic reaction the activation volume

Table 1.

$\Delta V^{\ddagger}$ (cm^3/mol)	Pressure (MPa)
5	366
10	183
25	76
40	44

falls within the range between +20 to -50 cm^3/mol. When the $\Delta V^{\ddagger}$ is negative the reaction proceeds from a more compact transition state, therefore it is favored by intensifying pressure. On the other hand, positive $\Delta V^{\ddagger}$ indicates an increase in molecular size and the reverse pressure effect is expected. It follows that one should apply pressure to a condensation reaction but decrease pressure in case of a dissociation process. Observed pressure effects on synthetic organic reactions are 5-40 kbar.* It is instructive to characterize several mechanistic features according to their contributions to $\Delta V^{\ddagger}$ [Asano, 1978] (Table 2).

1 bar= 0.9869 atm = 1.0197 kg/cm^{-1}
= 10^5 N/m^2 = 10^5 Pa = 0.1 MPa (N= newton; Pa = pascal).
While the correct unit is Pa, bar and kbar are most frequently reported units in the chemical literature.

Table 2.

Mechanistic Features	Contribution ($\Delta V^{\ddagger}$)
Bond cleavage	+10
Bond deformation	0
Bond formation	-10
Ionization	-20
Displacement	-5
Diels-Alder reaction	-25 ~ -50
Charge concentration	-5
Charge delocalization	+5
Charge neutralization	+20
Steric hindrance	(-)

Since pressure compensates for higher reaction temperature, in the preparation of heat sensitive substances the use of high pressure may be the last recourse. To carry out a high pressure reaction, a solution of the reactants in a vessel made of teflon or other pliable material is sealed and placed in the pressure chamber which is then filled with a fluid capable of transmitting pressure. For most preparative setup mechanically generated static pressure is applied externally by a piston-cylinder device.

It is extremely important to select an appropriate solvent because some of them solidify under pressure and cause damage to the apparatus. Usually the liquid-to-solid transition temperature of a solvent is raised by 15°-20°C per 1 kbar. Thus at 25°C the solidification pressures for cyclohexane (0.35 kbar), acetic acid (0.39 kbar), benzene (0.72 kbar), dioxane (1.2 kbar), and chloroform (5.5 kbar) indicate these to be of little use in high pressure reactions. However, the other common solvents such as methanol, ethanol, ether, acetone, dichloromethane, and *n*-hexane which remain liquid when pressurized at 20 kbar are quite suitable. For a more extensive list, see [Hamann, 1957].

Of foremost importance in carrying out chemical reactions under high pressure is safety. However, with proper equipment [Onodera, 1991] (For schematics of reaction vessel arrangements, see [Dauben, 1977]) high pressure reactions are very simple to perform. Normally all reactants and solvents are placed in the vessel and then pressurized. Standard workup procedure is followed after pressure is released at the end of the reaction.

4.1. SUBSTITUTION REACTIONS

Most nucleophilic substitutions are characterized by negative volumes of activation, they are facilitated by pressure. For Menschutkin reaction (alkylation of amines), $\Delta V^{\ddagger}$ becomes more negative in a less polar solvent, therefore to succeed in such a reaction (particularly of highly hindered reactants) the use of pressure is prescribed.

N-Tritylpyridinium chloride is by the conventional method but a yield of 60-70% is obtained from a pressure reaction (in dioxane, 4-5 kbar, 60-70°C, 10-15 h) [Okamoto, 1970]. Similarly, quaternization of hindered sparteine alkaloids such as α-isolupanine and 15-oxosparteine with iodomethane is practical only at high pressure [Jurczak, 1983].

+ Ph_3C-Cl → (4-5 kbar, dioxane, 60-70°) Cl^- 60-70%

α-isolupanine   15-oxosparteine

The synthesis of cryptands is operationally tedious, requiring high dilution techniques and the presence of a strong reducing agent. Such limitations are ameliorated by pressure which also seems to enhance the yields (often approaching quantitative), as shown in the smooth preparation of a chiral derivative [Newkome, 1981; Pietraszkiewicz, 1983]. Diazacoronands are available in an analogous manner [Jurczak, 1989].

8 kbar, 25°C/ 20 h; → Ph_3P, DMF, Δ

A successful synthesis of a [3]catenane in 11% yield [Ashton, 1991] indicates the necessity of high pressure, otherwise only the [2]catenane is obtainable. The process involves the second stage quaternization of 4,4-bipyridine with *p*-xylylene dibromide in the presence of the macrocycle tetra-*p*-phenylene-68-crown-20.

tetra-p-phenylene-68-crown-20, DMF, 10 kbar, 25°/5d

Some of the normally inert electrophiles undergo reaction with amines under high pressure conditions, for example, the deactivation of a benzylic tosylate by a trifluoromethyl group is counteracted by pressure [Pirkle, 1977]. Salt formation from triethylamine and dichloromethane to afford $Et_3N^+CH_2Cl\ Cl^-$ has been observed [Matsumoto, 1984b], and in its connection, a Mannich reaction occurs (93% yield) when a mixture of acetophenone, piperidine and dichloromethane is pressurized (8 kbar, 40°C) [Matsumoto, 1982].

NH_3/THF, 6 kbar, 130°/6h; > 95%

Formation of *p*-nitrophenylamines by displacement of the chloride can be accomplished by a pressure reaction [Ibata, 1987]. However, good yields of the products are obtained only with α-unsubstituted primary amines.

$PrNH_2$ + Cl–C_6H_4–NO_2 → THF, 7.2 kbar, 50°/20 h → 93%

In a synthesis of *cis*-trikentrin-A [Boger, 1991] the displacement of a mesyl group under pressure eliminates the problems due to either sluggishness (at room temperature) or epimerization (at higher temperatures).

SO_2Me + NH_2 — 13 kbar, 25°/3-4 d → SO_2Me, HN, 74% ⇒ HN, *cis*-trikentrin-A

Oxidation of aliphatic tertiary amines and thioethers to give the corresponding amine oxides and sulfoxides under high oxygen pressure has been reported [Riley, 1985a, 1985b].

Me_3N — H_2O / air, 7 kbar, 100° / 64 h → Me_3NO > 95%

Phosphonium salts are precursors of Wittig reagents. When a phosphonium salt contains thermally unstable functionality, its preparation at room temperature can be promoted by pressure [Dauben, 1984]. A strigol synthesis [Berlage, 1987] involving a Wittig reaction would not have been accomplished without the help of pressure in acquiring the phosphonium salt.

Br — PhH, PHMe, 15 kbar, 22° / 36 h → $\overset{+}{P}Ph_3$ Br^- 92%

COOMe — Ph_3P, 15 kbar, 70°; base → COOMe ⇒ strigol

O-Glycosylation by glycosyl bromides in the presence of mercury(II) cyanide in acetonitrile at room temperature is more highly stereoselective when under pressure (94% vs. 74%) [Kochetkov, 1987].

$Hg(CN)_2$ / MeCN, 12 kbar, 20°

47%

Because of steric hindrance, great difficulty is experienced in the generation of *t*-butyldimethylsilyl ether from tertiary alcohols. This reactivity problem is readily solved at high pressure, as exemplified by the silylation of linalool [Dauben, 1986a].

linalool + $tBuMe_2SiCl$ → (CH_2Cl_2 / py, 24 h)

9 bar /45°	5%
15 kbar / 20°	97%

The futile conversion of a neopentyl alcohol into a sulfide was resurrected on pressurization [Kotsuki, 1991]. Improved yields are generally obtained in other cases.

Phenylthioethers from Alcohols. [Kotsuki, 1991]

A mixture of the alcohol (0.25 mmol), diphenyl disulfide (164 mg, 0.75 mmol), tri-*n*-butylphosphine (202 mg, 1.0 mmol) in dry THF is placed in a Teflon reaction vessel, allowed to react at 10 kbar and 62°C. After the reaction, the mixture is diluted with ether, washed successively with 2N NaOH and brine. The crude product is purified.

$n\text{-}Bu_3P$ + PhS-SPh → $n\text{-}Bu_3\overset{+}{P}\text{-}SPh$ PhS^-

1 bar, 40 h	42%
10 kbar, 62°, 3 h	100%

Cyclic ethers undergo ring opening on admixture with acyl halides under 10 kbar pressure without the need for a catalyst [Kotsuki, 1988]. However, low conversions are attained by acyclic ethers.

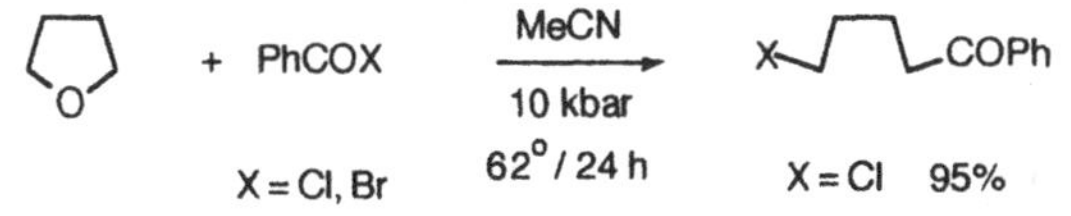

Epoxides (e.g., styrene oxide) are converted to 1,3-dithiolane-2-thiones as the major products when subjected to a high pressure reaction with excess carbon disulfide in the presence of a tertiary amine [Taguchi, 1988].

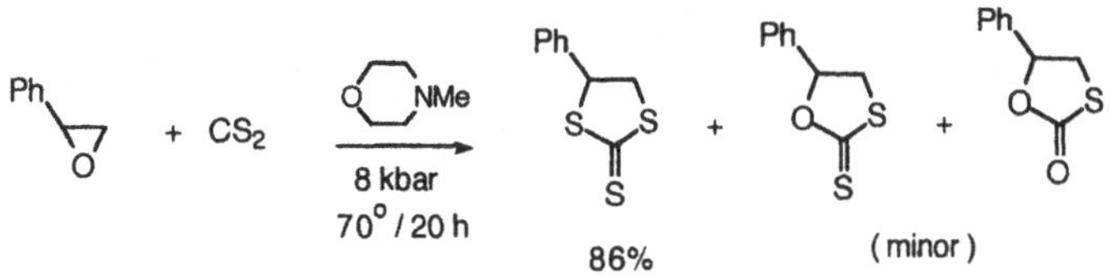

In the conventional peptide bond formation the first step is a bimolecular addition of the amine to the activated carboxylic acid derivative, hence it features a negative activation volume. The favorable effects of pressure have been noted [Yamada, 1984]. *N-t*-Butoxycarbonyllactams can be roused to react with amines by applying pressure [Kotsuki, 1992]. Furthermore, uncatalyzed aminolysis of ordinary esters is feasible at relatively low temperatures (e.g., 45°C) when 8 kbar pressure is applied [Matsumoto, 1986].

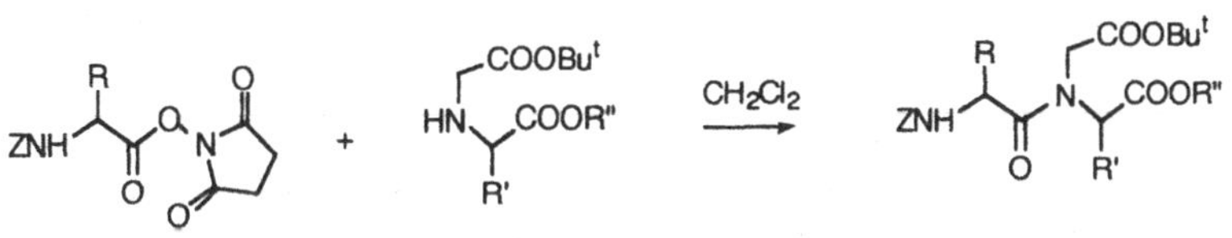

Hydrolysis of carboxylic esters (such as amino acid derivatives) without isomerization and racemization can be accomplished in the presence of *N*-methylmorpholine in aqueous acetonitrile under 10 kbar pressure [Yamamoto, 1991].

1,5-Dienes are accessible by coupling of allylmetals and allylic halides. Regioselectivity is the major concern of these reactions. It has been shown that the S_N2-type C-C bond formation can be directed predominantly at the γ-position of allyltrialkylstannanes at 10 kbar in dichloromethane [Yamamoto, 1984a; Golebiowski, 1986].

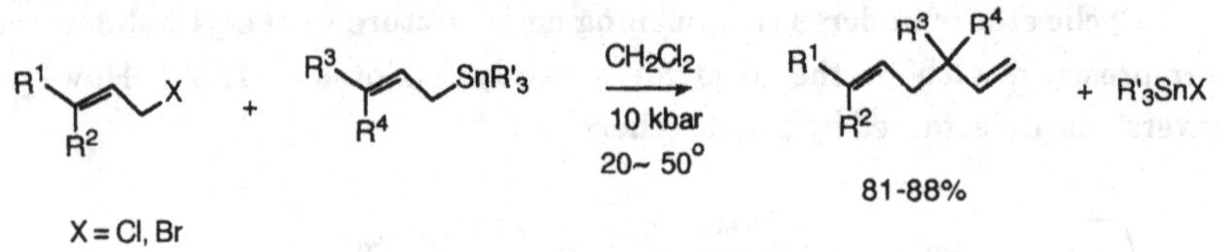

R¹ R² X = Cl, Br + R³ R⁴ SnR'₃ — CH_2Cl_2, 10 kbar, 20~50° → R¹ R² R³ R⁴ + R'₃SnX

81-88%

As expected, pressure influences electrophilic substitutions at a hindered site, although not many examples are known. However, a marked enhancement of reactivity of 1,3-bis(2-hydroxyhexafluoro-2-propyl)benzene toward hexafluoroacetone is evident [Jones, 1967]. No reaction occurs below 8 kbar.

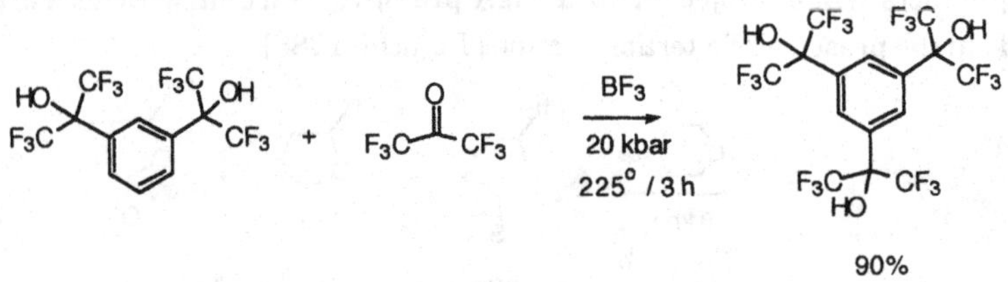

HO, CF_3, F_3C, OH, F_3C, CF_3 + F_3C–CO–CF_3 — BF_3, 20 kbar, 225° / 3 h →

90%

The substitution of five-membered heterocycles (furan, thiophene, *N*-methylpyrrole) with certain aldehydes and ketones in dichloromethane can be promoted by pressure [Jenner, 1984]. A synthesis of D-ribitol [Pikul, 1985] is based on this chemistry.

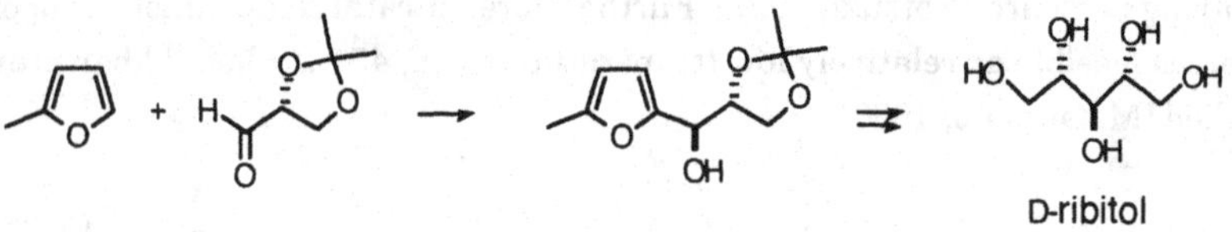

D-ribitol

The reaction of 2-methylfuran with *n*-butyl glyoxalate occurs as electrophilic substitution at C-5 as well as at the sidechain, when being conducted in toluene [Jenner, 1984]. 2,5-Dimethylfuran is restricted to the latter pathway.

Reduction of organic halides with tri-*n*-butyltin hydride without the addition of catalyst or initiator is achievable under pressure, although the reactions are slower [Rahm, 1989]. Undesirable side reactions are minimized.

$$n\text{-}C_8H_{17}Br + n\text{-}Bu_3SnH \xrightarrow[60^\circ / 24\,h]{14\ kbar} n\text{-}C_8H_{18} + n\text{-}Bu_3SnBr$$

100%

4.2. ADDITIONS

4.2.1. Addition to the Carbonyl Group

Acceleration of the asymmetric reduction of ketones with *B*-3-pinanyl-9-borabicyclo[3.3.1]nonane under high pressure has been observed [Midland, 1989]. Of great significance is also that the competing dehydroboration-reduction pathway is suppressed, erosion of enantioselectivity is avoided.

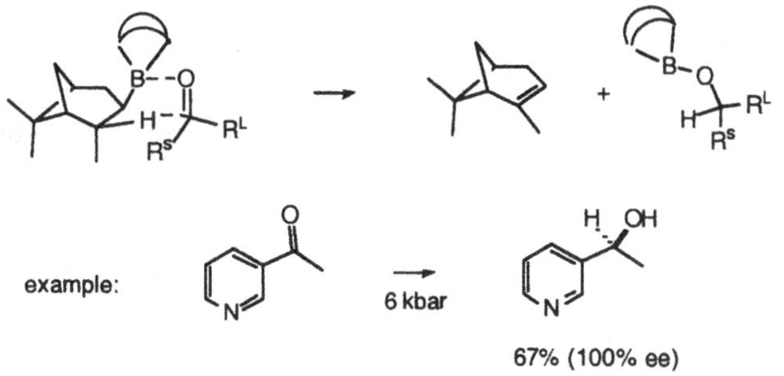

example: 6 kbar

67% (100% ee)

The smooth introduction of an α-hydroxyalkyl group to acrylic compounds CH_2=CHX, by reaction with a great number of aldehydes, is benefited by applied pressure [Hill, 1986]. The reactivity follows a trend according to X = CHO > COMe > CN > COOR > $CONH_2$. A β-substituent is detrimental. The catalytic effect of the tertiary amine seems to be related to its steric bulk, e.g., DABCO > quinuclidine > Me_3N >Et_3N.

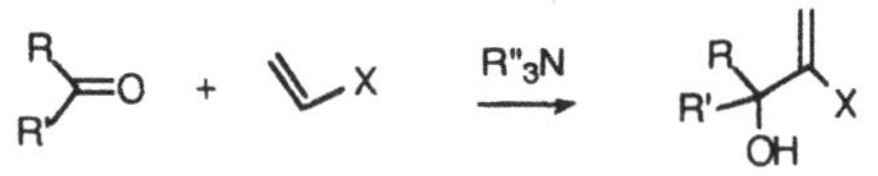

R"$_3$N

The Mukaiyama aldol condensation is catalyzed by Lewis acids. Interestingly, in the absence the catalyst the reaction proceeds under high pressure and manifests a different, and sometimes reverse diastereoselectivity [Yamamoto, 1983a]. The neutral conditions are also desirable, but there seems to be a serious limitation to aromatic aldehydes. On the other hand, *O* silyl ketene acetals react with ketones and aldehydes smoothly, but diastereoselectivity is generally low [Yamamoto, 1984b].

+ PhCHO → erythro + threo

	erythro		threo
1 bar, $TiCl_4$	1	:	3
10 kbar	3	:	1

The neopentyl situation of a ketone intermediate which must be converted to the unsaturated acid posed a serious problem near the end of an approach to retigeranic acid-A [Wright, 1988]. The forward progression was enabled by high pressure reaction with cyanotrimethylsilane in the presence of KCN and 18-crown-6.

Me_3SiCN, KCN, 18-c-6, 6.8 kbar

retigeranic acid

2-Methylcyclohexanone reacts with nitroalkanes in the presence of fluoride ion only under high pressure [Matsumoto, 1984a].

+ $MeNO_2$ → (n-Bu_4NF, 9 kbar, 30°/4 d) 50%

Pressurized ene reactions are accomplished at room temperatures [Gladysz, 1978]. It has also been demonstrated that slurries of the substrates in silica gel give products on pressurization [Dauben, 1992]. High pressure effectively increases the acidity of silica gel, yet on the other hand it is milder than conventional Lewis acids such as $ZnBr_2$ which can promote decomposition of sensitive products.

SiO_2/CH_2Cl_2, 15 kbar, 25°/53 h → 56% + 29%

Allylation of aldehydes with allyltributylstannane under pressure [Yamamoto, 1983b] is a concerted reaction via a cyclic transition state [Isaacs, 1992]. Under high pressure the boat transition state may be favored, as revealed by a study of the pressure dependence of the rate constant and diastereoselectivity for the condensation of octanal and crotyltributylstannane [Yamamoto, 1989]. The *(Z)*-stannane /threo product and *(E)*-stannane /erythro product correlation, with increasing trend in each series, argues for the interpretation.

H

Cl_3C O SnR_3 ‡

The reaction of 2-benzyloxypropanal with a 2-allyl-1,3,2-dioxaborolane under pressure (8.1 kbar), followed by treatment of the product with triethylamine, led to a cyclopropane derivative [Hoffmann, 1985].

CHO BnO + O B O OMe — 8.1 kbar, 45° /72 h ; Et_3N → OH OMe BnO

82%

Generally, mismatched double asymmetric reactions are slow. Such reactions are facilitated by high pressures, as exemplified in the allylation of the aldehyde with the dioxaborolane reagent which constitutes a key step toward synthesis of *(9S)*-dihydroerythronolide-A [Stürmer, 1993].

OMe ... CHO + Cy, Cy, B-O — 10 kbar, 3 d → OMe ... OH

Cy = cyclohexyl

4.2.2. Addition to Alkenes

The addition of ammonia and amines to simple alkenes is difficult to achieve, but it is promoted by high pressure [Howk, 1954]. Very highly hindered trialkylboranes are accessible by the high pressure technique [Rice, 1982a]. Unfortunately, these compounds tend to decompose or isomerize upon release of pressure, therefore their use as synthetic intermediates is voided. Hydrostannylation under pressure is a more promising reaction, as it delivers products unavailable otherwise (e.g., from β-pinene) [Rahm, 1982].

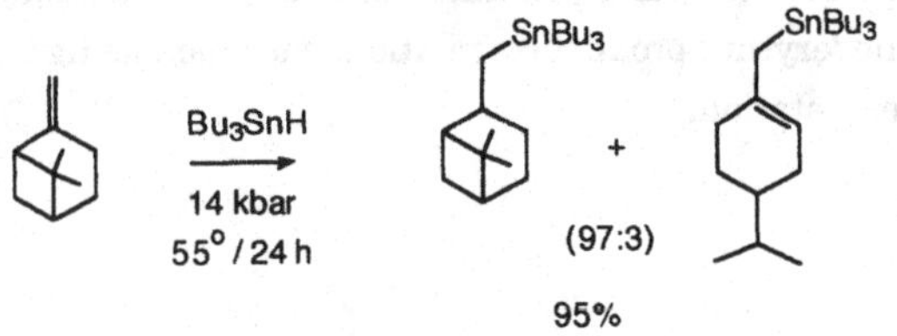

Although the Michael reaction has a broad scope and many catalysts have been developed to meet specific demands, from time to time chemists encounter recalcitrant addends. In many cases pressure is of considerable value for injecting reactivity [Matsumoto, 1980; Dauben, 1983b]. Of interest is that the normal adduct of 1-acetylcyclohexene and diethyl malonate is obtained by pressurizing the addends in acetonitrile at ambient temperature, but under more conventional conditions of base-catalysis (NaOEt, EtOH, reflux) the product is a bicyclic diketo ester.

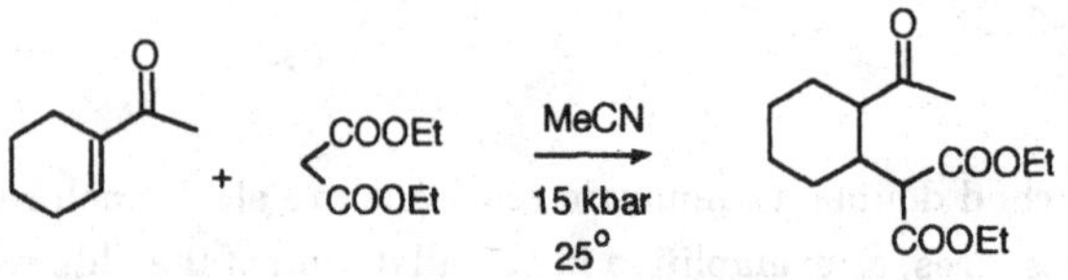

Quinine alkaloids act as chiral catalysts in Michael addition. However, no reaction occurs between chalcone and nitromethane in aprotic solvents unless high pressure is applied [Matsumoto, 1981]. Quantitative yield of the adduct with enantiomer excess of 24% and 51%, respectively, when (-)-quinine and (+)-quinidine are present in toluene solvent. Of significance is the observation that optical yields vary at different levels of pressure (higher pressure seems to lower asymmetric induction) when the reaction is catalyzed by quinine and they remain quite constant with quinidine.

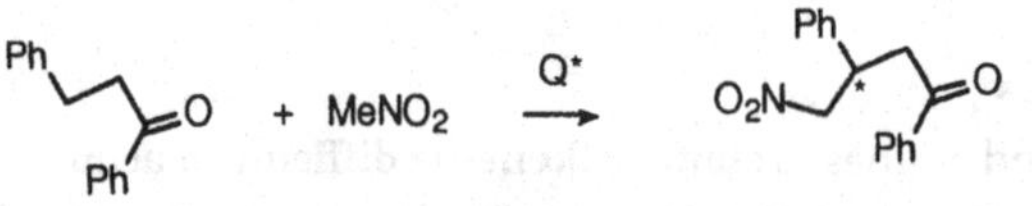

The condensation of a chiral methyl 2-aminocyclopentenecarboxylate with acrylic esters at 14 kbar affords adducts with high enantiomeric excess [Guingant, 1991]. The bulk of the acrylic ester (Me vs. *t*Bu) does not seem to affect the result.

(R)-β-Alanines are formed by addition of amines at 15 kbar to (-)-8-phenylmenthyl crotonate [d'Angelo, 1986]. The reaction proceeds with higher stereoselectivity (>99% vs. 60% in diastereomeric excess) but lower yields when the chiral auxiliary in the Michael acceptor is changed to the 8-(2-naphthyl)menthyl group.

50% (>99% de)

In connection with a synthesis of bruceantin, the introduction of an acetic ester chain to a tricarbocyclic enone intermediate is for its incorporation into the δ-lactone unit. This can be done by using a high pressure Michael addition [Kerwin, 1987] in which a bulky keteneacetal is the donor but no catalyst is required.

Acetylmanganation of an activated multiple bond is accomplished by reaction of the substrates with methylmanganese pentacarbonyl at 6 kbar [DeShong, 1988].

$MeMn(CO)_5$ + COOMe —6 kbar→ $Mn(CO)_4$ COOMe

$Me_3SiMn(CO)_5$ + O → Me_3SiO $Mn(CO)_5$ ⇌ $Mn(CO)_4$ Me_3SiO O

COOMe

COOMe ← Me_3SiO COOMe O- -$Mn(CO)_4$

OH O

4.3. CONDENSATION REACTIONS

The condensation of a tritylated sugar with a cyanoethylidene sugar derivative is subject to pressure effect. Preferential formation of *trans*-1,2-glycosides is maximized at 14 kbar [Kochetkov, 1987]. The reacting species are the dioxalenium ions instead of the glycosyl cations, due to smaller volumes of the bicyclic structures.

H O R CN H —Ph_3CClO_4→ H O +) R O H ⇌ O+ OCOR

↓ Ph_3C-OSugar (- Ph_3C^+)

O OSugar OCOR

Dehydrative condensation reactions made difficult or impossible by steric hindrance are best carried out under pressure. Thus the preparation of di-*t*-butyl ketoxime at 9 kbar [W.H. Jones, 1959], of 2-*t*-butylbenzimidazole from *o*-phenylene-diamine and pivalic acid at 8 kbar and 110°C [Holan, 1977] demonstrates the positive effects. Fenchone forms the ethyleneacetal in 82% yield [Dauben, 1986b].

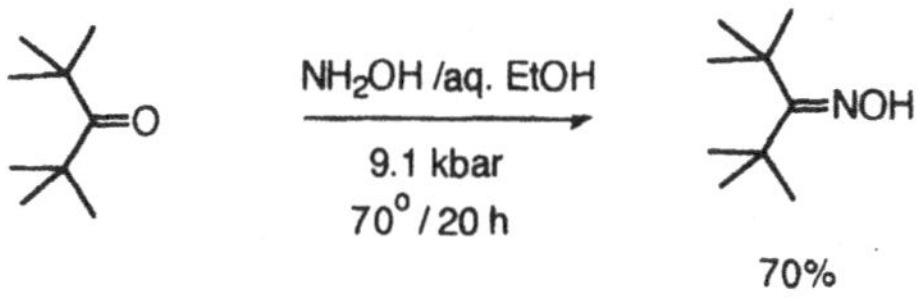

NH_2OH /aq. EtOH
9.1 kbar
70° / 20 h
=NOH
70%

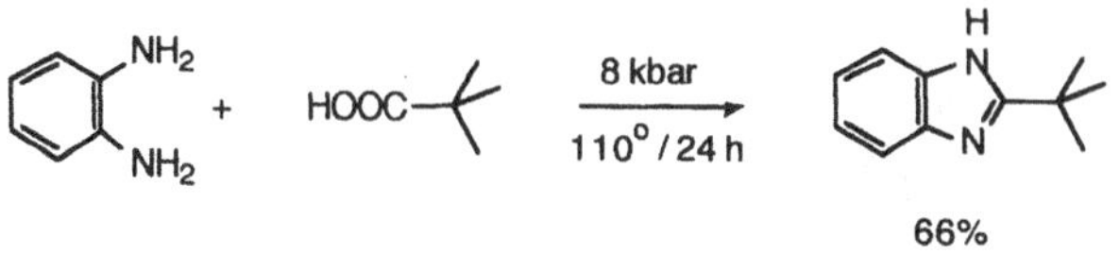

NH_2
NH_2
+ HOOC
8 kbar
110° / 24 h
H
N
N
66%

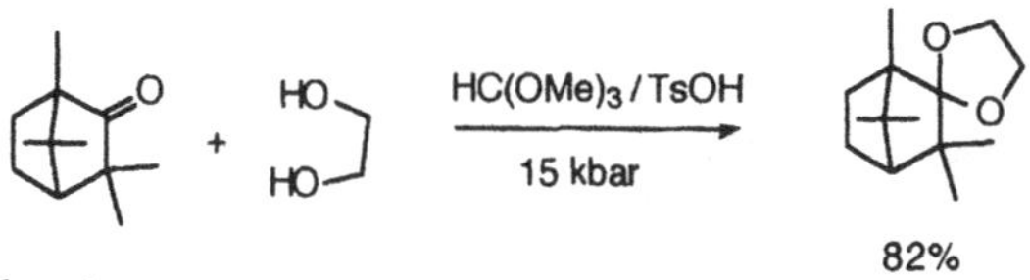

O
+ HO
HO
$HC(OMe)_3$ / TsOH
15 kbar
O
O
fenchone
82%

An increase in yield from 3% to quantitative of the Knoevenagel condensation product from cyclopentanone and ethyl cyanoacetate [Newitt, 1937] by changing the condition of piperidine catalysis to high pressure (15 kbar) is synthetically significant.

Robinson annulation is initiated by a Michael addition. The pressure effect is evident in that it enables addition to hindered substrates such as mesityl oxide [Dauben, 1983a]. Moreover, the aldol condensation leads to the more compact bridged ring systems.

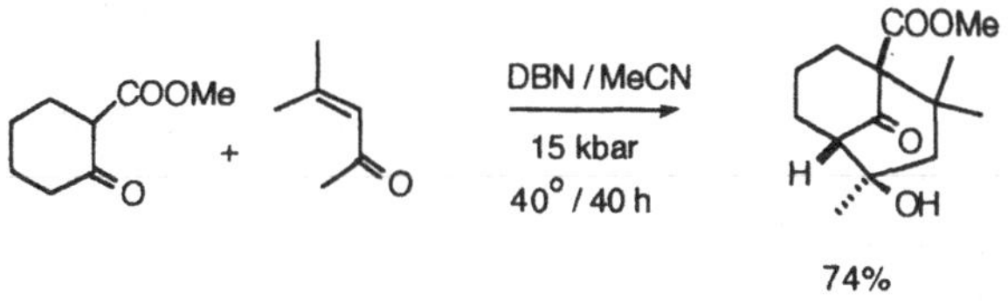

COOMe
O
+
O
DBN / MeCN
15 kbar
40° / 40 h
COOMe
O
H
OH
74%

Greatly improved efficiency is observed from pressure Wittig reactions of unreactive carbonyl compounds [Nonnenmacher, 1983]. For example, a 30% yield has been obtained from the reaction of 4-(*N,N*-dimethylamino)benzaldehyde with triphenylphosphoranylacetone at 10 kbar whereas there is no reaction at atmospheric pressure.

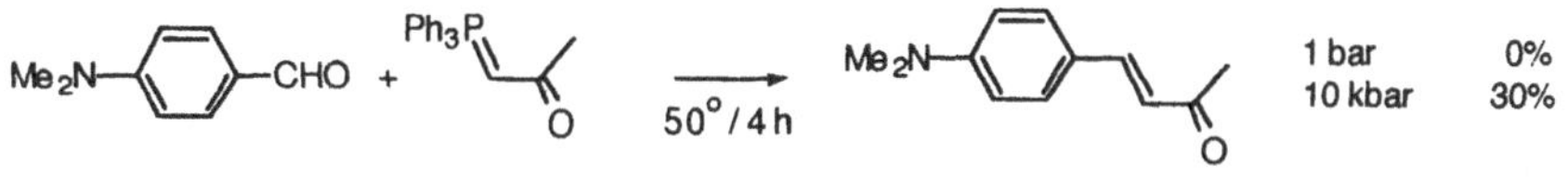

Me_2N CHO + Ph_3P O
50° / 4 h
Me_2N O

1 bar	0%
10 kbar	30%

Trimerization of nitriles to afford *sym*-triazines is generalized (to include nitrile with an α-hydrogen atom) [Cairns, 1952]. Phthalonitrile gives phthalocyanine quantitatively in the absence of water under 50 kbar at 410°C [Bengelsdorf, 1958], instead of 2,4,6-tri(*o*-cyanophenyl)-1,3,5-triazine which is obtained at a lower temperature in a sealed tube containing a little water.

3 RCN → (~7.5 bar)

4 (phthalonitrile) → (50 kbar, 410°/8 min, ~100%)

4.4. CYCLOADDITIONS

4.4.1. Diels-Alder Reactions

By far the most prominent reaction subject to effect of high pressures is the intermolecular Diels-Alder reaction. This is a natural consequence of the considerable negative volume of activation (-25 ~ -45 cm^3/mol) and of reaction. The effects include increase in yields, sometimes from 0%, lower reaction temperature, shortened reaction time, and changed regioselectivity. It is instructive to compare the thermal [Nozaki, 1968] and pressure reaction [Dauben, 1974] involving acrylonitrile and 1-piperidino-3,5,5-trimethyl-1,3-cyclohexadiene. The difficult thermal reaction caused a portion of the diene to isomerize before the proper condensation. Under such circumstances hydrolysis and apparent fragmentation of the adducts also occurred.

Δ	3%	10%	20%
pressure	90%	-	-

Increased reactivity due to pressure is apparent in the case of naphthalene which gives the maleic anhydride adduct in 1%. A mixture of endo-exo cycloadducts is produced in 78% yield at 9.5 kbar/100°C [Plieninger, 1969].

9.5 kbar

100° / 17 h

78%

A chiral cyclopentenone is obtainable from a $ZnCl_2$-catalyzed reaction of 4-acetoxy-2-cyclopentenone with a specially constructed cyclopentadiene under 6.5 kbar pressure [Borm, 1996]. Elimination of acetic acid from the Diels-Alder adduct occurred in situ. It seems important that pressure promoted the initial condensation before conversion of the dienophile to the very reactive cyclopentadienone which can give rise to its dimer in a competitive reaction pathway. The isolated adduct is an intermediate of (-)-methyl cucurbate and (-)-methyl jasmonate.

AcO

$ZnCl_2$

6.5 kbar

COOMe

OH

COOMe

COOMe

(-)-methyl jasmonate (-)-methyl cucurbate

A synthetic approach to the bakkenolide sesquiterpenes [Back, 1996] features a regioselective Diels-Alder reaction between a 5-spiroannulated cyclopentenone with piperylene. Both high pressure (ca. 16 kbar) and Lewis acid are required. Mixtures of β-exo, α-endo, and β-endo cycloadducts in variable ratios (depending on the Lewis acid) are obatainable.

The first reaction for a total synthesis of reserpine [Woodward, 1958] is the condensation of methyl 2,4-pentadienoate with *p*-benzoquinone. The low yield (28%) is due aromatization of the adduct in refluxing benzene, and improvement (to 64%) can be made by conducting a pressure (15 kbar) reaction at room temperature in dichloromethane [Dauben, 1982a]. By using the *(R)*-menthyl ester a chiral adduct (36% enantiomeric excess) is obtainable [Dauben, 1982b].

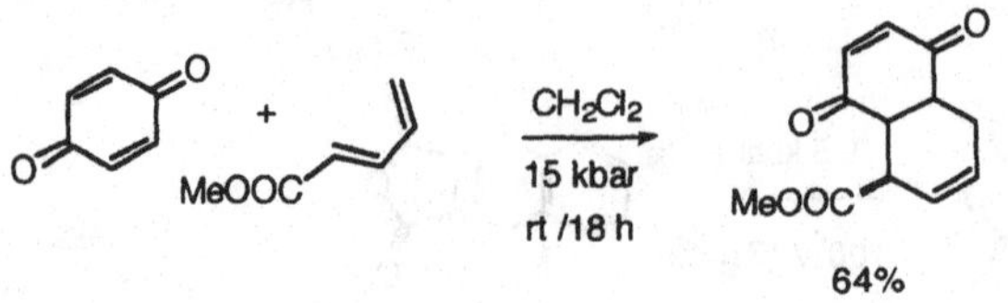

An analogous Diels-Alder adduct is suitable as precursor for aklavinone [Dauben, 1982a]. Reaction with a tricarbocyclic dienophile represents an alternative approach [Guingant, 1986].

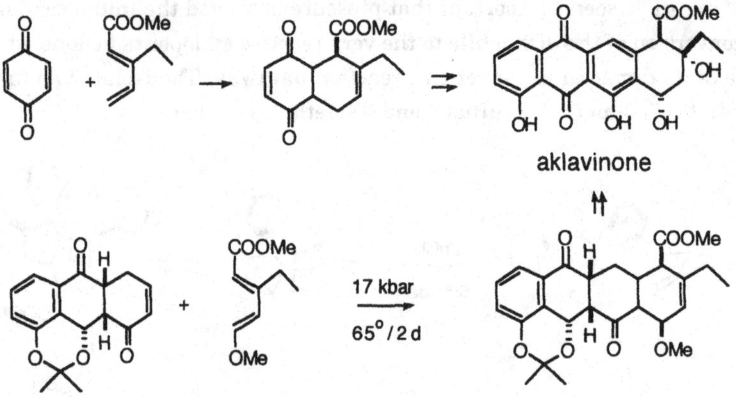

Good regioselectivity shown for the reaction of the highly hindered 1,3,3-trimethyl-2-vinylcyclohexene with 2-isopropyl-3-methoxy-1,4-benzoquinone at 12 kbar [Engler, 1989] makes it very useful for the synthesis of taxodione and royleanone.

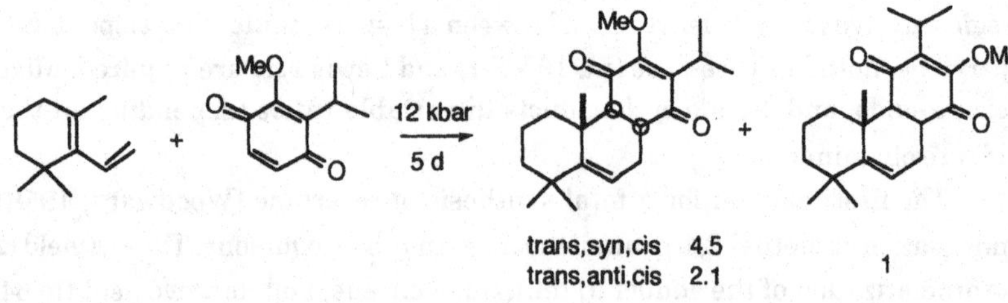

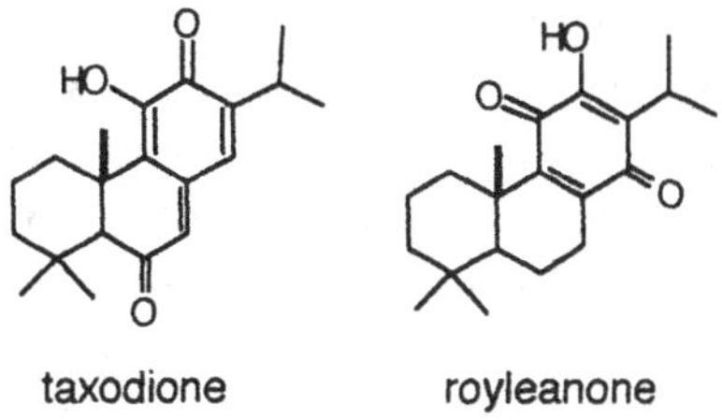

taxodione royleanone

For an approach to chlorothricolide a bicyclic compound was prepared by condensation of a butadiene derivative with 2-cyclohexenone [Hirsenkorn, 1990]. It appears that an exo transition state was adopted in the formation of the adduct. 3-Methyl-2-cyclopentenone and a *trans*-vinylhexalin which represents an AB-synthon for steroids indeed gives the desired skeleton, albeit in only 35% yield of a mixture of three compounds [Gacs-Baitz, 1994].

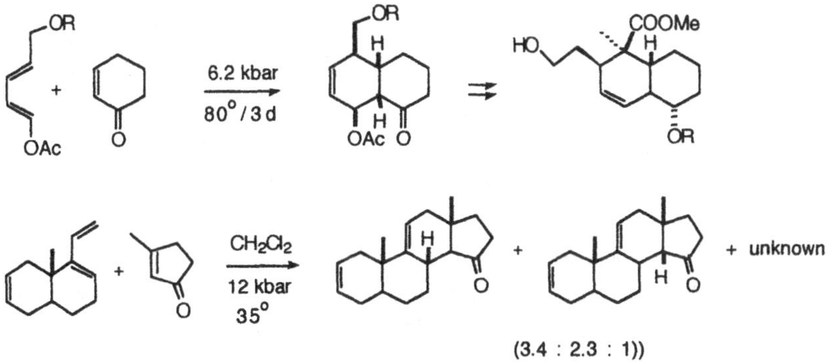

OR OAc + O — 6.2 kbar, 80°/3 d → OR H H OAc O ⇉ HO COOMe H OR

+ O — CH_2Cl_2, 12 kbar, 35° → H O + H O + unknown

(3.4 : 2.3 : 1))

Completion of the tetracyclic system of cerorubenic acid-III [Paquette, 1988] by a Diels-Alder reaction under pressure is straightforward. Unfortunately, the attachment of the missing side chain remains an insuperable hurdle.

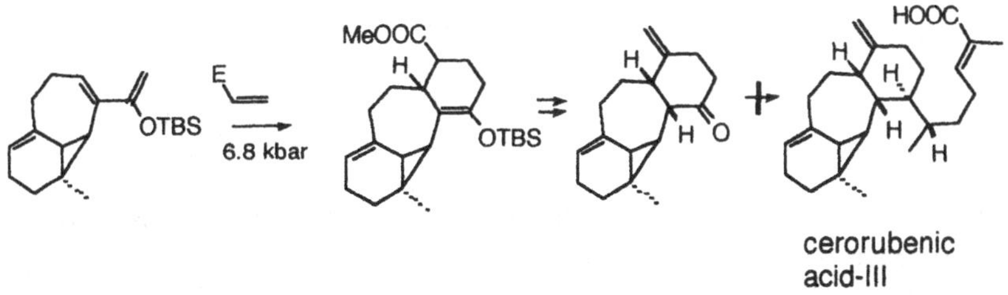

OTBS E — 6.8 kbar → MeOOC H OTBS ⇉ H H O + H H H HOOC H

cerorubenic acid-III

A method for synthesis of 4-substituted 2-cyclohexenones involves Diels-Alder reaction of 1-alkoxy-1,3-cyclohexadienes with enones or acrylic esters, followed by fragmentation of the bicyclo[2.2.2]octene system. In connection with a synthesis of the spirovetivane sesquiterpenes [Murai, 1981] observations indicate the Diels-Alder reaction at high pressure being *endo*-selective and furnishing a much higher ratio (3.2:1) ration of the *syn*-methyl adduct (vs. *anti*-methyl adduct), quite different from the purely thermal process. This result is favorable to elaboration of β-vetivone and hinesolone.

OMe COOMe 15 kbar 20° MeO COOMe + MeO COOMe (3.2 : 1)

β-vetivone hinesolone solavetivone

The Danishefsky diene used in a synthesis of erysotrine [Y. Tsuda, 1993] reacts normally only under pressure (10 kbar), whereby overcoming nonbonding interactions. Interestingly, the thermal process at atmospheric pressure selects the ketone group of the oxopyrrolinone ring as dienophile.

MeO COOMe MeO MeOOC Me_3SiO + OMe 10 kbar MeO COOMe MeO Me_3SiO E OMe MeO MeO MeO

(+)-erysotrine

For the construction of an octahydroisoquinolinedione to probe its potential as precursor of ircinal-B, a Diels-Alder reaction under high pressure is preferred because it is cleaner [Torisawa, 1996]. The cycloadduct was, however, withdrawn from service due to unsuccessful reduction of its nitrile function.

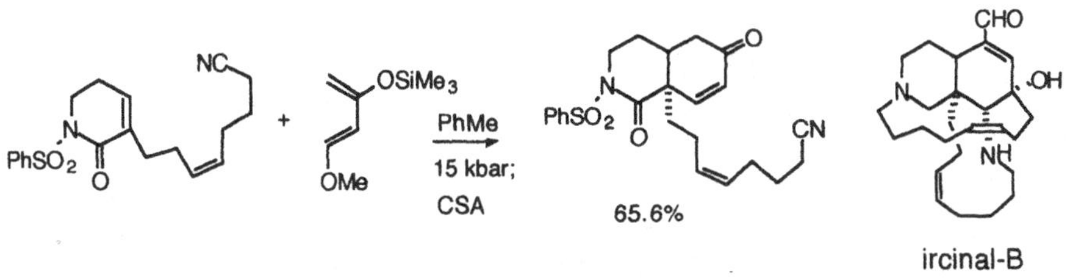

An undesirable effect due to pressure was encountered during a synthesis of actinobolin [Kozikowski, 1987]. While the Diels-Alder reaction (at 220°C) that provided a 1.7:1 ratio of diastereomers was unsatisfactory, the 6 kbar reaction (roome temperature) led to a 10:1 mixture in favor of the unwanted product.

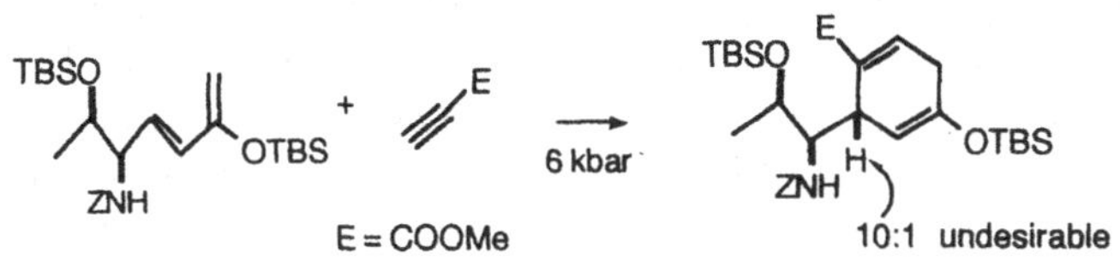

Novel compounds of the belt-type [Ashton, 1992] and cage-type [Ashton, 1993] owe their dramatic and smooth birth to high pressure.

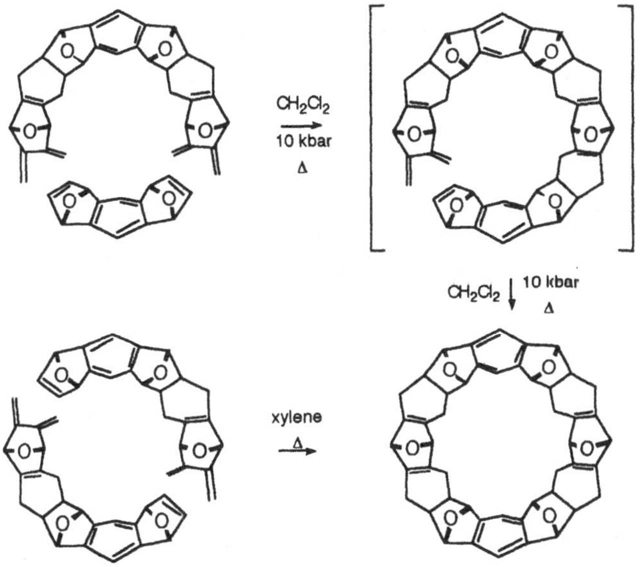

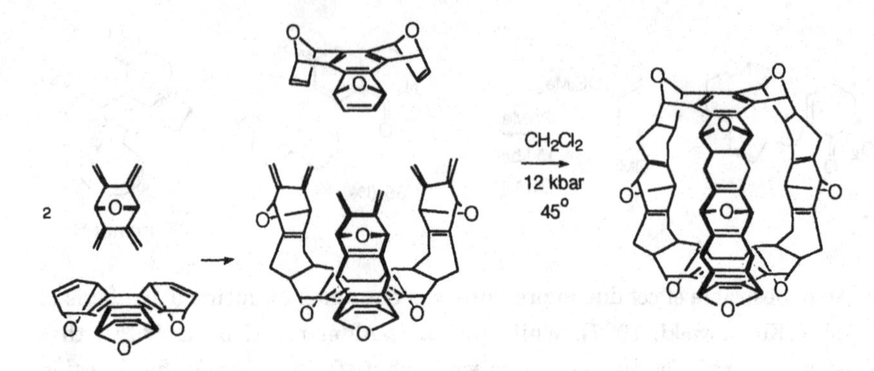

An expedient route to the skeleton of tigliane diterpenes employs a pressure reaction to condense methyl 3,3-dimethyl-1-cyclopropenecarboxylate with a diene which is constituted with a vinyl group and a hydrazulene nucleus [Rigby, 1989b]. Of interest are the results of a related model study [Rigby, 1989a] indicating opposite diastereoselectivity involving methyl 5,5-dimethyl-(5*H*)-pyrazole-3-carboxylate as the dienophile. The adducts from the latter reaction underwent photoelimination of dinitrogen to give the fused cyclopropane derivatives.

CHOMe; MeO; E; CH_2Cl_2; 8 kbar; 60%; MeO, COOMe, H; + ; (2 : 1)

OAc; 10 kbar; AcO, COOMe; +

	endo		exo
E (cyclopropene)	1	:	50
E (pyrazole, N=N) ; hv	50	:	1

A [4+2]cycloadduct is obtainable from tropone (as the diene) and C_{60} (as the dienophile) in toluene at 30 kbar [Takeshita, 1994]. It has been found to be stable until 140°C. The application of high pressure was meant to lower reaction temperatures to offset unfavorable addend/adduct equilibria.

A similarly prepared adduct from tropone and 2-cyclopentenone is a potential intermediate of ikarugamycin because its very characteristic 5:6:5-tricarbocyclic framework is readily accessible on photoinduced 1,3-acyl migration [Z.-H. Li, 1991].

A formal acetylene transfer to tropones which leads to the formation of homobarrelenones can be achieved by a two-step process involving high pressure condensation with dimethyl 7-oxabicyclo[2.2.1]hepta-2,5-diene-2,3-dicarboxylate which acts as the dienophile, and retro-Diels-Alder reaction [Tian, 1987]. Of course the latter step must be carried out separately.

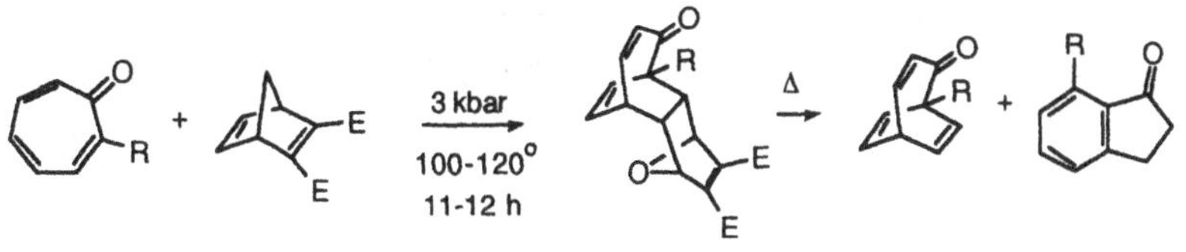

An unusual mode of [4+2]cycloaddition between azulene and dimethyl acetylenedicarboxylate was detected by carrying out the reaction at 50°C and high pressure [Klärner, 1982]. Thus formation of the heptalenedicarboxylic ester must consider this intermediate.

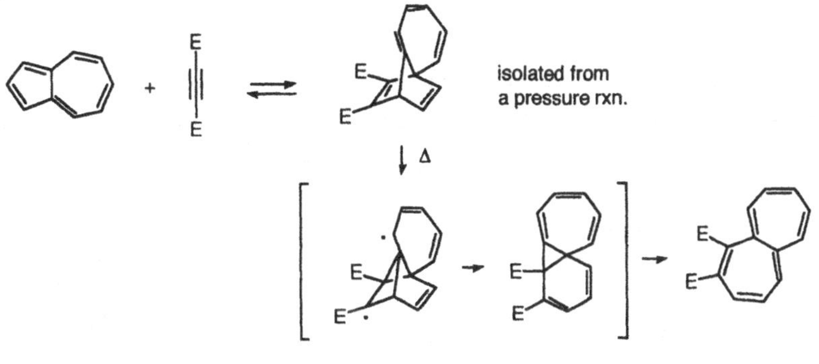

Due to thermal instability of Diels-Alder adducts of furans, great restriction is posed on their synthetic utility. In principle, the advance of pressure techniques effectively relieves such serious constraint. The adducts, formed at temperatures below their point of decomposition (retro-Diels-Alder fission), can be preserved for manipulation. Accordingly, the realization of previously unavailable adducts serves to magnify the scope of synthesis. The most eminent example must be the synthesis of cantharidin [Dauben, 1985], the tantalizingly direct access from a Diels-Alder adduct of furan and dimethylmaleic anhydride having been conceived long ago but thwarted until the development of the high pressure condensation. However, it must be emphasized that surrogation of the dienophile with 2,5-dihydrothiophene-3,4-dicarboxylic anhydride is a necessity. (Note that 3,4-dimethoxyfuran is capable of condensing with dimethylmaleic anhydride at 22 kbar at room temperature [Jurczak, 1982].)

Furan forms a Diels-Alder adduct with citraconic anhydride at 8 kbar, and the adduct can be hydrogenated to afford palasonin [Dauben, 1996] in 96% overall yield. The protocol involving reaction in lithium perchlorate in ether proves much inferior in this case.

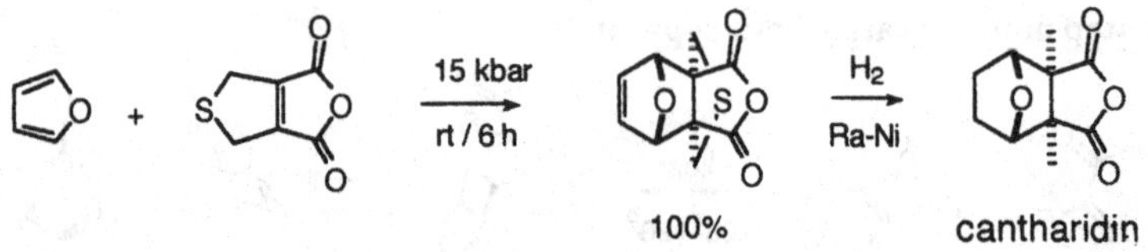

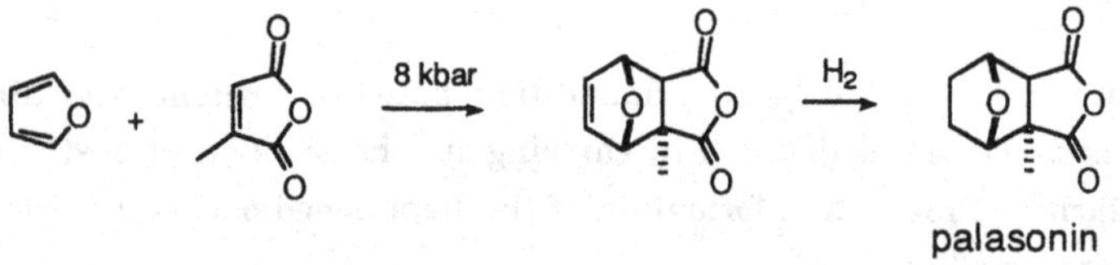

15 kbar, rt / 6 h; 100%; H_2, Ra-Ni; cantharidin

8 kbar; H_2; palasonin

Generally, the pressure reactions of furans and dienophiles are under kinetic control. Thus endo adducts are formed predominantly [Kotsuki, 1982]. Adducts derived from furan and dimethyl acetoxymethylenemalonate are promising synthons for *C*-nucleoside synthesis [Katagiri, 1988]. 2-Benzyloxymethyl-4-methylfuran forms an adduct with α-acetoxyacrylonitrile [Brown, 1992] which is useful for the synthesis of complex sesquiterpene lactones.

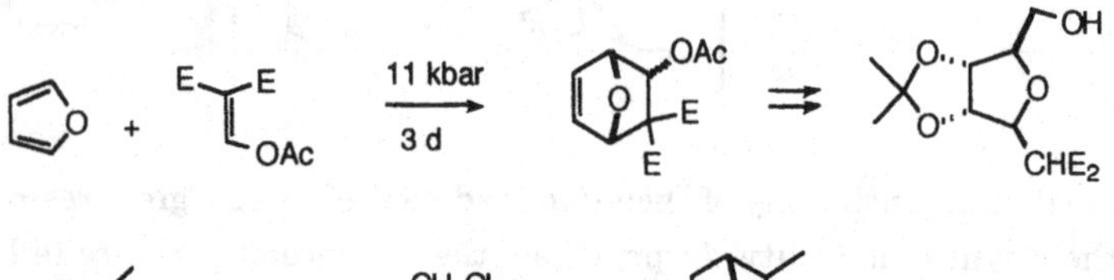

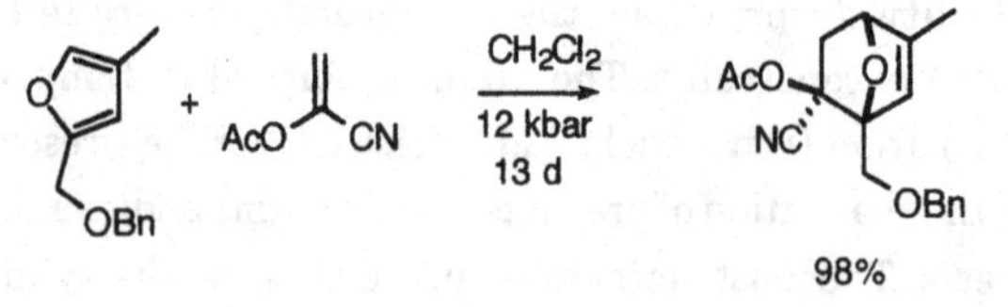

11 kbar, 3 d; OAc; E; OH; CHE_2

CH_2Cl_2, 12 kbar, 13 d; AcO; CN; NC; OBn; 98%

Another situation that requires high pressure is in a synthesis of jatropholone-A and jatropholone-B [Smith, 1986]. Here a fully substituted furan is caused to react with a cycloheptenone to provide an adduct for aromatization.

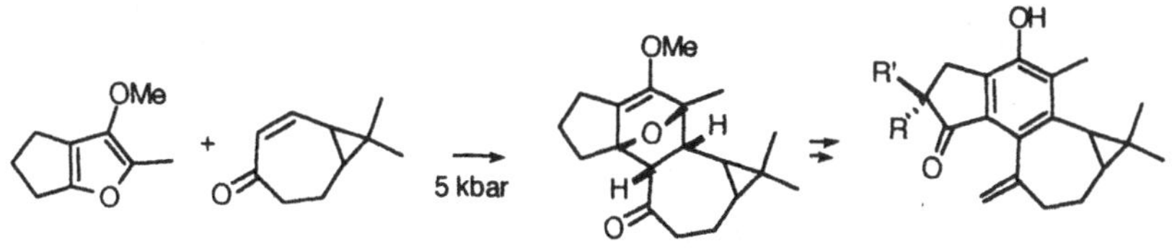

5 kbar

R=H, R'=Me jatropholone-A
R=Me, R'=H jatropholone-B

Furan fused to s sulfolane ring at $C_{(3)}/C_{(4)}$ loses sulfur dioxide on reaction with dienophiles under conventional conditions. Actually it is suspected that the cheletropic reaction provides the driving force to correct the unfavorable equilibrium. But to acquire the primary cycloadducts the employment of high pressure [Suzuki, 1994] is essential.

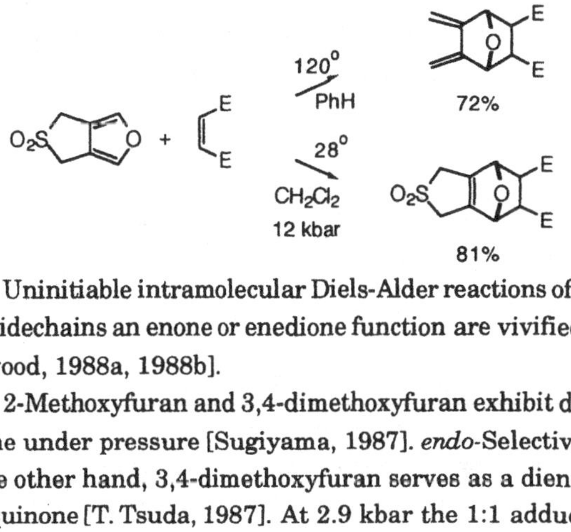

120° PhH 72%

28° CH_2Cl_2 12 kbar 81%

Uninitiable intramolecular Diels-Alder reactions of certain furans containing in their sidechains an enone or enedione function are vivified by high pressure (19 kbar) [Harwood, 1988a, 1988b].

2-Methoxyfuran and 3,4-dimethoxyfuran exhibit dienophilic character towards tropone under pressure [Sugiyama, 1987]. *endo*-Selectivity increases with pressure. On the other hand, 3,4-dimethoxyfuran serves as a diene in its condensation with *p*-tropoquinone [T. Tsuda, 1987]. At 2.9 kbar the 1:1 adduct reacts further with the α-diketone unit of another *p*-tropoquinone molecule.

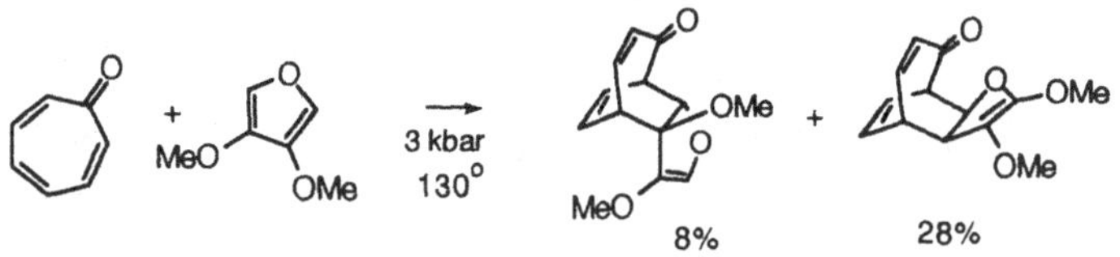

3 kbar 130° 8% 28%

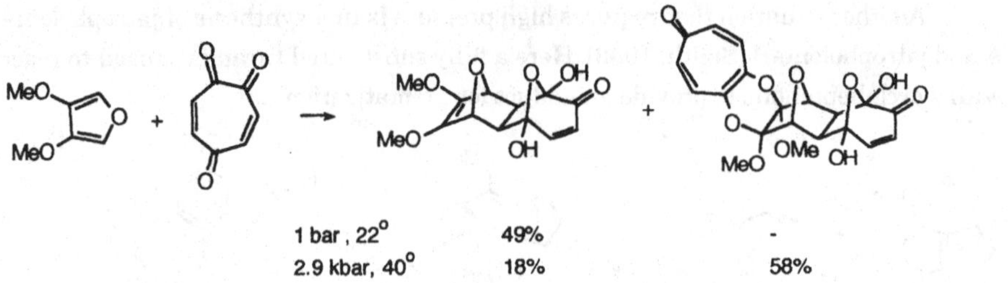

1 bar, 22°	49%	-
2.9 kbar, 40°	18%	58%

2-Vinylfuran prefers reaction with dienophiles involving the sidechain double bond, leading to benzofurans as major products [Kotsuki, 1981].

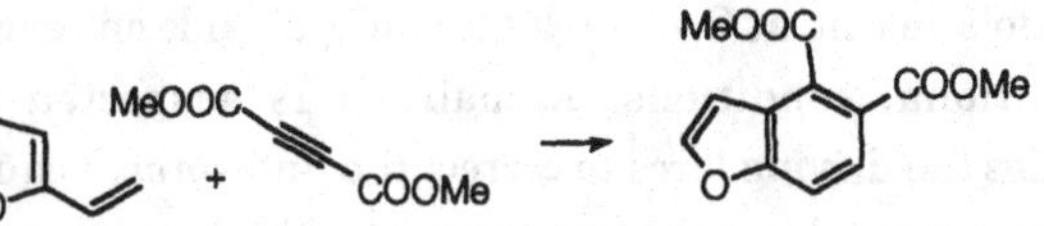

There is an intriguing solvent effect in the high pressure (10 kbar) reaction of 5-methoxyoxazoles with maleimides [Ibata, 1986], pertaining to elimination of methanol from the cycloadducts. It is surprising that the decomposition should occur, although it results in aromatic products.

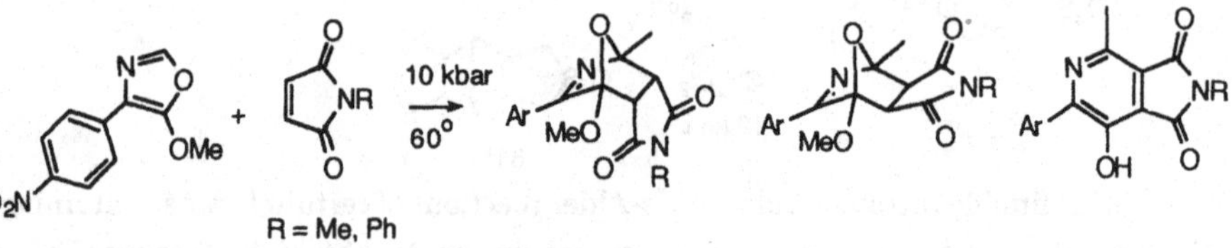

R = Me, Ph

The Diels-Alder adducts of α-pyrones often undergo decarboxylation in situ. It might be useful for the preparation of cyclohexadienes or formation of 1:2-adducts with reactive dienophiles, whereas preservation of the lactonic moiety could not be achieved until the forthcoming of high pressure techniques. In fact, the new option is very valuable, as it permits the execution of simplified synthetic schemes such as dealing with chorismic acid [Posner, 1987].

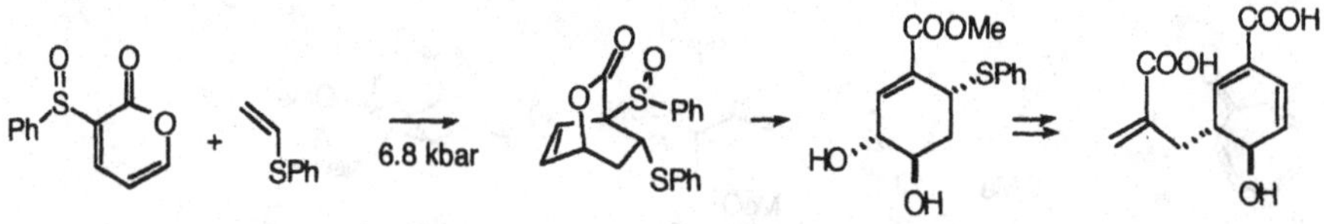

chorismic acid

A synthesis of sativene [Hatsui, 1994] was initiated by a high pressure Diels-Alder reaction, although the adduct must be decarboxylated subsequently. Pressure has the effect of suppressing isomerization of the dienophile (possibly by 1,5-sigmatropic rearrangement, cf. behavior of an oxapropellane under high pressure [Paquette, 1995]).

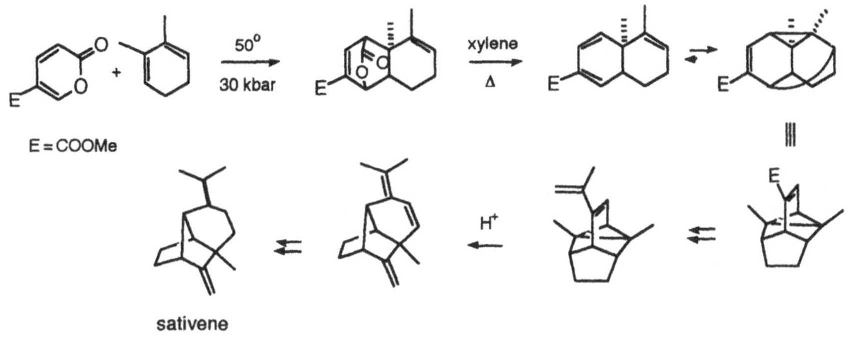

E = COOMe

50°, 30 kbar; xylene, Δ; H⁺

sativene

An approach to grandirubrine [Boger, 1995] features the union of a fused α-pyrones and the 2,2-dimethylpropyleneacetal of cyclopropenone at 12-13 kbar. The adduct is transformed into the tropone intermediate. Note that the pressure reaction could not be applied similarly to elaborate colchicine [Boger, 1985a].

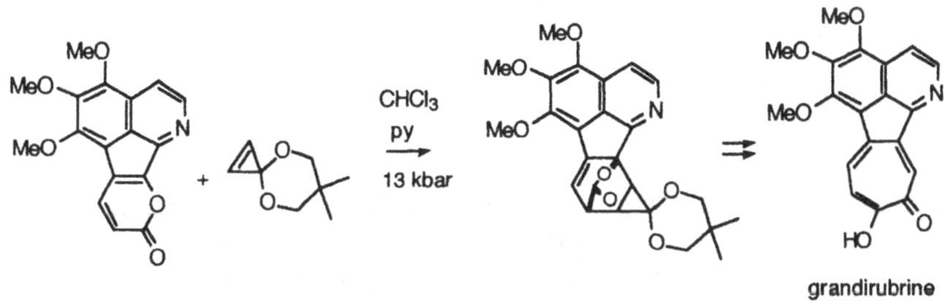

$CHCl_3$, py, 13 kbar

grandirubrine

Pyridazines can be used as dienes to generate benzene derivatives on reaction with dienophiles. With a 1,2,4-triazine precursor, the reaction with a propiophenone enamine led to formation of the pyridine ring while linking all the framework elements of streptonigrin [Boger, 1985b]. 1,2,4-Triazines also serve to provide the azadiene moiety of 3,8-methanoaza[10]annulenes upon reaction with benzocyclopropene (15 kbar, 55°C) [Maddox, 1980].

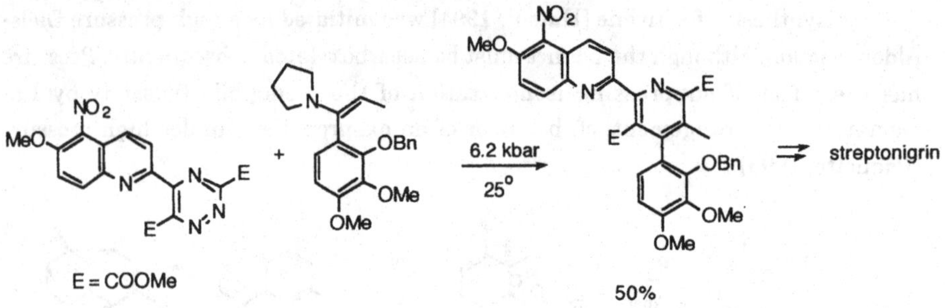

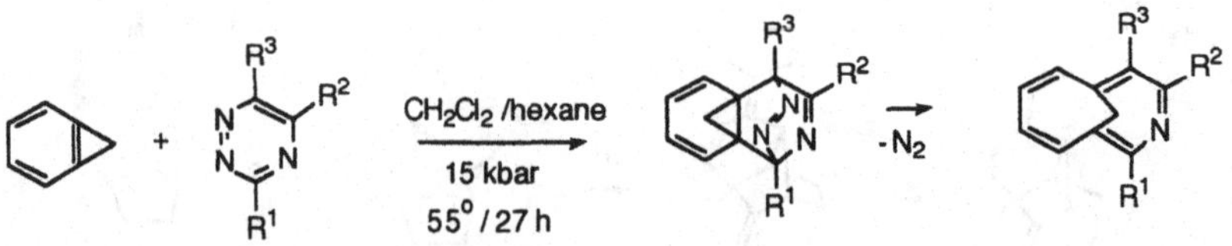

6.2 kbar
25°
streptonigrin
E = COOMe
50%

CH_2Cl_2 /hexane
15 kbar
55° / 27 h
- N_2

In addition to benzocyclopropene, it is significant that benzene participates in a Diels-Alder reaction, shown with hexachlorocyclopentadiene under pressure [Jarre, 1975].

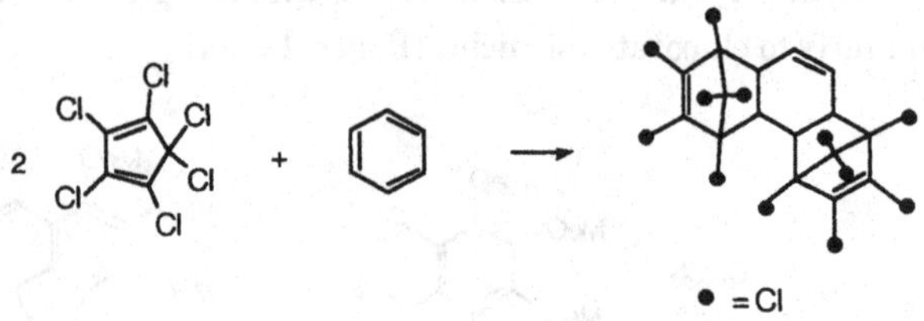

• = Cl

Other less common dienophiles made reactive to dienes by pressure include quinoneimine acetals [Kerr, 1995] and quinone monoacetals [Jarvo, 1996] . Pyranoid enolone esters also benefit from pressure regarding promotion of their reactivity [Dauben, 1990].

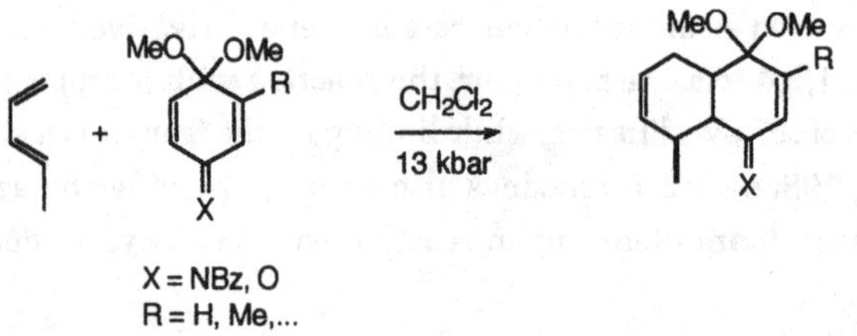

CH_2Cl_2
13 kbar
X = NBz, O
R = H, Me,...

CH_2Cl_2, 15 kbar

α-endo   α-exo

α > β
α-isomers: endo > exo

The Diels-Alder reactions of *N*-sulfinylimines may adopt different transition states when carried out in the presence of a Lewis acid catalyst or under pressure [Bell, 1988].

$TiCl_4$ (or $BF_3.OEt_2$)	100 :	0
CH_2Cl_2, 12 kbar	18 :	82

Only pressure brings about the reaction between 1,1,2,2-tetramethyl-1,2-disila-3.5-cyclohexadiene with phenyl vinyl sulfoxide [Sekiguchi, 1991].

Et_2O, 10 kbar, 100°

12%

Hetero-Diels-Alder reactions are affected by pressure in essentially the same way as the all-carbon versions. Consequently it facilitates the synthesis of pyranose sugars [Makin, 1974; Jurczak, 1979, 1986]. In certain circumstances, high pressure in conjunction with lanthanide catalysts generate superior results [Golebiowski, 1987]. For preparation of the precursors of 3-aminosugars, large enhancement of diastereoselectivity by pressure has been observed [Tietze, 1988].

Application of 20 kbar pressure enables formation of a tricyclic dihydropyran which is suitable for elaboration of ambreinolide [Daniewski, 1985].

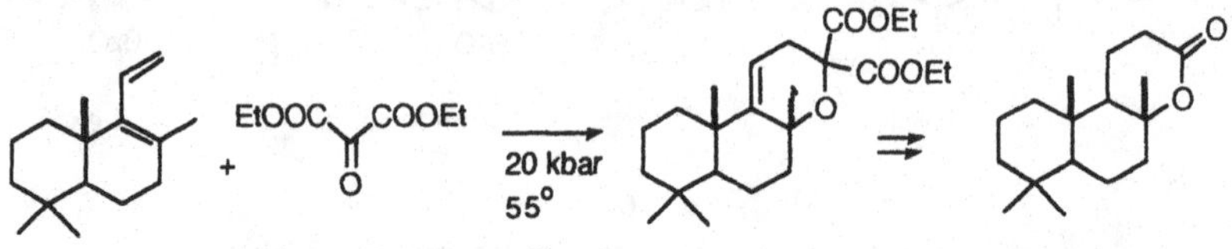

EtOOC COOEt + O → 20 kbar 55° → COOEt COOEt O ⇒ ambreinolide

Enhancement of enatioselectivity in chiral Lewis acid-catalyzed cycloadditions under high pressure [Tietze, 1993] is probably due to formation of complexes with different stoichiometry.

4.4.2. Other Cycloadditions

In comparison to the Diels-Alder reaction, far less studies on the other types of cycloadditions have been pursued. Despite the volumes of activation for 1,3-dipolar cycloadditions are highly negative (ca. -18 ~ -32 cm^3/mol) and their wide applicability, the intrinsic problems pertaining to 1,3-dipoles, such as their generation (sometimes by bond-breaking processes) and tendency to dimerize, counterplot the general utility. However, reactions involving alkenes with diazomethane [de Suray, 1974], certain nitrones [Dicken, 1982], and arenesulfonyl azides [Dauben, 1982] still profit enormously.

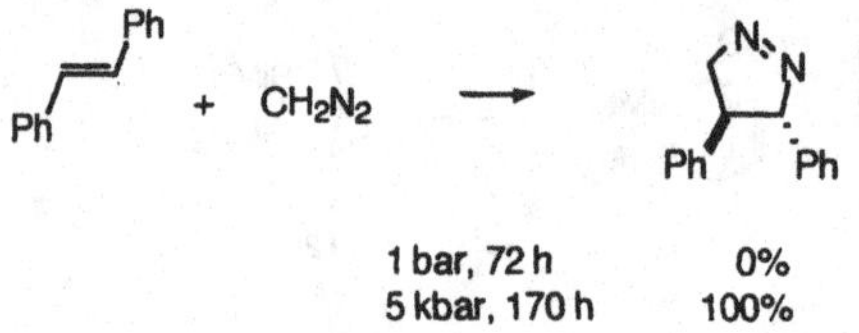

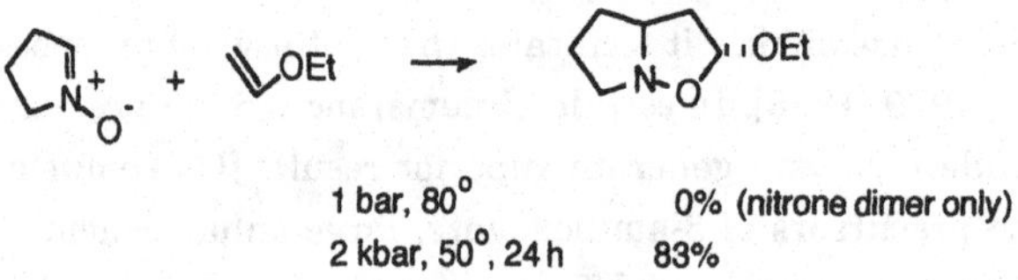

Ph Ph + CH_2N_2 → N N Ph Ph

1 bar, 72 h	0%
5 kbar, 170 h	100%

N+ O- + OEt → N-O OEt

1 bar, 80°	0% (nitrone dimer only)
2 kbar, 50°, 24 h	83%

OTBS ArSO$_2$N$_3$ ArSO$_2$ RO N–N N RO NSO$_2$Ar H$_2$O O NHSO$_2$Ar

Ar = 4-BrC$_6$H$_4$
R = TBS

1 bar, 75°, 96 h — 0%
15 kbar, 65°, 72 h — 79%

Regioisomerically pure Δ^1-1,2,3-triazolines can be obtained from high pressure reactions of alkyl azides and electron-deficient alkenes [Anderson, 1991]. The corresponding thermal process is usually nonregioselective.

The regioselectivity due to pressure effect on the cycloaddition involving unsymmetrical trimethylenemethane-PdL$_2$ complexes [Trost, 1995] has been ascribed to an increase in rate of the bimolecular addition relative to interconversion of the metal complexes. Kinetic trapping is made possible at high pressure.

Ph O + Me$_3$Si AcO → Ph O + Ph O

1 bar, Pd(OAc)$_2$, (iPrO)$_3$P	38%	1	:	3
15 kbar, (η^3-C$_3$H$_7$PdCl$_2$), tpdp	90%	3.6	:	1

Strained alkenes such as norbornene react with carbon disulfide under high pressure to give 2:2-adducts [Plieninger, 1972]. This reaction enables the accessibility of tetraselenafulvalenes in one step by using acetylenic esters and carbon diselenide [Rice, 1982b].

CS$_2$ 10 kbar 100° S S : 2 x S S S S

R R + CX$_2$ → R X X R R X X R

X = S, Se

Not many practical uses of pressured [2+2]-cycloadditions are known. Azetidine and β-lactam formation from imines [Aben, 1987] and isocyanates [Chmielewski, 1985], respectively, is facilitated by pressure. However, many of these products are unstable, ring opening occurs readily soon after the pressure is relieved.

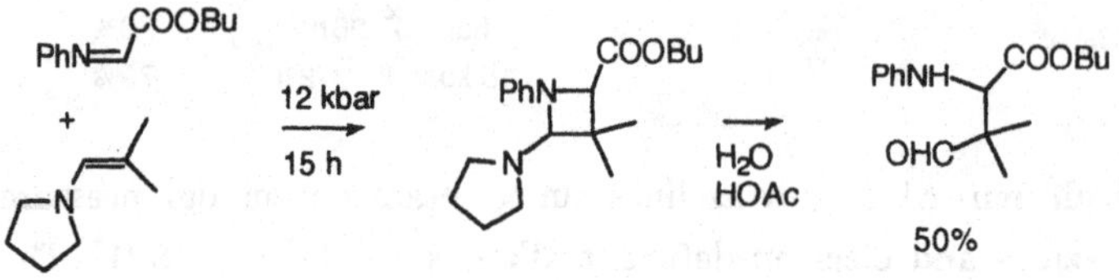

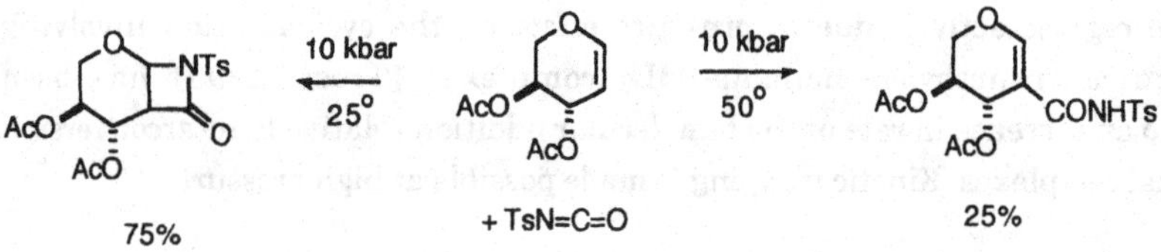

While norbornadiene involves C-2 and C-6 in its reaction with methyl propynoate, quadricyclane behaves differently [Jenner, 1987]. With excess quadricyclane a 2:1 adduct is obtained. Note that the monoadduct does not undergo further cycloaddition with quadricyclane.

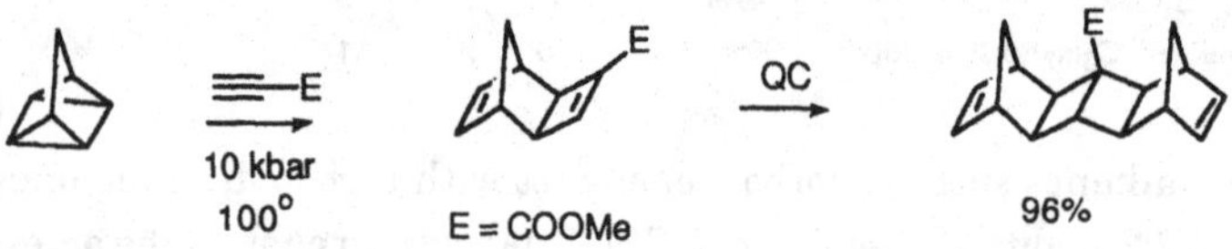

The use of pressure to coerce isomerization of 1,5,9-cyclododecatriyne to [Barkovich, 1977] and triquinacene to diademane [Gladysz, 1979] was unrewarding. Attempts at dimerizing triquinacene to give dodecahedrane [Gladysz, 1979] also met with failure (even at 400°C, 40 kbar).

2 x

dodecahedrane

Monographs and Reviews

Jenner, G. (1997) *Tetrahedron* **53:** 2669.

le Noble, W.J. [Ed.] (1988) *Organic High Pressure Chemistry*, Elsevier: Amsterdam.

Matsumoto, K.; Acheson, R.M., [Eds.] (1991) *Organic Synthesis at High Pressure*, Wiley: New York.

Matsumoto, K.; Sera, A.; Uchida, T. (1985) *Synthesis* 1.

Matsumoto, K.; Sera, A. (1985) *Synthesis* 999.

References

Aben, R.W.M.; Smith, R.; Scheeren, J.W. (1987) *J. Org. Chem.* **52:** 365.

Anderson, G.T.; Henry, J.R.; Weinreb, S.M. (1991) *J. Org. Chem.* **56:** 6946.

Asano, T.; le Noble, W.J. (1978) *Chem. Rev.* **78:** 407.

Ashton, P.R.; Brown, C.L.; Chrystal, E.J.T.; Parry, K.P.; Pietraszkiewicz, M.; Spencer, N.; Stoddart, J.F. (1991) *Angew. Chem. Int. Ed. Engl.* **30:** 1042.

Ashton, P.R.; Brown, G.R.; Isaacs, N.S.; Giuffrida, D.; Kohnke, F.H.; Mathias, J.P.; Slawin, A.M.Z.; Smith, D.R.; Stoddart, J.F.; Williams, D.J. (1992) *J. Am. Chem. Soc.* **114:** 6330.

Ashton, P.R.; Girreser, U.; Giuffrida, D.; Kohnke, F.H.; Mathias, J.P.; Raymo, F.M.; Slawin, A.M.Z.; Stoddart, J.F.; Williams, D.J. (1993) *J. Am. Chem. Soc.* **115:** 5422.

Back, T.G.; Gladstone, P.L.; Parvez, M. (1996) *J. Org. Chem.* **61:** 3806.

Barkovich, A.J.; Strauss, E.S.; Vollhardt, K.P.C. (1977) *J. Am. Chem. Soc.* **99:** 8321.

Bell, S.I.; Weinreb, S.M. (1988) *Tetrahedron Lett.* **29:** 4233.

Bengelsdorf, I.S. (1958) *J. Am. Chem. Soc.* **80:** 1442.

Berlage, U.; Schmidt, J.; Milkova, Z.; Welzel, P. (1987) *Tetrahedron Lett.* **28:** 3095.

Boger, D.L.; Brotherton, C.E. (1985) *J. Org. Chem.* **50:** 3425.

Boger, D.L.; Panek, J.S. (1985b) *J. Am. Chem. Soc.* **107:** 5745.

Boger, D.L.; Zhang, M. (1991) *J. Am. Chem. Soc.* **113:** 4230.

Boger, D.L.; Takahashi, K. (1995) *J. Am. Chem. Soc.* **117:** 12452.

Borm, C.; Winterfeldt, E. (1996) *Liebigs Ann. Chem.* 1209.

Brown, D.S.; Paquette, L.A. (1992) *J. Org. Chem.* **57:** 4512.

Cairns, T.L.; Larchar, A.W.; McKusic, B.C. (1952) *J. Am. Chem. Soc.* **74:** 5633.

Chmielewski, M.; Kaluza, Z.; Belzecki, C.; Salanski, P.; Jurczak, J.; Adamowicz, H. (1985) *Tetrahedron* **41:** 2441.

d'Angelo, J.; Maddaluno, J. (1986) *J. Am. Chem. Soc.* **108:** 8112.

Daniewski, W.M.; Kubak, E.; Jurczak, J. (1985) *J. Org. Chem.* **50:** 3963.

Dauben, W.G.; Kozikowski, A.P. (1974) *J. Am. Chem. Soc.* **96:** 3664.

Dauben, W.G.; Krabbenhoft, H.O. (1977) *J. Org. Chem.* **42:** 282.

Dauben, W.G.; Baker, W.R. (1982a) *Tetrahedron Lett.* **23:** 2611.

Dauben, W.G.; Bunce, R.A. (1982b) *Tetrahedron Lett.* **23:** 4875.

Dauben, W.G.; Bunce, R.A. (1982c) *J. Org. Chem.* **47:** 5042.

Dauben, W.G.; Bunce, R.A. (1983a) *J. Org. Chem.* **48:** 4642.

Dauben, W.G.; Gerdes, J.M. (1983b) *Tetrahedron Lett.* **24:** 3841.

Dauben, W.G.; Gerdes, J.M.; Bunce, R.A. (1984) *J. Org. Chem.* **49:** 4293.

Dauben, W.G.; Gerdes, J.M.; Smith, D.B. (1985) *J. Org. Chem.* **50:** 2576.

Dauben, W.G.; Gerdes, J.M.; Look, G.C. (1986a) *Synthesis* 532.

Dauben, W.G.; Gerdes, J.M.; Look, G.C. (1986b) *J. Org. Chem.* **51:** 4964.

Dauben, W.G.; Kowalczyk, B.A.; Lichtenthaler, F.W. (1990) *J. Org. Chem.* **55:** 2391.

Dauben, W.G.; Hendricks, R.T. (1992) *Tetrahedron Lett.* **33:** 603.

Dauben, W.G.; Lam, J.Y.L.; Guo, Z.R. (1996) *J. Org. Chem.* **61:** 4816.

DeShong, P.; Sidler, D.R.; Rybczynski, P.J.; Slough, G.A.; Rheingold, A.L. (1988) *J. Am. Chem. Soc.* **110:** 2575.

de Suray, H.; Leroy, G.; Weiler, J. (1974) *Tetrahedron Lett.* 2209.

Dicken, C.M.; DeShong, P. (1982) *J. Org. Chem.* **47:** 2047.

Engler, T.A.; Sampath, U.; Naganathan, S.; Vander Velde, D.; Takusagawa, F.; Yohannes, D. (1989) *J. Org. Chem.* **54:** 5712.

Gacs-Baitz, E.; Marrocchi, A.; Minuti, L.; Scheeren, H.W.; Taticchi, A. (1994) *Nat. Prod. Lett.* **5:** 165.

Gladysz, J.A.; Yu, Y.S. (1978) *Chem. Commun.* 599.

Gladysz, J.A. (1979) *Chemtech* 372.

Golebiowski, A.; Izdebski, J.; Jacobsson, U.; Jurczak, J. (1986) *Heterocycles* **24:** 1205.

Golebiowski, A.; Jacobsson, U.; Jurczak, J. (1987) *Tetrahedron* **43:** 3063.

Guingant, A.; d'Angelo, J. (1986) *Tetrahedron Lett.* **27:** 3729.

Guingant, A.; Hammami, H. (1991) *Tetrahedron: Asymmetry* **2:** 411.

Hamann, S.D. (1957) *Physico-Chemical Effects of Pressure*, Butterworth: London.

Harwood, L.M.; Leeming, S.A.; Isaacs, N.S.; Jones, G.; Pickard, J.; Thomas, R.M.; Watkin, D. (1988a) *Tetrahedron Lett.* **29:** 5017.

Harwood, L.M.; Jones, G.; Pickard, J.; Thomas, R.M.; Watkin, D. (1988b) *Tetrahedron Lett.* **29:** 5825.

Hatsui, T.; Hashiguchi, T.; Takeshita, H. (1994) *Chem. Lett.* 1415.

Hill, J.S.; Isaacs, N.S. (1986) *Tetrahedron Lett.* **27:** 5007.

Hirsenkorn, R.; Haag-Zeino, B.; Schmidt, R.R. (1990) *Tetrahedron Lett.* **31:** 4433.

Hoffmann, R.W.; Metternich, R. (1985) *Liebigs Ann. Chem.* 2390.

Holan, G.; Evans, J.J.; Linton, M. (1977) *J. Chem. Soc. Perkin Trans. I* 1200.
Howk, B.W.; Little, E.L.; Scott, S.L.; Whitman, G.M. (1954) *J. Am. Chem. Soc.* **76:** 1899.
Ibata, T.; Nakawa, H.; Isogami, Y.; Matsumoto, K. (1986) *Bull. Chem. Soc. Jpn.* **59:** 3197.
Ibata, T.; Isogami, Y.; Toyoda, J. (1987) *Chem. Lett.* 1187.
Isaacs, N.S.; Maksimovic, L.; Rintoul, G.B.; Young, D.J. (1992) *Chem. Commun.* 1749.
Jarre, W.; Bieniek, D.; Korte, F. (1975) *Angew. Chem. Int. Ed. Engl.* **14:** 181.
Jarvo, E.R.; Boothroyd, S.R.; Kerr, M.A. (1996) *Synlett* 897.
Jenner, G.; Papadopoulos, M.; Jurczak, J.; Kozluk, T. (1984) *Tetrahedron Lett.* **25:** 5747.
Jenner, G. (1987) *Tetrahedron Lett.* **28:** 3927.
Jones, E.S. (1967) *U.S. Pat.* 3304334.
Jones, W.H.; Tristram, E.W.; Benning, W.F. (1959) *J. Am. Chem. Soc.* **81:** 2151.
Jurczak, J.; Chmielewski, M.; Filipek, S. (1979) *Synthesis* 41.
Jurczak, J.; Kozluk, T.; Filipek, S.; Eugster, C.H. (1982) *Helv. Chim. Acta* **65:** 1021.
Jurczak, J.; Tkacz, M.; Majchrzak-Kuczynski, U. (1983) *Synthesis* 920.
Jurczak, J.; Bauer, T.; Jarosz, S. (1986) *Tetrahedron* **42:** 5045, 6477.
Jurczak, J.; Ostaszewski, R.; Salanski, P. (1989) *Chem. Commun.* 184.
Katagiri, N.; Akatsuka, H.; Kaneko, C.; Sera, A. (1988) *Tetrahedron Lett.* **29:** 5397.
Kerwin, S.M.; Paul, A.G.; Heathcock, C.H. (1987) *J. Org. Chem.* **52:** 1686.
Kerr, M.A. (1995) *Synlett* 1165.
Klärner, F.-G.; Dogan, B.; Roth, W.R.; Hafner, K. (1982) *Angew. Chem. Int. Ed. Engl.* **21:** 708.
Kochetkov, N.K.; Zhulin, V.M.; Klimov, E.M.; Malysheva, N.N.; Makarova, Z.G.; Ott, A.Ya. (1987) *Carbohyd. Res.* **164:** 241.
Kotsuki, H.; Nishizawa, H.; Ochi, M.; Matsuoka, K. (1982) *Bull. Chem. Soc. Jpn.* **55:** 496.
Kotsuki, H.; Ichikawa, Y.; Nishizawa, H. (1988) *Chem. Lett.*
Kotsuki, H.; Matsumoto, K.; Nishizawa, H. (1991) *Tetrahedron Lett.* **32:** 4155.
Kotsuki, H.; Iwasaki, M.; Nishizawa, H. (1992) *Tetrahedron Lett.* **33:** 4945.
Kozikowski, A.P.; Nieduzak, T.R.; Konoike, T.; Springer, J.P. (1987) *J. Am. Chem. Soc.* **109:** 5167.
Maddox, M.L.; Martin, J.C.; Muchowski, J.M. (1980) *Tetrahedron Lett.* **21:** 7.
Makin, S.M.; El'yanov, B.S.; Raifel'd, Yu.E. (1974) *Izv. Akad. Nauk SSSR, Ser. Khim.* 2654.
Matsumoto, K. (1980) *Angew. Chem. Int. Ed. Engl.* **19:** 1013.

Matsumoto, K.; Uchida, T. (1981) *Chem. Lett.* 1673.
Matsumoto, K. (1982) *Angew. Chem. Int. Ed. Engl.* **21:** 922.
Matsumoto, K. (1984a) *Angew. Chem. Int. Ed. Engl.* **23:** 617.
Matsumoto, K.; Hashimoto, S.; Ikemi, Y.; Otani, S. (1984b) *Heterocycles* **22:** 1417.
Matsumoto, K.; Hashimoto, S.; Otani, S. (1986) *Angew. Chem. Int. Ed. Engl.* **25:** 565.
Matsumoto, K.; Hashimoto, S.; Okamoto, T.; Otani, S.; Hayami, J. (1987) *Chem. Lett.* 803.
Midland, M.M.; McLoughlin, J.I.; Gabriel, J. (1989) *J. Org. Chem.* **54:** 159.
Murai, A.; Sato, S.; Masamune, T. (1981) *Tetrahedron Lett.* 1033.
Neuman, R.C. (1972) *Acc. Chem. Res.* 382.
Newitt, D.M.; Linstead, R.P.; Sapiro, R.H.; Boorman, E.J. (1937) *J. Chem. Soc.* 876.
Newkome, G.R.; Majestic, V.K.; Fronczek, F.R. (1981) *Tetrahedron Lett.* 3039.
Nonnenmacher, A.; Mayer, R.; Plieninger, H. (1983) *Liebigs Ann. Chem.* 2135.
Nozaki, H.; Yamaguchi, T.; Ueda, S.; Kondo, K. (1968) *Tetrahedron* **24:** 1445.
Okamoto, Y.; Shimakawa, Y. (1970) *J. Org. Chem.* **35:** 3752.
Onodera, A.; Amita, F. (1991) in Matsumoto, K.; Acheson, R.M., [Eds.] (1991) *Organic Synthesis at High Pressure*, Wiley: New York, pp. 77-131.
Paquette, L.A.; Poupart, M.A. (1988) *Tetrahedron Lett.* **29:** 273.
Paquette, L.A.; Branan, B.M.; Rogers, R.D. (1995) *J. Org. Chem.* **60:** 1852.
Pietraszkiewicz, M.; Salanski, P.; Jurczak, J. (1983) *Chem. Commun.* 1184.
Pikul, S.; Jurczak, J. (1985) *Tetrahedron Lett.* **26:** 4145.
Pirkle, W.H.; Hauske, J.R.; Eckert, C.A.; Scott, B.A. (1977) *J. Org. Chem.* **42:** 3101.
Plieninger, H.; Wild, D.; Westphal, J. (1969) *Tetrahedron* **25:** 5561.
Plieninger, H.; Heuck, C.C. (1972) *Tetrahedron* **28:** 73.
Posner, G.H.; Haces, A.; Harrison, W.; Kinter, C.M. (1987) *J. Org. Chem.* **52:** 4836.
Rahm, A.; Degueil-Castaing, M.; Pereyre, M. (1982) *J. Organomet. Chem.* **323:** C29.
Rahm, A.; Amardeil, R.; Degueil-Castaing, M. (1989) *J. Organomet. Chem.* **371:** C4.
Rice, J.E.; Okamoto, Y. (1982a) *J. Org. Chem.* **47:** 4189.
Rice, J.E.; Wojciechowski, P.S.; Okamoto, Y. (1982b) *Heterocycles* **18:** 191.
Rigby, J.H.; Kierkus, P.C. (1989a) *J. Am. Chem. Soc.* **111:** 4125.
Rigby, J.H.; Kierkus, P.C.; Head, D. (1989b) *Tetrahedron Lett.* **30:** 5073.
Riley, D.P.; Correa, P.E. (1985) *J. Org. Chem.* **50:** 1563.
Sekiguchi, A.; Maruki, I.; Ebata, K.; Kabuto, C.; Sakurai, H. (1991) *Chem. Commun.* 341.
Smith, A.B., III; Liverton, N.J.; Hrib, N.J.; Sivaramakrishnan, H.; Winzenberg, K. (1986) *J. Am. Chem. Soc.* **108:** 3040.
Stürmer, R.; Ritter, K.; Hoffmann, R.W. (1993) *Angew. Chem. Int. Ed. Engl.* **32:** 101.
Sugiyama, S.; Tsuda, T.; Mori, A.; Takeshita, H.; Kodama, M. (1987) *Bull. Chem. Soc.*

Jpn. **60:** 3633.
Suzuki, T.; Kubomura, K.; Takayama, H. (1994) *Heterocycles* **38:** 961.
Taguchi, Y.; Yanagiya, K.; Shibuta, I.; Shuhara, Y. (1988) *Bull. Chem. Soc. Jpn.* **61:** 921.
Takeshita, H.; Liu, J.-F.; Kato, N.; Mori, A.; Isobe, R. (1994) *J. Chem. Soc. Perkin Trans. I* 1433.
Tian, G.R.; Sugiyama, S.; Mori, A.; Takeshita, H. (1987) *Chem. Lett.* 1557.
Tietze, L.-F.; Hubsch, T.; Voss, E.; Buback, M.; Tost, W. (1988) *J. Am. Chem. Soc.* **110:** 4065.
Tietze, L.-F.; Ott, C.; Gerke, K.; Buback, M. (1993) *Angew. Chem. Int. Ed. Engl.* **32:** 1485.
Torisawa, Y.; Ali, M.A.; Tavet, F.; Kageyama, A.; Aikawa, M.; Fukui, N.; Hino, T.; Nakagawa, M. (1996) *Heterocycles* **42:** 677.
Trost, B.M.; Parquette, J.R.; Marquart, A.L. (1995) *J. Am. Chem. Soc.* **117:** 3284.
Tsuda, T.; Sugiyama, S.; Mori, A.; Takeshita, H. (1987) *Bull. Chem. Soc. Jpn.* **60:** 2695.
Tsuda, Y.; Hosoi, S.; Katagiri, N.; Kaneko, C.; Sano, T. (1993) *Chem. Pharm. Bull.* **41:** 2087.
Woodward, R.B.; Bader, F.E.; Bickel, H.; Frey, A.J.; Kierstead, R.W. (1958) *Tetrahedron* **2:** 1.
Wright, J.; Drtina, G.J.; Roberts, R.A.; Paquette, L.A. (1988) *J. Am. Chem. Soc.* **110:** 5806.
Yamamoto, Y.; Maruyama, K.; Matsumoto, K. (1983a) *J. Am. Chem. Soc.* **105:** 6963.
Yamamoto, Y.; Maruyama, K.; Matsumoto, K. (1983b) *Chem. Commun.* 489.
Yamamoto, Y.; Maruyama, K.; Matsumoto, K. (1984a) *Chem. Commun.* 548.
Yamamoto, Y.; Maruyama, K.; Matsumoto, K. (1984b) *Tetrahedron Lett.* **25:** 1075.
Yamamoto, Y.; Saito, K. (1989) *Chem. Commun.* 1676.
Yamamoto, Y.; Furuta, T.; Matsuo, J.; Kurata, T. (1991) *J. Org. Chem.* **56:** 5737.

5 FLASH VACUUM PYROLYSIS AND SOME OTHER THERMAL PROCESSES

Heat is the most convenient source of energy for promoting reactions. The popularity of the Diels-Alder reaction is at least partly attributable to its operational simplicity. Strong heating (pyrolysis) often causes disintegration of complex organic molecules into smaller fragments, and therefore it constitutes a method (with modifications such as admixture with S, Se, Zn...) for skeletal identification of natural products before the advent of modern spectroscopy. Actually, when benzene was a rare commodity it was prepared from the more readily available benzoic acid by Mitscherlich (in 1834) by distillation with lime. Berthelot obtained biphenyl on passing benzene vapor over a red-hot tube. These processes requiring extreme conditions fell out of favor for lack of control in selectivity and hence good yields of the desired products, although there are a few exceptions such as the preparation of ketene from acetone [Hurd, 1940]. However, in the second half of the twentieth century, because of the developments in petrochemical industry and the formulation of the Woodward-Hoffmann rules, revival of and furtherance of interests in purely thermal reactions become inevitable.

$$(CH_3)_2C{=}O \xrightarrow{700^{\circ}} CH_2{=}C{=}O + CH_4$$

Chemists have been inspired by patterns of electron-impact fragmentation in mass spectra of molecules and made some correlation and conjecture as to the semblance of thermal reactions. Synthetic transformations based on such patterns have been quite successful [Kametani, 1976]. In other words, molecules with particular constitutions can be designed and they are expected to undergo thermal reactions in analogous to the fragmentation patterns, and reactive species thus generated can be intercepted accordingly.

In this chapter selected thermal reactions which occur between 150°-1200°C are discussed in terms of their synthetic utility, with emphasis on flash vaccum pyrolysis [Hedaya, 1967, 1969a]. The Diels-Alder reaction and many other pericyclic reactions which have been reviewed extensively are mentioned only sporadically.

Static pyrolytic processes in a sealed tube or by reflux are easier to set up, but the product(s) may undergo secondary reactions on being subjected to continued heating. In a flow system compounds are exposed to heat in short durations ($10^{-3} \sim 1$ s) and the pyrolysates are cooled immediately to very low temperatures (-196°C), therefore bimolecular reactions are minimized. It is often noted that appreciably higher temperatures are required to complete flash pyrolysis to compensate for the short contact times. For example, bond cleavage leading to nonstabilized radicals is usually carried out at temperatures above 1000°C.

In flash vacuum pyrolysis (FVP) the flow is produced by either a high vacuum with the substrate into the reactor introduced by molecular distillation, or a stream of inert gas at 0.1 to 10 torr pressure. (Note the torr unit is generally reported, whereas in Chapter 4 high pressures are indicated by kbar. The relationship is 1 torr = 1.31×10^{-3} bar). It has been used to generate unusual, unstable, highly reactive species such as arynes, carbenes, nitrenes, and sulfenes. Species which decompose or polymerize in the liquid phase can be detected and analyzed by coupling the pyrolyzer to proper electro-optical instruments. Although some of the FVP motifs have not yet been applied to synthetic problems, one should always be reminded of the capability of generating very peculiar structures by the method. Accordingly, the design of synthesis can take advantage of FVP.

Conventional FVP is not without technical problems. But for the introduction of nonvolatile compounds into the hot zone on preparative scales the solution-spray technique [Rubin, 1991] has been developed. In the modified procedure a solution of the compound is converted to an aerosol spray within a heated quartz tube which is maintained under a moderate vacuum (1-2 torr).

Microwave oven is another convenient source of heat, although modifications are required to prevent explosion. The outstanding feature of microwave heating is that the fast rate of energy delivery minimizes unnecesary exposure of both substrates and products to high temperatures. For accomplishing an organic reaction the increase in reaction rate and decrease in reaction time are very desirable, and microwave heating provides the proper means [Abramovitch, 1991; Mingos, 1991; Caddick, 1995; Strauss, 1995]. Basically, dielectric heating by microwave is related to realignment of dipoles among molecules as influenced by an external electric field. The polar change in the capacitor causes molecules to absorb energy thereby raising the temperature in their surroundings. For example, it requires only two to three minutes to heat up *N,N*-dimethylformamide to 300°C by microwave.

5.1. ELIMINATION REACTIONS

5.1.1. α-Elimination

Pyrolytic decomposition of chloroform proceeds by α-elimination of hydrogen chloride to generate dichlorocarbene which dimerizes to afford tetrachloroethene. In the presence of an alkene the carbene is trapped and a *gem*-dichlorocyclopropane results. Particularly interesting is the formation of chlorobenzene (with <1% 6-chlorofulvene) and a mixture of 2- and 3-chloropyridine, respectively, on copyrolysis of chloroform with cyclopentadiene [Busby, 1971] and imidazole [Busby, 1969].

$CHCl_3$ —Δ, -HCl→ $Cl_2C:$; -HCl ; [10 : 1]

The generation of heptafulvalene on pyrolysis of diazocycloheptatriene at 600°C or lower [Joines, 1969; Schissel, 1970] apparently involves elimination of dinitrogen to form a carbene intermediate, while higher temperatures favor the formation of stilbenes. [7]Paracyclophane arising from the tosylhydrazone of a spirocyclic dienone [Wolf, 1973] must be considered as the rearrangement of a carbene intermediate. Aryl azides give nitrenes under similar conditions, and the primary products stabilize themselves through ring contraction to afford cyanocyclopentadienes.

600°, $-N_2$; 2x ; heptafulvalene

=N-NTs Li^+ ; 380°, $-N_2$; [7]paracyclophane

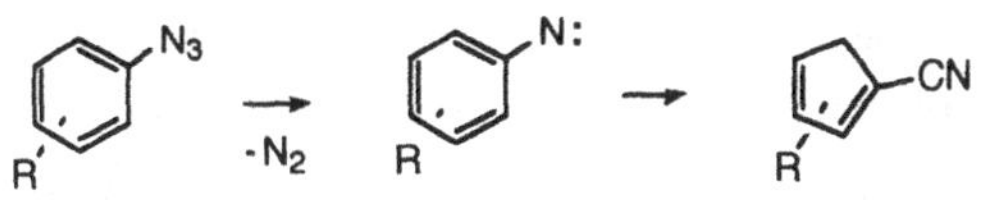

α-Diazoketones undergo pyrolytic α-elimination of dinitrogen and Wolff rearrangement. Despite the popularity of the photolysis, when the substrate contains other light-sensitive functionalities pyrolysis offers an alternative. A case in point is the preparation of 2-oxobenzocyclobutenecarboxylic esters from 2-diazo-1,3-indandione [Spangler, 1977]. Photolysis leads to the homophthalic esters.

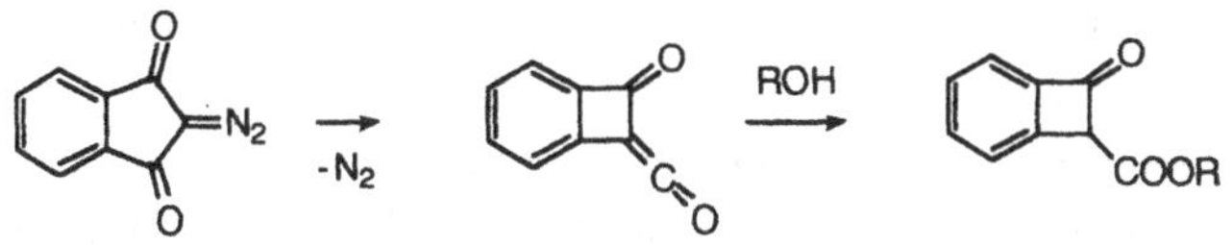

When the structure of a carboxylic ester does not permit customary β-elimination, pyrolysis may cause α-elimination instead. Slightly higher temperatures may also be required. The process is involved in the formation of benzocyclobutenedione, 4-methylene-2-cyclobutenone [Trahanovsky, 1974, 1979] and acenaphthenequinone [R.F.C. Brown, 1976].

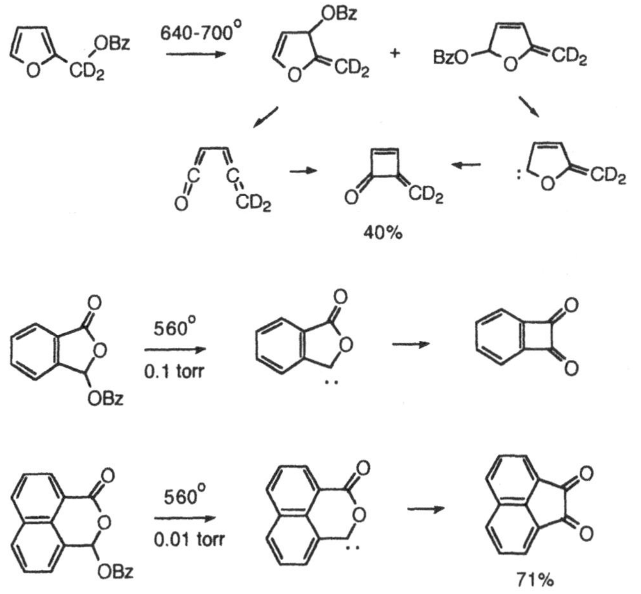

The conversion of hexachlorocyclopentadiene to octachlorofulvalene in the presence of iron at 500°C [Ginsberg, 1962] is a case of 1,1-dechlorination.

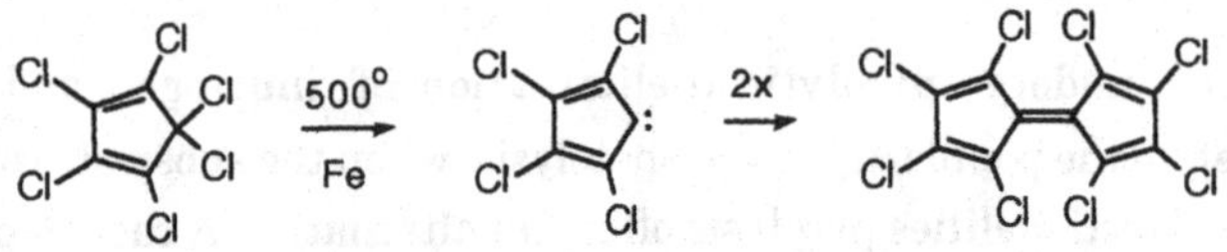

5.1.2. β-Elimination and Related Cycloreversions

The best known and synthetically useful thermal β-elimination deals with carboxylic esters. It is a reasonably good method for the generation of alkenes while avoiding strong acidic or alkaline conditions. The steric course of ester pyrolysis is well established, the passage through a six-centered transition state means detachment of a *cis* hydrogen atom from the β-carbon atom. Statistically, elimination favors the formation of a less substituted double bond, but this trend can be subverted by the generation of a conjugated system.

In addition to carboxylic esters, alkyl carbonates, carbamates, xanthates, thionocarboxylates, and thiocarboxylates behave similarly. In fact, many of these exters have lower decomposition temperatures [DePuy, 1960].

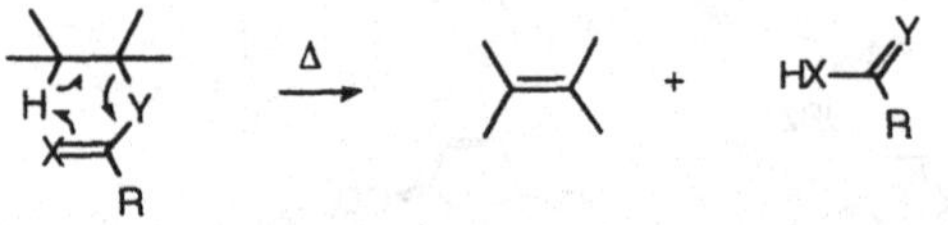

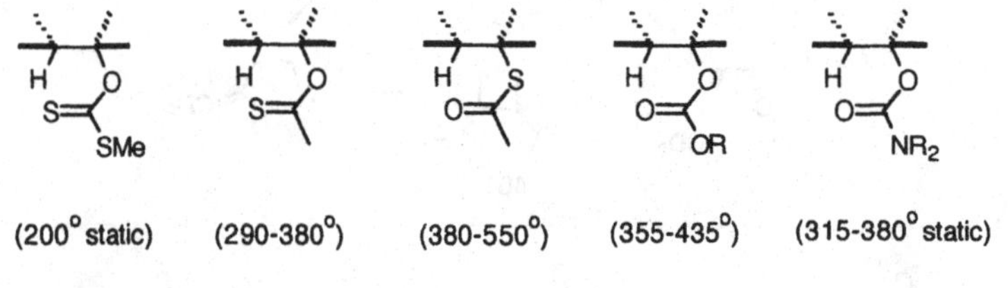

Two steps of a sinularene synthesis [Collins, 1979] entailed pyrolysis of an acetate at 450°C, the first for the creation of an isopropenyl group (to be hydrogenated) and its repeated employment appeared at the conclusion of the synthesis.

OAc 450° ⇄ OAc 450°

sinularene

Elimination of two molecules of acetic acid provided a diene for the synthesis of [1.1.1.1]paracyclophane [Miyahara, 1983]. Interestingly, the double bonds were not at the benzylic positions presumably conjugation with the benzene rings would induce much steric strain.

OAc OAc 550° PhSO Δ

[1.1.1.1]paracyclophane

In most applications the ester pyrolysis is for alkene synthesis, acetates and benzoates are the most common substrates. However, the need for temporary protection of a carboxyl group may be gainfully fulfilled by conversion into the *t*-butyl ester, by taking advantage of the thermal decomposition of *t*-butyl esters into isobutene and the carboxylic acids. Thus arylacetaldehydes may be prepared by the homologation protocol of Darzens reaction and pyrolysis of the epoxy esters when *t*-butyl chloroacetate is used in the condensation step [Blanchard, 1963]. This alternative method completely avoids saponification and hence the possible exposure of any inadvertently generated arylacetaldehydes (due to premature protonation and decarboxylation) to bases.

Ph O O O 350-360° N_2 Ph O H O O + $-CO_2$ Ph CHO

2-Phenylpropanal. [Blanchard, 1963]

t-Butyl β-methyl-β-phenylglycidate (23.4 g, 0.1 mol) is passed with a stream of nitrogen (5 mL/min) through a hot tube (350-360°C) filled with glass helices. The pyrolysate which is collected in a cold trap is taken up in ether, and the solution is washed with aq. sodium carbonate. The dried ether solution is filtered and evaporated. The residue is distilled to give 2-phenylpropanal (8.62 g, 62.8%).

Unstable molecules prepared by ester pyrolysis are exemplified by the preparation of highly strained bridgehead alkenes [Becker, 1979], and several other unsaturated compounds, including benzocyclopentadienone and isoindole [Bailey, 1955; Marvel, 1954; Bonnet, 1973].

AcO — 400° →

OAc, OAc — 400°, N_2 → 47% + OAc 30%

O, OAc — 550° → O 20%

O, OMe, N–O — 550°, 0.01 torr → NH + CO_2 + MeOH

Certain lactones undergo thermal decomposition to give ω-alkenoic acids [Bailey, 1977].

O, O, $(CH_2)_n$ — Δ → O, OH, $(CH_2)_n$

n = 4, 6

O, O — Δ → O, OH

Although acrylic anhydride is not an ester, its pyrolysis follows a similar and expected course; a mixture anhydride of butadienoic acid also gives an acylcumulene [Blackman, 1978; R.D. Brown, 1979].

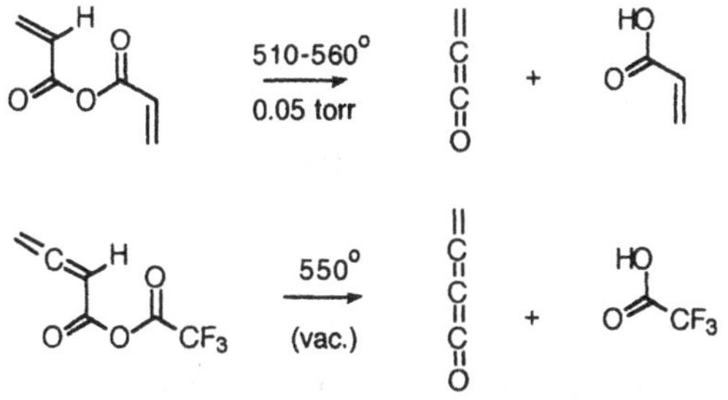

510-560° 0.05 torr

550° (vac.)

Ketenes which cannot be formed by dehydrochlorination of acyl chlorides with a tertiary amine may be obtained by pyrolysis (400-600°C) of the proper derivatives of Meldrum's acid. Cogenerated with the ketenes in this reaction are carbon monoxide and acetone. While ethyleneketene is such an example [Baxter, 1975], interesting secondary reactions may lead to other products. Acylcumulenes are known to decarbonylate, giving alkenylcarbenes which then undergo rearrangement [Baxter, 1978].

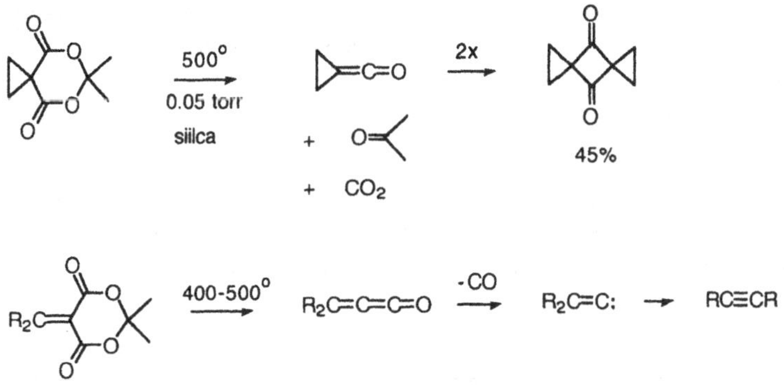

500° 0.05 torr silica; 2x; + CO_2; 45%

400-500°; $R_2C{=}C{=}C{=}O$ $\xrightarrow{-CO}$ $R_2C{=}C{:}$ → $RC{\equiv}CR$

In certain esters the β-elimination pathway may not dominate [Carlson, 1967; Sonnenberg, 1976]. Decomposition of patchouli acetate gives α- and γ-patchoulene which have a different skeleton [Büchi, 1961]. Unwitting of rearrangement undergone by the patchouli alcohol derivative during the pyrolysis caused its structural misidentification.

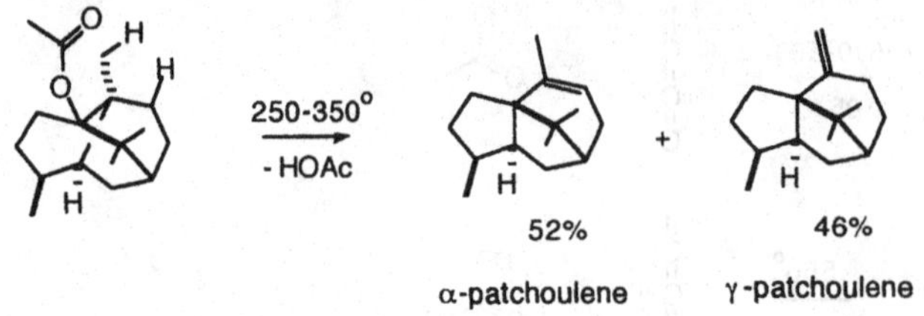

250-350°
- HOAc

52% α-patchoulene + 46% γ-patchoulene

Preclusive of β-elimination by ring strain (which must provide an anti-Bredt alkene) 2-adamantyl sulfonates elect to display reactivity for 1,3-elimination [Boyd, 1972].

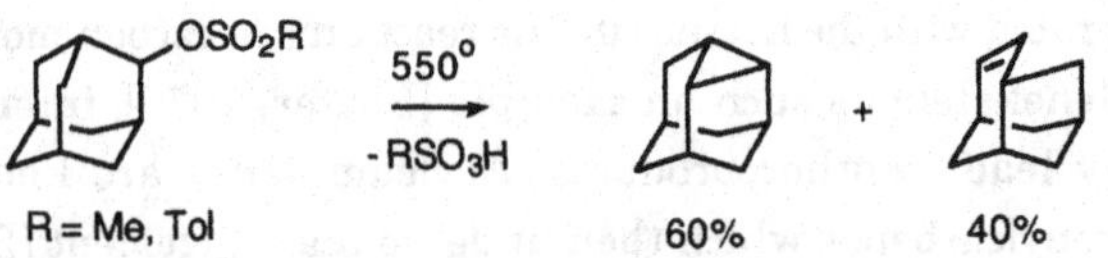

OSO_2R ; R = Me, Tol
550°
$- RSO_3H$
60% + 40%

Pyrolysis of 2-Adamantyl Mesylate. [Boyd, 1972]

2-Admanantyl mesylate (100 mg) is placed in a boat inside a silica tube which is wrapped with a heating tape (maintained at 150°). The whole system is evacuated while a stream of helium carries the sublimate into the pyrolysis chamber (550°C) at 0.5 torr. The product from the cold trap is recovered with ether and analyzed.

Some esters undergo thermal elimination of alcohol to afford ketene products. These species may in turn participate in further reactions such as decarbonylation-cycloaddition, [2+2]cycloaddition/ cycloreversion [Leyendecker, 1976].

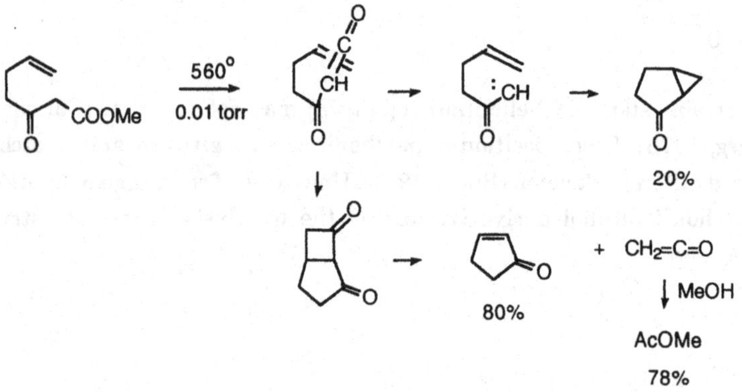

COOMe
560°
0.01 torr
20%
80%
+ CH_2=C=O
MeOH
AcOMe
78%

Pyrolysis of organic halides usually suffer from nonregioselective elimination, therefore this method is of little synthetic value.

5.1.3. Remote Eliminations

Bridged diazolines have been identified as convenient precursors of 1,3-diyls [Little, 1986]. Elimination of nitrogen from these compounds can be performed photochemically or thermally. The thermal decomposition requires moderate temperatures only. However, intramolecular trapping of the diyls leading to carbocycles constitutes a useful synthetic method.

Aromatic compounds such as *o*-substituted toluenes and phenols which contain a benzylic leaving group are amenable to elimination on pyrolysis, yielding *o*-quinodimethanes or *o*-quinomethanes as primary products. Thus α-chloro-*o*-xylene suffers dehydrochlorination at 630°C to give benzocyclobutene [Loudon, 1969, 1970]. Transannular bond formation that results in superphane [Sekine, 1979] from a bis-*o*-quinodimethane generated in the same manner is energetically favorable.

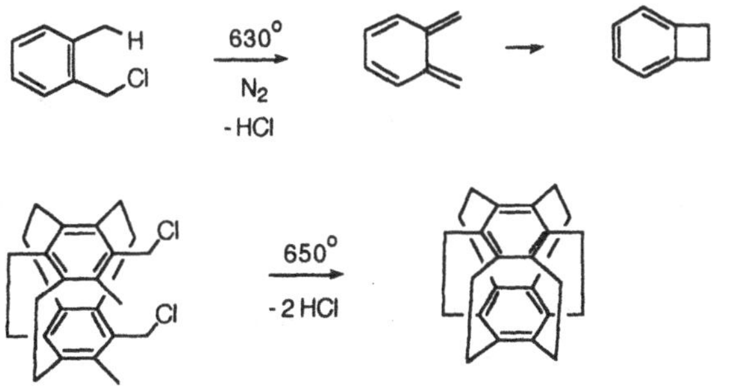

H
Cl
630°
N_2
-HCl

Cl
Cl
650°
-2 HCl
superphane

A convenient preparation method for [6]radialene is based on a threefold 1,4-elimination of hydrogen chloride on short contact time (0.01 s) [Harruff, 1978; Schiess, 1978].

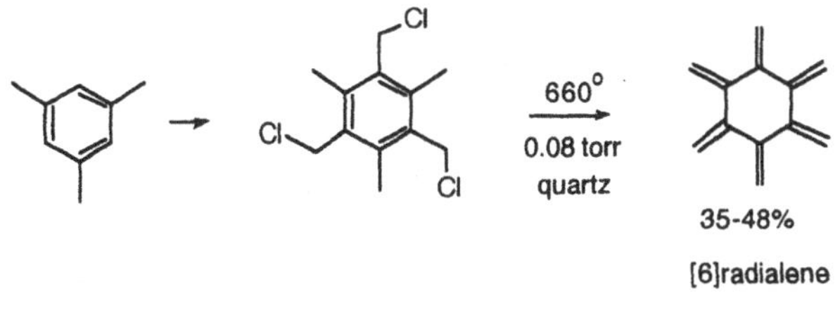

Cl
Cl
Cl
660°
0.08 torr
quartz
35-48%
[6]radialene

While *o*-toluoyl chloride forms benzocyclobutenone [Loudon, 1969a; Schiess, 1977] but [1.5]-hydrogen shift intervenes when homologous *o*-alkylbenzoyl chlorides are subjected to the same thermal conditions. The ketene intermediates clearly prefer conversion to the *o*-alkenylbenzaldehydes [Schiess, 1977].

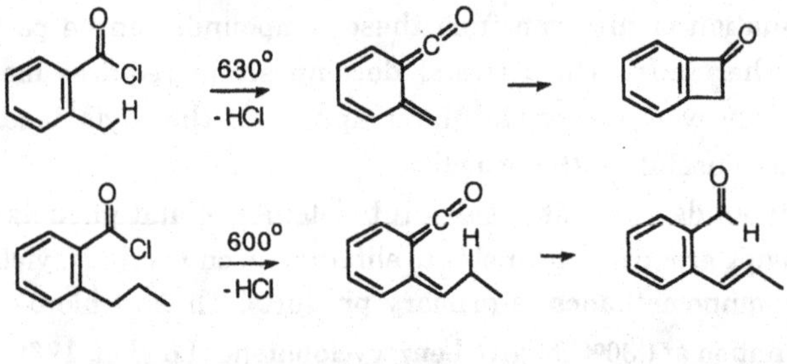

Elimination of methanol in an analogous manner is represented by the following examples [Cavitt, 1962; Mamer, 1974; De Champlain, 1976]:

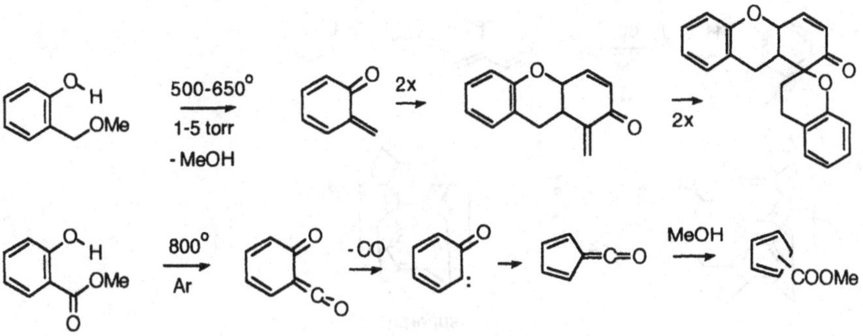

A synthesis of azulene from 6-(dialkylaminopentadienyl)fulvene [Ziegler, 1955] is a classic example of remote elimination by pyrolysis.

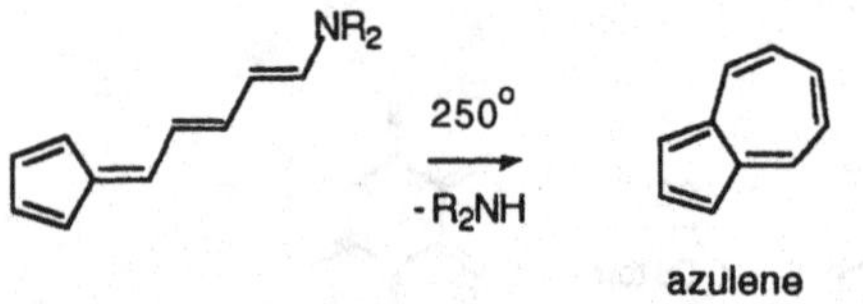

azulene

Two FVP pathways to corannulene by remote dehydrobromination have been developed [Borchardt, 1992; Scott, 1992].

1000° 18% 1000° 40%

corannulene

5.2. CHELETROPIC EXTRUSIONS AND OTHER RING FISSIONS

Cheletropic reactions are those involving concerted formation or cleavage of two σ bonds terminating at a single atom. Selection rules for two subclasses (linear and nonlinear) of cheletropic reactions have been defined.

Perfluorodicyclopentadiene behaves quite unusually as thermal extrusion of difluorocarbene competes with the retro-Diels-Alder fission [Banks, 1966; R.W. Hoffmann, 1971].

F_{12} 650-660° F_6 F_{10} + F_2C: 2x $F_2C{=}CF_2$

Decarbonylation and extrusion of sulfur dioxide are the most common cheletropic reactions exploited in synthesis. In the absence of proper nucleophiles cyclopropanones can undergo dissociation into carbon monoxide and alkenes. In connection with the synthesis of *cis*-jasmone two different routes [Büchi, 1971; Berkowitz, 1972] have been developed, each apparently proceeding via a bicyclo-[3.1.0]hexane-2,5-dione intermediate which promptly lost carbon monoxide to give the cyclopentenone.

xylene Na_2CO_3 -CO dehydrojasmone

COOEt 540°

[2_3](1,3,5)cyclophane may be obtained from a triketone by a temperature-dependent, stepwise decarbonylation process [Breitenbach, 1992]. The intermediate ketones could be isolated.

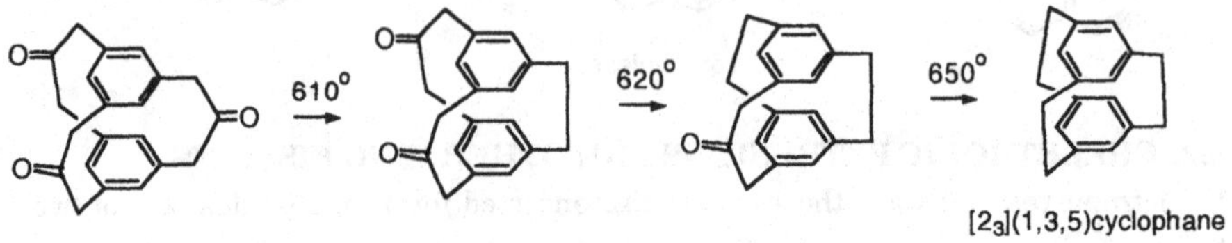

610° 620° 650°

[2_3](1,3,5)cyclophane

Cyclopentadienone is an antiaromatic compound. Its existence as a transient species has been detected by infrared spectroscopy in the pyrolysate (550°C) of *o*-benzoquinone, when collected at liquid nitrogen temperature [O.L. Chapman, 1971]. Pyrolysis of *p*-benzoquinone at 850°C leads predominantly to 3-buten-1-yne, the doubly decarbonylated product [Hageman, 1972]. The best method for access to dibenzopentalene seems to be that involving pyrolytic decomposition of chrysene-6,12-quinone [Schaden, 1977].

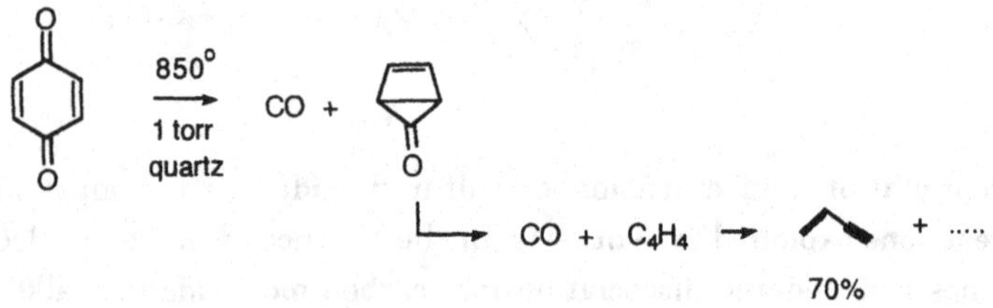

850° 1 torr quartz → CO + ... → CO + C_4H_4 → ... + ... 70%

An expedient synthesis of linear polyynes with odd numbers if triple bonds involves thermal decomposition of 3,4-dialkynyl-3-cyclobutene-1,2-diones. The difficulties pertaining to the lack of volatility of high molecular weight substrates have been solved by the development of a solution-spray technique [Rubin, 1991].

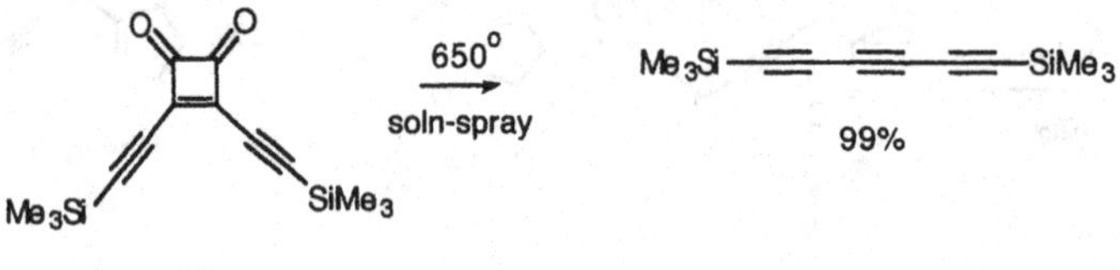

Me_3Si ... $SiMe_3$ → 650° soln-spray → Me_3Si—≡—≡—≡—$SiMe_3$ 99%

The cyclic sulfite and carbonate esters of catechol behave differently on exposure to high temperatures. The sulfite loses sulfur monoxide to generate *o*-benzoquinone and thence to cyclopentadienone, which dimerizes at the lower temperature and in a condensed phase [De Jongh, 1972]. On the other hand, the first step of thermal decomposition for catechol carbonate [De Jongh, 1970] results in the loss of carbon dioxide.

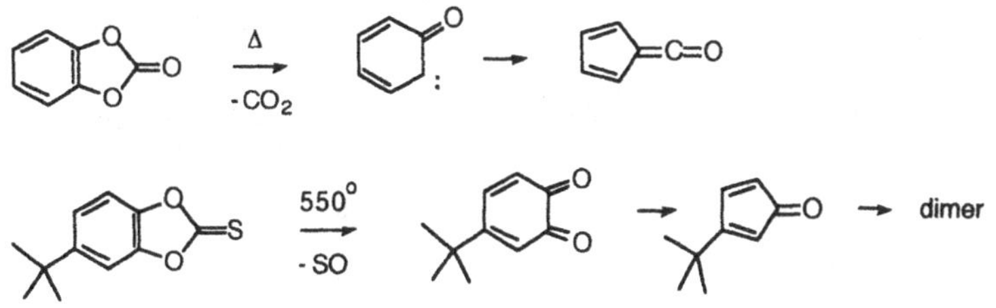

Cheletropic extrusion of sulfur dioxide from 3-sulfolenes requires relatively mild conditions. Accordingly, sulfolenes are usually considered as latent 1,3-dienes. Accordingly, chain extension at one or both termini of a conjugated diene is quite readily achieved via the sulfur dioxide adduct. The sulfone group serves to activate its α-position and departs on subsequent thermolysis [Bloch, 1983]. Further synthetic utility based on the theme of tandem cheletropic extrusion and intramolecular Diels-Alder reaction is now very well established. Highly efficient routes to complex molecules, e.g., elaeokanine-A [Schmitthenner, 1980] and aspidospermine [S.F. Martin, 1980] have evolved. Other applications of 3-sulfolenes to synthesis are evident [Chou, 1989].

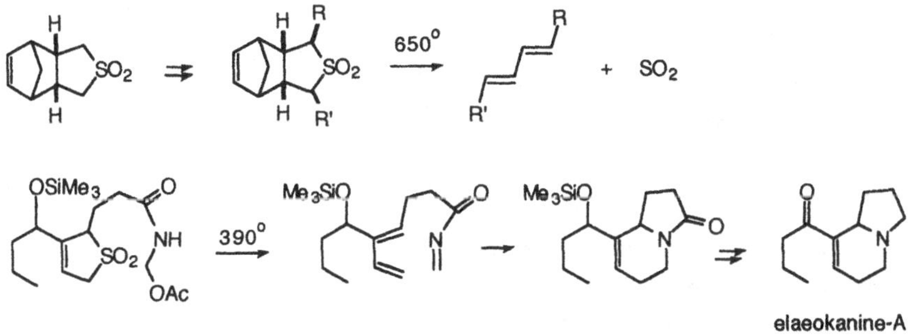

elaeokanine-A

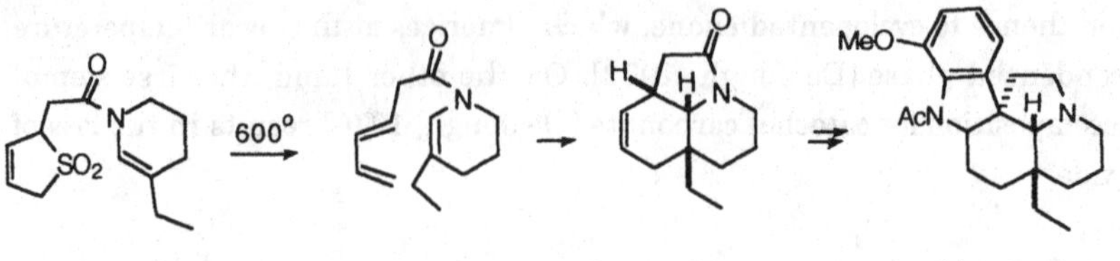

600°

aspidospermine

[3]Dendralene has been synthesized from 3-vinyl-3-sulfolene [Cadogan, 1991] by the cheletropic extrusion method in excellent yield (87%) and milder conditions (550°C, 0.001 torr) than pyrolysis of the acyclic triacetate (860-900°C, 10^{-4} torr) [Trahanovsky, 1992].

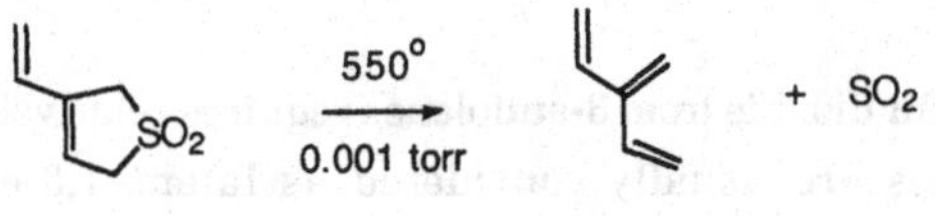

550°, 0.001 torr; + SO_2

[3]dendralene

Dihydroisobenzothiophene *S,S*-dioxides are a good source of *o*-quinodimethanes. In the absence of trapping agents the *o*-quinodimethane products in the pyrolysates are isomerized to benzocyclobutenes [Cava, 1960; Giovannini, 1977].

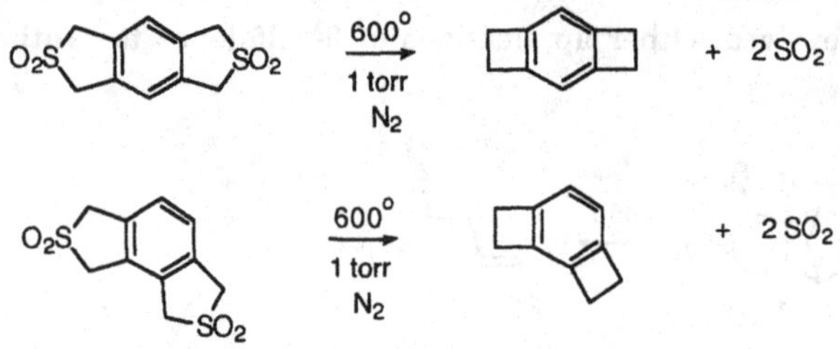

600°, 1 torr, N_2; + 2 SO_2

600°, 1 torr, N_2; + 2 SO_2

The high reactivity of *o*-quinodimethanes toward dienophiles makes them excellent precursors of tetralins. That an *o*-quinodimethane can be trapped intramolecularly even by an unactivated double bond opens a short pathway to the A-aromatic steroids [Oppolzer, 1980].

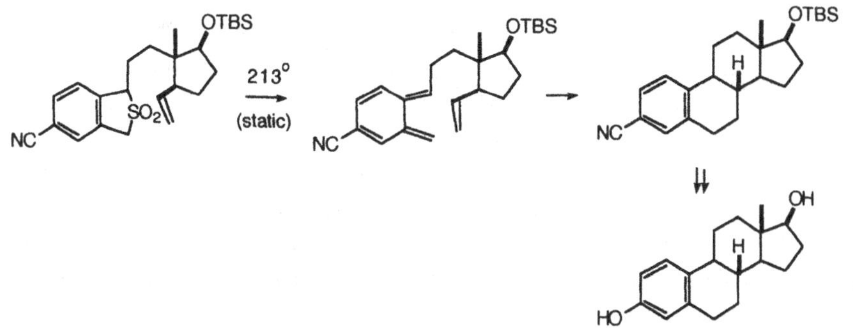

OTBS 213° (static) NC SO2 OTBS NC OTBS H NC OH H HO

estradiol

Benzylic sulfones undergo homolytic C-S bond cleavage at high temperatures, therefore bibenzyls are readily available by pyrolysis of dibenzyl sulfones. This process is particularly expedient for the preparation of cyclophanes [Vögtle, 1969, 1979] including the highly strained adamantanocyclophanes [Lemmerz, 1993]. (Note a synthesis of all-carbon paddlane [Vögtle, 1978] based on a triptycene skeleton by pyrolysis of a disulfone which is not even benzylic.)

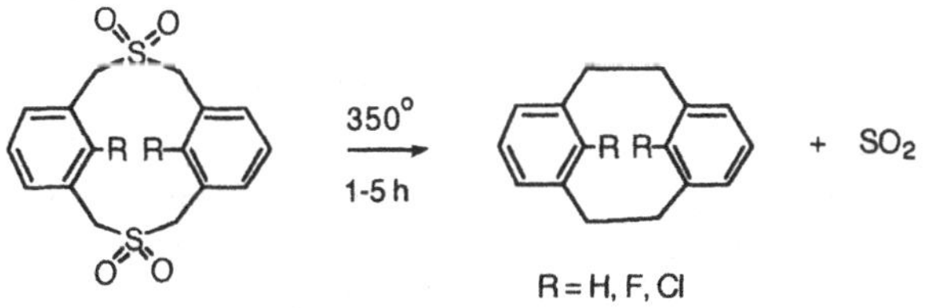

350° 1-5 h + SO_2

R = H, F, Cl

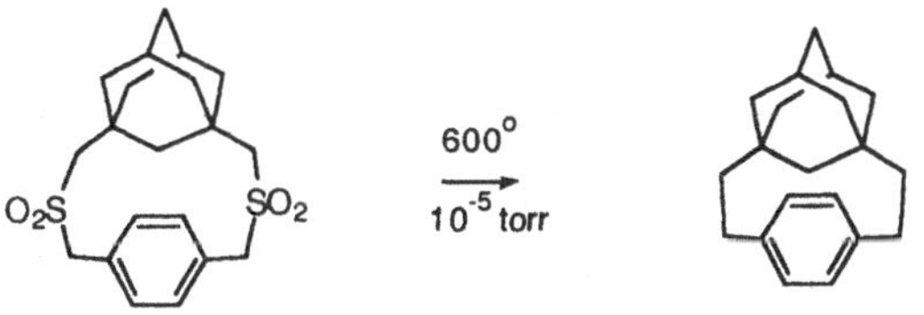

600° 10^{-5} torr

[2.2](1,3)adamantoparacyclophane

It must emphasized that the thermal reaction of benzylic sulfones of the type aforementioned is not cheletropic. Another motif is represented by the extensive bond scission of thiabicyclo[3.2.0]heptane *S,S*-dioxides to give 1,5-dienes [Williams, 1981].

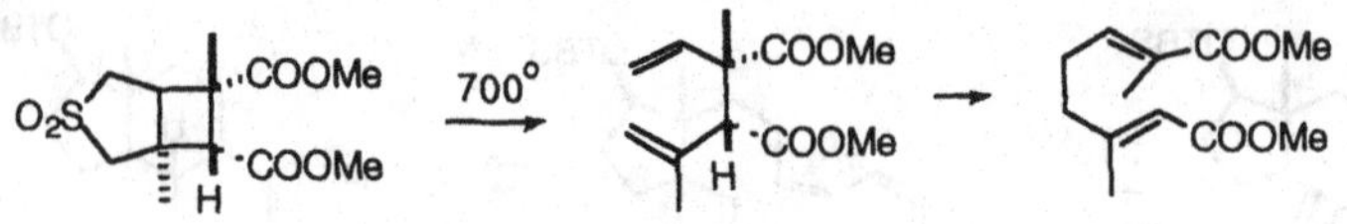

The very uncommon thermal elimination of O_2 from a diketone formed an azuleno-azulene has been observed [Vogel, 1984]. Apparently, maintenance of the aromatic 14π-electron system was so favorable as to virtually preclude tautomerization and thereby to allow decarbonylation. Anthracene was found to be the minor product of the pyrolysis.

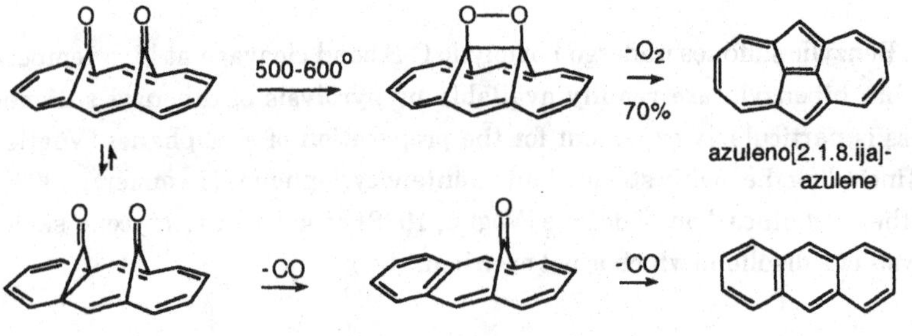

azuleno[2.1.8.ija]-azulene

Extrusion of dinitrogen from aromatic heterocycles embodying two adjacent nitrogen atoms inside the ring is a favorable course of thermal decomposition [MacBride, 1972]. Thus indole, isoindoles, azacyclobutadienes, and diazaphenylene are accesible by this method [Lawrence, 1970; Gilchrist, 1975; Seybold, 1973; MacBride, 1974]. Thioketenes are formed in 60-70% yield from 1,2,3-thiadiazoles [Seybold, 1977].

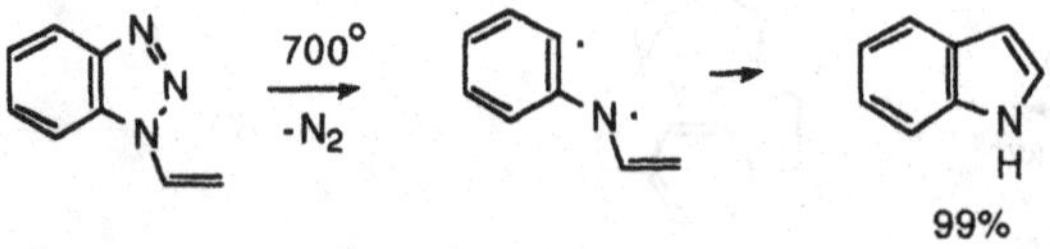

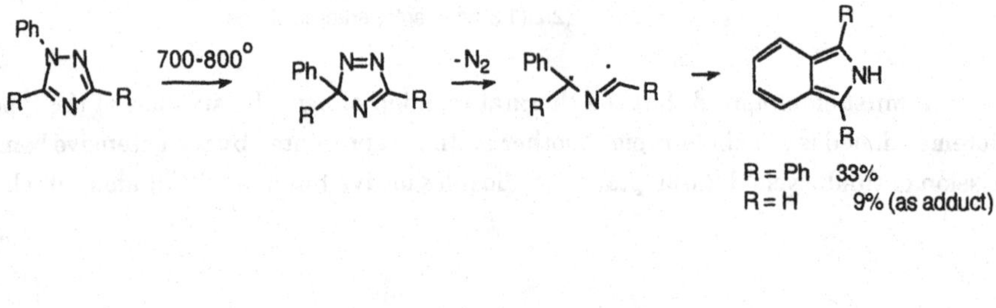

R = Ph 33%
R = H 9% (as adduct)

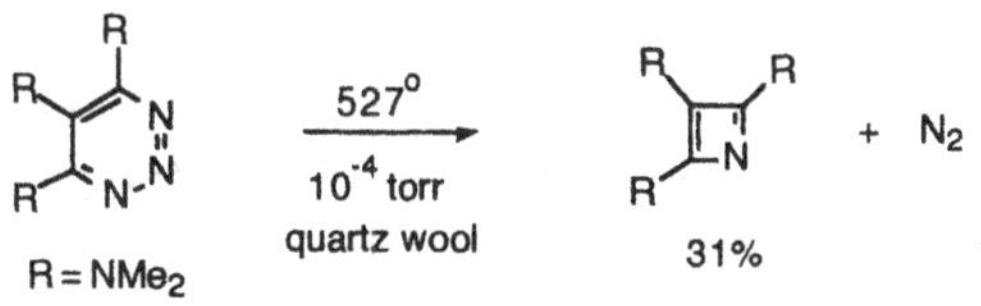

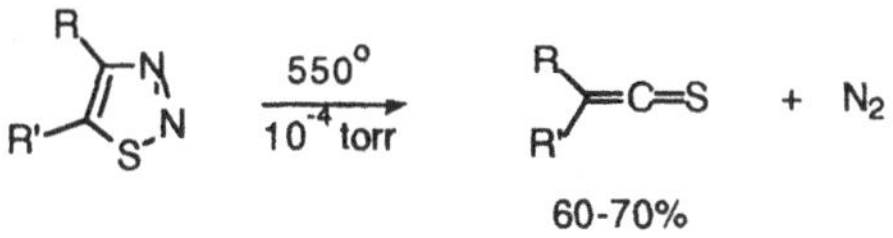

Benzotriazines give benzyne (which dimerizes to afford biphenylene) upon pyrolytic elimination of dinitrogen and a nitrile [Adger, 1975; Forster, 1971].

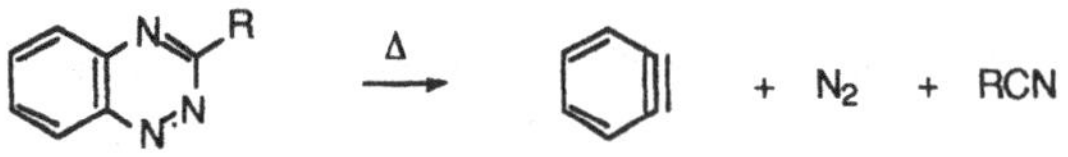

A convenient way to prepare benzocyclobutenedione in 88% yield [Forster, 1971] is by a tandem retro-Diels-Alder elimination of cyclopentadiene and dinitrogen from a phthalhydrazide at 500°C/0.01 torr. On the other hand, the extrusion of dinitrogen from a heterocycle may trigger elimination of another smaller molecule.

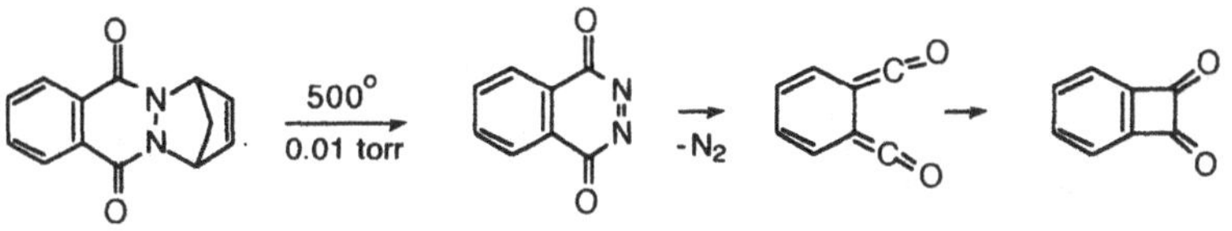

5.3. THERMAL CLEAVAGE OF FOUR-MEMBERED RINGS

Strain relief is the driving force for the thermal lability of the four-membered ring (relative to five- and six-membered homologues). However, the cleavage reaction still requires high temperatures.

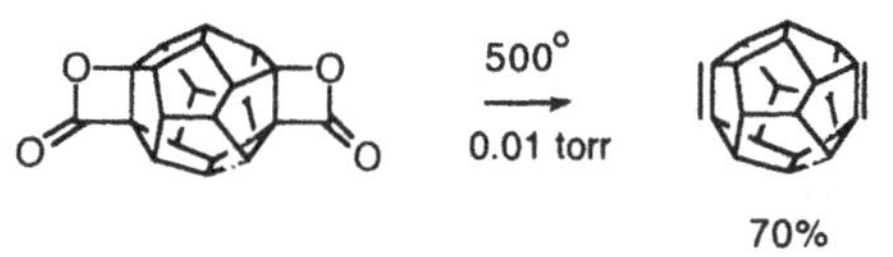

Thermal decomposition of β-lactones generates alkenes regioselectively [Krapcho, 1974; Adam, 1993]. One of the most significant utility of this process is in the preparation of a dodecahedradiene [Melder, 1990]. It is also remarkable that the valence isomer of α-pyrone reverts to the monocyclic compound on heating to 400°C, but it decarboxylates to furnish cyclobutadiene at low pressure and higher temperatures (835°C) [Hedaya, 1969b]. Remotely related is the thermal decarboxylation of phthaloyl peroxide to furnish benzyne [Wittig, 1961].

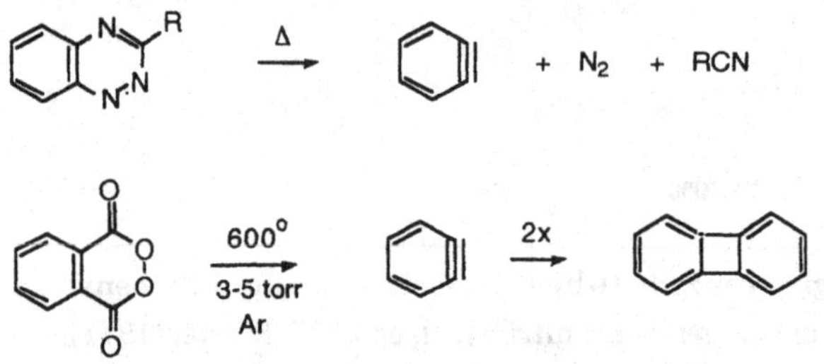

Δ + N_2 + RCN

600° 3-5 torr Ar 2x

2,2-Dimethyl-3-isopropylidene-β-propiolactone undergoes decarboxylation [J.C. Martin, 1964] whereas other dimers of ketenes decompose into their monomer components.

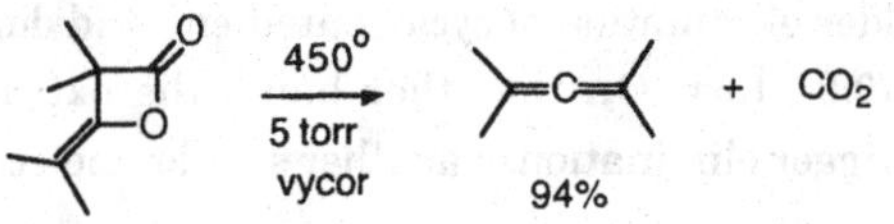

450° 5 torr vycor + CO_2 94%

Some very reactive species are accessible by thermal cycloreversion of other heteracyclobutanes. These include sulfene [Block, 1976] and methylenedimethylsilane [Nametkin, 1966; Gusel'kov, 1989].

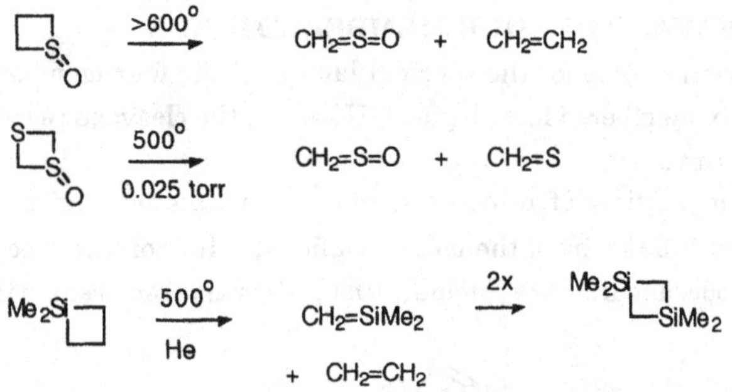

>600° CH_2=S=O + CH_2=CH_2

500° 0.025 torr CH_2=S=O + CH_2=S

Me_2Si 500° He CH_2=$SiMe_2$ + CH_2=CH_2 2x Me_2Si $SiMe_2$

Oxetanes are quite readily available from aldehydes and alkenes by the Paterno-Büchi reaction. As synthetic intermediates they show the potential for inserting an alkeno link to the α-bond of an aldehyde when cycloalkenes are used as the photoaddend, on the basis of pyrolytic cycloreversion. Thus the precursor of *(E)*-6-nonenol, a sex attractant of the Mediterranean fruit fly, has been assembled in that manner [G.Jones, 1975].

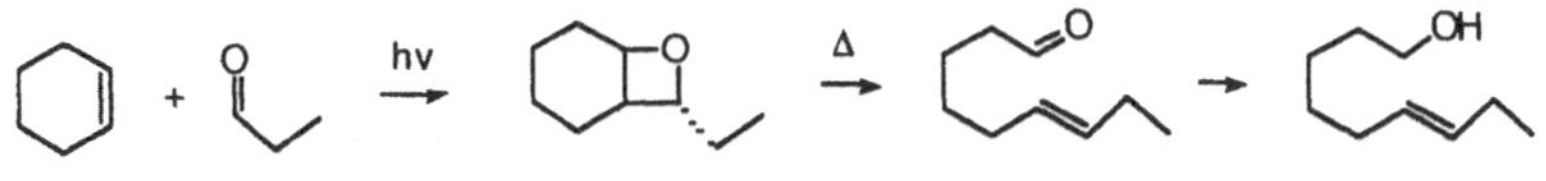

The formal all-carbon metathesis involving photocycloaddition and pyrolytic cycloreversion is illustrated in the approach to hirsutene and $\Delta^{9(12)}$-capnellene [Mehta, 1986]. The regioselectivity of the ring cleavage is due to generation of two conjugated ketone units.

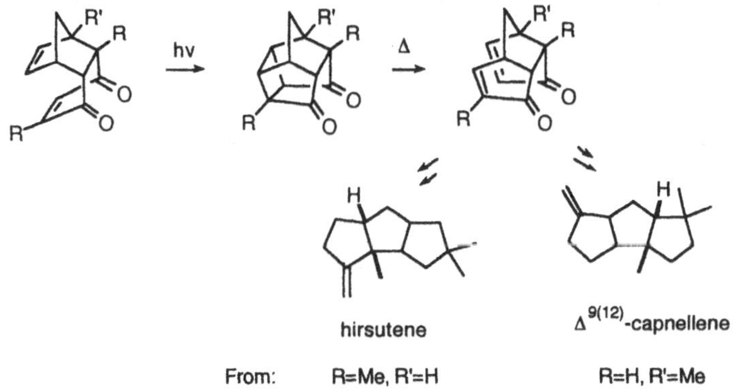

hirsutene $\Delta^{9(12)}$-capnellene

From: R=Me, R'=H R=H, R'=Me

A tricarbocyclic dilactone, which was also originated from an intramolecular photocycloaddition, proved to be an excellent precursor of byssochlamic acid [White, 1992]. Because of the higher strain its thermolysis requires a much lower temperature.

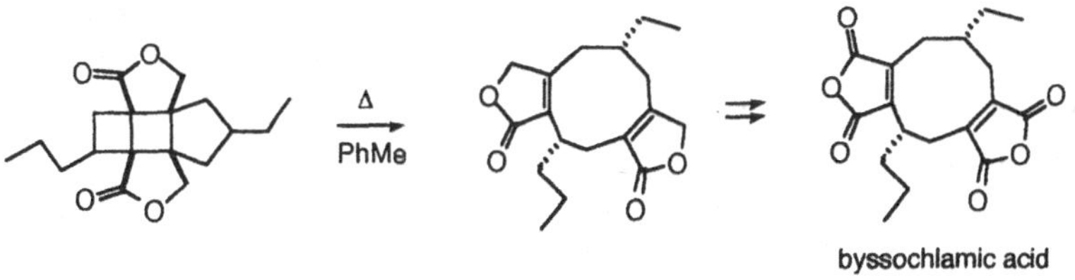

byssochlamic acid

In a flow system the photocycloadduct of piperitone and 1-methylcyclobutene was transformed into shyobunone and stereoisomers, as well as isoacoragermacrone

[Williams, 1980a]. The secondary intramolecular ene reaction was probably arrested during the relatively short thermolysis, as only a cadinane derivative could be isolated from a condensed phase reaction (250°C, 0.5 h).

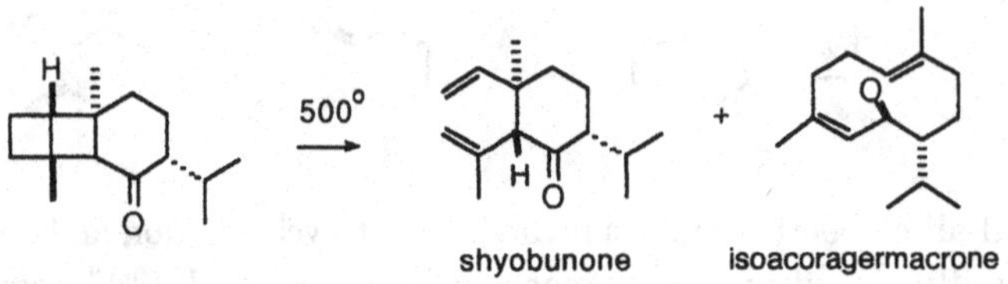

500°

shyobunone isoacoragermacrone

Of industrial significance is the pyrolysis of pinane derivatives. Many versatile and valuable acyclic intermediates for use in manufacture of flavor and fragrance materials as well as vitamin-A and vitamin-E are obtained. The simplest of the series of compounds are myrcene, ocimene [Goldblatt, 1941] and linalool [Coxon, 1972]. Nopinone likewise is converted to a dienone [Mayer, 1970], which has been exploited in a synthesis of β-bisabolene [Ho, 1980]. An epoxide of β-farnesene is now readily available from caryophyllene in two steps [Giersch, 1994], i.e., epoxidation and flash vacuum pyrolysis.

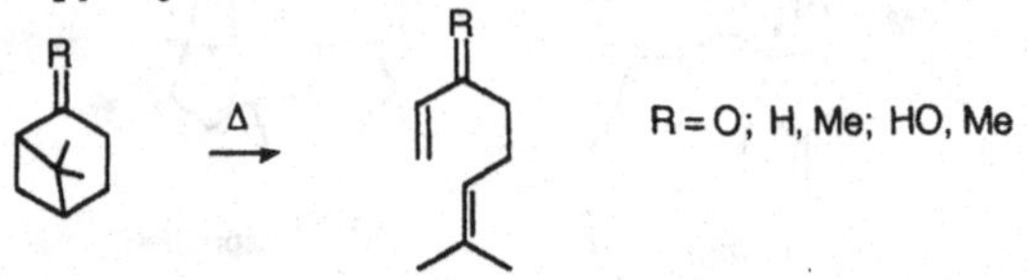

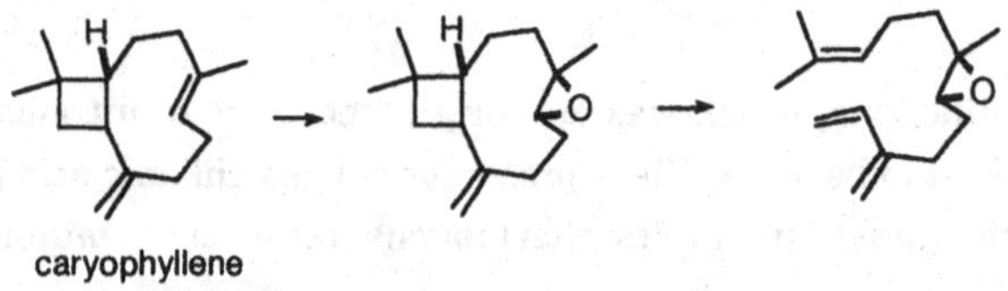

Δ

R = O; H, Me; HO, Me

caryophyllene

Carvonecamphor is a tricyclic product resulted from intramolecular [2+2]-photocycloaddition [Büchi, 1957]. On exposure to high temperature this compound is converted to a cyclopentenone which finds utility as a presursor of methylenomycin [Sternbach, unpubl.].

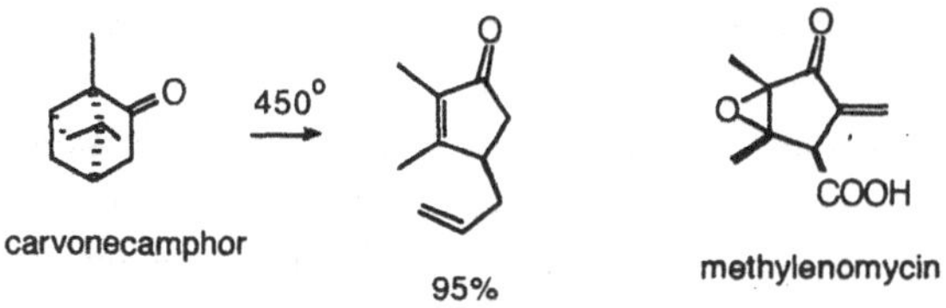

450°

carvonecamphor 95% methylenomycin

Strain relief is an important factor for the dissociation of [2,2]paracyclophane to give two molecules of *p*-xylylene. As *p*-xylylene prefers polymerization in the absence of trapping agents a method for forming a protective coating of various items is by spraying them with the pyrolysate of [2,2]paracyclophane [Gorham, 1966]. Another precursor of *p*-xylylene is the propellano Dewar benzene [Landheer, 1974]. However, its utility is extremely uneconomical.

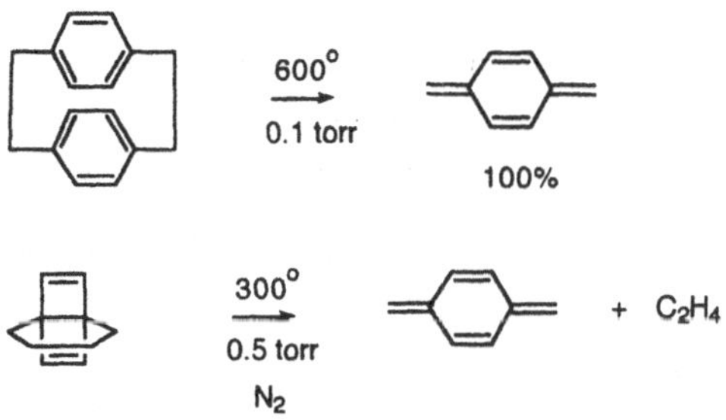

600° 0.1 torr 100%

300° 0.5 torr N_2 + C_2H_4

It may be mentioned here also the thermal cleavage of aziridine-2-carboxylic esters. Generation of 1,3-dipolar species is indicated by the formation of proline lactones from alkenoxy esters [DeShong, 1985]. Reaction in a sealed tube led to extensive decomposition, which augments the argument for the advantages of FVP.

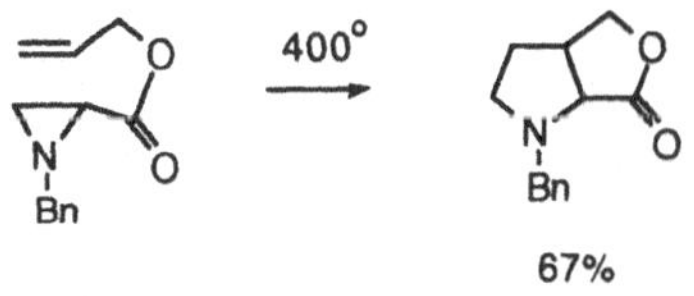

400° 67%

5.4. ELECTROCYCLIC REACTIONS

The classification of this class of reactions (thermal and photo versions) began with the formulation of the Woodward-Hoffmann rules [Woodward, 1970]. Previously the few examples known were considered under valence tautomerism. The selection rules for the thermal electrocyclic processes of open-chain π-electron systems are:

conrotatory movement of the terminal groups for $4n$ electrons, but disrotatory of the groups for $4n+2$ electrons.

Many substituted butadienes are more expediently accessible from cyclobutenes. As on most occasions the dienes are used in Diels-Alder reactions, direct thermolysis of the cyclobutenes with the proper dienophiles accomplishes the synthetic purpose. A case in point is the preparation of 1-acetoxy-2-arylthio-cyclohexene derivatives [Trost, 1980].

AcO, ArS, 150°, AcO, ArS, X, Y, AcO, ArS, X, Y

Ar = —C₆H₄—OMe

Electrocyclic opening of benzocyclobutenes is favored in terms of molecular strain factor. Interestingly, when the resulting *o*-quinodimethanes are well juxtaposed to undergo intramolecular dimerization, multibridged [2_n]phanes may be formed [Boekelheide, 1978; Aalbersberg, 1979].

700°

750°
0.01 torr

Trapping of *o*-quindimethanes by dienophiles is mentioned in a section above. It seems benzocyclobutenes are more readily accessible in different substitution patterns and therefore conducive to more stringent and variegated synthetic demands. For steroid synthesis [Kametani, 1981b] the small ring section may supply portion of the B-ring or C-ring. Because of the very high stereoselectivity observed in the intramolecular Diels-Alder reactions the adducts need only few more steps to be elaborated into the target compounds.

Cobalt-promoted [2+2+2]-cycloaddition is an alternative method to provide the benzocyclobutene intermediate. In fact at temperature above 100°C the subsequent reactions occur in tandem, resulting in the steroid skeleton *in toto* [Funk, 1977, 1979].

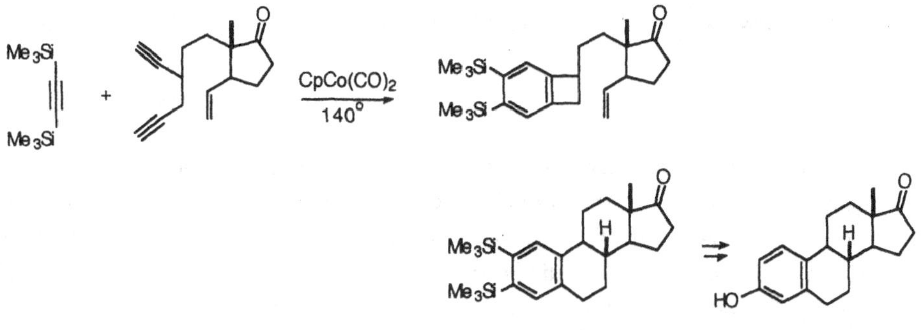

Me$_3$Si

Me$_3$Si

CpCo(CO)$_2$

140°

estrone

The theme have been repeatedly and effectively disseminated. A remarkable demonstration is the assembly of a pentacyclic precursor of triterpenes which contains four contiguous asymmetric centers [Kametani, 1978]. Also of high efficiency are processes for the formation of BCD-system of steroids using a B-aromatic precursor [Kametani, 1981a], and the expedient synthesis of chelidonine [Oppolzer, 1971].

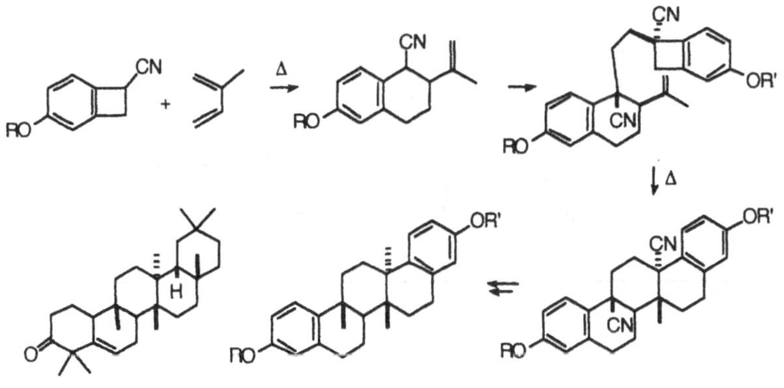

Δ

alnusenone

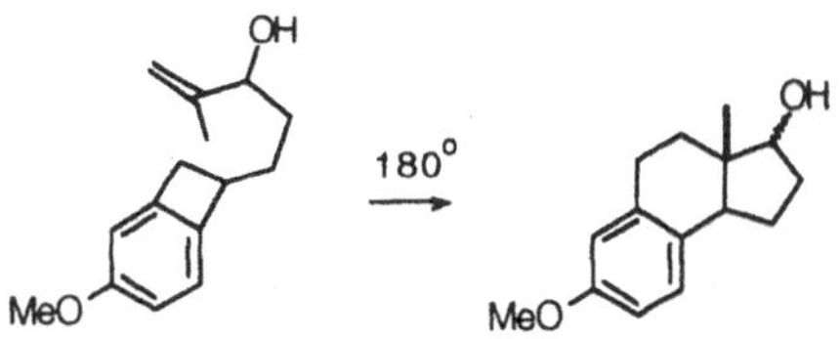

180°

85.5%

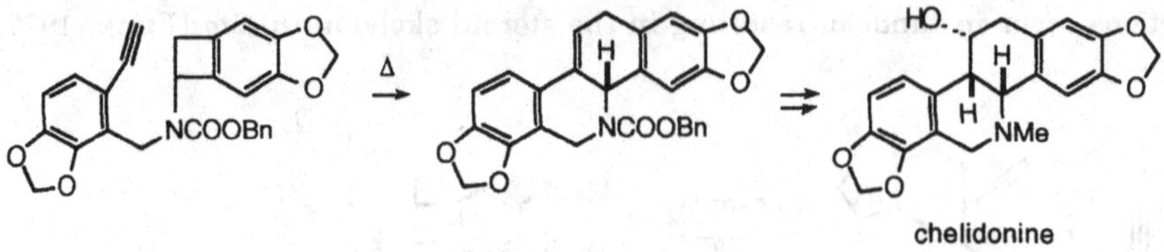

chelidonine

When the four-membered ring of the *o*-quinodimethane precursor bears an unsaturated side chain, the ring cleavage may be followed by other pericyclic reactions in tandem. Thus the creation of an intermediate for atisine [Shishido, 1989] by such a method is quite significant in terms of synthetic design. For synthesis of heterocyclic systems reference can be made to synthetic approaches to xylopinine [Kametani, 1973, 1975], sendaverine [Kametani, 1979], and physostigmine [Shishido, 1986]. From the scheme outlining several key intermediates of physostigmine, it can be seen that a lactone emerged from pyrolysis of the benzocyclobutene is the final product of sequential 4e-electrocycloreversion, 6e-electrocyclization, and Claisen rearrangement.

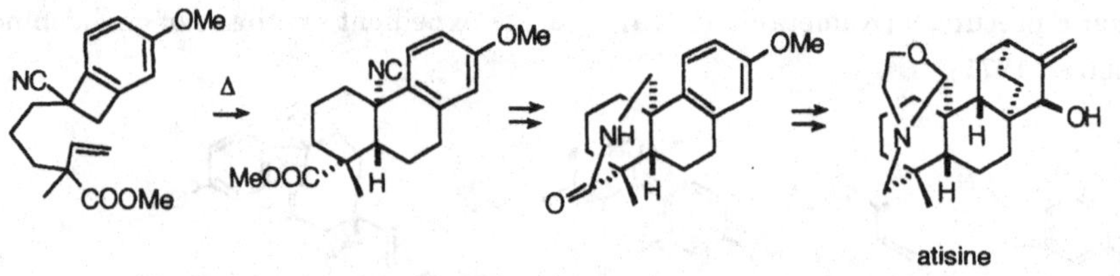

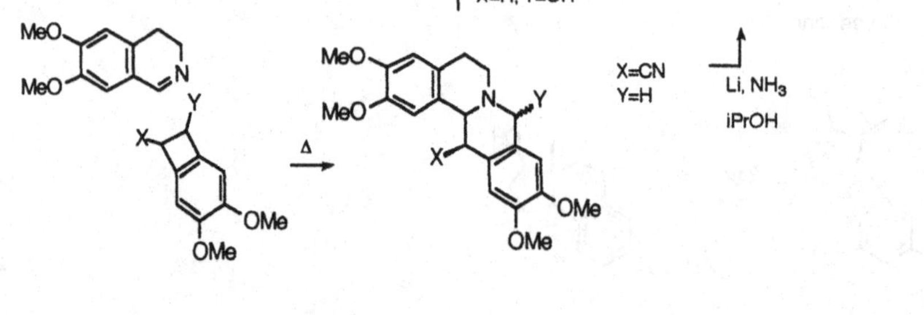

atisine

xylopinine

MeO BnO COOH Δ MeO BnO O O MeO HO N OMe

sendaverine

MeO O O Δ MeO O O MeO O O

O MeNH O NMe N H Me MeO O N COOMe MeO O O

physostigmine

7-Benzyloxy-6-methoxyisochroman-3-one. [Kametani, 1979]

4-Benzyloxy-5-methoxybenzocyclobutene-1-carboxylic acid (2.97 g) is heated at 170°C for 1 h under a nitrogen atmosphere. After cooling the product is chromatographed over silica gel to give the isochroman-3-one (1.24 g, 41.7%). This compound is a precursor of sendaverine.

N-Benzylidene-2-aminocyclobutenecarboxylic esters give 3,4-dihydroisoquinolines by way of electrocyclic opening and reclosure, and 1,5-hydrogen migration [Wessjohann, 1993].

E N R 500-550° 10^{-4} torr E N R H E N R H E N R

E = COOMe

50-68%

Benzocyclobutenediones undergo bisdecarbonylation and electrocyclic opening leading to bisketenes which can recyclize to phthalide carbenes [R.F.C. Brown, 1988; Scott, 1988].

-2 CO

$-CO_2$

:CH

(shown by isotope label)

The cycloethenologous Bergman rearrangement [Fritch, 1978] is a fascinating process. Its extension enables the synthesis of tetraalkynylcyclobutadiene complexes [Altmann, 1997].

Me_3Si CoCp

550°

$SiPr^i_3$

85%

Azetines serve as precursor of azadienes. When the nitrogen atom is substituted with an unsaturated side-chain the pyrolysis of azetines can lead directly to bicyclic azacycles [Jung, 1991].

δ-coniceine

5.5. RETRO-DIELS-ALDER REACTIONS

Although the Diels-Alder reaction cannot be treated in this book, the inclusion of its somewhat less well known reverse reaction [Ripoll, 1978; Ichihara, 1987] is proper. Retro-Diels-Alder reactions require higher temperatures, and they are better performed by the flow pyrolysis technique. Such conditions disfavor recombination of the dissociated components and minimize exposure of the products to high temperatures.

Highly reactive species which have been accessed by the retro-Diels-Alder reaction include: benzocyclopropene [Vogel, 1965], 2-cyclohexene-1,4-dione [D.D. Chapman, 1969], 3,4-dimethylenecyclobutene [H.-D. Martin, 1975], butatriene [Roth, 1975], pentatetraene [Ripoll, 1976, 1977], vinylidenecyclopentadiene [Spangler, 1972], alkylideneketenes [Clough, 1973; R.F.C. Brown, 1977], isobenzofuran [Wege, 1971; Wiersum, 1972], isoindole [Bornstein, 1972], and tetramethylsilylene [Barton, 1976].

COOMe, COOMe — 400°, 0.1 torr → + COOMe, COOMe

175° → +

425° → +

/ / — Δ → $CH_2=C=C=CH_2$

C — 700° → $CH_2=C=C=C=CH_2$ +

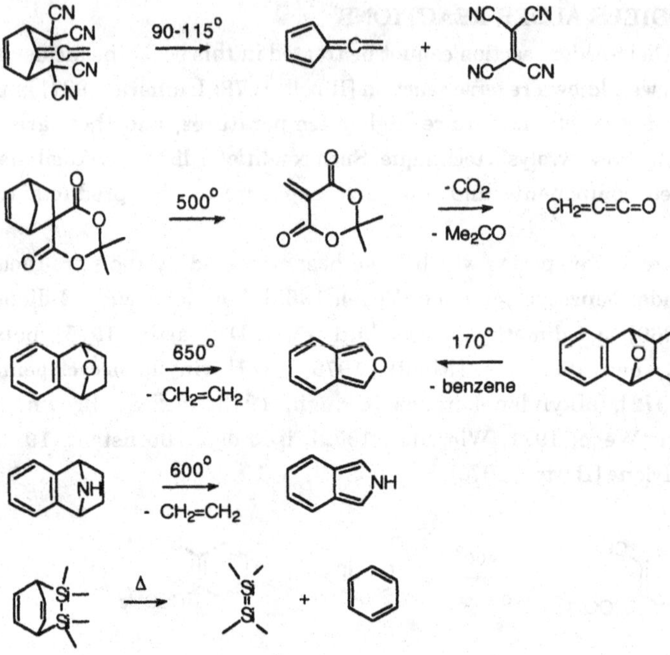

There are also quite a number of extremely unstable compounds which are accessible only by the FVP technique. Thioxoethanal thus generated [Bourdon, 1990] is confirmed to have an open-chain structure which is different from ethandithial. The latter prefers to exist in the cyclic dithiete form.

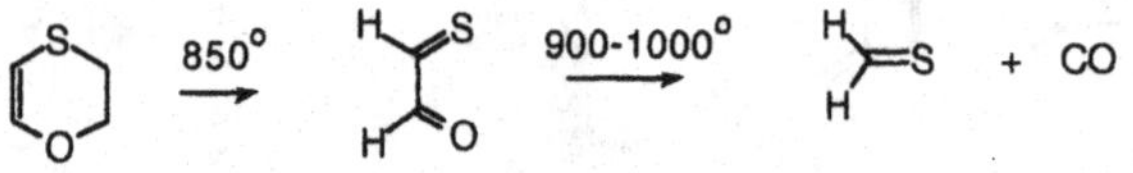

A primary enamine [Ripoll, 1980] and the parent sulfene [King, 1972; Fisher, 1973] are obtainable from thermal cracking processes.

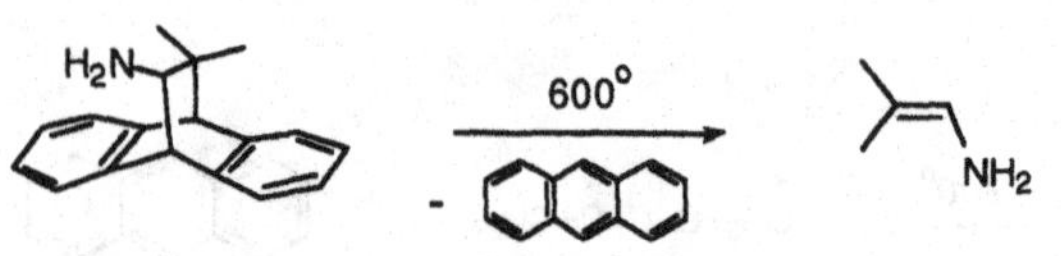

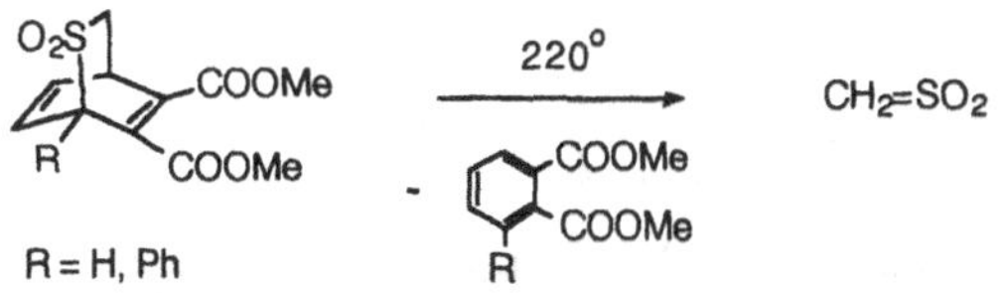

O_2S COOMe COOMe R; R = H, Ph — 220° → $CH_2{=}SO_2$ − (R, COOMe, COOMe)

The excellent geometry of certain bridged azo compound to undergo retro-Diels-Alder reaction that elimination of dinitrogen occurs at or below room temperature. A synthesis of semibullvalene [Paquette, 1970] by oxidation of the corresponding hydrazine took advantage of the situation.

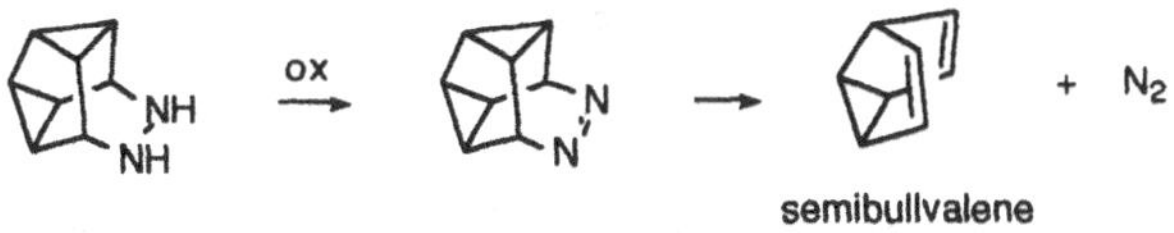

NH NH — ox → N N → + N_2

semibullvalene

Regarding synthetic applications to more traditional target molecules it may be mentioned an improved preparation [Stork, 1972] of the Nazarov ketoester and the release of *ar*-turmerone [Ho, 1974] involve thermolysis of 5-acyl-2-norbornene derivatives. Cyclic ketones fused to the norbornene skeleton also decompose to give conjugated cycloalkeneones which for various reasons are difficult to generate by other methods. Another advantage is the stereocontrol afforded by such a framework during the synthetic process. Theus the syntheses of methyl jasmonate [Ducos, 1973], maleimycin [Lee, 1992], terrein [Klunder, 1981], pentenomycin [Verlaak, 1982], pyrenolide-B [Asaoka, 1985], mycorrhizin-A [R.F.C. Brown, 1985] and laburnine [Arai, 1994] have been devised on such a basis.

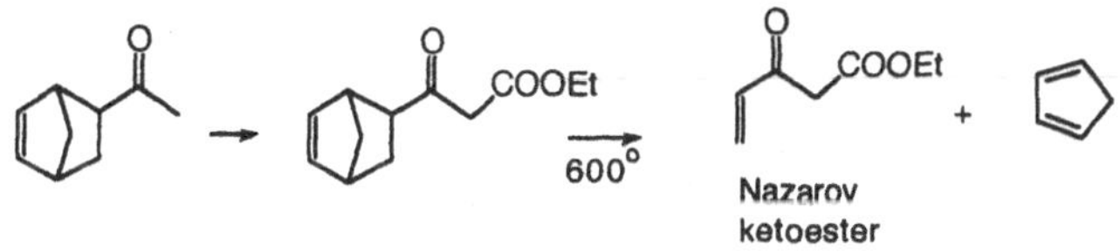

O → O COOEt — 600° → O COOEt +

Nazarov ketoester

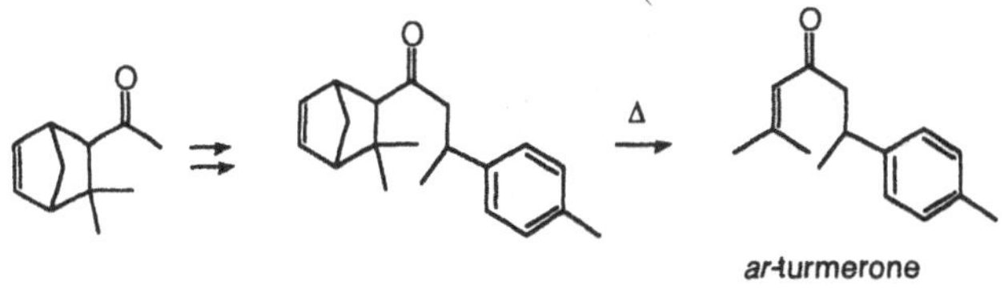

O ⇉ O — Δ → O

ar-turmerone

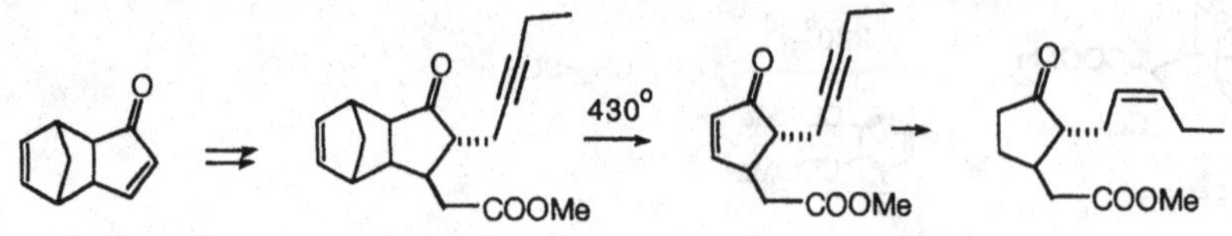

430°

COOMe

COOMe

COOMe

methyl jasmonate

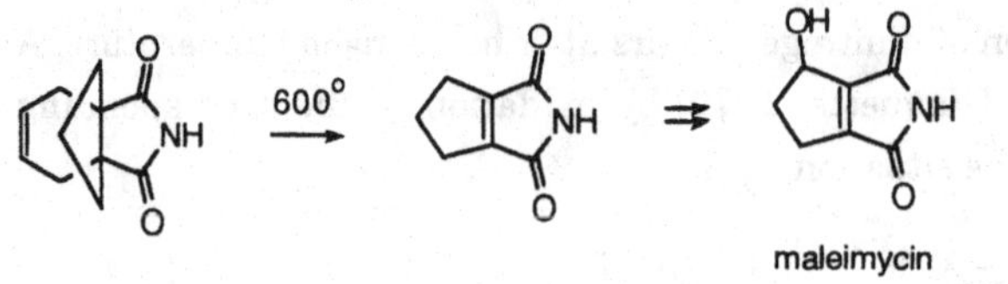

600°

NH

NH

OH

NH

maleimycin

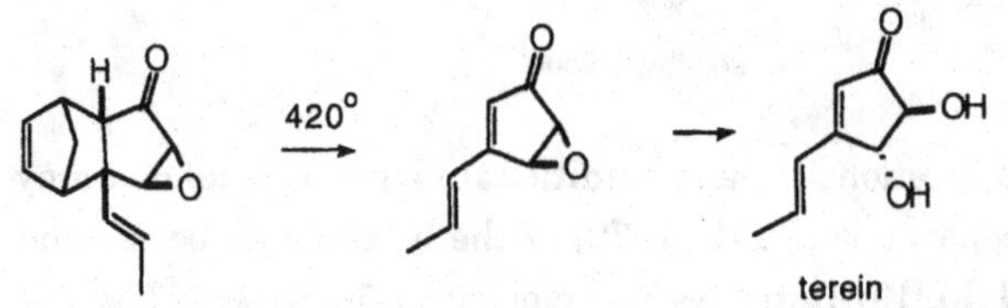

H

420°

OH

OH

terein

525°

OH

OH

OH

OH

OH

OH

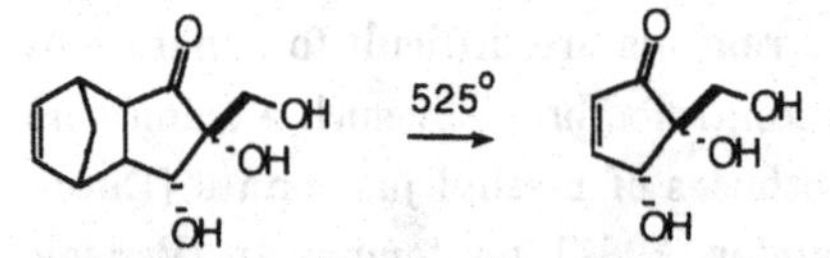

pentenomycin

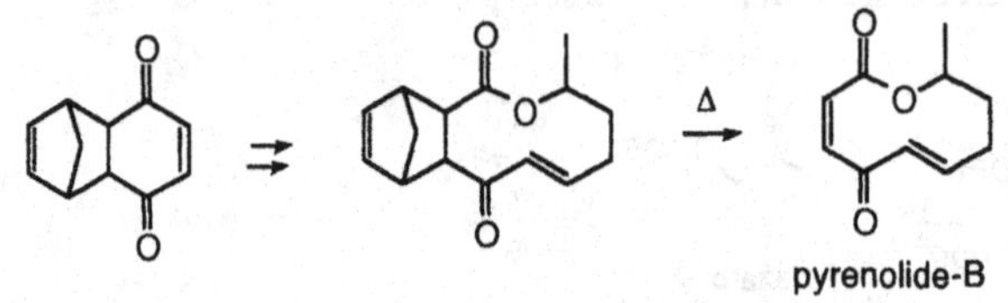

Δ

pyrenolide-B

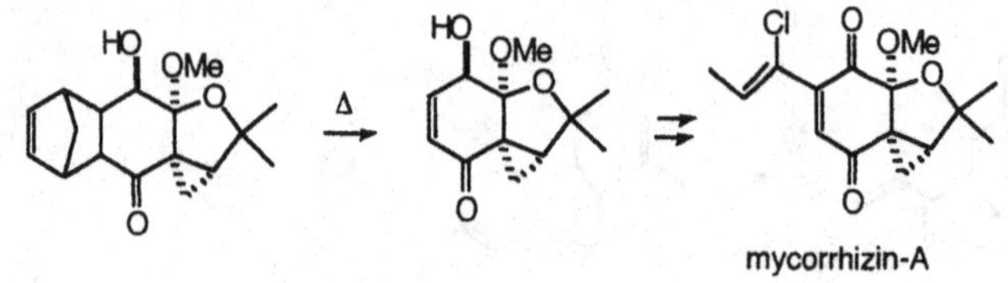

HO OMe

Δ

HO OMe

Cl OMe

mycorrhizin-A

450-500°

- C_5H_6

laburnine

It must be emphasized that a carbonyl group conjugated to the emergent alkene is not a necessity, although its presence should lower the reaction temperatures. Thus multifidene was liberated from a hydrocarbon precursor [Boland, 1978], and an approach to ipsenol [Haslouin, 1977] illustrates the possible service of many other modified norbornenes as synthetic intermediates.

450°

multifidene

450°

ipsenol

Elimination of cyclopentadiene from an azanorbornene derivative with a dienoyl substituent at the nitrogen atom paved the way to izidine compounds [Lasne, 1982]. The pyrolysis generated the required dienophile for an intramolecular Diels-Alder reaction.

450°

δ-coniceine

A useful preparation of 2-cyclopentenones consists of pyrolysis of 2-siloxynorbornenes and acid hydrolysis [Bloch, 1979].

600°

Terminal functionalization of a double bond can also be accomplished after derivatization with anthracene, with subsequent pyrolytic regeneration in the gas phase. A synthesis of methylenomycin-B [Siwapinyoyos, 1982] was carried out on the basis of this method.

COOMe ⇛ O —450°, - anthracene→ O

methylenomycin-B

Flash vacuum pyrolysis achieves the conversion of 1-naphthylmethyl propynoate into a γ-lactone [Anderson, 1990]. It involves a Diels-Alder reaction followed by elimination ethyne.

O O —750°, 0.02 torr→ O O → O O

A very useful synthetic tactic for the generation of cyclic dienes is to conduct a Diels-Alder/retro-Diels-Alder reaction tandem using dienes tethered by units which can become stable molecules (e.g., CO_2, HCN). The many applications to natural product synthesis include those of occidentalol [Watt, 1972], digitopurpone [Cano, 1983], α-yohimbine [S.F. Martin, 1985], reserpine [S.F. Martin, 1987], podophyllotoxin [D.W. Jones, 1989], staurosporine aglycone [Moody, 1990], involving decarboxylation, and an approach to furanoterpenes including paniculide-A [Jacobi, 1984a], ligularone [Jacobi, 1984b], and gnididione [Jacobi, 1984c], in which hydrogen cyanide was eliminated. The thermal condensation of 1,3-cyclohexadienes with dienophiles followed by elimination of ethylene constitutes a useful method for preparation of benzene derivatives, a modern application is in the elaboration of daunomycinone [Krohn, 1979].

Daunomycinone and Isodaunomycinone. [Krohn, 1979]

The mixture of cycloadducts from a tricyclic quinone and 1-methoxy-1,3-cyclohexadiene is heated at 140°C for 30 min. Products isolated by preparative thick layer chromatography are identified as daunomycinone (17% yield) and isodaunomycinone (32% yield).

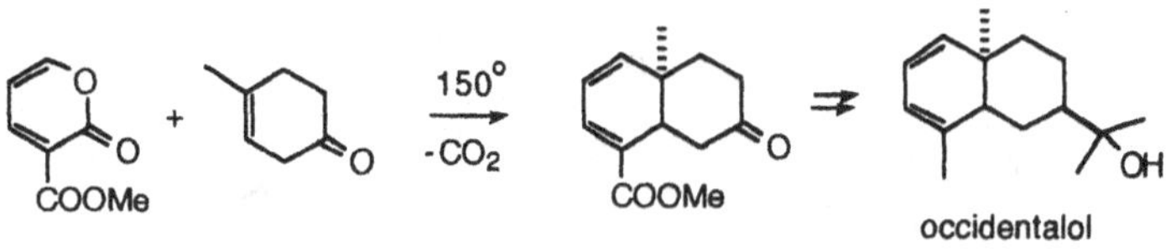

150°
$-CO_2$
COOMe
occidentalol

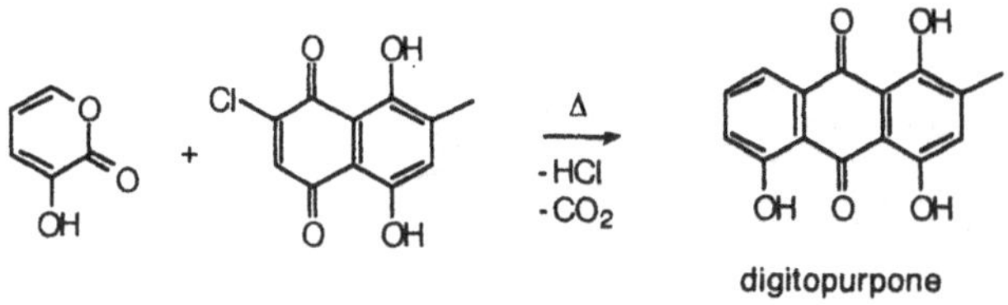

Δ
$-HCl$
$-CO_2$
digitopurpone

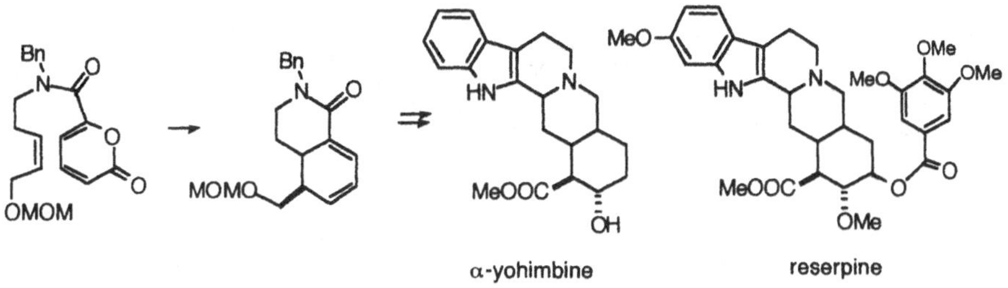

Bn
OMOM
MOMO
α-yohimbine
reserpine

Δ
$-CO_2$
podophyllotoxin

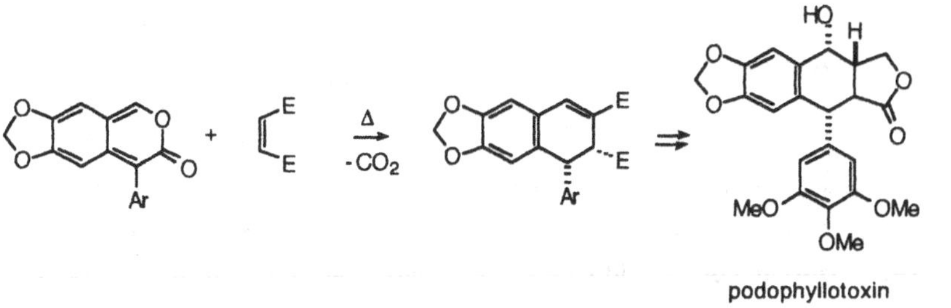

Δ
$-CO_2$
$(EtO)_3P$
staurosporine aglycone

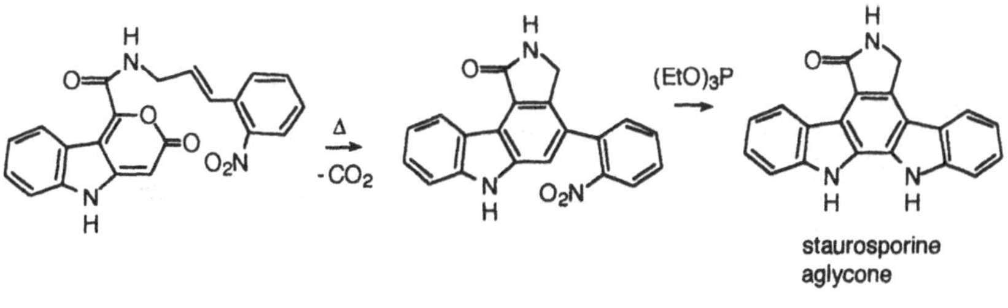

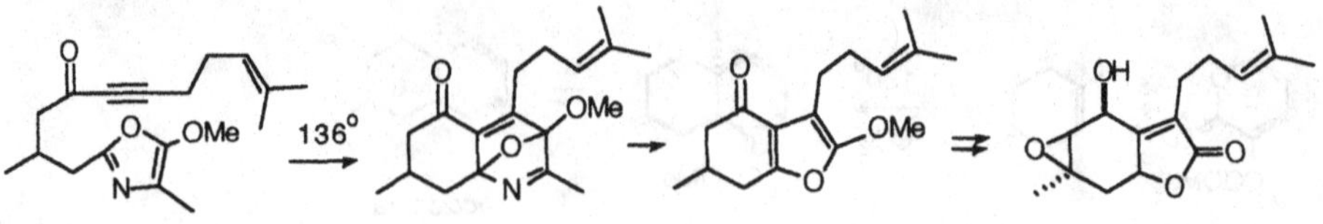

136°

paniculide-A

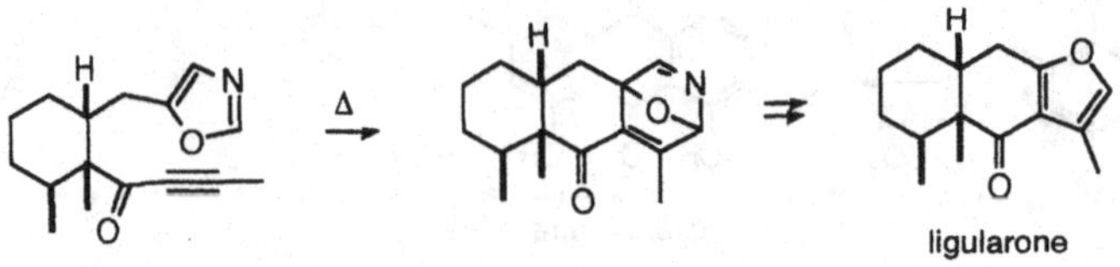

Δ

ligularone

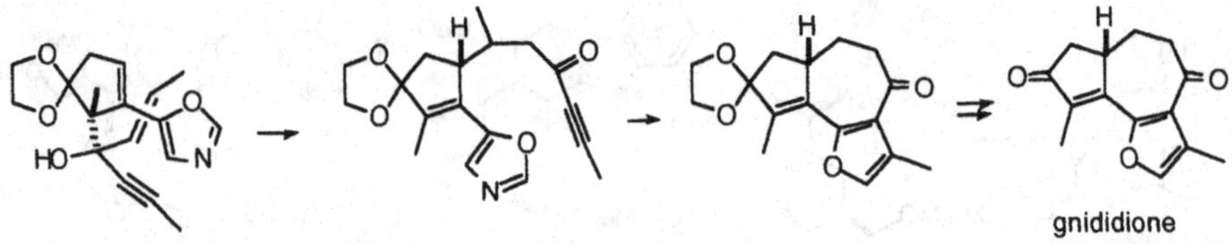

gnididione

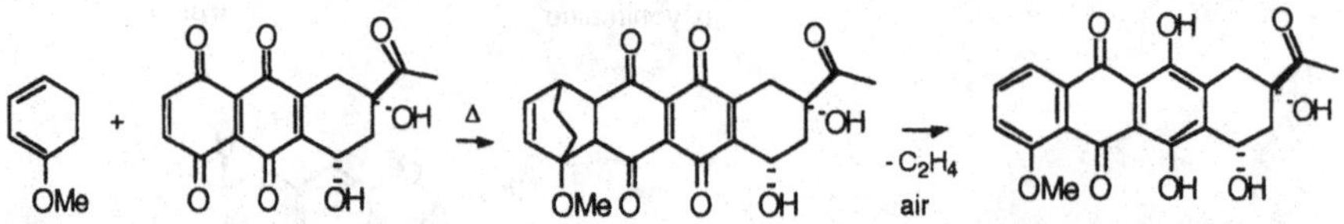

Δ

$-C_2H_4$

air

daunomycinone

The access of a methoxyisobenzofuran from α-pyrone and a benzo-oxabicyclo[2.2.1]heptadiene played a key role in another synthesis of daunomycinone [Kende, 1977]. The Diels-Alder adduct was decomposed at 140°C by way of two very favorable retro-Diels-Alder reactions: decarboxylation and elimination of benzene.

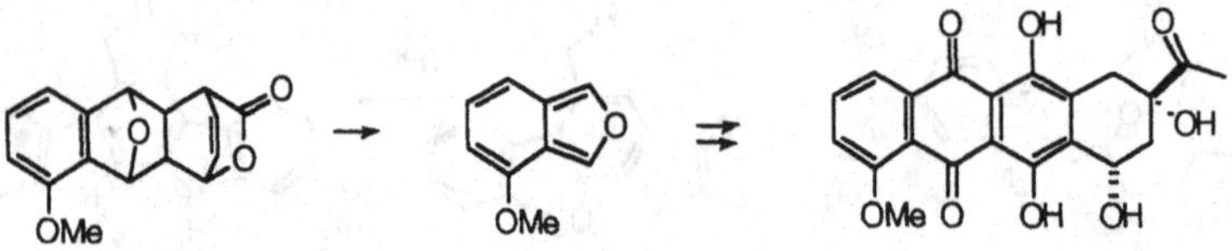

daunomycinone

5.6. THE ENE REACTION AND ITS REVERSION

The ene reaction [H.M.R. Hoffmann, 1969] unites two unsaturated centers while an allylic hydrogen in the ene component is being transferred to the end of the multiple bond of the enophile. It proceeds through the interaction of the ene-HOMO and enophile-LUMO. The synthetic potential of this concerted thermal process is narrower than those mentioned above, due to the relatively high temperatures required and several other factors, but some unique applications have evolved. Like the Diels-Alder reaction, the ene reaction is subject to catalysis by Lewis acids [Snider, 1980], when the enophile contains a polarizable group.

The ene reaction might become an adverse side reaction to destroy a molecular skeleton constructed by another thermal reaction. For example, the pyrolytic scission of the cyclobutane unit of pinanols not only generates linalool, but also substituted cyclopentanols arising from an ene reaction of the primary product [Ikeda, 1936; Strickler, 1967]. In this case the ene reaction cannot be avoided because its energy of activation is lower than that required for the ring cleavage step. However, the plinol has been used as starting material for a synthesis of cyclonerodiol [Nozoe, 1971].

The intramolecular version of the ene reaction is proven much more useful synthetically [Oppolzer, 1978]. Acorane sesquiterpenes [Oppolzer, 1977], isocomene [Oppolzer, 1979], modhephene [Oppolzer, 1981b,c] have been synthesized in a way that relied on the ene reaction to create a new ring with stereoselective establishment of two contiguous asymmetric centers, one of which bearing a methyl group.

Pyrolysis of Ethyl 2-(1-Cyclohexenyl)-5-hexenoate. [Oppolzer, 1977]

The ester (10 g, 0.45 mol) is dissolved in dry toluene (54 mL) is heated in a Pyrex ampoule at 290°C for 72 h. Chromatography of the evaporated reaction mixture (8 g) gives the "trans" spiro ester (2.54 g, 25%) and the "cis" spiro ester (4.31 g, 43%).

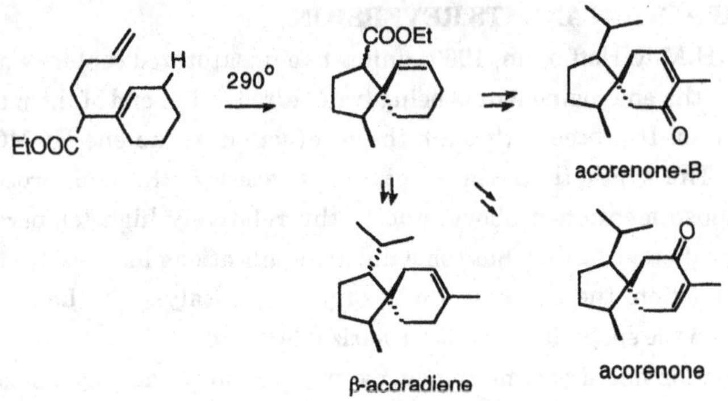

COOEt

H

290°

EtOOC

acorenone-B

β-acoradiene

acorenone

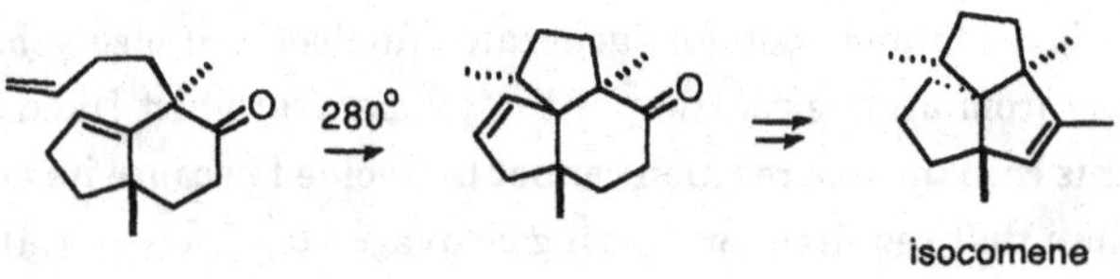

280°

isocomene

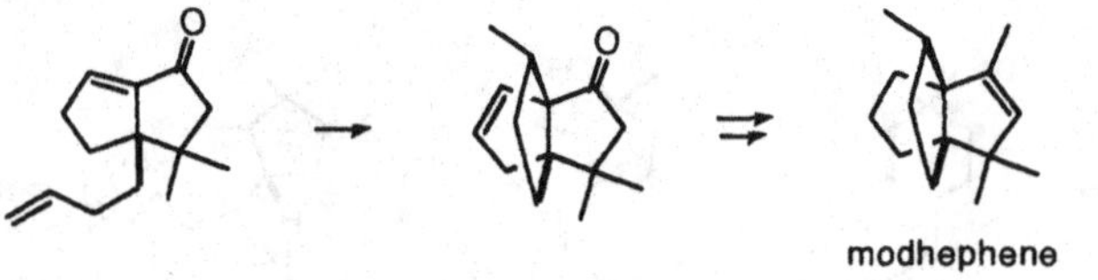

modhephene

The access to (+)-α-allokainic acid [Oppolzer, 1984] demonstrates the mild conditions (-35°C) of catalyzed reactions and an effective 1,5-stereoinduction by a chiral auxiliary as well as template effect. A 90% diastereomeric excess was observed, and the trans:cis ratio >95:5.

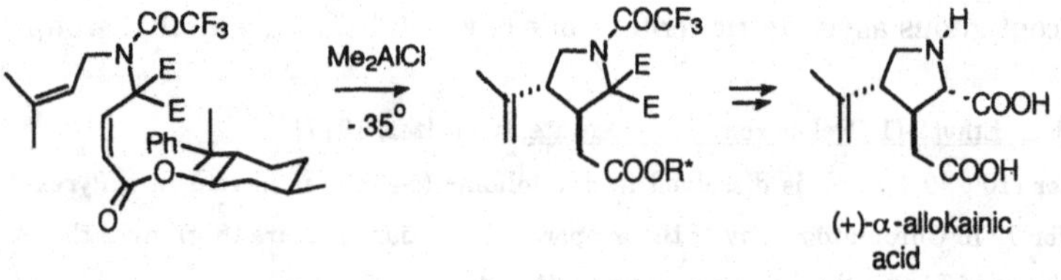

COCF$_3$

Me_2AlCl

- 35°

COOR*

COOH

(+)-α-allokainic acid

The intramolecular ene reaction of a 1,6-enyne served to form a methylenecyclopentane precursor of the prostaglandins [Stork, 1976]. By this reaction the original double bond was shifted tothe required position between $C_{(13)}$ and $C_{(14)}$. The suprafacial nature of the reaction pathway also rendered the *(E)*-configuration intact.

Such a stereochemical consequence is generally recognized and it formed the basis for an elegant route to chiral acetic acid [Townsend, 1975].

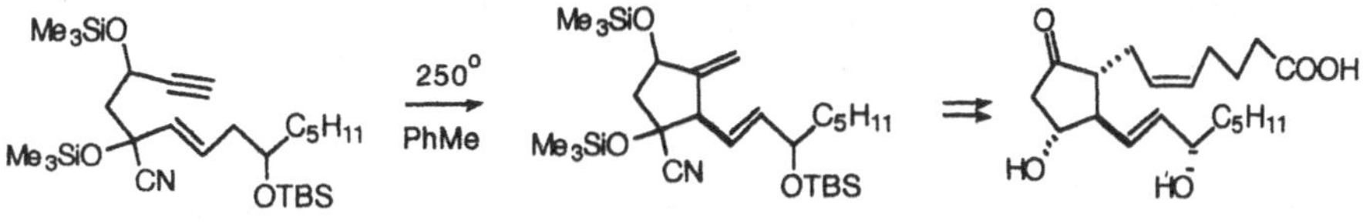

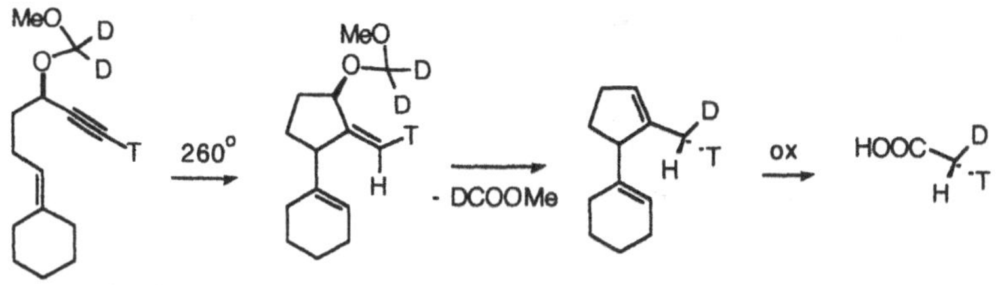

It is of interest to note the the regiochemical difference between an ene reaction effected by FVP and catalyzed with palladium(II) acetate [Trost, 1985].

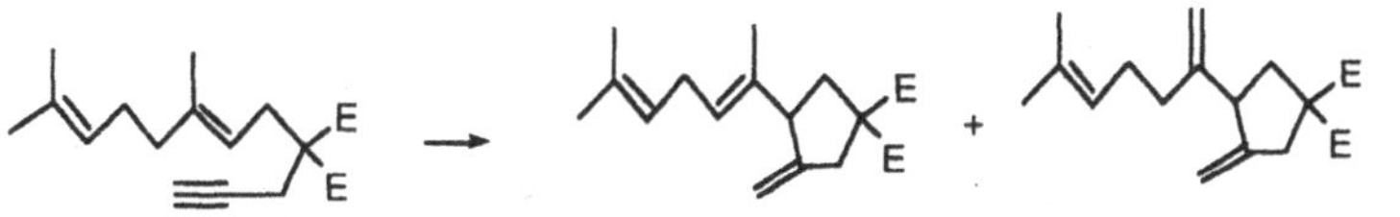

FVP	(83%)	1	:	0
$Pd(OAc)_2$	(80%)	1	:	16

1,2-Stereoinduction by a substituent at the other allylic position of the ene component effectively enabled the elaboration of a steroid CD-ring synthon [Takahashi, 1988]. For the synthesis of (+)-estrone, the ene reaction proceeded via an endo transition state and in tandem with a Claisen rearrangement. Thus a C-seco intermediate was rapidly constructed, which on simple manipulation afforded a precursor of the hormone [Mikami, 1990b].

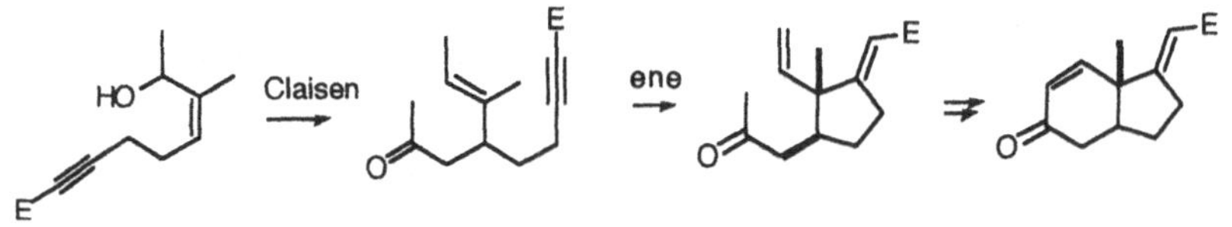

E = COOMe

Far less obvious in skeletal transformation by the ene reaction are found in the synthesis of sesquiterpenes which contain a decalin nucleus [Wender, 1980a,b, 1982; Williams, 1980b]. In these cases the intramolecular ene reaction occurred from 10-membered ring intermediates which were formed by pyrolysis of linear tricarbocycles composed of 6/4/4 subunits. The terpene candidates include calameon, β-eudesmol, atractylon, isoalantolactone, and warburganal.

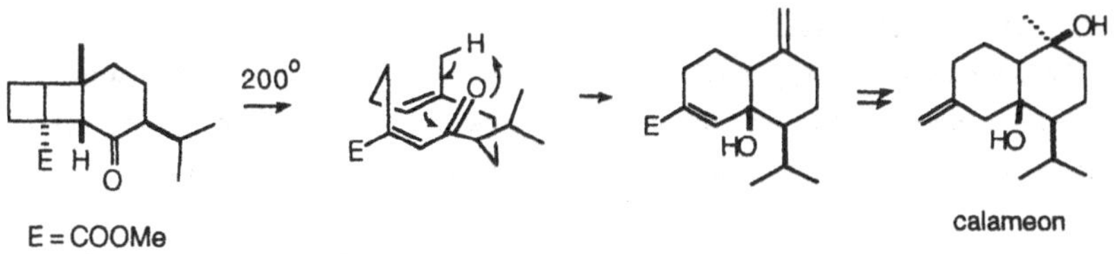

E = COOMe

calameon

The Conia reaction is a special intramolecular ene reaction which features an enol component. Under selected conditions the reverse reaction is much less favorable, therefore even if the enol content from an ordinary ketone (the "ene" component) is small the Conia reaction can be of preparative value [Conia, 1975]. Significantly, condensed rings, spirocycles, and bridged systems can be created. In a

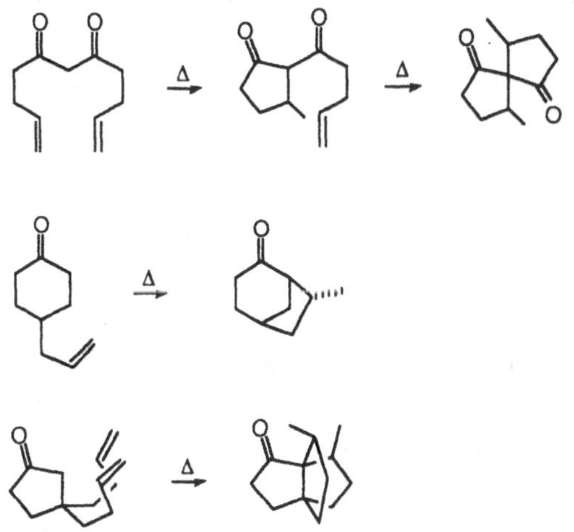

synthesis of modhephene [Schostarez, 1981] the propellane skeleton was formed by this reaction. Note that the enophile partner was an alkyne chain because the alkene analogue furnished an epimer at the new stereocenter that bears the methyl group and it is impossible to readjust its configuration.

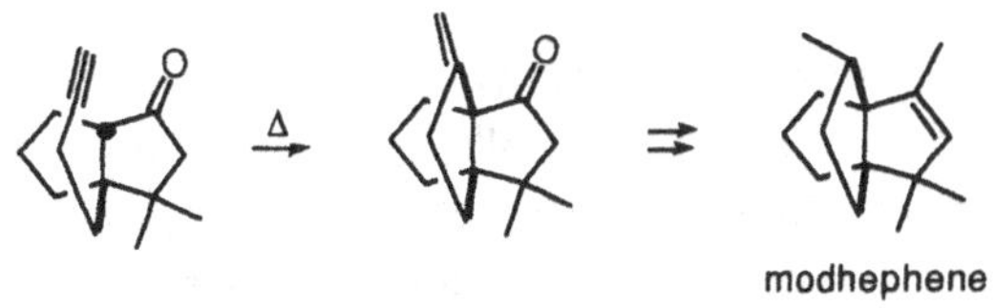

modhephene

cis,cis-8,9-Dimethyl-[3.3.3]propellan-2-one. [Drouin, 1980]

A solution of 3,3-di(but-3-en-1-yl)cyclopentanone (100 mg) in decalin (0.9 mL) is divided into two portions and sealed in Pyrex ampoules. The ampoules are heated at 335°C for 2 h. The product is isolated from silica gel chromatography and identified as a meso ketone (50 mg, 50%).

Certain aldehydes act as enophiles, and by virtue of the ene reaction such molecules are legitimized as sources of homoallylic alcohols. A more important aspect is the attendant diastereoselection or diastereofacial selection because the carbon chain assembly is given excellent control. Moreover, by using α-oxy aldehydes such control can be finely tuned on judicious admittance of a Lewis acid with preference for either monodentate or bidentate ligands. Thus it has been demonstrated that the steroid side-chain having a configuration characteristic of the ecdysones or the brassinosteroids can be extended from a *(Z)*-$\Delta^{17(20)}$ precursor [Mikami, 1990a, 1992].

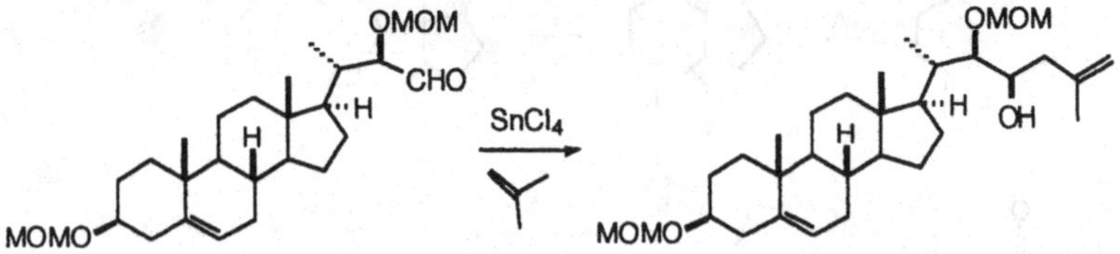

OMOM CHO H H MOMO SnCl4 OMOM H OH H MOMO

Some acylnitroso compounds undergo intramolecular ene reaction to give *N*-hydroxylactams. When the acylnitroso substrates are generated thermally a tandem process gives rise to the heterocyclic products. Construction of the hydroindolone skeleton of mesembrine has been reported [Keck, 1982].

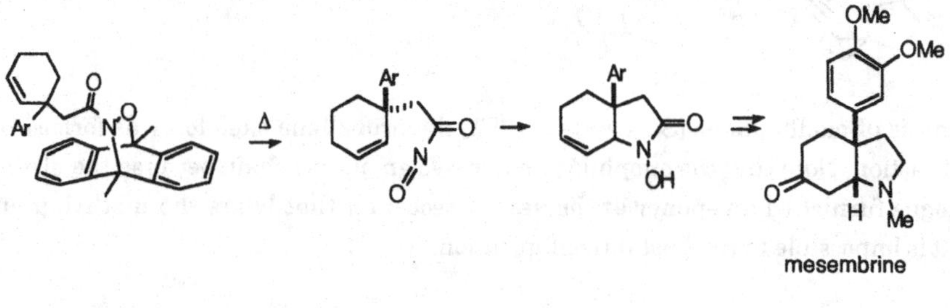

O Ar N O Δ Ar O N O Ar O N OH OMe OMe O N H Me

mesembrine

The retro-ene reaction also proceeds thermally. Decomposition of *cis*-2-alkyl-1-alkenylcyclopropanes in such a manner to give *(Z)*-alkenes is favorable due to strain relief. Thus the transformation of 2-carene (in thermal equilibrium with 3-carene) to *trans*-2,8-menthadiene afforded a useful intermediate for the synthesis of qinghaosu [Ravindranathan, 1990], and deconvolution of 6-vinylbicyclo[3.1.0]hexan-2-one to give 3-(1-propenyl)-2-vinylcyclopentanone paved the way to sarkomycin [Hudlicky, 1980a].

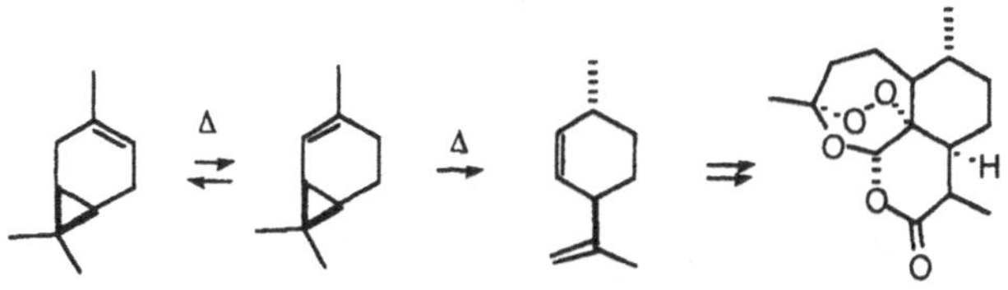

3-carene 2-carene qinghaosu

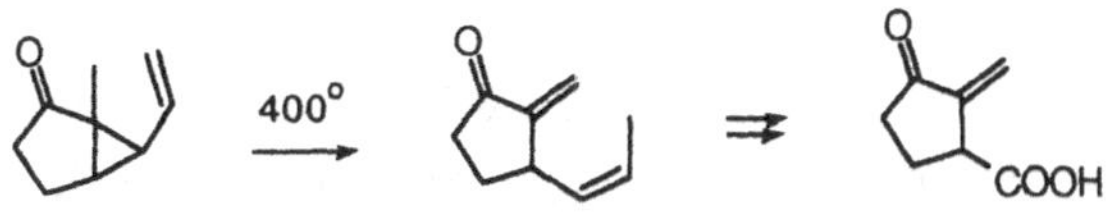

sarkomycin

A 2-alkenylcyclopropyl carbinol has been converted to a building block in an elaboration of the cecropia juvenile hormone JH-I [Corey, 1970]. When 2,2-dimethylcyclopropyl cyanide was transformed into the cyclopropyl *n*-hexyl ketone, followed by pyrolytic ring cleavage and ozonolysis, the precursor of of dihydrojasmone was obtained [Ho, 1977].

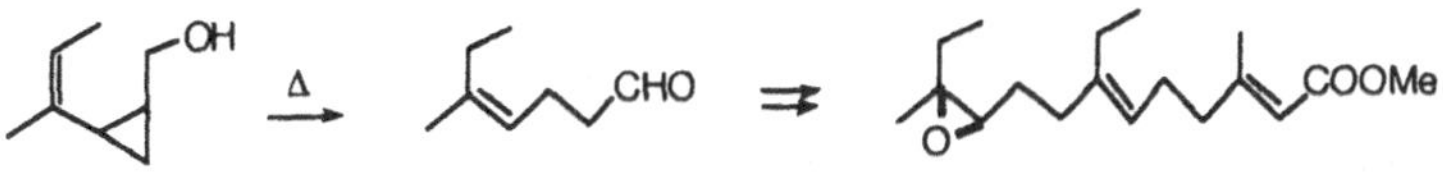

juvenile hormone JH-1

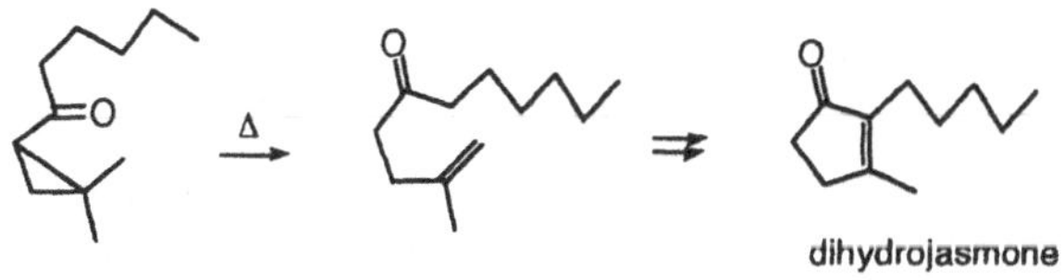

dihydrojasmone

2-[(*Z*)-2-pentenyl]cyclopent-2-enone, a well-known precursor of *cis*-jasmone and methyl jasmonate, may be prepared from a spirocyclic ketone [Bahurel, 1974]. It is also possible to couple the retro-ene reaction to a retro-Diels-Alder reaction in tandem, permitting the use of a different substrate [Ho, 1996].

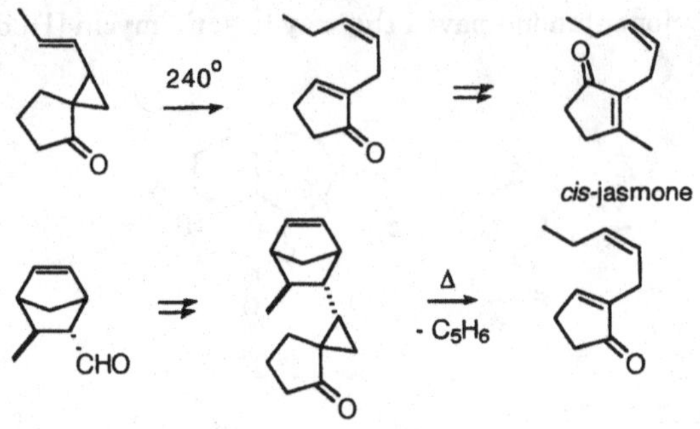

cis-jasmone

The pyrolytic formation of allyl aroates from the diallyl acetals of araldehydes [Ho, 1975] and of 3-silacyclobutenes from diallylsilanes [Block, 1978] involves elimination of propene by a retro-ene reaction. In the case of vinyldiallylphosphine the pyrolysis proceeds with retro-ene reaction, electrocyclization and dehydrogenation, resulting in phosphabenzene [Le Floch, 1993].

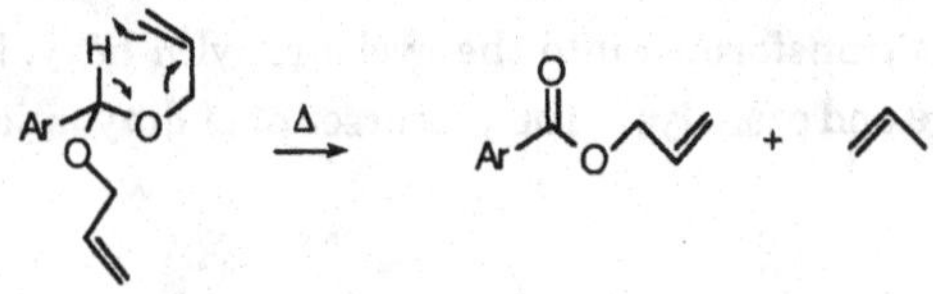

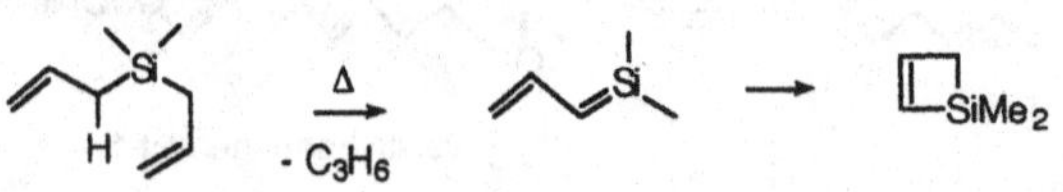

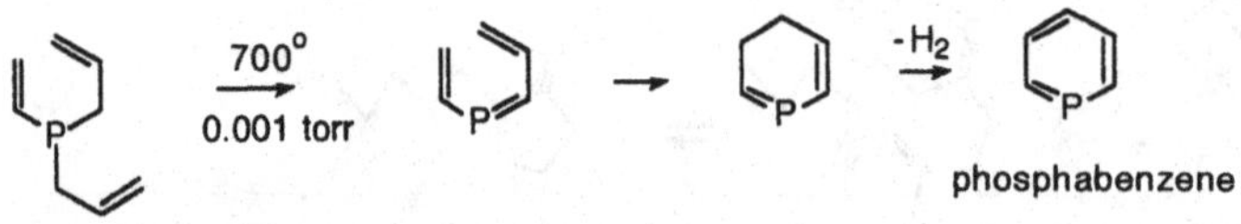

phosphabenzene

A retro-ene reaction may generate structures which are thermally labile. Useful transformations include the Diels-Alder reaction and Cope rearrangement, as exemplified by the formation of a dehydroindolizidinone [Earl, 1982] and a macrocyclic ketone [Karpf, 1977], respectively.

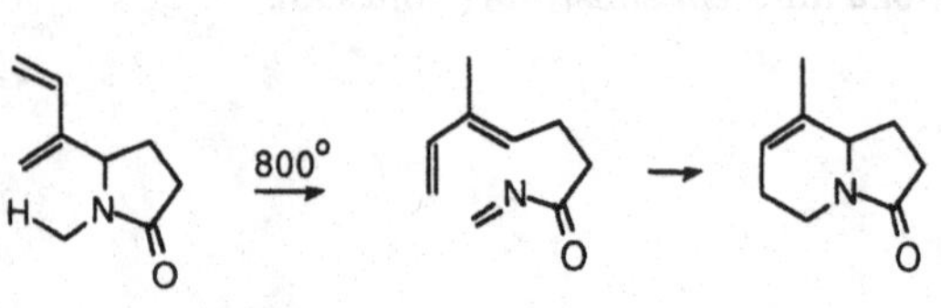

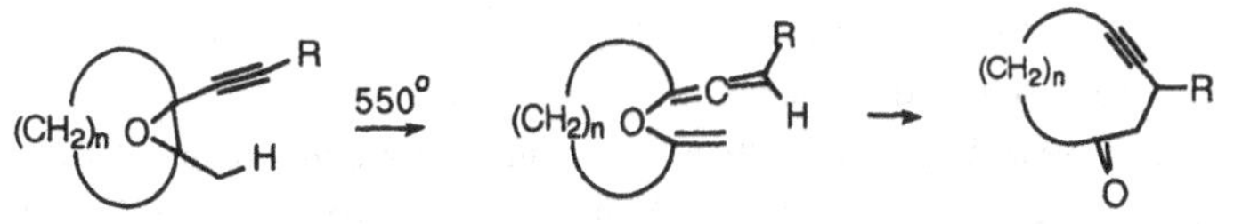

5.7. REARRANGEMENTS

Many thermal rearrangements are useful for the establishment of molecular skeletons from more readily accessible species. Because of the existence of extensive reviews on sigmatropy such as the Cope, Claisen, and Wittig rearrangements, their discussion is largely omitted.

5.7.1. 1,2-Rearrangements

The most useful 1,2-rearrangements that are promoted thermally involve the transformation of an alkyne to an alkenylcarbene [R.F.C.Brown, 1974], with intramolecular insertion to follow, resulting in the formation of a new C-C bond. An excellent example is the quantitative conversion of 7,10-diethynylfluoranthine to corannulene at 1000°C in a high vacuum [Scott, 1991]. (Note that a silylated compound also gives the same product [Zimmermann, 1994].)

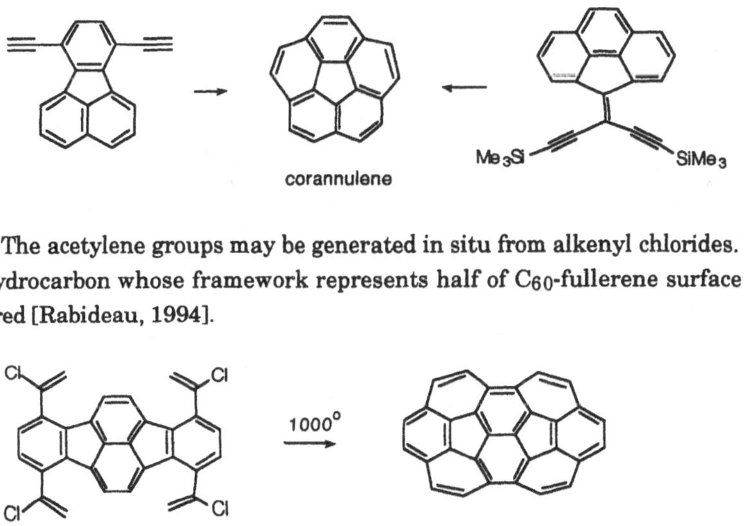

corannulene

The acetylene groups may be generated in situ from alkenyl chlorides. Thus a C_{30} hydrocarbon whose framework represents half of C_{60}-fullerene surface can be prepared [Rabideau, 1994].

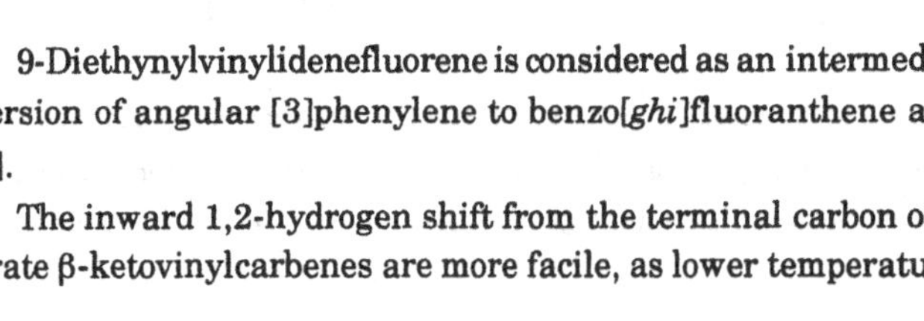

9-Diethynylvinylidenefluorene is considered as an intermediate in the pyrolytic conversion of angular [3]phenylene to benzo[*ghi*]fluoranthene at 1000°C [Matzger, 1997].

The inward 1,2-hydrogen shift from the terminal carbon of ethynyl ketones to generate β-ketovinylcarbenes are more facile, as lower temperatures are required. As

such carbene species readily undergo intramolecular insertion to form cyclopentenones, incorporation of this step to the synthesis of several terpenes was investigated, resulting in the novel elaboration of modhephene [Karpf, 1981], albene [Manzardo, 1983], clovene [Ackroyd, 1984]. The approach to $\Delta^{9(12)}$-capnellene [Huguet, 1982] employed the method on two occasions.

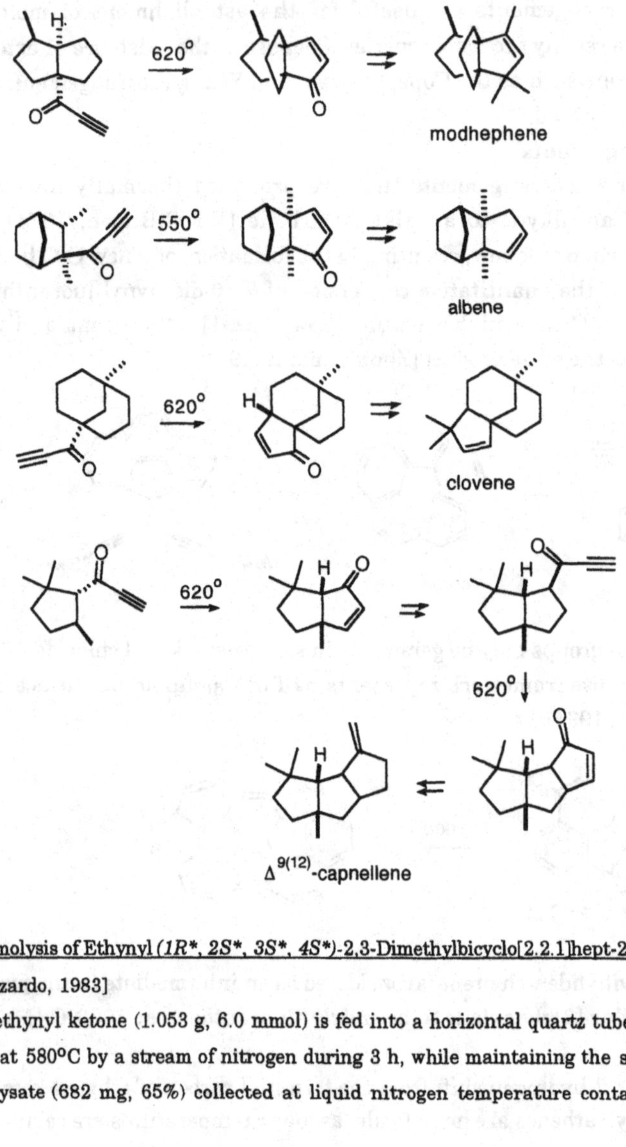

620°

modhephene

550°

albene

620°

clovene

620°

620°

$\Delta^{9(12)}$-capnellene

Thermolysis of Ethynyl *(1R*, 2S*, 3S*, 4S*)*-2,3-Dimethylbicyclo[2.2.1]hept-2-yl Ketone.

[Manzardo, 1983]

The ethynyl ketone (1.053 g, 6.0 mmol) is fed into a horizontal quartz tube (1.6x43 cm) which is held at 580°C by a stream of nitrogen during 3 h, while maintaining the system at 14 torr. The pyrolysate (682 mg, 65%) collected at liquid nitrogen temperature contains albenone and two cyclopentenone isomers in a ratio of 91:6:3. These isomers originate from carbene insertion at the endo-methyl group and the bridgehead C-H bond.

2-Ethynylphenol and its *C*-trimethylsilyl derivative form benzofurans by virtue of rearrangement and carbene insertion into the O-H bond [Bloch, 1981; Barton, 1985].

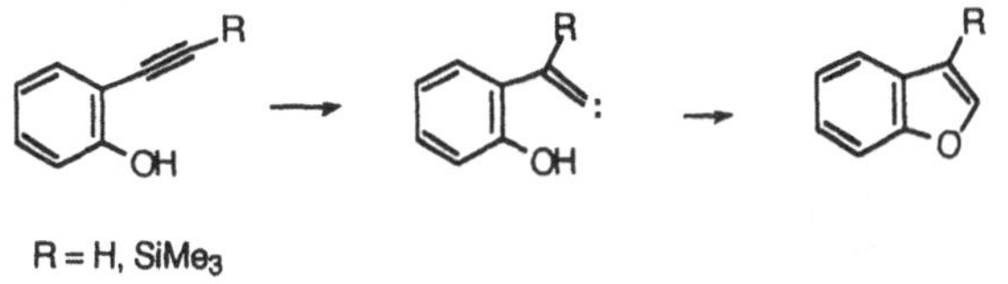

R = H, $SiMe_3$

Some internal alkynes also undergo the pyrolytic rearrangement. Thus the formation of 1-ethynylacenaphthylene and 3-ethynylacenaphthylene from a FVP experiment of phenanthrene-1,2-dicarboxylic anhydride [R.F.C. Brown, 1993a] likely proceeded via the 1,2-phenanthryne, from which a ring contraction to give a vinylidene carbene followed.

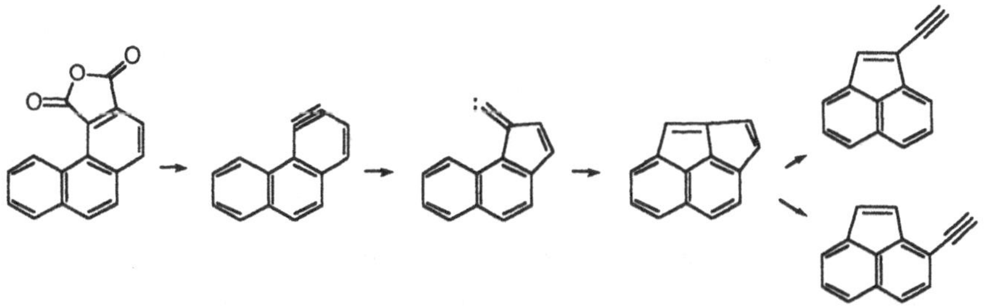

3-Phenylnaphthalene-1,2-dicarboxylic anhydride was shown to give indeno[2,1-*a*]indene in 95% yield [R.F.C. Brown, 1993b], which is also a minor product (19%) in the pyrolysate of 1,4-diphenylbutadiyne.

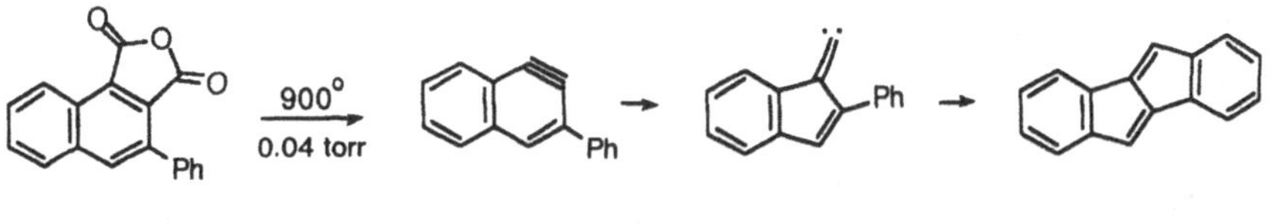

900° 0.04 torr

A formal 1,2-rearrangement of two C-C bonds was observed in an apparent attempt to generate 3-methylenecyclopropene in 56% yield [Mühlebach, 1993]. Previously, the formation of oxirene was also redirected to a dibenzocycloheptatrienecarboxaldehyde [Lewars, 1976].

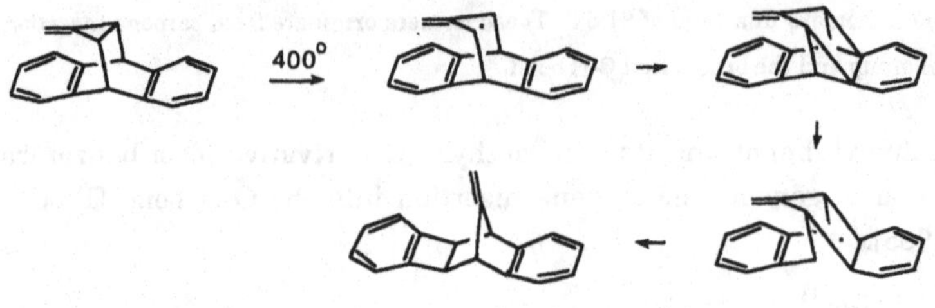

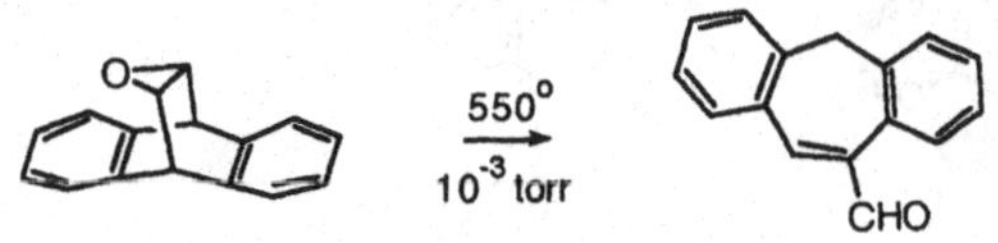

A bowl-shape hydrocarbon $C_{32}H_{12}$ has been acquired on FVP of a tribromide [Hagen, 1997]. It formation probably involved homolysis of the Ar-Br bonds, hydrogen atom migration, and cyclization.

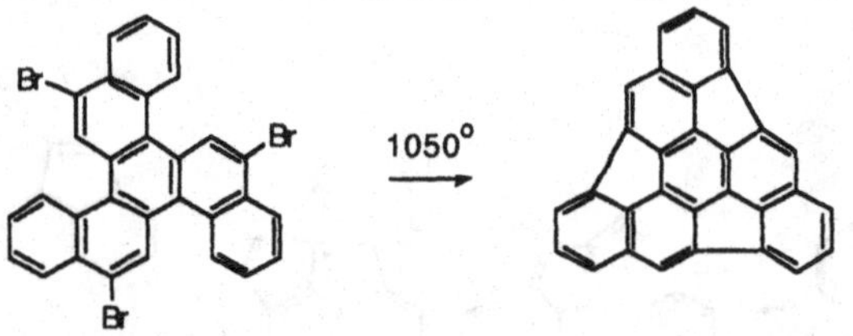

A smooth conversion of indigo under FVP conditions to dibenzonaphthyridindione [Haucke, 1995] involves α-cleavage of the carbonyl units and reclosure of two rings which constitutes formal 1,2-rearrangements.

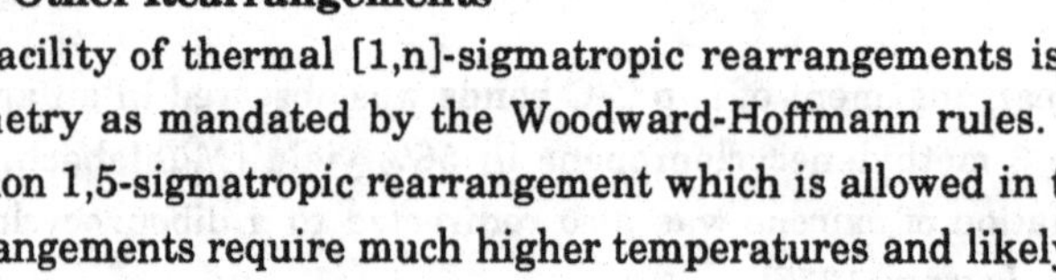

5.7.2. Other Rearrangements

The facility of thermal [1,n]-sigmatropic rearrangements is correlated to orbital symmetry as mandated by the Woodward-Hoffmann rules. Contrary to the more common 1,5-sigmatropic rearrangement which is allowed in the ground state, 1,3-rearrangements require much higher temperatures and likely proceed via diradical intermediates. Special among these is the vinylcyclopropane-to-cyclopentene rearrangement [Hudlicky, 1985] which benefits from relief of ring strain of the substrates. The design in the synthesis of several natural products incorporating such a step was well rewarded: *cis*-sativenediol [McMurry, 1976], longifolene [Schultz, 1985], zizaene [Piers, 1982], hirsutene [Hudlicky, 1980b], isocomene [Short, 1983, Ranu, 1984], retigeranic acid-A [Hudlicky, 1989], α-vetispirene [Yan, 1982], β-vetivone [Piers, 1977; Barnier, 1984], and aphidicolin [Trost, 1979].

Preparation of 1-Methyl-4,5,6,7-tetrahydroindan-4-one. [Piers, 1982]

A solution of 3-(1-methylcyclopropyl)-2-cyclohexen-1-one (12.3 g) in hexane (30 mL) is added dropwise to a vertically held Pyrex tube (1.2x40 cm) which is filled with glass helices and placed in a tubular furnace held at 450°C over 1.5 h, while a slow stream of nitrogen sweeps through the system. After the addition the column is rinsed by hexane (30 mL). The condensate collected at -78°C is evaporated and treated with NaOMe in methanol at 0°C for 1 h to convert the unconjugated ketone in the mixture to the conjugated ketone (10.7 g, 87%).

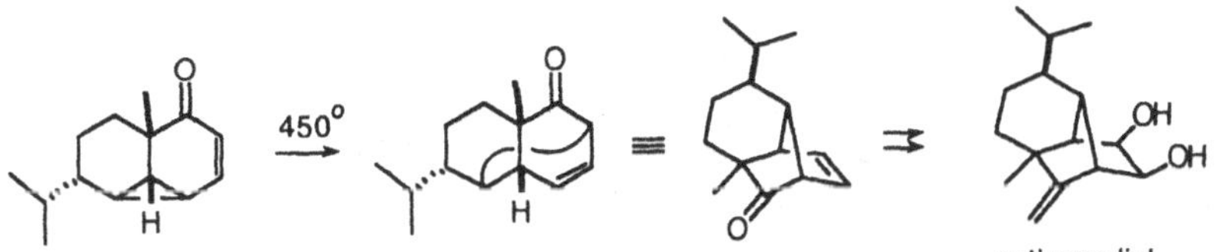

450°

sativenediol

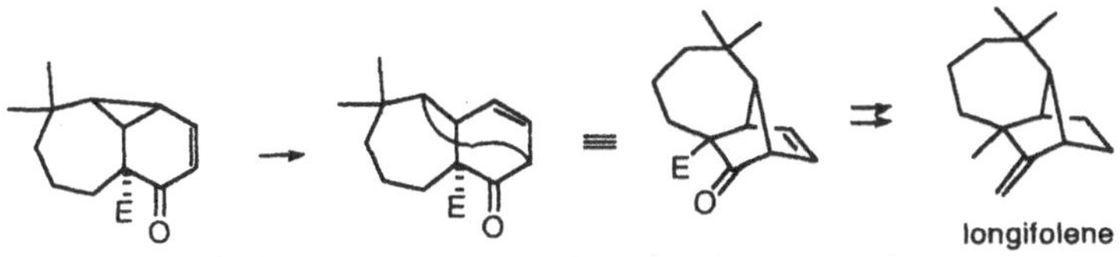

longifolene

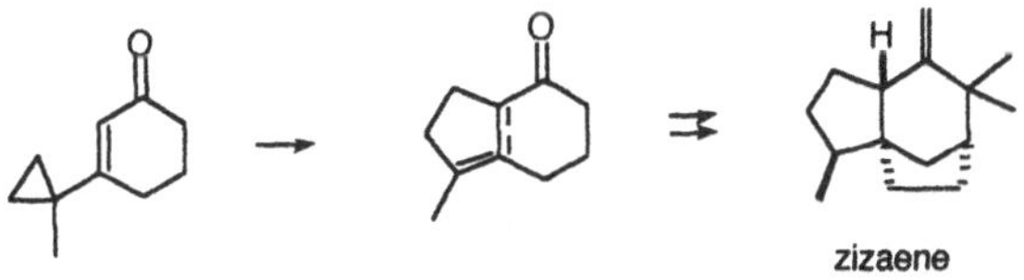

zizaene

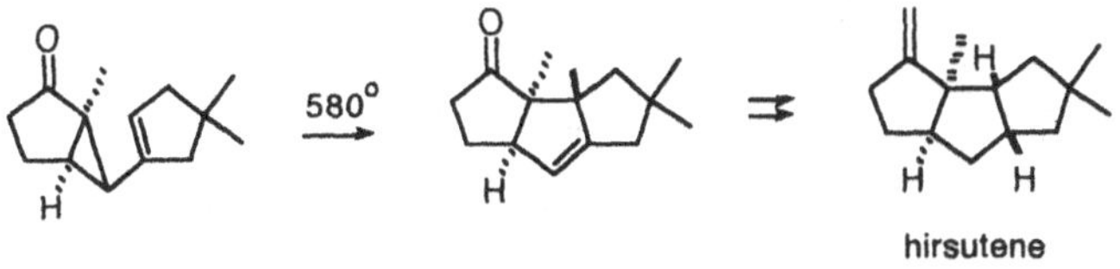

580°

hirsutene

EtOOC → EtOOC → isocomene

585° PbCO$_3$ EtOOC → EtOOC → HOOC retigeranic acid-A

OMe → OMe → α-vetispirene

OSiMe$_3$ H → (1 : 3) → β-vetivone

OSiMe$_3$ OSiR$_3$ → OSiMe$_3$ OSiR$_3$ → OSiR$_3$ → β-vetivone

Me$_3$SiO 610° → Me$_3$SiO → OH OH H HO HO aphidicolin

Norbornadiene undergoes thermal isomerization, to cycloheptatriene at 475°C and thence to toluene at higher temperatures (>500°C) [Woods, 1958; Klumpp, 1963]. Thus the isomerization makes cycloheptatriene readily available for many interesting investigations.

475° >500°

The conversion of β-himachalene to cuparene [Subba Rao, 1968] is quite remarkable as optical activity was retained to the extent of 20%.

490°

(+)-β-himachalene (+)-cuparene

The thermal behavior of diazabasketene is unusual. It splits off hydrogen cyanide to form azocine [McNeil, 1971], probably by way of two retro-Diels-Alder reactions interjected by a 1,3-bond migration and followed by electrocyclic reversion.

560° 10^{-6} torr -HCN

While the intramolecular Diels-Alder reaction was preempted by a 1,3-acyl shift of an enol ester, the β-diketone product underwent partial cycloaddition via the enol form [Shea, 1982]. The transformation failed under static conditions.

>360°

Spirocyclic isoxazolidines, obtained from nitrones and methylenecycloalkanes by the 1,3-dipolar cycloaddition, undergo thermal rearrangement to give azacyclo-alkanones [Brandi, 1993]. For volatile isoxazolidines the rearrangment is best carried out under FVP conditions, although others furnish products in the condensed phase and in acceptable yields, as exemplified by the reaction leading to an ipecac alkaloid intermediate [Brandi, 1988]. The process can be rationalized as involving N-O bond homolysis, formation of the oxoalkyl radical and ring closure.

MeO MeO Δ MeO MeO 61%

The thermal transformation of a polysilabicyclic allene involves a formal 1,3-dyatropic isomerization [Petrich, 1993].

520° N_2

Monographs and Reviews

Brown, R.F.C. (1980) *Pyrolytic Methods in Organic Chemistry*, Academic Press: New York.

Karpf, M. (1986) *Angew. Chem. Int. Ed. Engl.* **25:** 414.

McNab, H. (1996) *Contemp. Org. Synth.* **3:** 373.

Seybold, G. (1977) *Angew. Chem. Int. Ed. Engl.* **16:** 365.

Wiersum, U.E. (1984) *Aldrichimica Acta* **17:** 31.

References

Aalbersberg, W.G.L.; Vollhardt, K.P.C. (1979) *Tetrahedron Lett.* 1939.

Abramovitch, R.A. (1991) *Org. Prep. Proc. Int.* **23:** 683.

Ackroyd, J.; Karpf, M.; Dreiding, A.S. (1984) *Helv. Chim. Acta* **67:** 1963.

Adam, W.; Salgado, V.O.N.; Peters, E.-M.; Peters, K.; von Schnering, H.G. (1993) *Chem. Ber.* **126:** 1481.

Anderson, M.R.; Brown, R.F.C.; Coulston, K.J.; Eastwood, F.W.; Ward, A. (1990) *Aust. J. Chem.* **43:** 1137.

Arai, Y.; Kontani, T.; Koizumi, T. (1994) *J. Chem. Soc. Perkin Trans. 1* 15.

Asaoka, M.; Naito, S.; Takei, H. (1985) *Tetrahedron Lett.* **26:** 2103.

Altmann, M.; Roidl, G.; Enkelmann, V.; Bunz, U.H.F. (1997) *Angew. Chem. Int. Ed.*

Engl. **36:** 1107.
Bahurel, Y.; Cottier, L.; Descotes, G. (1974) *Synthesis* 118.
Bailey, W.J.; Rosenberg, J. (1955) *J. Am. Chem. Soc.* **77:** 73.
Bailey, W.J.; Bird, C.N. (1977) *J. Org. Chem.* **42:** 3895.
Banks, R.E.; Harrison, A.C.; Haszeldine, R.N. (1966) *J. Chem. Soc. C.* 2102.
Barnier, J.P.; Salaün, J. (1984) *Tetrahedron Lett.* **25:** 1273.
Barton, T.J.; Kilgour, J.A. (1976) *J. Am. Chem. Soc.* **98:** 7746.
Barton, T.J.; Groh, B.L. (1985) *J. Org. Chem.* **50:** 158.
Baxter, G.J.; Brown, R.F.C.; Eastwood, F.W.; Harrington, K.J. (1975) *Tetrahedron Lett.* 4283.
Baxter, G.J.; Brown, R.F.C. (1978) *Aust. J. Chem.* **31:** 327.
Becker, K.B.; Pfluger, R.W. (1979) *Tetrahedron Lett.* 3713.
Berkowitz, W.F. (1972) *J. Org. Chem.* **37:** 341.
Berkowitz, W.F.; Ozorio, A.A. (1975) *J. Org. Chem.* **40:** 527.
Blackman, G.L.; Brown, R.D.; Brown, R.F.C.; Eastwood, F.W.; McMullen, G.L.; Robertson, M.L. (1978) *Aust. J. Chem.* **31:** 209.
Blanchard, E.P.; Büchi, G. (1963) *J. Am. Chem. Soc.* **85:** 955.
Bloch, R. (1979) *Tetrahedron Lett.* 3945.
Bloch, R.; Orvane, P. (1981) *Tetrahedron Lett.* **22:** 3597.
Bloch, R.; Abecassis, J. (1983) *Tetrahedron Lett.* **24:** 1247.
Block, E.; Penn, R.E.; Olsen, R.J.; Sherwin, P.F. (1976) *J. Am. Chem. Soc.* **98:** 1264.
Block, E.; Revelle, L.K. (1978) *J. Am. Chem. Soc.* **100:** 1630.
Bonnet, R.; Brown, R.F.C.; Smith, R.G. (1973) *J. Chem. Soc. Perkin Trans. I* 1432.
Boekelheide, V.; Ewing, G. (1978) *Tetrahedron Lett.* 4245.
Boland, W.; Jaenicke, L. (1978) *Chem. Ber.* **111:** 3262.
Borchardt, A.; Fuchicello, A.; Kilway, K.V.; Baldridge, K.K.; Siegel, J.S. (1992) *J. Am. Chem. Soc.* **114:** 1921.
Bornstein, J.; Remy., D.E.; Shields, J.E. (1972) *Chem. Commun.* 1149.
Bourdon, F.; Ripoll, J.-L.; Vallee, Y.; Lacombe, S.; Pfister-Guillouzo, G. (1990) *J. Org. Chem.* **55:** 2596.
Boyd, J.; Overton, K.H. (1972) *J. Chem. Soc. Perkin Trans. I* 2533.
Brandi, A.; Garro, S.; Guarna, A.; Goti, A.; Codero, F.M.; De Sarlo, F. (1988) *J. Org. Chem.* **53:** 2430.
Brandi, A.; Codero, F.M.; De Sarlo, F.; Goti, A.; Guarna, A. (1993) *Synlett* 1.
Breitenbach, J.; Ott, F.; Vögtle, F. (1992) *Angew. Chem. Int. Ed. Engl.* **31:** 307.
Brown, R.D.; Brown, R.F.C.; Eastwood, F.W.; Godfrey, P.D.; McNaughton, D. (1979) *J. Am. Chem. Soc.* **101:** 4705.
Brown, R.F.C.; Eastwood, F.W.; Harrington, K.J.; McMullen, G.L. (1974) *Aust. J.*

Chem. **27:** 2393.

Brown, R.F.C.; Eastwood, F.W.; Lim, S.T.; McMullen, G.L. (1976) *Aust. J. Chem.* **29:** 1705.

Brown, R.F.C.; Eastwood, F.W.; McMullen, G.L. (1977) *Aust. J. Chem.* **30:** 179.

Brown, R.F.C.; Caldwell, K.B.; Gatehouse, B.M.; Teo, P.Y.T. (1985) *Aust. J. Chem.* **38:** 1339.

Brown, R.F.C.; Eastwood, F.W.; Kissler, B.E. (1988) *Tetrahedron Lett.* **29:** 6861.

Brown, R.F.C.; Eastwood, F.W.; Wong, N.R. (1993a) *Tetrahedron Lett.* **34:** 1223.

Brown, R.F.C.; Eastwood, F.W.; Wong, N.R. (1993b) *Tetrahedron Lett.* **34:** 3607.

Büchi, G.; Goldman, I.M. (1957) *J. Am. Chem. Soc.* **79:** 4741.

Büchi, G.; Erickson, R.E.; Wakabayashi, N. (1961) *J. Am. Chem. Soc.* **83:** 927.

Büchi, G.; Egger, B. (1971) *J. Org. Chem.* **36:** 2021.

Busby, R.E.; Iqbal, M.; Parrick, J.; Shaw, C.J.G. (1969) *Chem. Commun.* 1344.

Busby, R.E.; Iqbal, M.; Langston, R.J.; Parrick, J.; Shaw, C.J.G. (1971) *Chem. Commun.* 1293.

Caddick, S. (1995) *Tetrahedron* **51:** 10403.

Cano, P.; Echavarren, A.; Prados, P.; Farina, F. (1983) *J. Org. Chem.* **48:** 5373.

Carlson, R.G.; Bateman, J.H. (1967) *J. Org. Chem.* **32:** 1608.

Cava, M.P.; Deana, A.A.; Muth, K. (1960) *J. Am. Chem. Soc.* **82:** 2524.

Cavitt, S.B.; Sarrafizadeh, R.; Gardner, P.D. (1962) *J. Org. Chem.* **27:** 1211.

Chapman, D.D.; Musliner, W.J.; Gates, J.W. (1969) *J. Chem. Soc. [C]* 124.

Chapman, O.L.; McIntosh, C.L. (1971) *Chem. Commun.* 770.

Chou, T.-s.; Tso, H.H. (1989) *Org. Prep. Proc. Int.* **21:** 259.

Clough, S.C.; Kang, J.C.; Johnson, W.R.; Osdene, T.S. (1973) *Chem. Ind.* 323.

Collins, P.A.; Wege, D. (1979) *Aust. J. Chem.* **32:** 1819.

Conia, J.M.; LePerchec P. (1975) *Synthesis* 1.

Corey, E.J.; Yamamoto, H.; Herron, D.K.; Achiwa, K. (1970) *J. Am. Chem. Soc.* **92:** 6635.

Coxon, J.M.; Garland, R.P.; Hartshorn, M.P. (1972) *Aust. J. Chem.* **25:** 353.

De Champlain, P.; Luche, J.L.; Marty, R.A.; de Mayo, P. (1976) *Can. J. Chem.* **54:** 3749.

De Jongh, D.C.; Brent, D.A. (1972) *J. Org. Chem.* **35:** 4204.

De Jongh, D.C.; Van Forssen, R.Y. (1972) *J. Org. Chem.* **37:** 1129.

DePuy, C.H.; King, R.W. (1960) *Chem. Rev.* **60:** 431.

DeShong, P.; Kell, D.A.; Sidler, D.R. (1985) *J. Org. Chem.* **50:** 2309.

Drouin, J.; Leyendecker, F.; Conia, J.M. (1980) *Tetrahedron* **36:** 1203.

Ducos, P.; Rouessac, F. (1973) *Tetrahedron* **29:** 3233.

Earl, R.A.; Vollhardt, K.P.C. (1982) *Heterocycles* **19:** 265.
Fisher, N.M.; Lin, H.N. (1973) *J. Org. Chem.* **38:** 3073.
Forster, D.L.; Gilchrist, T.L.; Rees, C.W.; Stanton, E. (1971) *Chem. Commun.* 695.
Fritch, J.R.; Vollhardt, K.P.C. (1978) *J. Am. Chem. Soc.* **100:** 3643.
Funk, R.L.; Vollhardt, K.P.C. (1977) *J. Am. Chem. Soc.* **99:** 5483.
Funk, R.L.; Vollhardt, K.P.C. (1979) *J. Am. Chem. Soc.* **101:** 215.
Giersch, W.K.; Boschung, A.F.; Snowden, R.L.; Schulte-Elte, K.H. (1994) *Helv. Chim. Acta* **77:** 36.
Gilchrist, T.L.; Rees, C.W.; Thomas, C. (1975) *J. Chem. Soc. Perkin Trans. I* 12.
Ginsberg, A.E.; Paatz, R.; Korte, F. (1962) *Tetrahedron Lett.* 779.
Giovannini, E.; Vuilleumier, H. (1977) *Helv. Chim. Acta* **60:** 1452.
Goldblatt, L.A.; Palkin, S. (1941) *J. Am. Chem. Soc.* **63:** 3517.
Gorham, W.F. (1966) *J. Polym. Sci. (A-1)* **4:** 3027.
Grütze, J.; Vögtle, F. (1977) *Chem. Ber.* **110:** 1978.
Gusel'nikov, L.E.; Avakyan, V.G.. (1989) *Sov. Sci. Rev. (B), Chem.* **13:** 39.
Hageman, H.J.; Wiersum, U.E. (1972) *Angew. Chem. Int. Ed. Engl.* **11:** 333.
Hagen, S.; Bratcher, M.S.; Erickson, M.S.; Zimmermann, G.; Scott, L.T. (1997) *Angew. Chem. Int. Ed. Engl.* **36:** 406.
Harruff, L.G.; Brown, M.; Boekelheide, V. (1978) *J. Am. Chem. Soc.* **100:** 2893.
Haslouin, J.; Rouessac, F. (1977) *Bull. Soc. Chim. Fr.* 1242.
Haucke, G.; Graness, G. (1995) *Angew. Chem. Int. Ed. Engl.* **34:** 67.
Hedaya, E.; McNeil, D.W. (1967) *J. Am. Chem. Soc.* **89:** 4213.
Hedaya, E. (1969a) *Acc. Chem. Res.* **2:** 367.
Hedaya, E.; Miller, R.D.; McNeil, D.W.; D'Angelo, P.F.; Schissel, P. (1969b) *J. Am. Chem. Soc.* **91:** 1875.
Ho, T.-L. (1974) *Synth. Commun.* **4:** 189.
Ho, T.-L..; Wong, C.-M. (1975) *Synth. Commun.* **5:** 213.
Ho, T.-L. (1977) *Synth. Commun.* **7:** 351.
Ho, T.-L.; Liu, S.-H. (1980) *Synth. Commun.* **10:** 603.
Ho, T.-L.; Hsu, K.F. (1996) *Unpublished results.*
Hoffmann, H.M.R. (1969) *Angew. Chem. Int. Ed. Engl.* **8:** 556.
Hoffmann, R.W. (1971) *Angew. Chem. Int. Ed. Engl.* **10:** 529.
Hudlicky, T.; Koszyk, F.J. (1980a) *Tetrahedron Lett.* **21:** 2487.
Hudlicky, T.; Kutchan, T.M.; Koszyk, F.J.; Sheth, J.P. (1980b) *J. Org. Chem.* **45:** 5020.
Hudlicky, T.; Kutchan, T.M.; Naqvi, S.M. (1985) *Org. React.* **33:** 247.
Hudlicky, T.; Fleming, A.; Radesca, L. (1989) *J. Am. Chem. Soc.* **111:** 6691.
Huguet, J.; Karpf, M.; Dreiding, A.S. (1982) *Helv. Chim. Acta* **65:** 2413.
Hurd, C.D. (1940) *Org. Syn.* **Coll. 1:** 330.

Ichihara, A. (1987) *Synthesis* 207.

Ikeda, T.; Wakatsuki, K. (1936) *J. Chem. Soc. Jpn.* **57:** 425.

Jacobi, P.A.; Kaczmarek, C.S.R.; Udodong, U.E. (1984a) *Tetrahedron Lett.* **25:** 4859.

Jacobi, P.A.; Craig, T.A.; Walker, D.G.; Arrick, B.A.; Frechette, R.F. (1984b) *J. Am. Chem. Soc.* **106:** 5585.

Jacobi, P.A.; Selnick, H.G. (1984c) *J. Am. Chem. Soc.* **106:** 3041.

Joines, R.C.; Turner, A.B.; Jones, W.M. (1969) *J. Am. Chem. Soc.* **91:** 7754.

Jones, D.W.; Thompson, A.M. (1989) *Chem. Commun.* 1370.

Jones, G., II; Acquadro, M.A.; Carmody, M.A. (1975) *Chem. Commun.* 206.

Jung, M.E.; Choi, Y.M. (1991) *J. Org. Chem.* **56:** 6729.

Kametani, T.; Ogasawara, K.; Takahashi, T. (1973) *Tetrahedron* **29:** 73.

Kametani, T.; Kajiwara, M.; Takahashi, T.; Fukumoto, K. (1975) *J. Chem. Soc. Perkin Trans. I* 737.

Kametani, T.; Fukumoto, K. (1976) *Acc. Chem. Res.* **9:** 319.

Kametani, T.; Hirai, Y.; Shiratori, Y.; Fukumoto, K.; Shiratori, Y.; Satoh, S. (1978) *J. Am. Chem. Soc.* **100:** 554.

Kametani, T.; Enomoto, Y.; Takahashi, K.; Fukumoto, K. (1979) *J. Chem. Soc. Perkin Trans. I* 2836.

Kametani, T.; Matsumoto, H.; Honda, T.; Nagai, M.; Fukumoto, K. (1981a) *Tetrahedron* **37:** 2555.

Kametani, T.; Nemoto, H. (1981b) *Tetrahedron* **37:** 3.

Karpf, M.; Dreiding, A.S. (1977) *Helv. Chim. Acta* **60:** 3045.

Karpf, M.; Dreiding, A.S. (1981) *Helv. Chim. Acta* **64:** 1123.

Keck, G.E.; Webb, R.R., II (1982) *J. Org. Chem.* **47:** 1302.

Kende, A.S.; Curran, D.P.; Tsay, Y.; Mills, J.E. (1977) *Tetrahedron Lett.* 3537.

King, J.F.; Lewars, E.G. (1972) *Chem. Commun.* 700.

Klumpp, K.N.; Chesick, J.P. (1963) *J. Am. Chem. Soc.* **85:** 130.

Klunder, A.J.H.; Bos, W.; Zwanenburg, B. (1981) *Tetrahedron Lett.* **22:** 4557.

Krapcho, A.P.; Jahngen, E.G.E. (1974) *J. Org. Chem.* **39:** 1322.

Krohn, K.; Tolkiehn, K. (1979) *Chem. Ber.* **112:** 3453.

Landheer, I.J.; de Wolf, W.H.; Bickelhaupt, F. (1974) *Tetrahedron Lett.* 2813.

Lasne, M.-C.; Ripoll, J.L.; Thuillier, A. (1982) *J. Chem. Res. Synop.* 214.

Lawrence, R.; Waight, E.S. (1970) *Org. Mass Spectrom.* **3:** 367.

Lee, C.-J.; Mundy, B.P.; Jun, J.-G. (1992) *Synth. Commun.* **22:** 803.

Le Floch, P.; Mathey, F. (1993) *Chem. Commun.* 1295.

Lemmerz, R.; Nieger, M.; Vögtle, F. (1993) *Chem. Commun.* 1168.

Lewars, E.; Morrison, G. (1976) *Chem. Ind.* 488.

Leyendecker, F. (1976) *Tetrahedron* **32:** 349.

Little, R.D. (1986) *Chem. Rev.* **86:** 875.

Loudon, A.G.; Maccoll, A.; Wong, S.K. (1969a) *J. Am. Chem. Soc.* **91:** 7577.

Loudon, A.G.; Maccoll, A.; Wong, S.K. (1969) *J. Chem. Soc. (B)* 1733.

MacBride, J.A.H. (1972) *Chem. Commun.* 1219.

MacBride, J.A.H. (1974) *Chem. Commun.* 359.

McMurry, J.E.; Silvestri, M.G. (1976) *J. Org. Chem.* **41:** 3953.

McNeil, D.W.; Kent, M.E.; Hedaya, E.; D'Angelo, P.F.; Schissel, P.O. (1971) *J. Am. Chem. Soc.* **93:** 3817.

Mamer, O.A.; Rutherford, R.G.; Seidewand, R.J. (1974) *Can. J. Chem.* **52:** 1983.

Manzardo, G.G.G.; Karpf, M.; Dreiding, A.S. (1983) *Helv. Chim. Acta* **66:** 627.

Martin, J.C. (1964) *U.S. Pat.* 3131234 (*Chem. Abstr.* **61:** 2969).

Martin, S.F.; Desai, S.R.; Phillips, G.W.; Miller, A.C. (1980) *J. Am. Chem. Soc.* **102:** 3294.

Marvel, C.S.; Hinman, C.W. (1954) *J. Am. Chem. Soc.* **76:** 5435.

Martin, H.-D.; Kagabu, S.; Schiwek, H.J. (1975) *Tetrahedron Lett.* 3311.

Martin, S.F.; Rüeger, H. (1985) *Tetrahedron Lett.* **26:** 5227.

Martin, S.F.; Grzejszczak, S.; Rüeger, H.; Williamson, S.A. (1987) *J. Am. Chem. Soc.* **109:** 6124.

Matzger, A.J.; vollhardt, K.P.C. (1997) *Chem. Commun.* 1415.

Mayer, C.F.; Crandall, J.K. (1970) *J. Org. Chem.* **35:** 2688.

Mehta, G.; Murthy, A.N.; Reddy, D.S.K.; Reddy, A.V. (1986) *J. Am. Chem. Soc.* **108:** 3443.

Melder, J.-P.; Pinkos, R.; Fritz, H.; Prinzbach, H. (1990) *Angew. Chem. Int. Ed. Engl.* **29:** 95

Mikami, K.; Terada, M.; Nakai, T. (1990a) *J. Am. Chem. Soc.* **112:** 3949.

Mikami, K.; Takahashi, K.; Nakai, T. (1990b) *J. Am. Chem. Soc.* **112:** 4035.

Mikami, K.; Shimizu, M. (1992) *Chem. Rev.* **92:** 1021.

Mingos, D.M.P.; Baghurst, D.R. (1991) *Chem. Soc. Rev.* **20:** 1.

Miyahara, Y.; Inazu, T.; Yoshino, T. (1983) *Tetrahedron Lett.* **24:** 5277.

Moody, C.J.; Rahimtoola, K.F. (1990) *Chem. Commun.* 1667.

Mühlebach, M.; Neuenschwander, M.; Engel, P. (1993) *Helv. Chim. Acta* **76:** 2089.

Nametkin, N.S.; Vdorin, V.M.; Gusel'nikov, L.E.; Za'yadov, V.I. (1966) *Izv. Akad. Nauk SSSR, Ser. Khim.* 584.

Nozoe, S.; Goi, M.; Morisaki, N. (1971) *Tetrahedron Lett.* 3701.

Oppolzer, W.; Keller, K. (1971) *J. Am. Chem. Soc.* **93:** 3836.

Oppolzer, W.; Mahalanabis, K.K.; Bättig, K. (1977) *Helv. Chim. Acta* **60:** 2388.

Oppolzer, W.; Snieckus, V. (1978) *Angew. Chem. Int. Ed. Engl.* **17:** 476.

Oppolzer, W.; Bättig, K.; Hudlicky, T. (1979) *Helv. Chim. Acta* **62:** 1493.

Oppolzer, W.; Roberts, D.A. (1980) *Helv. Chim. Acta* **63:** 1703.
Oppolzer, W. (1981a) *Pure Appl. Chem.* **53:** 1181.
Oppolzer, W.; Bättig, K. (1981b) *Helv. Chim. Acta* **64:** 2489.
Oppolzer, W.; Marazza, F. (1981c) *Helv. Chim. Acta* **64:** 1575.
Oppolzer, W.; Robbiani, C.; Bättig, K. (1984) *Tetrahedron* **40:** 1391.
Paquette, L.A. (1970) *J. Am. Chem. Soc.* **92:** 5765.
Petrich, S.A.; Pang, Y.; Young, V.G.; Barton, T.J. (1993) *J. Am. Chem. Soc.* **115:** 1591.
Piers, E.; Lau, C.K. (1977) *Synth. Commun.* **7:** 495.
Piers, E.; Banville, J. (1982) *Can. J. Chem.* **60:** 2965.
Rabideau, P.W.; Abdourazak, A.H.; Folsom, H.E.; Marcinow, Z.; Sygula, A.; Sygula, R. (1994) *J. Am. Chem. Soc.* **116:** 7891.
Ranu, B.C.; Kavka, M.; Higgs, L.A.; Hudlicky, T. (1984) *Tetrahedron Lett.* **25:** 2447.
Ravindranathan, T.; Kumar, M.A.; Menon, R.B.; Hiremath, S.V. (1990) *Tetrahedron Lett.* **31:** 755.
Rienäcker, R.; Ohloff, G. (1961) *Angew. Chem.* **73:** 240.
Ripoll, J.L. (1976) *Chem. Commun.* 235.
Ripoll, J.L.; Thuillier, A. (1977) *Tetrahedron* **33:** 1333.
Ripoll, J.L.; Rouessac, A.; Rouessac, F. (1978) *Tetrahedron* **34:** 19.
Ripoll, J.L.; Lebrun, H.; Thuillier, A. (1980) *Tetrahedron* **36:** 2497.
Roth, W.R.; Humbert, H.; Wegener, G.; Erker, G.; Exner, H.-D. (1975) *Chem. Ber.* **108:** 1655.
Rubin, Y.; Lin, S.S.; Knobler, C.B.; Anthony, J.; Boldi, A.M.; Diederich, F. (1991) *J. Am. Chem. Soc.* **113:** 6943.
Schaden, G. (1977) *Angew. Chem. Int. Ed. Engl.* **16:** 50.
Schiess, P.; Heitzmann, M. (1977) *Angew. Chem. Int. Ed. Engl.* **16:** 469.
Schiess, P.; Heitzmann, M. (1978) *Helv. Chim. Acta* **61:** 844.
Schissel, P.; Kent, M.E.; McAdoo, D.J.; Hedaya, E. (1970) *J. Am. Chem. Soc.* **92:** 2147.
Schmitthenner, H.F.; Weinreb, S.M. (1980) *J. Org. Chem.* **45:** 3372.
Schostarez, H.; Paquette, L.A. (1981) *Tetrahedron* **37:** 4431.
Schultz, A.G.; Puig, S. (1985) *J. Org. Chem.* **50:** 915.
Scott, L.T.; Roelofs, N.H. (1988) *Tetrahedron Lett.* **29:** 6857.
Scott, L.T.; Hashemi, M.M.; Meyer, D.T.; Warren, H.B. (1991) *J. Am. Chem. Soc.* **113:** 7082.
Scott, L.T.; Hashemi, M.M.; Bratcher, M.S. (1992) *J. Am. Chem. Soc.* **114:** 1920.
Sekine, Y.; Brown, M.; Boekelheide, V. (1979) *J. Am. Chem. Soc.* **101:** 3126.
Seybold, G.; Jersak, U.; Gommper, R. (1973) *Angew. Chem. Int. Ed. Engl.* **12:** 847.
Seybold, G.; Heibl, C. (1977) *Chem. Ber.* **110:** 1225.
Shea, K.J.; Wada, E. (1982) *Tetrahedron Lett.* **23:** 1523.

Shishido, K.; Shitara, E.; Komatsu, H.; Hiroya, K.; Fukumoto, K.; Kametani, T. (1986) *J. Org. Chem.* **51:** 3007.

Shishido, K.; Hiroya, K.; Fukumoto, K.; Kametani, T.; Kabuto, C. (1989) *J. Chem. Soc. Perkin Trans. I* 1443.

Short, R.P.; Ranu, B.C.; Revol, J.M.; Hudlicky, T. (1983) *J. Org. Chem.* **48:** 4453.

Siwapinyoyos, T.; Thebtaranonth, Y. (1982) *J. Org. Chem.* **47:** 599.

Snider, B.B. (1980) *Acc. Chem. Res.* **13:** 426.

Sonnenberg, F.M.; Stille, J.K. (1976) *J. Org. Chem.* **41:** 3441.

Spangler, R.J.; Kim, J.H. (1972) *Tetrahedron Lett.* 1249.

Spangler, R.J.; Kim, J.H.; Cava, M.P. (1977) *J. Org. Chem.* **42:** 1697.

Sternbach, D.D.; Mane, M. (unpubl.) priv. commun. to Crimmins, M.T. (1988) *Chem. Rev.* **88:** 1453.

Stork, G.; Guthikonda, R.N. (1972) *Tetrahedron Lett.* 2755.

Stork, G.; Kraus, G.A. (1976) *J. Am. Chem. Soc.* **98:** 6747.

Strauss, C.R.; Trainor, R.W. (1995) *Aust. J. Chem.* **48:** 1665.

Strickler, H.; Ohloff, G.; Kovats, E.s. (1967) *Helv. Chim. Acta* **50:** 759.

Subba Rao, H.N.; Damodaran, N.P.; Dev, S. (1968) *Tetrahedron Lett.* 2213.

Takahashi, K.; Mikami, K.; Nakai, T. (1988) *Tetrahedron Lett.* **29:** 5277.

Townsend, C.A.; Scholl, T.; Arigoni, D. (1975) *Chem. Commun.* 921.

Trahanovsky, W.S.; Emeis, S.L.; Lee, A.S. (1976) *J. Org. Chem.* **41:** 4043.

Trahanovsky, W.S.; Alexander, D.L. (1979) *J. Am. Chem. Soc.* **101:** 142.

Trahanovsky, W.S.; Koeplinger, K.A. (1992) *J. Org. Chem.* **57:** 4711.

Trost, B.M.; Nishimura, Y.; Yamamoto, K. (1979) *J. Am. Chem. Soc.* **101:** 1328.

Trost, B.M.; Vladuchick, W.C.; Bridges, A.J. (1980) *J. Am. Chem. Soc.* **102:** 3554.

Trost, B.M.; Lautens, M. (1985) *Tetrahedron Lett.* **26:** 4887.

Verlaak, J.M.J.; Klunder, A.J.H.; Zwanenburg, B. (1982) *Tetrahedron Lett.* **23:** 5463.

Vögtle, F. (1969) *Angew. Chem. Int. Ed. Engl.* **8:** 274.

Vögtle, F.; Mew, P.K.T.. (1978) *Angew. Chem. Int. Ed. Engl.* **17:** 60.

Vögtle, F.; Rossa, L. (1979) *Angew. Chem. Int. Ed. Engl.* **18:** 515.

Vogel, E.; Grimme, W.; Korte, S. (1965) *Tetrahedron Lett.* 3625.

Vogel, E.; Markowitz, G.; Schmalstieg, L.; Ito, S.; Breuckmann, R.; Roth, W.R. (1984) *Angew. Chem. Int. Ed. Engl.* **23:** 719.

Watt, D.S.; Corey, E.J. (1972) *Tetrahedron Lett.* 4651.

Wege, D. (1971) *Tetrahedron Lett.* 2337.

Wender, P.A.; Hubbs, J.C. (1980a) *J. Org. Chem.* **45:** 365.

Wender, P.A.; Letendre, L.J. (1980b) *J. Org. Chem.* **45:** 367.

Wender, P.A.; Eck, S.L. (1982) *Tetrahedron Lett.* **23:** 1871.

Wessjojann, L.; Giller, K.; Zuck, B.; Skattebol, L.; de Meijere, A. (1993) *J. Org. Chem.*

58: 6442.

White, J.D.; Dillon, M.P.; Butlin, R.J. (1992) *J. Am. Chem. Soc.* **114:** 9673.

Wiersum, U.E.; Mijs, W.J. (1972) *Chem. Commun.* 347.

Williams, J.R.; Callahan, J.F. (1980a) *J. Org. Chem.* **45:** 4475.

Williams, J.R.; Callahan, J.F. (1980b) *J. Org. Chem.* **45:** 4479.

Williams, J.R.; Lin, C. (1981) *Chem. Commun.* 752.

Wittig, G.; Ebel, H.F. (1961) *Liebigs Ann. Chem.* **650:** 20.

Wolf, A.D.; Kane, V.V.; Levin, R.H.; Jones, M. (1973) *J. Am. Chem. Soc.* **95:** 1680.

Woods, W.G. (1958) *J. Org. Chem.* **23:** 110.

Woodward, R.B.; Hoffmann, R. (1970) *The Conservation of Orbital Symmetry*, Verlag Chemie: Weinheim.

Yan, T.-H.; Paquette, L.A. (1982) *Tetrahedron Lett.* **23:** 3227.

Ziegler, K.; Hafner, K. (1955) *Angew. Chem.* **67:** 301.

Zimmermann, G.; Nuechter, U.; Hagen, S.; Nuechter, M. (1994) *Tetrahedron Lett.* **35:** 4747,

6 ELECTROORGANIC TRANSFORMATIONS

The study of organic electrochemistry has a long history which dates back to the era of Michael Faraday. Based on those results Hermann Kolbe in 1849 purposefully devised a method of alkane synthesis from carboxylic acids. Fifty years later F. Haber laid down the theoretical foundation of electrochemical transformations by the investigation of nitrobenzene reduction; he found the dependence of product formation on electrode potentials. The development of the potentialstat then turned electrochemistry into a common tool for organic synthesis. Given the procedural simplicity, absence of side products derived from reagents, and high potential for accomplishing selective oxidoreductions under very mild conditions, electrosynthesis still appears to be undervalued, even though important industrial scale works such as the Monsanto adiponitrile process, the Nalco tetraethyllead process, and about one hundred others have demonstrated its appealing features.

The apparatus required for electrosynthesis can be as simple as a beaker placed with a pair of electrodes connected to a DC-voltage source. The electrodes are fixed at a distance of 1-5 mm to minimize resistance and achieve the desired current densities. Of course a stirrer, thermometer, cooling jacket, inert gas inlet, or any combination thereof can be added. For some reactions the separation of the electrodes by a diaphragm is mandatory to prevent product(s) from one electrode to diffuse to the other and be destroyed. An ideal diaphragm is only permeable to ions but not to other chemical species including the solvent; the material is should also be chemically inert. Such a set-up has higher resistivity and energy consumption.

Two common conditions for electrochemical processes are called constant current and controlled potential. The constant current process is often preferred as it is completed much faster, easily scaled up, and does not require a potentiometer.

Electrode materials for preparative purposes should be sturdy in both physical and chemical terms, low in cost, easily shaped, able to induce selective and fast reactions, and relatively nontoxic. Not all the above criteria can be met by one type of material and compromises must be made in the choice. A useful guide for the selection is the hydrogen overvoltage of metals, from those approaching zero (vs. SCE): Pt, Ni, Ag, Cu (hydrogenation catalysts) to those having a high value: Pb, Hg, Cd. For electrooxidation the most common electrodes are made of platinum, carbon, and lead dioxide. Generally, Pt electrodes limit one-electron oxidation (to product radicals or cation radicals) whereas carbon electrodes can be used to generate cations. Platinum electrodes are durable and their surfaces can be refreshed by washing in hot nitric acid, but carbon electrodes are laible to damage by perchlorates if these salts are used as electrolytes. Comparing to platinum, lead dioxide electrodes are much cheaper; they can also be prepared by coating on various other materials including titanium, iron, nickel, and graphite.

Solvents are important for electrochemical processes. Unfortunately the most suitable solvent, water, is only seldom used in organic electrosynthesis on account of incompatible solvent property toward most substrates and/or products. The solvents to be considered in electrooxidation are acetic acid, pyridine, nitromethane, and acetonitrile. Acetonitrile has the best attributes except toxicity. Owing to the high volatility dichloromethane can be used at low temperatures. For electroreduction alcohols (MeOH, EtOH, etc.), ethers (THF, 1,2-DME, diglyme), amides (DMF, DMA), acetone and pyridine are suitable solvents. Besides mixed solvents an often promising solution to the solubility problem is found in phase transfer.

The presence of a supporting electrolyte is required if the substrate itself does not conduct electricity, as usually the case in organic electrochemistry. Quaternary ammonium salts are preferred because in aprotic media potentials up to ca. -2.9 V can be reached. The counter ion should be reistant to oxidation (in anodic processes) therefore hexafluorophosphate, hexafluoroborate, perchlorate and tosylate are preferred. Perchlorate salts are less expensive and do not require rigorously dry conditions but they are explosion prone (e.g., during purification).

It should be noted that some reactions carried out in electrolytic cells are indirect processes. They involve active chemical species which are generated continuously at one of the electrodes. The mediators before electrochemical activation need only be present in small (catalytic) amounts.

6.1. ELECTROOXIDATION AND FUNCTIONALIZATION

6.1.1. Of Hydrocarbons

The C-H bond undergoes oxidation only at very high positive potential: CH_3 and CH_2 groups at above 3.4 V, whereas CH betwee 3.0 to 3.4 V (vs SCE). An exception is the adamantane skeleton which permits functionalization at one or two bridgehead positions at much lower potentials [Koch, 1973; Bewick, 1977]. Very strained cyclic compounds such as tricyclo[4.1.0.0^{2,7}]heptane undergoes oxidative cleavage [Laurent, 1976].

OH 83% — 1.8 V — CF_3COOH / $(CF_3CO)_2O$ (Pt) — 2.3 V — OH, OH 70%

MeCN (Pt) — $LiClO_4$ 2.35 V — NHAc 90%; Bu_4NBF_4 — NHAc, NHAc 58%

<u>1-Acetamidoadamantane</u>. [Koch, 1973]

A three-compartment cell in which the chambers are separated by glass frits is equipped with a platinum sheet as anode, and a reference electrode system of Ag/0.1 N $AgNO_3$ in acetonitrile. Adamantane is added to the acetonitrile medium which contains lithium perchlorate as supporting electrolyte. After purging the cathode and anode compartments with nitrogen, electrolysis at 2.3 V begins at room temperature. Initial currents at 50-100 times the background of ca. 0.1 mA/cm^2. When the current drops back to the background level, the reaction is discontinued. The anolyte is concentrated, extracted with water and ether. The ether layer yields ~90% of the product.

- 2e, MeOH — [+ —OMe] — —OMe

Functional group can influence the oxidizability of C-H. Particularly the use of electrogenerated oxidants obviate the difficulties. The oxidative addition of acetic acid to butadiene being an example [Coleman, 1990].

$HOAc + Mn(OAc)_2 \longrightarrow Mn(OAc)_3 + H^+ + e^-$

$Mn(OAc)_3 + HOAc \longrightarrow Mn(OAc)_2 + HOAc + \cdot CH_2COOH$

$CH_2{=}CH{\cdot}CH{=}CH_2 + \cdot CH_2COOH \longrightarrow CH_2{=}CHC^{\cdot}H(CH_2)_2COOH$

$Mn(OAc)_3 + CH_2{=}CHC^{\cdot}H(CH_2)_2COOH \longrightarrow CH_2{=}CHC^{+}H(CH_2)_2COOH + Mn(OAc)_2 + HOAc$

$CH_2{=}CHC^{+}H(CH_2)_2COOH + HOAc \longrightarrow CH_2{=}CHCH(OAc)(CH_2)_2COOH + AcOCH_2CH{=}CH(CH_2)_2COOH +$ γ-vinyl-γ-butyrolactone

Alkenes are much more easily oxidized than alkanes. The attachment of an electron-donating substituent to the double bond further facilitates the process, while an electron-withdrawing group has the opposite effect. Several types of products may arise, depending on the alkene structure and the supporting electrolyte. As expected, the Birch reduction products are very readily rearomatized [Shono, 1975a; Birch, 1978].

Epoxidation of perfluoropropene by electrolytic method is an industrially viable process. Under specified conditions highly selective epoxidation can be achieved.

HOAc-H_2O, HNO_3, (PbO_2 anode)

65- 75% conversion
90% selectivity

A rose oxide synthesis involves anodic oxidation and intramolecular trapping of the allylic cation to form a tetrahydropyran ring [Shono, 1971]. Among other modes of reaction is oxidative addition of nucleophiles, which, when applied to a cyclopentadiene derivative gives an important intermediate of allethrolone [Shono, 1976d].

- 2e, Et_4N OTs, MeCN — 26% — rose oxide

- 2e, HOAc, Et_3N — allethrolone

<u>3-5-Diacetoxy-1-allyl-2-methylcyclopentene.</u> [Shono, 1976d]

A solution of 2-allyl-3-methylcyclopentadiene (0.02 mol) and triethylamine (0.02 mml) in acetic acid (30 mL) is introduced into an undivided electrolysis cell (50 mL) which is equipped with carbon rod electrodes and cooled in a water bath. A constant current (0.1A) is supplied until 1.0~1.2 equiv. of electricity is consumed, the reaction mixture is poured into satd. NaCl solution. Workup furnishes the diacetate (46% current efficiency).

Enol derivatives are more easily oxidized on the anode. The radical cation intermediates may undergo dimerization or reaction with nucleophiles. Formation of conjugated carbonyl compounds is favored when enol acetates are tetrasubstituted [Shono, 1974, 1976c, 1983a]. By varying the reaction parameters, double bond cleavage may be effected [Torii, 1982].

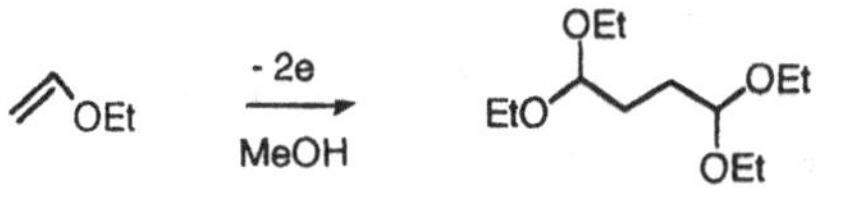

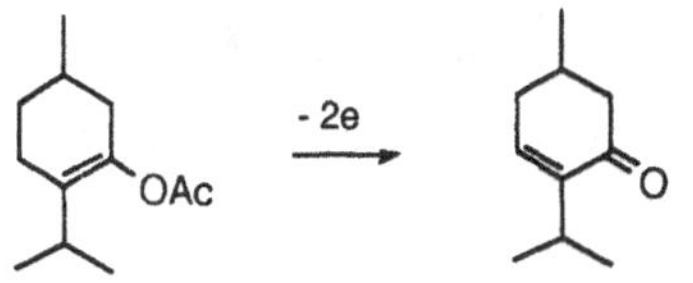

4-*p*-Menthen-3-one. [Shono, 1983a]

A constant current (0.1 A) is passed through an undivided cell which is charged with 3-acetoxy-3-*p*-menthene (9.8 g, 50 mmol), tetraethylammonium tosylate (1.5 g) and acetic acid (50 mL) while cooling externally with water. The reaction is terminated when 2.7-3.0 F/mol of electricity is consumed. Thew mixture is diluted with water (100 mL) and the aqueous solution is extracted with ether. Distillation gives the enone in 90-97% yield, b.p. 77°C/10 torr.

Oxidative chlorination of ethene to give vinyl chloride can be performed on a graphite anode using hydrochloric as source of chlorine. Allylic chlorides produced electrochemically [Torii, 1981a, b] are transposed. Of course in these processes the electrooxidation does not occur directly at the double bond.

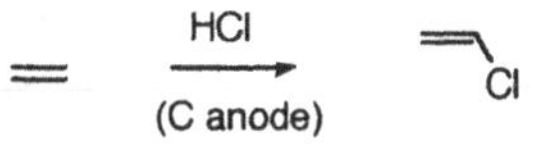

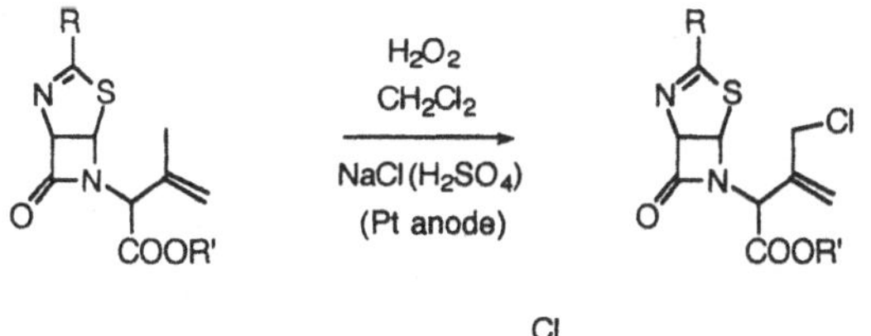

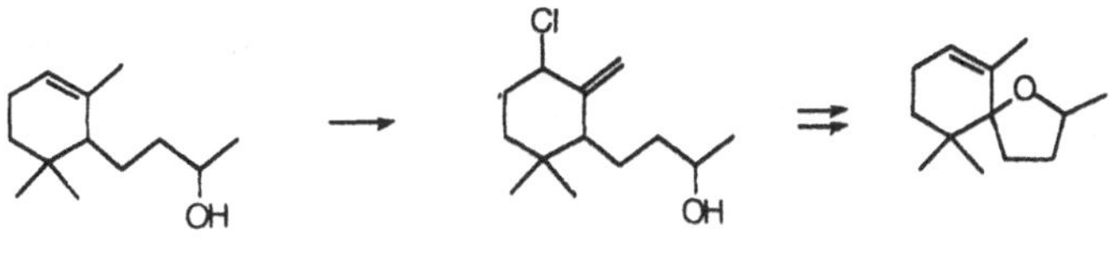

A regioselective and stereoselective electrochlorination at the unactivated 9α position of chlostestan-3α-ol via the *m*-iodobenzoate [Breslow, 1976] is a case of sophisticated maneuver.

6.1.2. Of Aromatic Systems

Simple aromatic compounds have high oxidation potentials therefore their direct electrooxidation (leading to dimers) is relatively difficult. Compounds bearing electron-donating groups are better substrates. Cross-coupling of phenols is not only feasible, it often gives products in excellent yields [Grujic, 1976].

- 4e
H_2O
NaOAc
95%

Intramolecular reactions such as those leading to *O*-methylflavinanthine from laudanosine [Miller, 1978] and to a 4,5,9,10-tetrahydropyrene from a dimethoxymetacyclophane [Kerr, 1979] are particularly favorable.

laudanosine
- 2e
$MeCN-H_2O$
$NaHCO_3$
$Me_4N\,BF_4$
O-methylflavinanthine

MeO OMe
- 2e
$-2H^+$
90%

Electrooxidation of benzene in aqueous sulfuric acid on a lead dioxide anode can give benzoquinone which is reducible to hydroquinone on the lead cathode. However, since the conversion is below 10% the method is not competitive in the economic sense. For synthesis of phenols the route involving trifluoroacetoxylation is preferred (vs. acetoxylation) due to the much higher stability of the products [So, 1976].

PbO_2 anode: H_2SO_4, H_2O

Pb cathode: H_2SO_4, H_2O

-e, CF_3COONa, CF_3COOH; $OCOCF_3$; NaOH; OH; 65%

p-Benzoquinone is the anodic oxidation product of phenol, hydroquinone mono- and dimethyl ethers. When methanol is used as solvent the monoacetal or bisacetal is formed [Nilsson, 1978].

OH; OR; OR'; MeO, OMe

R = R' = H
R = H, R' = Me
R = R' = Me

4,4-Dimethoxycyclohexa-2,5-dienone. [Nilsson, 1978]

Constant-current (500 mA) electrolysis of a stirred solution of *p*-methoxyphenol (10 g) in methanol (60 mL) containing $LiClO_4$ (4 g) is conducted using a Pt anode (50 cm^2) and tungsten wire (0.5 mm dia.) as cathode in an open vessel with water cooling. At 95-100% consumption (nmr monitoring) of starting material the reaction is stopped, mixture poured to a phosphate buffer (pH 6, 300 mL) and extracted with CH_2Cl_2. Solvent evaporation (< 30°C) affords the product in 97% yield.

Monoprotected quinones are synthetically useful, as shown by a route to colchicine [Evans, 1981]. An alternative method of generating such intermediates is by a similar electrolysis of *p*-methoxyarenoxyethanols followed by selective hydrolysis of the dimethylacetal unit [Nilsson, 1978].

OMe OMe OMe - 2e / MeOH; H_3O^+ O MeO OMe O MeO OMe OMe MeO MeO OH MeO MeO OMe OMe

CF_3COOH; DDQ

MeO MeO MeO NHAc O OMe ⇄ MeO MeO MeO OMe O

colchicine

OMe R O OH - 2e MeOH KOH MeO OMe R O O HOAc THF O R O O

The quinone bisacetals are also highly valued intermediates for the synthesis of, inter alia, complex polycyclic quinones [Swenton, 1983].

Nuclear 1,4-dimethoxylation of isoeugenol (with subsequent elimination of methanol from the adduct) enables a simple synthesis of γ-asarone [Vargas, 1989].

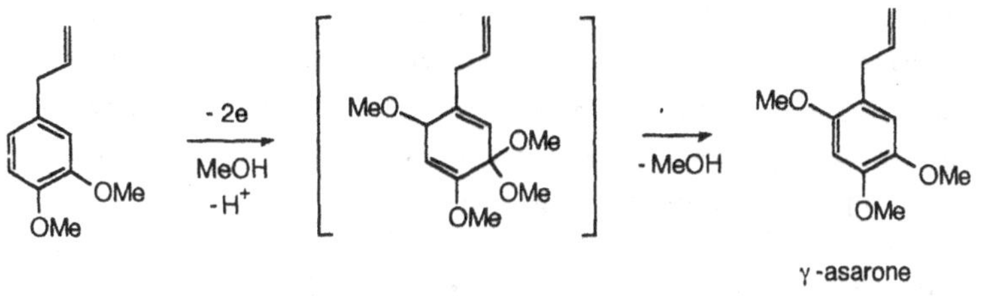

γ-asarone

The cationic intermediates from anodic oxidation can be intercepted by weaker nucleophiles such as alkenes. A novel approach to complex natural products based on the bicyclo[3.2.1]oct-3-en-2,8-diones thus formed by either an intermolecular or intramolecular reaction has been developed [Yamamura, 1991]. Applications include helminthosporal, 8.14-cedraneoxide and silphinene.

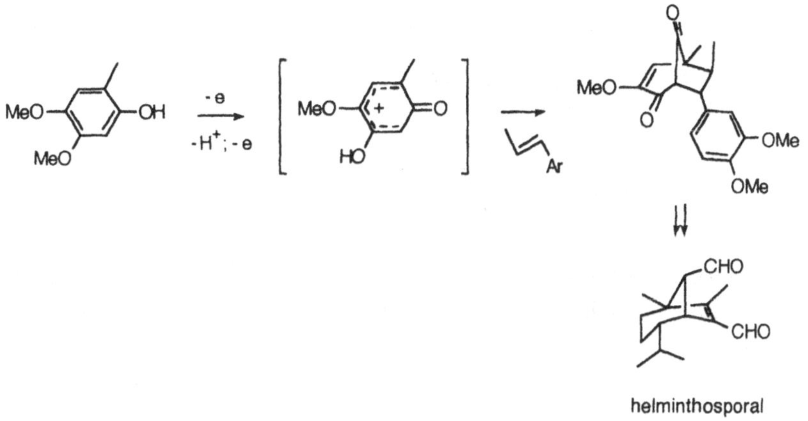

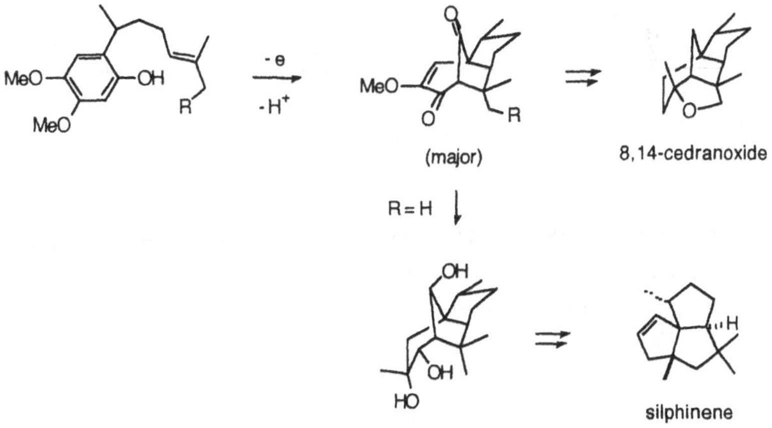

helminthosporal

(major)

8,14-cedranoxide

silphinene

Electrolysis of 6-(2-hydroxy-4,5-dimethoxyphenyl)-2-methylhept-2-ene. [Yamamura, 1991]
A stirred solution of the phenol (9.0 mg) in acetic anhydride (10 mL) containing n-Bu_4NBF_4 (150 mg) is electrolyzed at a constant current (1.84 mA) and stopped at ca. 2.0 F/mol. After concentration of the solution in vacuo the product is isolated by preparative TLC to furnish the tricyclic product (5.0 mg, 59%). This compound is a precursor of silphinene.

Alkylaromatic compounds possessing a benzylic hydrogen are subjected to anodic oxidation at the side chain as well as at a nuclear position. Sometimes it is possible to revert one of the products to the starting substrate in situ. Thus in the presence of a hydrogenolysis catalyst the benzylic acetate from an acetoxylation process is being removed to allow accumulation of only the aryl acetate(s) [Eberson, 1981]. However, this method is uneconomical.

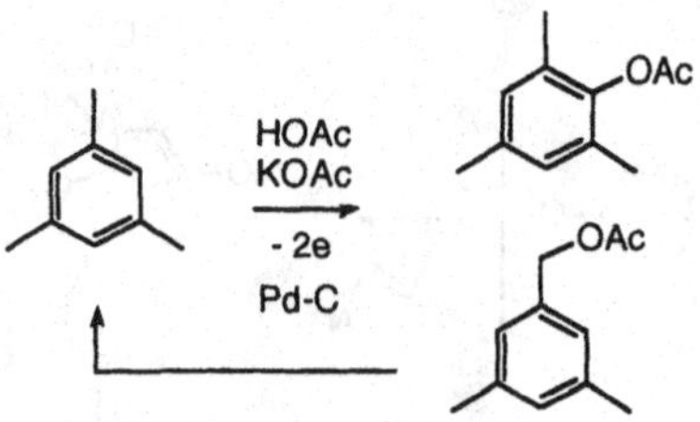

HOAc
KOAc
- 2e
Pd-C
OAc
OAc

Introduction of a $\Delta^{9(11)}$-double bond to estrone methyl ether is achieved under mild conditions by benzylic methoxylation and elimination of methanol [Ponsold, 1979].

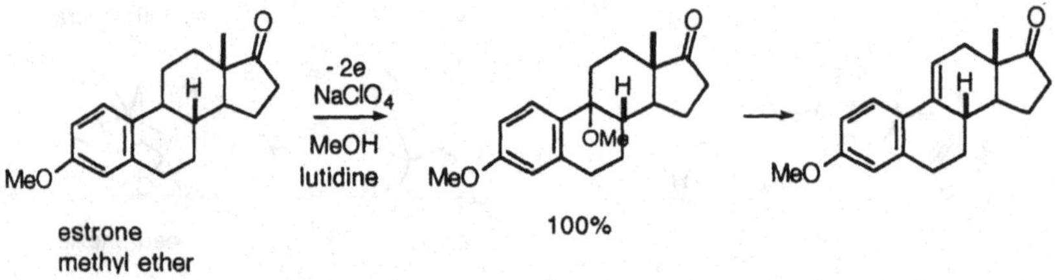

- 2e
$NaClO_4$
MeOH
lutidine
MeO
H
OMe
100%
estrone
methyl ether

Oxidation of a side chain methyl group can be controlled by varying potential of the working electrode. As benzyl ethers are readily converted to araldehydes the use of the p-methoxybenzyl group to protect alcohols acquires a higher value [S.M. Weinberg, 1975].

$$\text{cyclodecyl benzyl ether} \xrightarrow[\text{MeCN - H}_2\text{O (Pt)}]{-2e,\ \text{LiClO}_4} \text{cyclodecanol (89\%)} + \text{PhCHO}$$

Oxidation of electroregenerated chemical species is more practical. An outstanding method for saccharin preparation (80% yield) is by anodic oxidation of *o*-toluenesulfonamide in the presence of Cr(III) under acidic conditions [Kuhn, 1979]. Ruthenium(IV) complexes can be used to mediate oxidation of arylalkanes to aroic acids, e.g., *p*-xylene to terephthalic acid in excellent current yields [Moyer, 1980].

H_2SO_4, e, Cr(VI) / Cr(III); SO_2NH_2 → saccharin

In the presence of acetonitrile, ammonium nitrate or N_2O_4, alkali cyanide the anodic reaction of arenes gives aryl acetamides [Hammerich, 1974], nitroarenes [Eberson, 1980], and aryl cyanides [K. Yoshida, 1979] in good yields, respectively.

- 2e / MeCN, $Bu_4N\,BF_4$, $(CF_3CO)_2O$ → NHAc, 85%

- 2e / MeCN, $Bu_4N\,PF_6$, N_2O_4 → NO_2, 91%

- 2e / MeOH, NaCN → CN, 74%

6.1.3. Of Heterocycles

Furans are a good source of 1,4-dicarbonyl compounds on account of their clean transformation by anodic oxidation [Clauson-Kaas, 1952]. Actually the 1,4-

dimethoxyl adducts generated in methanol are themselves valuable for various synthetic purposes: pyridoxine [Elming, 1955], maltol [Shono, 1976a], rethrolones [Shono, 1976b], cyclotene [Shono, 1977c].

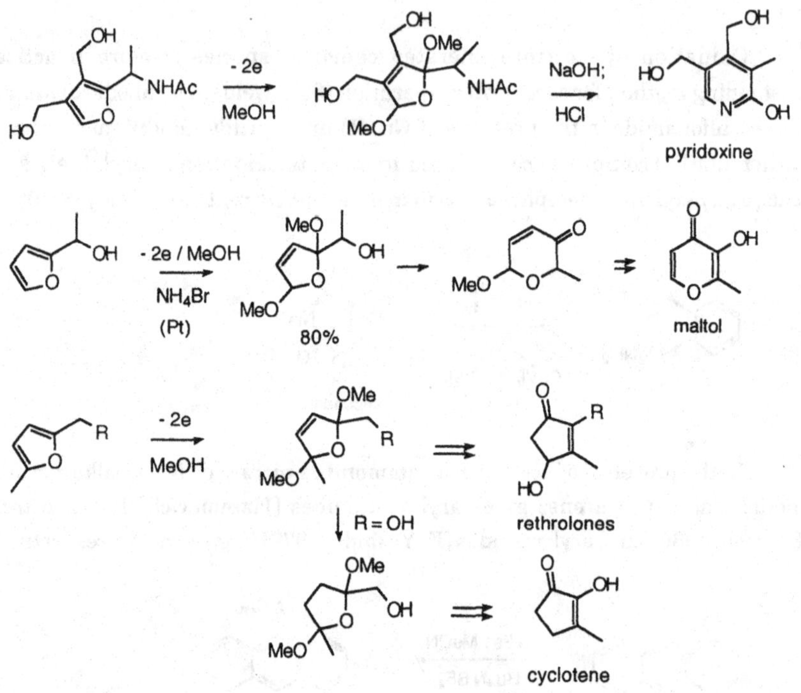

The analogous formation of 2,5-diacetoxy-2,5-dihydrofuran [Shono, 1981] makes available of 2-acetoxyfuran which readily undergoes electrophilation to produce γ-substituted 2-butenolides. However, it should be noted that 2,5-dimethylfuran yields 2,5-bisacetoxymethylfuran.

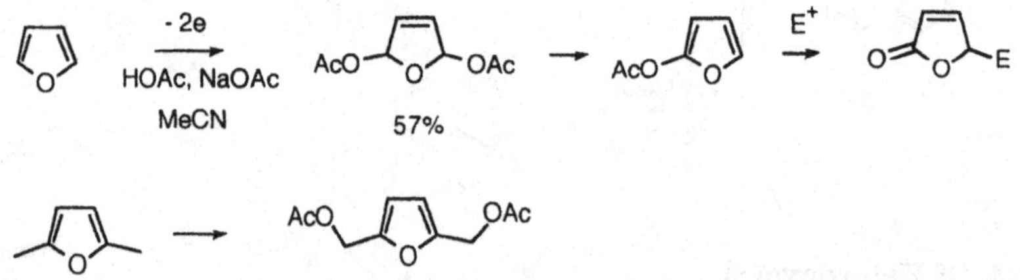

Furfuryl alcohol and 2-furoic acid undergo ring opening and degradation at high current densities [Tanaka, 1976; Iwasaki, 1982].

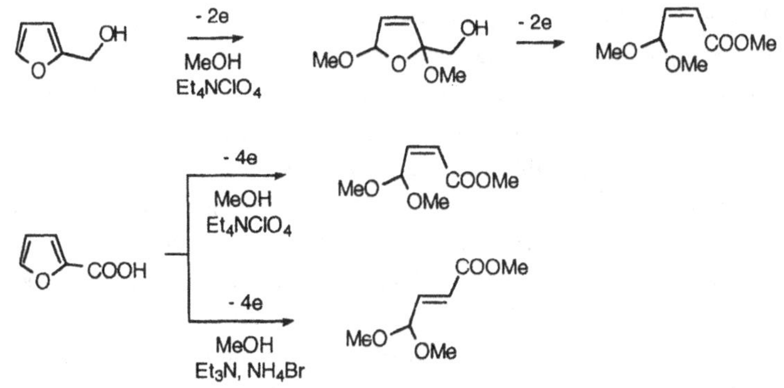

Methyl (Z)-4,4-dimethoxy-2-butenoate. [Tanaka, 1976]

A solution of furfuryl alcohol (3.0 g) and tetraethylammonium perchlorate (0.1 g) in methanol (20 mL) is electrolyzed at 15-16°C in a compartment cell equipped with Pt foil electrodes. After passage of 8 F/mol electricity the reaction mixture is subjected to workup and distillation to give the ester in 75% yield.

Pyrrole derivative generally behave similarly to furans under electrolytic conditions, although pyrrole itself undergoes polymerization. Actually the yellow polypyrrole film thus prepared in an acetonitrile solution of Et_4NBF_4 at 0.8 V is semiconducting; in a highly colored doped form it exhibits electric conductivity of 30-100 S/cm [Tourillon, 1982; Pickup, 1984].

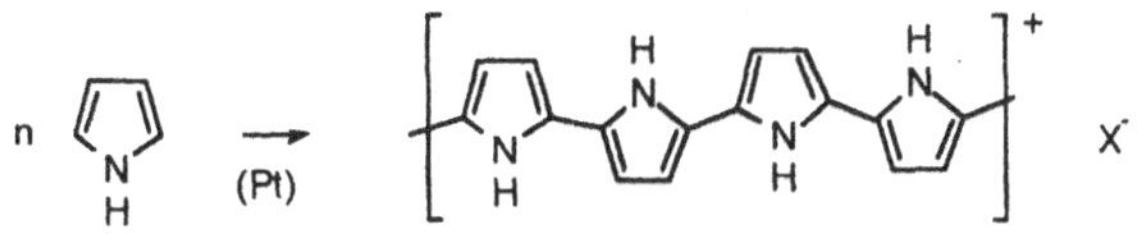

Azacyclooctane and azacyclononane undergo transannular closure to afford pyrrolizidine (52%) and indolizidine (71%), respectively, on electrolysis (graphite electrode) in the presence of lithium bromide [Elofson, 1985]. This process is reminiscent of the Hofmann-Löffler-Freytag reaction.

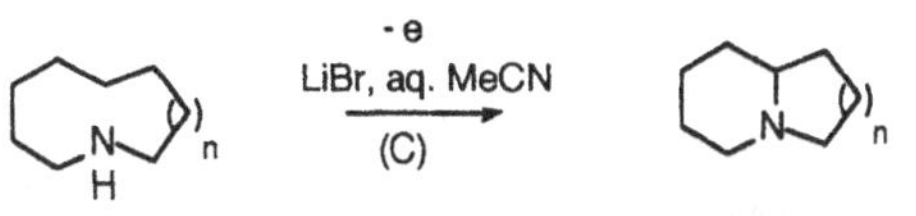

6.1.4. Of Heterofunctional Compounds

Many n-donor compounds are oxidizable under electrolytic conditions. Except for a few cases such as 2-butyne-1,4-diol , the anodic oxidation of which being a significant method for the preparation of acetylenedicarboxylic acid [Jäger, 1978], the high oxidation potentials of simple alcohols cause their direct oxidation difficult, thus the more effective method involves generation of an oxidizing agent on anode at a lower potential to promote the reaction. Only catalytic amount of the mediating agent is required. It should be noted that electroreduction can also be carried out based on the same principle. For oxidation of benzylic alcohols and ethers, trisbromoarylamines can serve as catalysts by virtue of their formation of stable aminium ions covering a wide range of potential (0.76 to 2.0 V vs SCE). To obtain the aldehydes a base is usually added, otherwise (when methanol is used as solvent) the dimethyl acetals are formed [Brinkhaus, 1983].

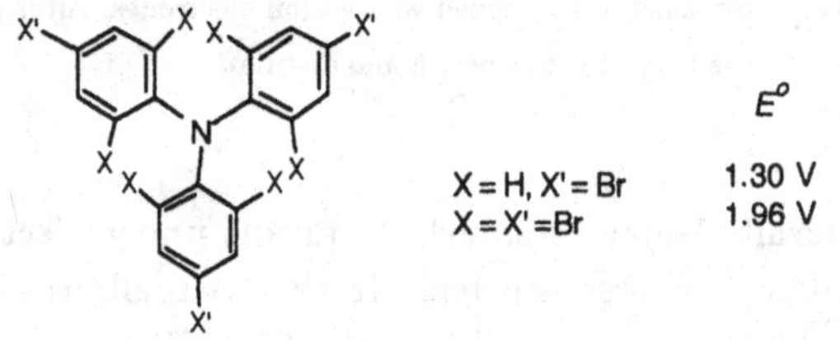

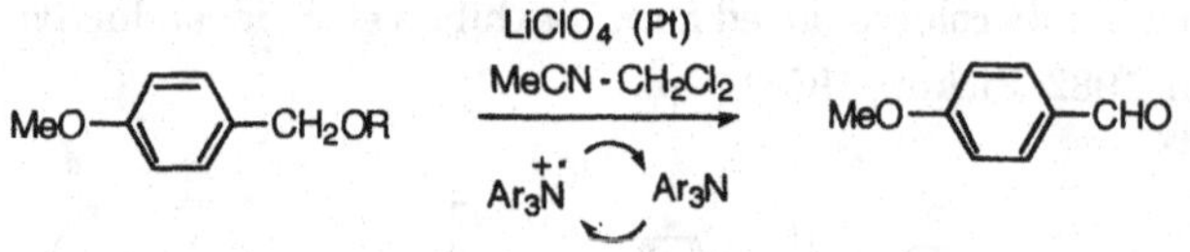

Controlled deblocking of benzyl ethers is illustrated in the following reaction sequence. Note that electrolysis in the presence of tris(4-bromophenyl)amine takes off the 4-methoxybenzyl group only. It requires tris(2,4-dibromophenyl)amine to mediate the removal of the unsubstituted benzyl residue.

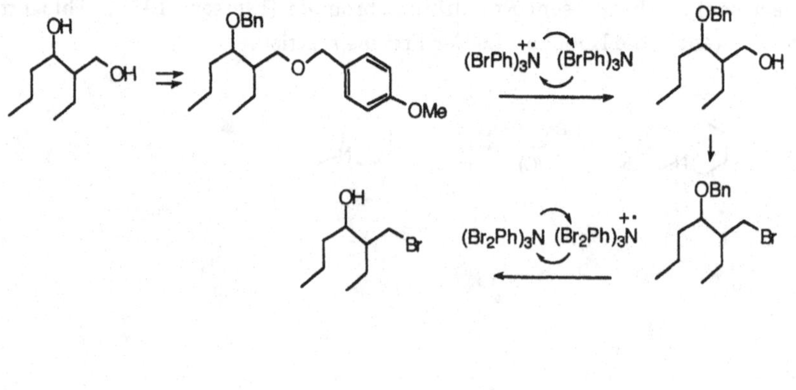

N-Hydroxyphthalimide has been introduced as an electrocatalyst for oxidation [Masui, 1983]. In the present of a base it forms the anion which undergoes oxidation to the oxy radical to mediate alcohol oxidation. A number of substrates susceptible to hydrogen atom abstraction are readily oxidized in this manner.

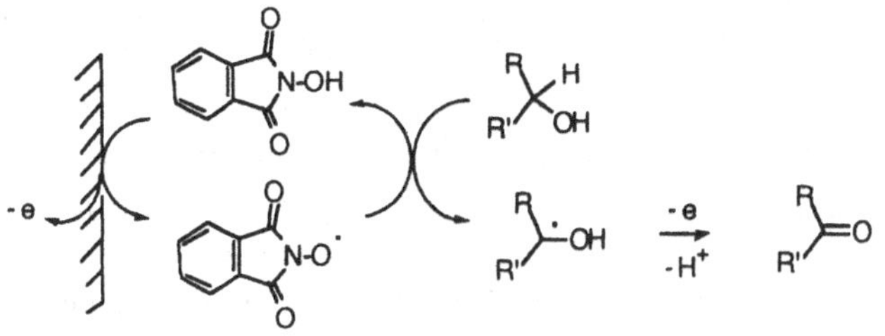

Interestingly, 2,2,6,6-tetramethylpiperidinoxyl is converted to the cation and reacts by hydride abstraction [Semmelhack, 1983]. In contrast to *N*-hydroxyphthalimide which promotes selective oxidation of a secondary alcohol, the oxoammonium ion acts on primary alcohols preferentially.

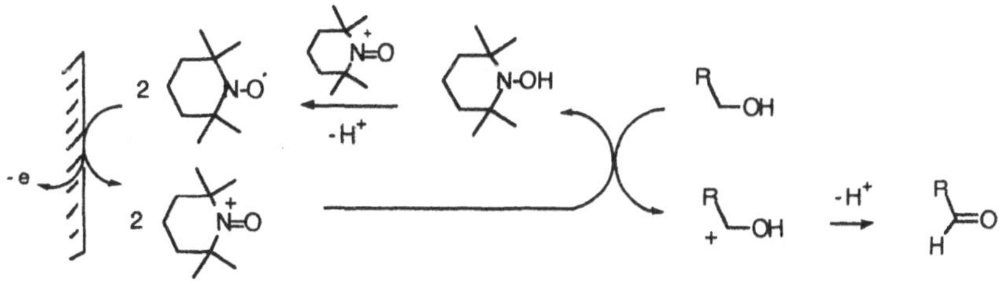

Electrooxidation of alcohols is mediated by halide ions [Shono, 1979]. It should be noted that a combination of a 2,2,6,6-tetramethylpiperidinoxyl derivative and bromide ion forms an efficient double mediatory system which operates in a biphasic medium [Inokuchi, 1991]. The bromide is involved in the elctrochemical process while oxidation of an alcohol occurs in the organic layer with the oxoammonium species which is regenerated continuously.

Oxidation of Undecanol. [Inokuchi, 1991]

Undecanol (17.23 g, 100 mmol) and 4-benzoyloxy-2,2,6,6-tetramethylpiperidinoxyl (0.28 g, 1.0 mmol) are placed in a beaker-type cell (1.5 L) which contains dichloromethane (200 mL) and 25% NaBr (400 mL, pH 8.6). With two graphite plates (3x6 cm^2) as electrodes the electrolysis is conducted with constant current of 900 mA (current density 50 mA/cm^2) until 2.8 F/mol of electricity is passed. The reaction is quenched with ethanol (2 mL) and the mixture extracted with dichloromethane (2x50 mL). Distillation furnishes the aldehyde product in 88% yield.

Many other electrocatalytic systems are now known, including those with double mediators. For example, a buffered (pH 4) NaCl solution containing small amount of RuO_2 to oxidize alcohols in carbon tetrachloride works by the mechanism of chlorine generation to oxidize the ruthenium [Torii, 1986].

Direct electrooxidative cleavage of glycols [Shono, 1975b] is relatively facile. It is superior to chemical method, being environmentally friendly and unrestricted by stereochemical features. However, the mono- and dimethyl ether of glycols are also affected (note the behavior of furfuryl alcohol mentioned above). A useful synthetic sequence for the monoacetal of 6-oxoalkanals from cyclohexanone involves enoletherification, electromethoxylation, organometallic reaction and electrolytic cleavage of the 2-methoxycyclohexanols.

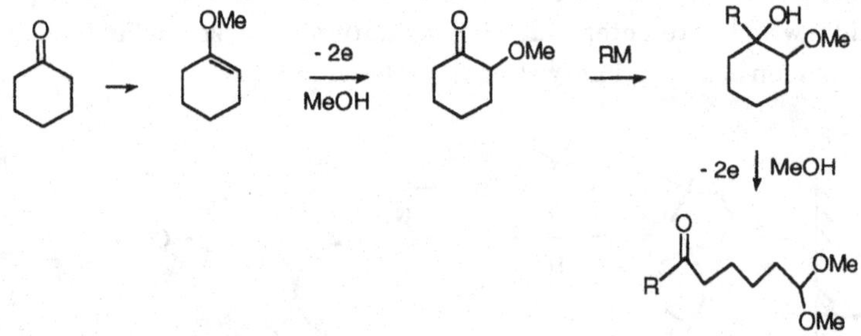

OMe
- 2e
MeOH
OMe
RM
R OH
OMe
- 2e MeOH
O
R
OMe
OMe

The behavior of a 1,2-aminocyclohexane derivative is shown below. The imine apparently undergoes further oxidation to afford a nitrile [Torii, 1987].

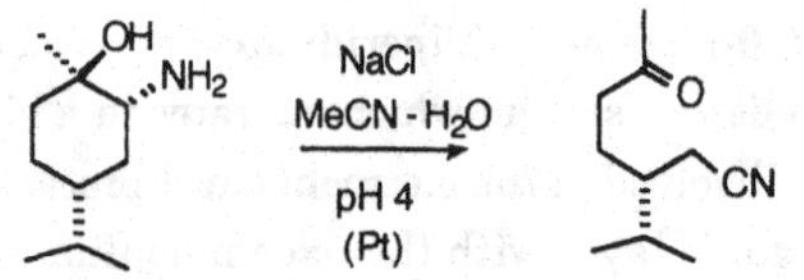

OH
NH2
NaCl
MeCN - H2O
pH 4
(Pt)
O
CN

Degradation of α,β-epoxy ketones by the electrochemical method is of some synthetic utility. For example, from one enantiomeric carvone routes can be developed to acces either optical isomer of *trans*-chrysanthemic acid [Torii, 1983b].

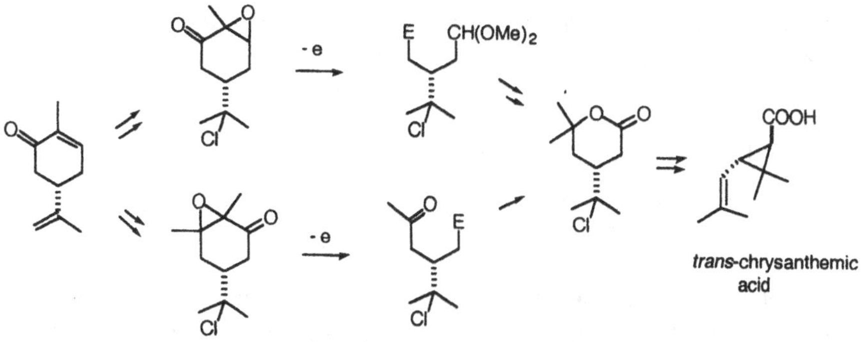

1-Alkoxy-2,2-dichlorocyclopropanes also suffer ring cleavage on electrolysis [Klehr, 1975]. This behavior reflects the relief of ring strain to be a contributing factor.

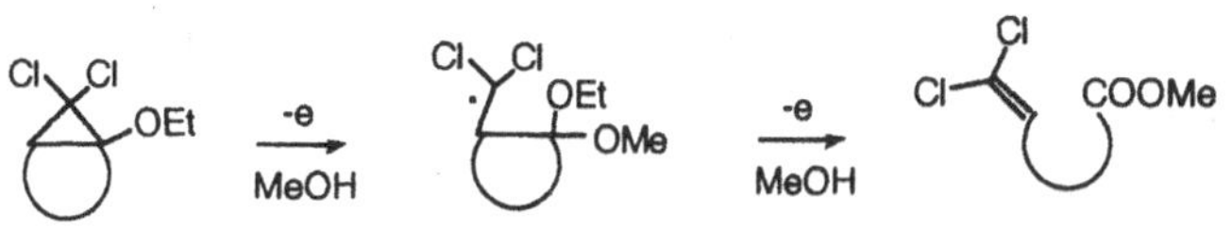

Under acidic condtions ethers (such as THF) are converted to acetals [Shono, 1987], but acetals of aldehydes undergo anodic methoxylation under basic conditions [Scheeren, 1978].

Methoxylation of 2-Propyl-1,3-dioxolane. [Scheeren, 1978]

A 2.5 M solution of the acetal (0.3 ~ 1 mol) in methanol containing 2% KOH is electrolyzed at room temperature in an undivided cell (Pt anode, Ni cathode, applied voltage 15 V) until 6 F/mol electricity is passed. Workup gives the cyclic orthoester in 72% yield.

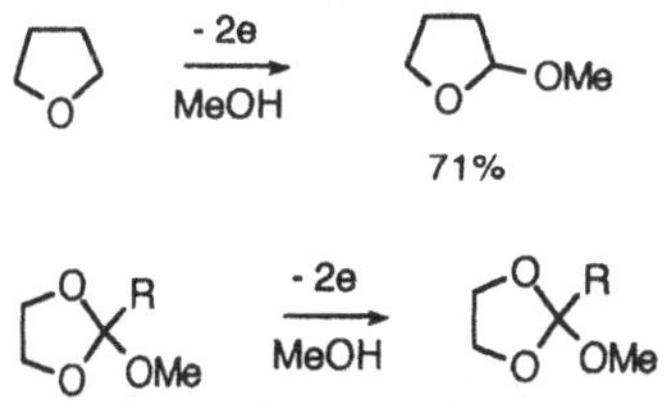

The oxidation potential of an ether is lowered by an α-stannyl or α-silyl group. Electrolysis induces fragmentation to give an oxonium ion which can be intercepted by a remote double bond to form a tetrahydropyran [J. Yoshida, 1992]. Carbamyl α-cations are similarly created. The reaction is not suitable for generating five-membered ring products.

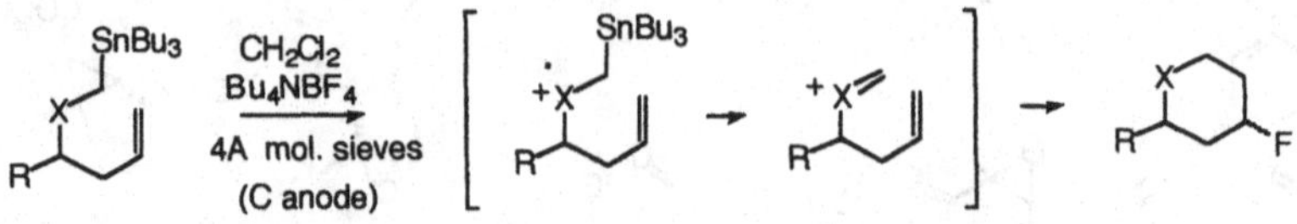

Oxidation performed on nickel-hydroxide anode (which is prepared by deposition of Ni(O)OH to a Pt or Ni on electrolyzing an alkaline solution of a Ni(II) salt) is particularly suitable for primary alcohols to acids. For obvious reasons the method is far superior to those employing stoichiometric amounts of potassium permanganate and dichromate. Accordingly, it has been applied to the conversion of sorbose diacetonide to gulonic acid diacetonide in the production of vitamin C [Robertson, 1983]. Lactones afford keto acids due to in situ saponification to expose the oxidizable hydroxyl group [Ruholl, 1987]. In these transformations abstraction of a hydrogen α to the hydroxyl group takes place.

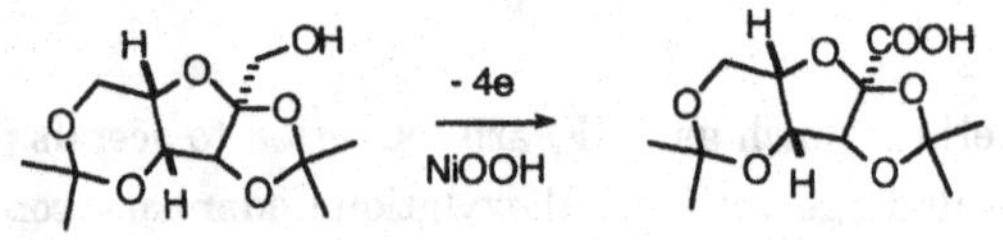

It must not be construed that the nickel hydroxide electrode is unsuitable for the electrooxidation of secondary alcohols. Actually a very selective oxidation of a steroid triol has been reported [Kaulen, 1982].

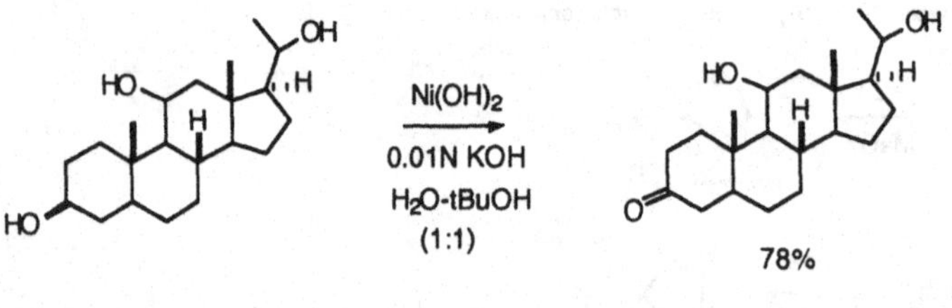

Amines are dealkylated when electrolyzed in the presence of water. In the presence of a better nucleophile in the solvent α-functionalization of amines results. Introduction of a methoxy or cyano group represents the more useful process, with methoxylation showing excellent regioselectivity at the N-methyl substituent [Shono, 1982a].

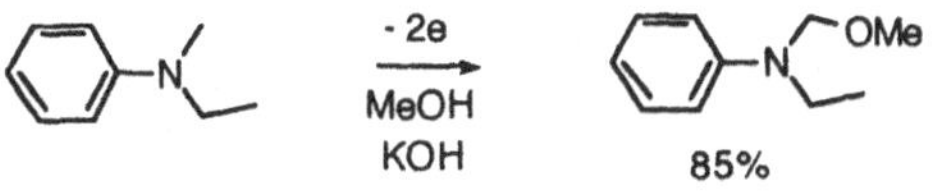

The electrooxidation products, including the α-hydroxylated compounds, of amides and carbamates are more amenable to use in synthesis [Shono, 1984a]. Interestingly, prolonged reaction of *N*-carbomethoxypiperidine in different media leads to either the 2,6-dimethoxylated product [Shono, 1982b] or (in HOAc with KOAc as supporting electrolyte) the 2,3-diacetoxylated derivative [Shono, 1984b].

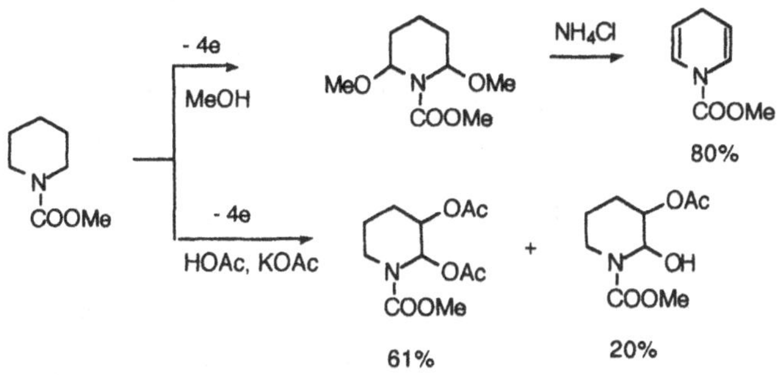

The susceptibility of the acyliminium intermediates to interception by various nucleophiles coupled with the mild conditions of electrolysis conspire to evoke synthetic exploitation. Thus its involvement in the *N*-debenzylation of a highly reactive α-methylene-β-lactam derivative [Mori, 1985] serves to confirm the high

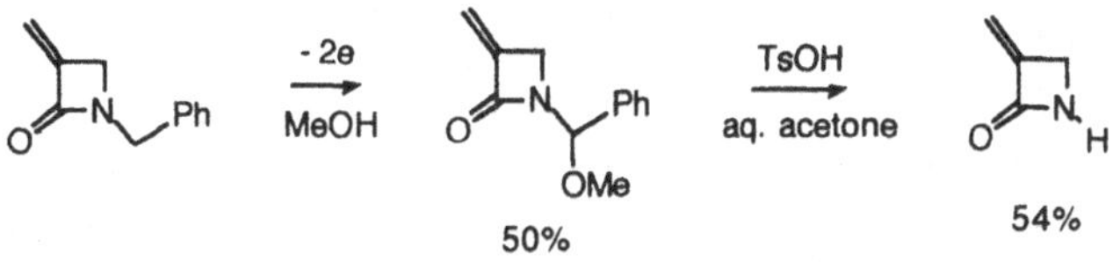

expectation. Carbon-carbon bond formation by trapping the intermediates with external or internal nucleophiles is now well-established, it constitutes the key step in the synthesis of many alkaloids: coniine [Shono, 1983c], elaeokanine-A [Shono, 1984c], trachelanthamidine and isoretronecine [Blum, 1984; Shono, 1984c]. By means of an intramolecular alkylation of the δ-methoxylated δ-valerolactam a quinolizidinone precursor of lupinine and epilupinine is readily obtained [Okita, 1983].

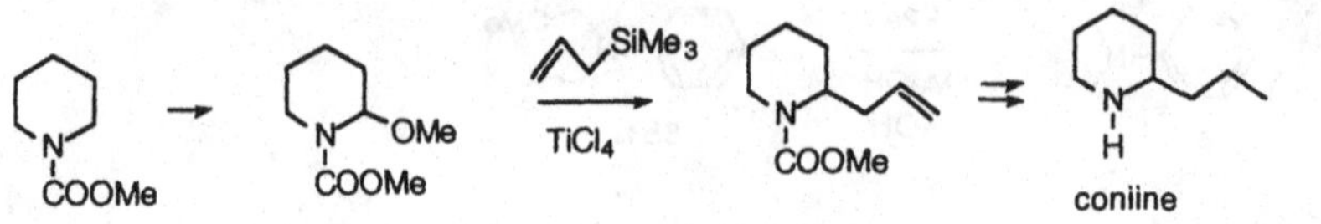

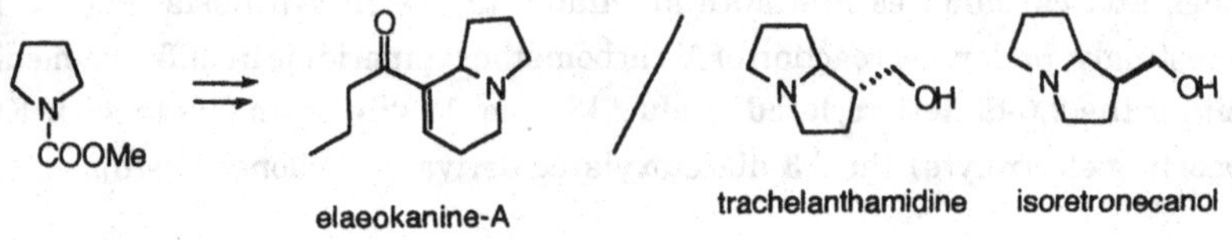

COOMe COOMe OMe SiMe3 TiCl4 COOMe coniine

COOMe elaeokanine-A trachelanthamidine isoretronecanol

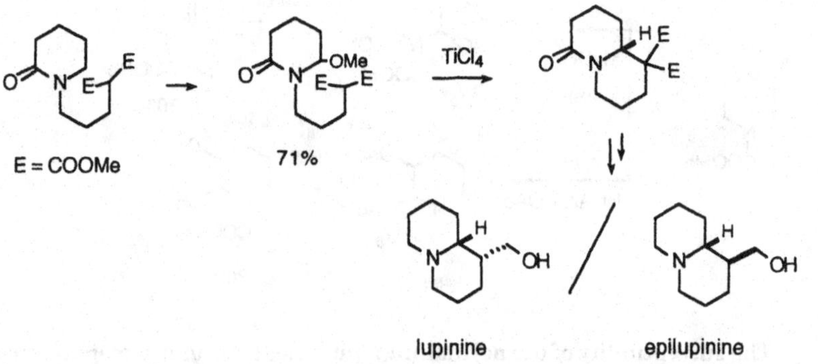

E = COOMe 71% OMe TiCl4 lupinine epilupinine

The electrooxidation products of amides and carbamates derived from cyclic amines have other synthetic uses. For example, they can be converted to ω-amino-α-amino acids by Strecker synthesis [Warning, 1978]. ω-(*N*-Alkoxycarbonyl)amino-α,β-unsaturated esters can be synthesized by a Wittig reaction of the α-hydroxy analogues.

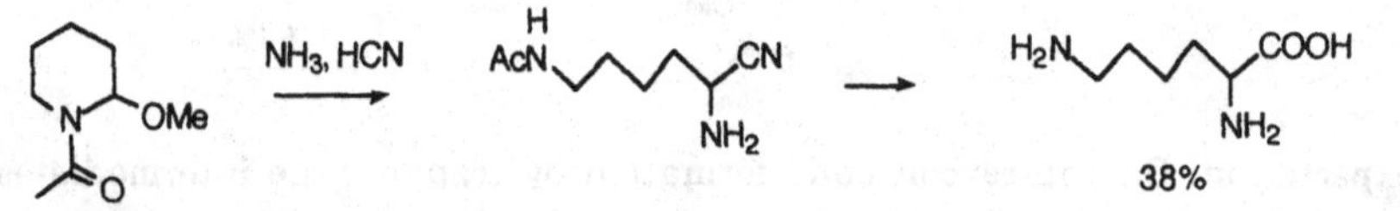

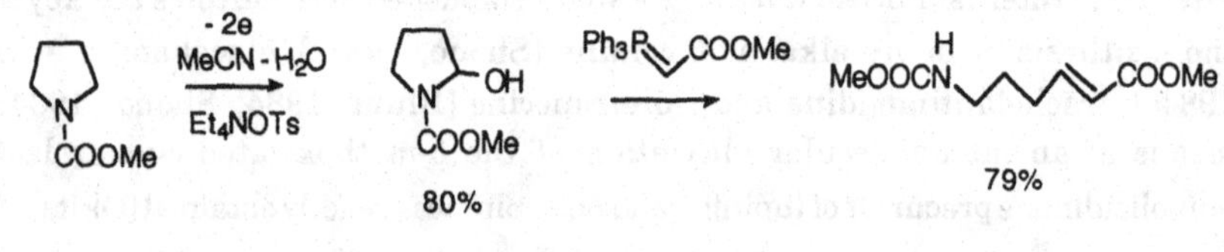

OMe NH_3, HCN AcNH CN NH_2 H_2N COOH NH_2 38%

COOMe -2e MeCN-H_2O Et_4NOTs OH COOMe 80% Ph_3P COOMe MeOOCNH COOMe 79%

Cyanation at the α-position of an amino nitrogen can also be accomplished electrochemically. As the cyano group of such products is readily replaced, e.g., on

Grignard reaction, the cyanation constitutes a useful synthetic process. In serving an elaboration of gephyrotoxin-223AB [Yang, 1994] two reaction sequences converted *N*-benzylpiperidine to a *cis*-2,6-dialkyl derivative.

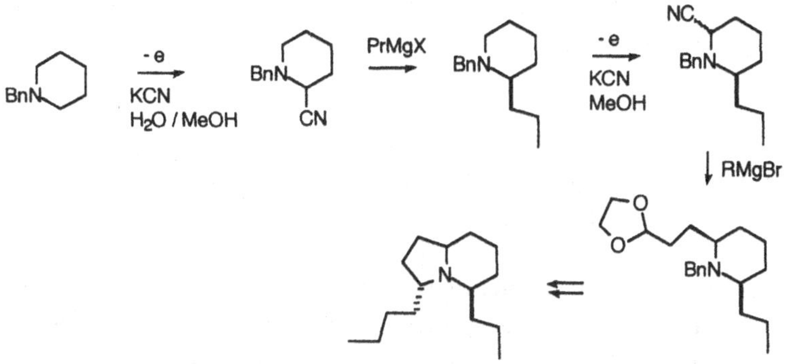

gephyrotoxin-223AB

Oxidative desulfurization of sulfides in an alcohol leading to acetals [Torii, 1980a] is not possible with diaryl sulfides. However, the process can be utilized to convert a chiral secondary alcohol into the chloride [Shono, 1983b]. The results may be rationalized by an S_N2 displacement of the alkoxydiphenylsulfonium species with chloride ion.

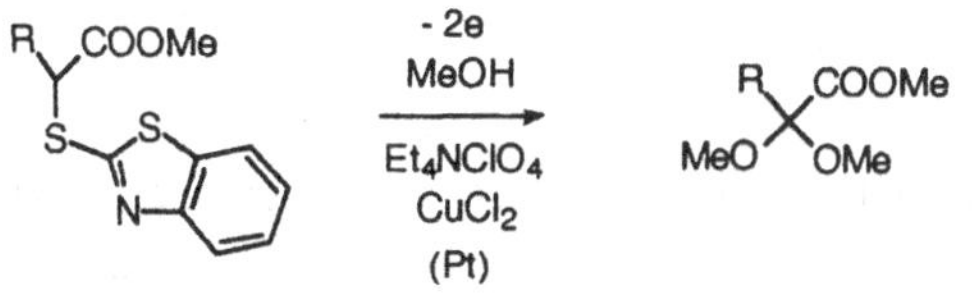

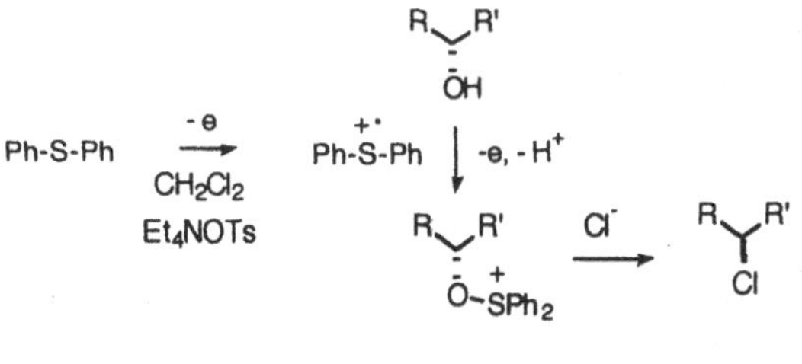

Synthesis of Chlorides from Alcohols.

A solution of diphenyl sulfide (15 mmol) and tetraethylammonium tosylate (5 mmol) in dichloromethane (15 mL) is charged into a water-cooled undivided cell equipped with two graphite electrodes (dia. 8 mm) A constant current of 0.2 A is passed until 3-5 F/mol is consumed. The product is then isolated (55-95% yield).

Cleavage of dithioacetals in aqueous acetonitrile containing sodium carbonate is a mild method [Platen, 1984]. The same transformation can also be effected with mediation of a triarylamine.

MeCN - H_2O, $NaClO_4$ (Pt) → PhCHO 77%

Reactive species generated by anodic oxidation of halogen or halide ions can be used to functionalize alkenes and halogenation of arenes. In acetonitrile the iodide ion is converted to $[CH_3C{=}N{-}I]^+$ which is very reactive toward compounds such as anisole [Miller, 1976], but deactivated aromatic compounds (e.g., nitrobenzene, benzonitrile) can only be attacked by species generated in the presence of trifluoroacetic acid [Lines, 1980].

I_2 - 2e, $2I^+$, CF_3COOH

X = NO_2, CN, CHO, Ac

Diphenylselenide can be activated to participate in selenoalkoxylation of alkenes by anodic oxidation of the bromide ion of a supporting electrolyte. The adducts then enter an anodic oxidation and thermal elimination sequence an overall transformation of the alkenes to transposed allylic ethers is accomplished [Torii, 1980b]. A simple approach to rose oxide is based on this method.

Ph-Se-Se-Ph —(- 2e, ROH)→ Ph-Se-OR; OAc, RO, SePh, -2 e, PhSe=O

Radical cations generated by anodic oxidation of trialkyl phosphites can be trapped by arenes [Ohmori, 1979]. The products are converted to arylphosphonates by treatment with NaI.

$$(RO)_3P \xrightarrow{-e} (RO)_3\overset{+\cdot}{P} \xrightarrow{ArH} [(RO)_3P\text{-}ArH]^{+\cdot} \xrightarrow[-H^+]{-e} (RO)_3\overset{+}{P}\text{-}Ar \xrightarrow{NaI} (RO)_2\overset{O}{\overset{\|}{P}}\text{—}Ar$$

Diethyl phenylphosphonate. [Ohmori, 1979]

A suspension of K_2CO_3 (3 g) in a solution of triethyl phosphite (5 mmol) and benzene (25 mmol) in acetonitrile (40 mL) containing 0.2 M $NaClO_4$ is electrolyzed under nitrogen at 1.7 V in the anodic chamber of a divided cell equipped with a glassy carbon anode. When the current is decreased to less than 2% of the initial value (ca. 10 h) the mixture is filtered, concentrated in vacuo, and extracted with chloroform. The extracts are evaporated, and the residue is treated with NaI (0.9 g) in refluxing acetone (100 mL) for 6 h. Product is obtained on workup and distillation. Yield 59%.

Trialkylboranes (R_3B) are Lewis acids and they cannot be oxidized. However, in the presence of hydroxide or alkoxide ion, the ensuing ate complexes are susceptible to electrooxidation to generate alkyl radicals which afford dimers and alcohols when they suffer oxidation to the carbocations [Schäfer, 1972].

A more intriguing series of transformation involves electrolysis of a mixture of trialkylborane, iodide ion, an electrophile [Takahashi, 1978]. Alkyl radical split from the borane reacts with the electrophile.

Ethyl 2-methylhexanoate. [Takahashi, 1978]

A solution of tripropylborane (4.9 mmol), ethyl mathacrylate (4 mL) and n-Bu_4NI (0.5 g, 1.35 mmol) in acetonitrile (17 mL) is placed in an undivided cell equipped with two Pt electrodes (400 mm^2) and a magnetic stirrer. Electrolysis is conducted under nitrogen at a constant current (0.5 Acm^{-2}) for 5 h, then the mixture is treated with 3N NaOH (5.0 mL) and 30% H_2O_2 (5.0 mL) for 2 h. The ester product is obtained in in 73% yield.

6.1.5. Of Anions

Oxidation of anionic species leads to dimers via free radical intermediates. The reaction pattern is followed by carbanions of malonate esters and organometallic compounds, but it is generally of little synthetic value. However, the electrocatalytic oxidative coupling of 2-naphthol in the presence of (-)-sparteine on a graphite felt electrode which is modified with 2,2,6,6-tetramethylpiperidin-1-yloxyl is most interesting, as *(S)*-2,2'-binaphthol is obtained in 99.5% enantiomeric excess [Osa, 1994].

$$(ROOC)_2CH{-}R' \xrightarrow{-H^+} \left[(ROOC)_2C^{-}{-}R' \xrightarrow{-e} (ROOC)_2C^{\cdot}{-}R' \right] \xrightarrow{2x} (ROOC)_2C(R'){-}C(R')(COOR)_2$$

$$Ph{-}C{\equiv}CLi \xrightarrow{-e} \left[Ph{-}C{\equiv}C\cdot \right] \xrightarrow{2x} Ph{-}C{\equiv}C{-}C{\equiv}C{-}Ph$$

$$\text{2-naphthol} \xrightarrow[\text{(-)-sparteine}]{-2e} \text{(S)-2,2'-binaphthol} \quad 99.5\%\ ee$$

As will be seen, halide ions often mediated reactions such as oxidation of organic compounds through their own transformation into active halogens (X^+). However, the electrochemical process is a particularly useful technique of fluorination [Fuchigami, 1997] as the handling of reactive and often dangerous fluoringating agents can be avoided. Fluorination such as desulfurative fluorination of dithioacetals can also be carried out using 4-methoxyiodobenzene difluoride which is formed in situ by trapping of the electrooxidized iodoarene with fluoride ion [Fuchigami, 1994].

6.1.5.1. Kolbe Electrolysis

By far the most important electrochemical reaction involving organic anions is the Kolbe electrolysis. Carboxyl radicals generated from the anions undergo decarboxylation which may dimerize or be further oxidized to cations. Various factors contribute to the preference of the two pathways.

$$R{-}COO^- \rightleftharpoons R{-}COO^-_{ads} \xrightarrow{-e} R{-}COO^{\cdot}_{ads} \xrightarrow{-CO_2} R^{\cdot}_{ads} \rightleftharpoons R^{\cdot} \xrightarrow{-e} R^+ \rightarrow \text{alcohols, ethers, esters, alkenes}$$

$$R{-}COO^{\cdot}_{ads},\ R^{\cdot}_{ads} \rightarrow R{-}H + R{-}R$$

Correlation of the electrolytic products of various carboxylates with the ionization potentials of the intermediate radicals has led to the conclusion that alkyl radicals with gas phase potentials below 8 eV give mainly the carbenium ions [Eberson, 1963]. Thus electronic effects exerted by an α-substituent can favor one type of intermediate or the other. Bulky groups hinder coupling of the free radicals so that they suffer further oxidation more readily.

The classical Kolbe electrolysis is identified with decarboxylative dimerization. The process is favored by the use of a platinum electrode, high current density, as well as a high concentration of the carboxylate ion but absence of strong nucleophiles. One should also select neutral or weakly acidic conditions (by neutralizing 2-5% of the carboxylic acid to be electrolyzed with alkali) and keep the anion concentration constant in the process. This condition is easily achieved in an undivided cell as consumption of carboxylate on the anode is always balanced by the generation of an equivalent amount of base on the cathode which serves to deprotonate the acid. At the point when the acid is depleted the pH of the electrolyte abruptly rises to the alkaline region. Anions other than the carboxylate should be excluded from the system so that interference of the carboxylate layer formation at the anode is minimized.

High temperatures (>50°C) should be avoided as disproportionation of the free radicals and/or esterification of the acid are prominent under such conditions. When methanol is used as solvent high temperatures also promote esterification of the carboxylic acid. Methanol is probably the most suitable medium although aqueous acetonitrile can be used. In water the current yield of the product is low; insufficient aqueous solubilities of many carboxylic acids can cause problems.

For electrolysis in nonaqueous media one may use platinized titanium, gold, and nonporous graphite as anode; generally there is no restriction on the cathode except when unsaturated carboxylic acids are being electrolyzed. Hydrogenation of the substrate/product can be avoided by using a steel cathode.

<u>General Procedure for Electrolysis of Carboxylic Acids</u>.

An acid (0.2 mol) is dissolved in methanol (100 mL) containing sufficient amount of NaOMe to neutralize 2% of the acid in a cell equipped with smooth platinum plates (4x2.5 cm) which are placed a few cm apart. Electrolysis is conducted with a constant current of 1.5-2.0 A while maintaining the temperature at 40-50°C by external cooling. When the mixture becomes alkaline, electricity is turned off and the solvent is evaporated. The product is extracted into ether and purified.

When aqueous methanol is used as the electrolytic medium a layer of alkanes is added to free the electrodes of nonpolar products.

Many functional groups are unaffected during the mild electrolysis, therefore steps associated with their protection are saved. This is an attractive attribute of the method. A synthesis of (+)-α-onocerin acetate [Stork, 1963] enlisted the Kolbe electrolysis as a key step.

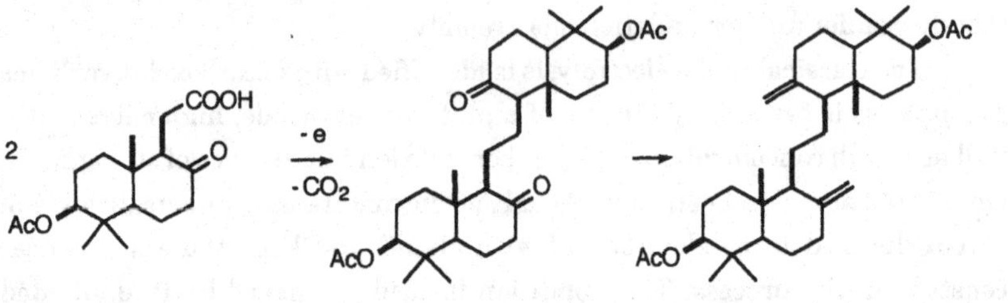

α-onocerin acetate

Cross coupling product was observed by Wurtz on electrolysis of a mixture of two carboxylic acids. Usually an equimolar mixture gives rise to a 1:2:1 distribution of products consisting of two homodimers in the minor ratio. In order to maximize the unsymmetrical product one can use the less expensive acid in excess to divert the formation of only two major products. According to the equation %=100n/(1+n), the calculated yield of R-R' from RCOOH and R'COOH is 50, 66, 80, 86, 91%, respectively when n=1, 2, 4, 6, 10, where n is the mole ratio of the two acids. Actual results indicate each value is factored by about 0.7, due to efficiency and other reasons contributing to the yield loss. Some acids which have low solubility in water must be totally converted into the salt in the mixed coupling. In such circumstances a mercury cathode is used, the electrolyte remains neutral because the alkali metal cations are amalgamated.

Examples illustrating synthetic applications of the Kolbe electrolysis of mixed carboxylic acids include muscalure [Gribble, 1973; Yadav, 1984], disparlure [Klünenberg, 1978], *exo*-brevicomin [Knolle, 1975], and the vitamin-E side chain [Takahashi, 1979]. In the formulae each wavy line indicates the C-C bond created by mixed coupling of two acids. In-chain components are conveniently assembled from the monoester of a dicarboxylic acid, which after the first cross-couping the ester group of the product is saponified to reveal a new acid for the second stage electrolysis.

muscalure

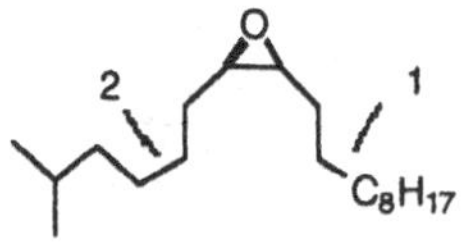

disparlure

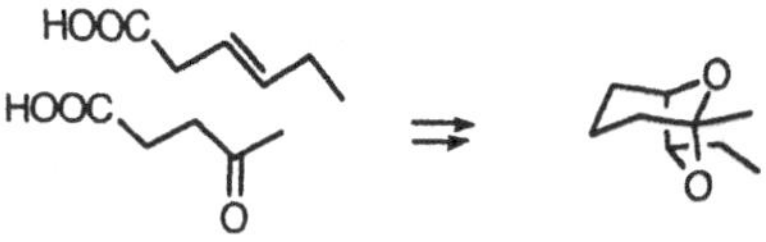

exo-brevicomin

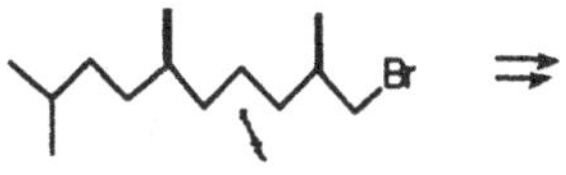

vitamin-E

6.1.5.2. Non-coupling Processes of Carboxylic Acids

Structural features can divert the the free radicals derived from electrooxidation of carboxylate ions to other more favorable reactions than coupling. For example, the radical readily undergoes intramolecular addition to a double bond to form a five membered ring, as shown in a preparation of a prostaglandin intermediate [Becking, 1988]. This approach of radical generation is better than the customary route involving debromination by toxic organotin hydride reagents.

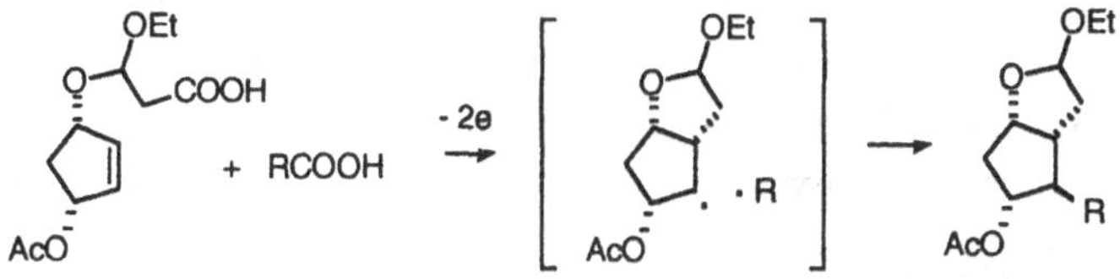

Available to β,γ-epoxy carboxylic acids is a reaction pathway leading to α,β-enones [Torii, 1979a]. The epoxy radical intermediates are prone to cleavage to relieve ring strain, and further oxidation of the resulting oxygen radical furnishes the observed products. Fragmentation of γ-hydroxy acids [Corey, 1959; Wharton, 1963] may occur from the radical intermediates or after rapid oxidation to the cations. γ-Lactones suffer the same fate under basic conditions [Michaelis, 1987]. Cyclic acetals of γ-keto acids are simultaneously oxidized and fragmented [Wuts, 1986]. Carboxylic acids with a δ-substituent which can stabilize a positive charge at the γ-position are also prone to fragmentation [Schäfer, 1985].

COOH; -e, MeOH, Et_3N (Pt); 84%

COOH, H, OH; -2e, $-CO_2$, $-2H^+$

R, COOH; -e, $-H^+$, $-CO_2$; R, +; R, +; H_2O; R, O, OH

COOH, Me_3Si; -e, MeCN - EtOH (C anode); 85%

Several other features favor non-Kolbe processes. Both steric hindrance and stabilization of the free radical encourage its oxidation to the cation. Selected examples are degradation of disubstituted malonic esters [Nokami, 1979], of β-lactam carboxylic acids [Mori, 1988], and bisdecarboxylation of succinic acids [Westberg, 1968; Radlick, 1968]. When acetonitrile is used as solvent the electrolysis of 1-methylcyclohexanecarboxylic acid gives *N*-acetyl-1-methylcyclohexylamine [Kornprobst, 1968].

HOOC, COOH; -e, Et_3N - HOAc, tBuOH (C); NaOH; *cis*-jasmone

RO, H, COOH, NH; -e, HOAc - NaOAc, MeCN; RO, H, OAc, NH

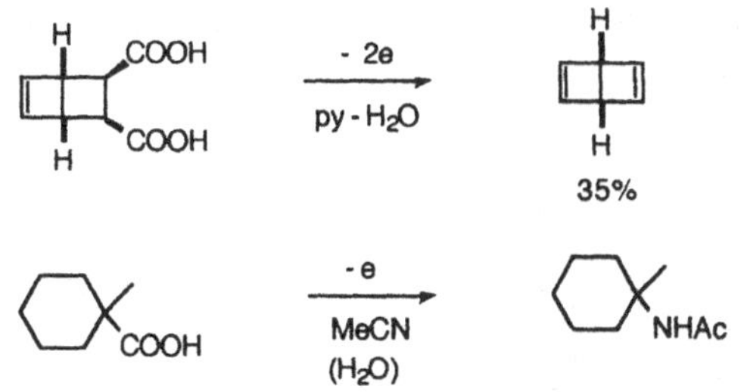

Skeletal rearrangement resulting from electrolysis of certain carboxylic acids can be correlated to the intervension of cations. Advantage has been taken to utilize such reactions in synthesis, e.g., of methyl jasmonate [Torii, 1975] and muscone [Shono, 1977b].

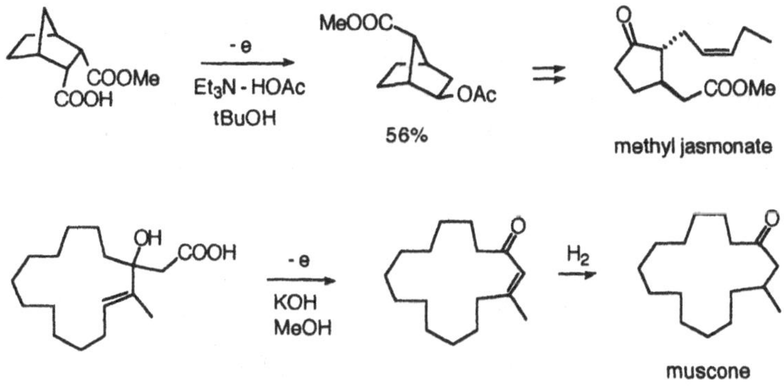

methyl jasmonate

muscone

Pseudo-Kolbe electrolysis occurs when the β-position of a carboxylic acid is oxidized faster. Decarboxylation is not preceded by oxidation of the carboxylate anion. This phenomenon is observed in the electrolysis of 4-methoxyphenylacetic acid [Coleman, 1974].

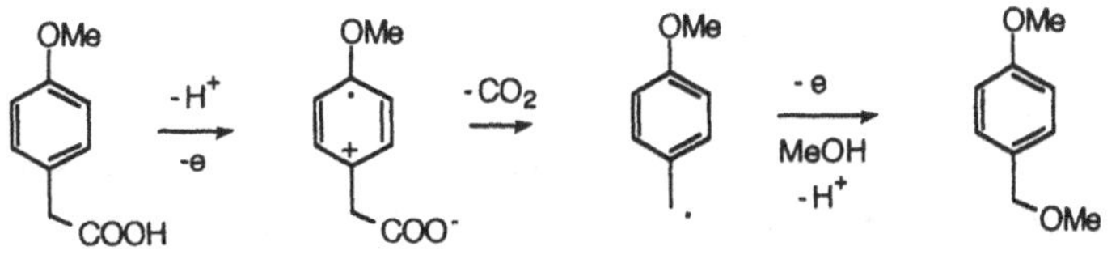

The oxidative degradation of α-sulfenyl acids [Nokami, 1978] to norketones may proceed by the same mechanism. The decarboxylative rearomatization of methoxylated dihydrobenzoic acids [Slobbe, 1977] to afford *m*-substituted anisoles represents the vinylogous examples.

R, COOH; MeO; R' —[$-H^+$, -e; MeOH, NaOMe]→ R; MeO; R' (60-70%)

6.2. ELECTROREDUCTION

In this section not only reduction of functional groups is covered, further reactions of the electroreduced species will also be discussed.

6.2.1. Of CC Multiple Bonds

Alkenes, alkynes, and arenes may be reduced by direct electron transfer or electrocatalytic method. The former involves generation and protonation of radical anions followed by a second-step reduction to form carbanions which are protonated. In the electrocatalytic method protons are reduced at the cathode and the adsorbed hydrogen is the reducing agent. The mechanistic difference of the two is reflected in the reduction of 8,9-dehydroestradiol 3-methyl ether which gives a product with either the *cis*- or *trans*-B/C ring junction [Junghans, 1973, 1974].

+ 2e, Pd; $2H^+$ / + 2e, $MeNH_2$; Et_4NOTs; - 20°C

It is remarkable that the side chain double bond of cinnamic acid is selectively hydrogenated but allylbenzene gives 1-allylcyclohexa-1,4-diene [Benkeser, 1969]. The electroreduction of phthalic anhydride to 1,2-dihydrophthalic acid is an industrial process [Wegenknecht, 1983].

+ 2e

+ 2e
LiCl
$MeNH_2$

+ 2e
H_2O, H_2SO_4
(Pb)

Co-electrolysis of a conjugated diene and an ester (except ArCOOMe) using a magnesium electrode generates a 3-cyclopentenol [Shono, 1992]. Magnesium not only supplies electron for the reduction of the diene, it also participates in reaction to form a magnesiacyclopentene intermediate. Under similar conditions cyclopropanols are formed from styrenes.

+ R"COOMe
$LiClO_4$, THF
(Mg cathode
Cd anode)

+ AcOMe

Dialkylalkynes are reduced to either the *(Z)*- or *(E)*-alkenes by varying the electrolytic conditions [Campbell, 1943; Benkeser, 1968].

+ 2e
EtOH - HCl
(Ni)
80%

+ 2e
LiCl, $MeNH_2$
(Pt)
undivided cell

The radical carbanions produced from styrenes are readily intercepted by acetonitrile or DMF which are used as solvent [Engels, 1978]. Bisenones as well as acrylonitrile undergo reductive coupling at the β-carbon [Mandell, 1976; Danly, 1984]. The production of adiponitrile, which is a starting material for Nylon 66, represents a prominent application of electrochemical process in industry.

Ph, + 2e, $LiClO_4$-DMF (Pt), Ph, OHC, CHO, 82%

2-Phenylsuccinaldehyde. [Engels, 1978]

A solution of styrene (20 mmol) and lithium perchlorate (40 mmol) in anhydrous DMF (80 mL) is electrolyzed at -10°C with a current density of ca. 25 mA/cm^2 in a diaphragm cell equipped with Pt mesh cathode (surface 18 cm^2) until 20 mF of electricity is passed. The product is isolated as the oxime methyl ether (82% yield).

+ 2e, H_2O, MeCN, O, O, O, O, 81%

2 CN, + 2e, Et_4NOTs, H_2O, NC, CN

When a conjugated system juxtaposed with a leaving group (X) undergoes electroreduction, intramolecular alkylation occurs at the β-carbon [Gassman, 1981; Nugent, 1982]. The reaction indeed involves reduction of the conjugated carbonyl system and displacement instead of reductive cleavage of the C-X bond followed by Michael addition has been confirmed using a stereochemical probe [Gassman, 1989b].

TsO, O, + 2e, O

An analogous pathway ia apparently followed during the generation of 2-phenyl-4-oxopentanoic acid upon electrolysis of benzalacetone in the presence of carbon dioxide [Harada, 1984]. Only that the C-C bond formation involves an external electrophile.

(Hg)
MeCN - Et_4NClO_4
CO_2

Electrohydrodimerization [Baizer, 1980] is well known for its application to adiponitrile preparation from acrylonitrile [Baizer, 1964]. It is generally considered that the radical anion is protonated and adds to a second alkene. The reaction is terminated by reduction and protonation of the new radical. Alkenes with electron-withdrawing substituents undergo this process is very useful for reductive cyclization of appropriate substrates to form 3-, 5-, and 6-membered rings. The method plays an important role in a synthesis of sterpurene [Moëns, 1986].

+ 2e
MeCN - Bu_4NBr
$CH_2(COOEt)_2$
$CeCl_3$
(Hg)

sterpurene

Intramolecular reductive cyclization involving a butenolide and a cyclopentylidenemalononitrile provides a key intermediate of pentalenolactone-E methyl ester [Bode, 1990]. The desired diastereomer is obtained predominantly (ratio 11:1) with magnesium perchlorate as supporting electrolyte. It is also crucial that the substrate is an alkylidenemalononitrile (or a malonic ester) because the corresponding acrylonitrile fails to cyclize (reduction of the unsaturated lactone moiety only).

- 1.6 V
MeCN - Bu_4NBr
$CH_2(COOMe)_2$
$Mg(ClO_4)_2$

cis-syn-cis

pentalenolactone-E methyl ester

It is interesting that in a synthesis of quadrone [Sowell, 1990], formation of the two component rings to complete the bridged system relied on such intramolecular alkylation method using an aldehyde acceptor.

CHO COOMe —(+ 2e, MeCN - Bu_4NBr, $CH_2(COOMe)_2$;)→ HO COOMe 89% + lactone ⇒ HCCN OR CHO —(+ 2e, MeCN - Bu_4NBr, $CH_2(COOMe)_2$;)→ NC OR HO H ⇒ quadrone

quadrone

Esters conjugated to an allene behave in the expected manner. When another conjugated ester group is tethered to the allene at suitable length, an intramolecular reaction leads to cyclic diester [Torii, 1992].

COOEt R COOMe R' —(+ 2e, DMF - Et_4NOTs, (C))→ EtOOC R COOMe R' 69-96%

Reduction of a β-keto ester enol phosphate at a mercury anode seems to generate a vinyl radical at the β-carbon which can add initiate a cycloaddition to a remote double bond of the proper substrate. Cyclopropane formation is witnessed [Gassman, 1989a].

EtOOC OPO(OEt)$_2$ —(+e, DMF - Bu_4NClO_4, (Hg))→ EtOOC 58%

$(EtO)_2P(O)$ COOEt —[+e; DMF - Bu_4NClO_4; (Hg)]→ EtOOC 63%

6.2.2. Of C-X Bonds

For electrochemical reduction of the C-Hal bond the ease follows the order of R-I > R-Br > R-Cl >> R-F and tertiary > secondary > primary halides. A halogen atom geminal to another similar atom suffers cleavage more readily, i.e., $R\text{-}CX_3 > R\text{-}CHX_2 > R\text{-}CH_2X$. Intermediates for the reduction are either free radicals or radical anions.

Direct anodic oxidation of an organohalide occurs by way of electron transfer from the nonbonding orbital of the halogen atom to the electrode. Subsequent scission of the C-Hal bond liberates a carbocation which, depending on its nature, may undergo Wagner-Meerwein rearrangment or be trapped by a nucleophile. The species generated from electrolysis of cyclopropylmethyl iodide apparently opts for the rearrangement route, as *N*-acetylcyclobutylamine is formed in 90% yield when the electrolysis is carried out in acetonitrile [Laurent, 1976].

I —[-e; MeCN; $LiClO_4$; (Pt)]→ [+ → +] —[MeCN; H_2O]→ NHAc 90% + NHAc

The demarcation of reactivities between bromides, iodides and chlorides and fluorides is clearly indicated by the electrochemical behaviors of 1-haloadamantanes [Koch, 1973; Vincent, 1976].

X —[MeCN; $LiClO_4$]→ X=Cl, 91%; X=F, 65%: X, NHAc; X=Br, 89%: NHAc

Aromatic halides are also reduced relatively easily, with formation of radical anions of varied lifetimes. From a deuterium incorporation study it can be concluded that long-lived species diffuse into solution before ejecting halide and abstracting a hydrogen atom from the solvent, whereas short-lived radical anions lose halide immediately upon formation and the resulting aryl radicals undergo rapid second reduction on the electrode surface to furnish the anions which then protonate (deuterate) [Grimshaw, 1975]. However, a radical cation can also act as an electrophile for the unoxidized substrate and the substitution product undergoes further oxidation [Miller, 1977].

$$\text{Ph-I} \xrightarrow[\text{LiClO}_4\ (\text{Pt})]{-e,\ \text{MeCN}} \left[\text{Ph-I}^{+\bullet}\right] \xrightarrow[-e,\ -H^+]{\text{Ph-I}} \text{Ph-}\overset{+}{\text{I}}\text{-C}_6\text{H}_4\text{-I}\ \ \text{ClO}_4^-$$

Interception of an iodoarene radical cation with chloride ion forms a chloroiodoarene radical which is a source of chlorine atom. Hydrogen abstraction from a proximal C-H bond then generates a radical site which can react to give the chloride. Within the steroid framework the reaction sequence serves to functionalize an unactivated carbon site. Thus the electrochemical procedure can be developed into a highly regioselective, remote functionalization method [Breslow, 1976], with site selection being controllable by varying the chain tethering the iodoarene group and the steroid skeleton. The following example shows the introduction of a 9α-chlorine atom.

C_8H_{17} H H O O I; -e, MeCN, Bu_4NX (Pt,C); C_8H_{17} H H O O I Cl; C_8H_{17} H O O I; .Cl; C_8H_{17} H Cl O O I

An arylzinc bromide used for Pd-catalyzed coupling with another bromoarene can be generated in situ by electrochemical method [Sibille, 1993]. Actually the organozinc is formed by transmetallation with the organonickel species. Thus reaction takes place at the nickel cathode gives rise to the latter while Zn(II) ion is formed at the anode.

Biaryl. [Sibille, 1993].

A mixture of freshly distilled DMF (35 mL), dried $ZnBr_2$ (8 mmol), $Ni(BF_4)_2bpy_3$ (1 mmol), 2,2'-bipyridine (2 mmol), and Bu_4NBr (1 mmol) is introduced into a cylindrical undivided cell fitted with Zn anode and carbon fiber cathode. Under argon at room temperature a 2 F/mol electricity (constant current density 0.15 A) is passed. The solution is delivered via syringe to another DMF solution (10 mL) of $(Ph_3P)_2PdCl_2$ (0.5-2%) and the haloarene (10 mmol) to be coupled. Stirring continues for 1-2 h to allow completion of reaction. Usual workup followed by silica gel chromatography furnishes the product.

Zn(0) anode --(-2e)--> Zn(II)

Ar-X → Ar-NiX → Ar-ZnX

Ni(0) ←(+ 2e, cathode)— Ni(II) + Ar-ZnX

Ar-ZnX --(Ar'-X, Pd catalyst)--> Ar-Ar'

Selective electrohydrogenolysis of polyhalogen compounds is synthetically more interesting [Fry, 1970; Pletcher, 1980]. Stereoselectivity is also observed, for example in the formation of *endo*-7-chlorobicyclo[4.1.0]heptane from the *gem*-dichloro compound [Fry, 1968].

(7,7-dichlorobicyclo[4.1.0]heptane) --(+ 2e, MeOH - HCl)--> *endo*-7-chlorobicyclo[4.1.0]heptane, 100%

Carbanions formed by electroreduction can be used to form ketones [Shono, 1977d]. For example benzyl ethyl ketone is derived from benzyl chloride and acetyl chloride in 69% yield. Note that benzyl chloride affords the carbanion more cleany than the bromide and iodide because less negative potential required for the latter two allows accumulation of the free radical which tends to dimerize.

$PhCH_2Cl \xrightarrow[MeCN - Et_4NOTs]{+ 2e} [PhCH_2^-] \xrightarrow{EtCOCl} PhCH_2COEt$ 69%

Allyl bromide is converted into the organometallic reagents in electrocatalytic systems containing small amount of tin [Uneyama, 1984b] or bismuth [Wada, 1987] ion. These reagents react with aldehydes readily. Selective 1,2-addition to conjugated aldehydes is observed. Another advantage of this method is that the active hydrogen atoms of hydroxy carbonyl compounds (e.g., glyceraldehyde) do not require protection; highly enolizable ketones such as methyl acetoacetate also react.

COOMe → (+ 2e; Sn / allyl Br; MeOH) → OH, COOMe

(Note that allylic acetates are converted by electroreductively generated Pd(0) to π-allylpalladium species which are further reduced and react with a zinc halide to give allylzinc reagents. In situ attack on carbonyl compounds results in the formation of homoallylic alcohols [Qiu, 1989].)

α-Halocarbonyl compounds are usually reduced to the carbanion stage because the α-acyl radical accepts a second electron very readily. Trapping of the enolate with an aldehyde (aldol condensation) is possible when a lanthanide salt is added [Fry, 1987]. The electroreduction of 2,2-dibromo-1,3-cyclohexanedione in the presence of an alkene affords a dihydrofuran [J. Yoshida, 1985].

Ph, COOMe, Br → (+ 2e; $LaBr_3$; PhCHO) → Ph, O, O, Ph

O, Br, Br, O → (+ 2e; R-alkene) → O, O, R

α-Bromoalkyl cyclic β-keto esters undergo ring expansion [Shono, 1990]. In the absence of a Lewis acid the reaction proceeds in low yield.

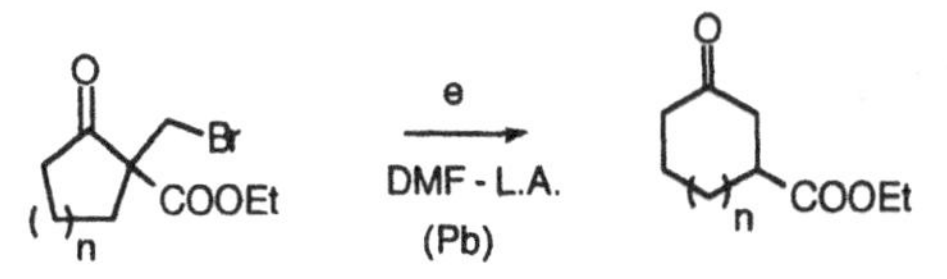

Reduction of carbon tetrachloride in DMF proceeds to the trichloromethide ion stage which can be used to form C-C bond with aldehydes and acrylonitrile [Baizer, 1972; Karrenbrock, 1978]. In the presence of chloroform the current yields reach up to 1000% owing to regeneration of Cl_3C^- by proton exchange. Tetroses are formed from the trichloromethyl adducts of D-glyceraldehyde acetonide on electroreduction, followed by methoxide displacement and hydrolysis [Shono, 1982c].

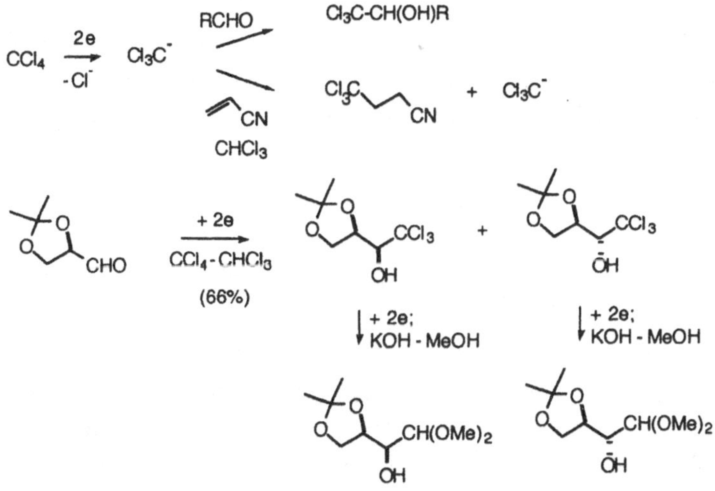

The exclusion of electrophile to react with the trichloromethide ion allows it to eject a chloride ion to form dichlorocarbene. Dichlorocyclopropanation of electron-rich alkenes can then be accomplished [Fritz, 1978]. Difluorocarbene is similarly accessible [Fritz, 1979]. On the other hand, *gem*-dichloro compounds can be dechlorinated to form carbenes [Fry, 1972].

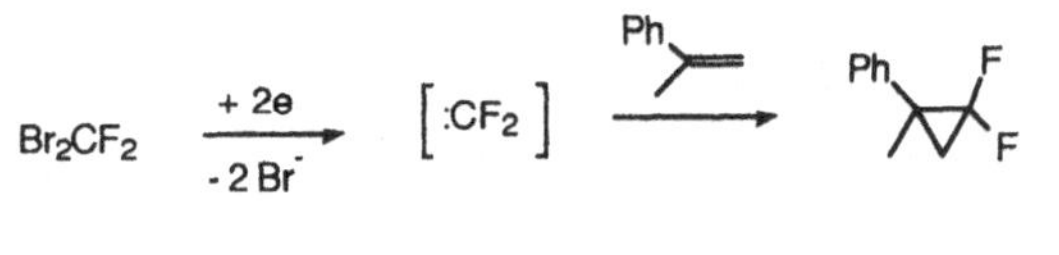

+ 2e
DMF

Generation of Difluorocarbene. [Fritz, 1979]

In a divided glass cell mounted with a ceramic diaphragm and provided with Pt electrodes, inert gas inlet and outlet and a mercury safety valve is charged in the cathode with dibromodifluoromethane (4 mL) and a dichloromethane (55 mL) solution of tetrabutylammonium bromide (0.35 M) and 2-phenylpropene (0.7 M, as trapping agent), while the anode is added the solution (35 mL) of the supporting electrolyte and some olefin (2 mL, for removal of bromine which is to be generated from oxidation of the bromide ion). Electrolysis is carried out at 18°C under dry nitrogen. The current yield is 57% from the catholyte, 4% from the anolyte.

Perhaps it should be mentioned that tosylnitrene is formed cathodically from the *N,N*-dichlorosulfonamide [Fuchigami, 1976]. *N*-Chlorocarbamates can be reduced to the anions and used as Michael donors for acrylic esters [Berube, 1975].

$$Ts\text{-}NCl_2 \xrightarrow[-2Cl^-]{+2e} [Ts\text{-}N:]$$

$$EtOOC\text{-}NHCl + CH_2{=}CHCOOMe \xrightarrow[MeOH\text{-}CHCl_3]{+2e} EtOOC\text{-}NH\text{-}CH_2CH_2COOMe \quad 61\%$$

vic-Dihaloalkenes undergo electrochemical elimination in the same manner as that induced by chemical reagents. Note that the side chain dibromide of 4-vinylcyclohexene is obtainable by electrodebromination of the tetrabromide [Husstedt, 1979], the cyclic dibromide is the product of controlled bromination. Electrochemical method produces no side products that could be detrimental to the alkenes. Thus, bicyclo[2.2.0]hex-1(4)-ene is produced quantitatively [Casanova, 1974]. 1-Adamantene is similarly acquired from 1,2-dihaloadamantane [Lenoir, 1983].

Br Br Br Br
+ 2e
- 2 Br⁻
Br Br
2 Br_2
Br_2
Br Br

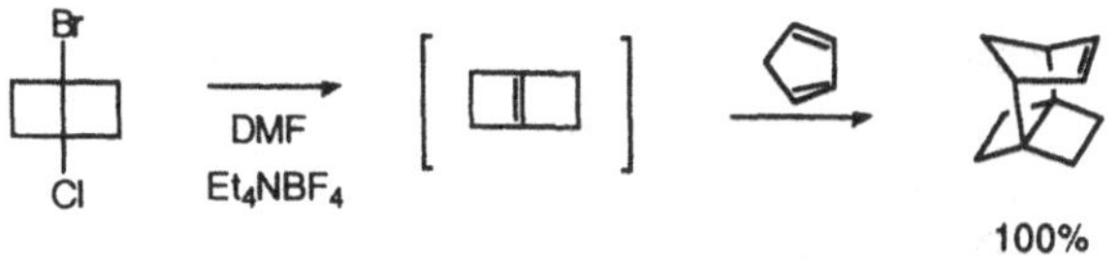

Br, Cl; DMF, Et_4NBF_4; 100%

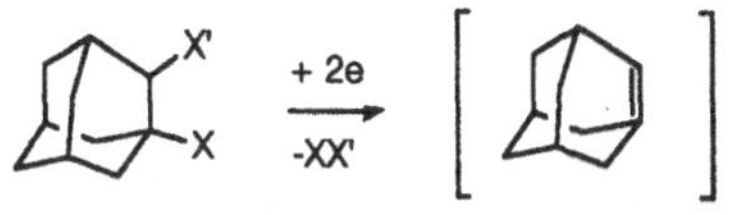

X', X; + 2e; -XX'

A related reaction is the removal of the 2,2,2-trichloroethyl group from protected alcohols (and acids), amines, and thiols [Semmelhack, 1972].

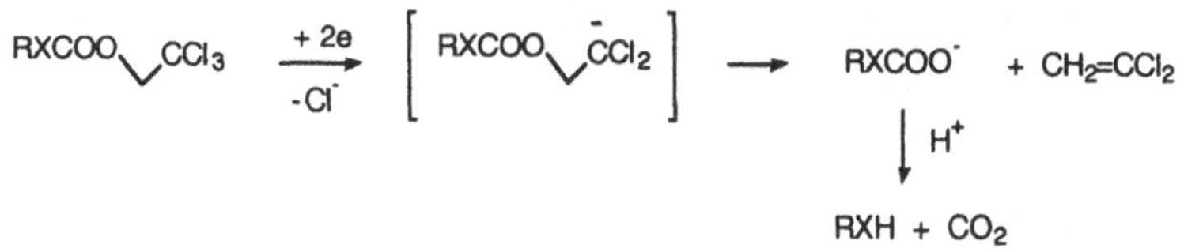

$RXCOO\diagdown\!\diagup CCl_3 \xrightarrow[-Cl^-]{+2e} [RXCOO\diagdown\!\diagup \bar{C}Cl_2] \longrightarrow RXCOO^- + CH_2{=}CCl_2$

$RXCOO^- \xrightarrow{H^+} RXH + CO_2$

Reductive elimination of 1,n-dihalide (n = 3,4,6) to form carbocycles of various sizes is shown in the following examples [Rifi, 1971; Caroll, 1980; Satoh, 1978].

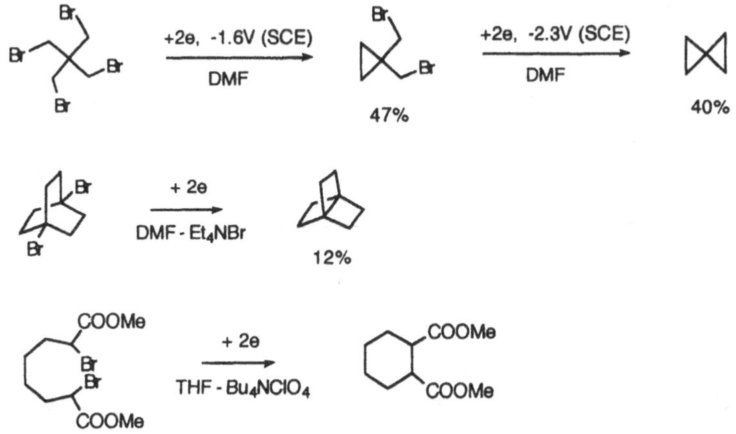

Br; +2e, -1.6V (SCE); DMF; 47%; +2e, -2.3V (SCE); DMF; 40%

Br; + 2e; DMF - Et_4NBr; 12%

COOMe, Br; + 2e; THF - Bu_4NClO_4; COOMe

The cyclization of bromoalkenes mediated by stoichiometric organotin hydride is undesirable in many aspects. A much improved method involves a nickel(II)-catalyzed electroreductive process [Ihara, 1996]. Not only the noxious reagent is avoided, it does not require high dilution.

EtO O Br COOMe — - 1.5 V (Ag/AgCl), Ni(cyclam)(ClO_4)$_2$, NH_4ClO_4, DMF → EtO O COOMe 88% ⇛ ipecac / corynanthe alkaloids

Reduction of halides in the presence of vitamin-B_{12} generates radical-type intermediates. Ring formation by intramolecular interception of such species is possible. Interestingly, by this method a bicyclic lactol ether derived from dihydro-α-ionone is stereoisomeric to the one obtained from tin hydride promoted cyclization [Hutchinson, 1987]. The reaction of acyl halides results in the formation of acylcobalt(III) complexes. In the presence of an electron-deficient alkene and assistance by visible light irradiation, transfer of the acyl group to the alkene acceptor occurs [Scheffold, 1983].

OEt Br O O — e, vit-B_{12}, MeOH → OEt O O

R X O, X⁻, R O, Co^{+}, Co^{3+}, + e, cathode, Co^{2+}, Z, R O Z

Quaternary phosphonium and ammonium salts undergo C-P and C-N bond cleavage, repectively [Horner, 1977b, 1988]. On a lead cathode electrochemical cleavage occurs at the intercyclic C-N bond of 1-oxoquinuclidine, the ketone then suffers further reduction to result in 5-azacyclodecanol [Leonard, 1952].

$n = 1\text{-}3$ $\xrightarrow[H_2O]{+ 2e}$ $(CH_2)_nNMe_2$ 88-90%

$Ph-P^{+}-Bn$ X^- $\xrightarrow[H_2O]{+ 2e}$ Ph–P 72%

The C-S bond of a sulfone is susceptible to scission. Synthetically more useful processes are selective removal of allylic sulfonyl group (mediated by anthracene), cleavage of arenesulfonamides in the presence of other protecting groups [Horner, 1969], and reductive elimination of β-substituted sulfones to give alkenes, in which even a hydroxide ion can act as leaving group [Shono, 1978a].

$PhSO_2$ ... SO_2Ph $\xrightarrow[\text{anthracene, DMF - } Bu_4NI]{+ 2e}$ $PhSO_2$ ... 60%

MeO ... N H ... COOH, N(H)SO_2Ph (-2.4 V) $\xrightarrow[EtOH,\ Me_4NCl]{- 2.15 V}$ MeO ... N H ... COOH, NH_2 90%

$RCOOMe + ArSO_2CH_2MgI \rightarrow RCOCH_2SO_2Ar \rightarrow RCH(OH)CH_2SO_2Ar \xrightarrow{+ 2e,\ DMF} R\text{-}CH{=}CH_2$

6.2.3. Reduction and Reductive Alkylation of X=Y Bonds

The electrochemical reduction of carbonyl compounds in the presence of some other functional groups makes it a valuable synthetic method. A particularly intriguing and economically attractive process is the cathodic reduction of glucose to sorbitol while at the anode glucose is being oxidized to gluconic acid [Park, 1985].

sorbitol $\xleftarrow[\text{(Ra-Ni)}]{+ 2e}$ glucose $\xrightarrow[\text{(C)}]{- 2e,\ CaBr_2}$ gluconic acid

80-100% 100%

An aldehyde can be reduced selectively in the presence of a ketone at lead cathode. Thus after ozonolysis of 1-methylcyclohexene in aqueous acetic acid electroreduction of the mixture (NaOAc added) proceeds by O-O bond cleavage and elimination of acetic acid to give 6-oxoheptanal which then undergoes the selective reduction [Gora, 1982]. The aldehyde reduction can be avoided if a graphite electrode is used.

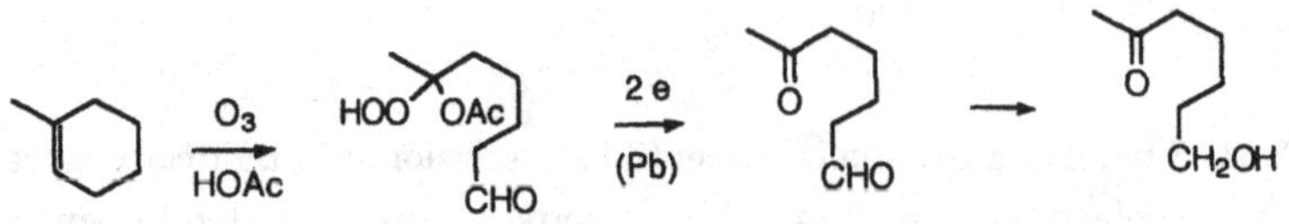

A pH dependence has been observed in the reduction course of aryl alkyl ketones. Secondary alcohols are formed in acid media, whereas *vic*-diols are the major products when the solution is basic.

Esters and acids which cannot undergo Kolbe electrolysis can be reduced cathodically [Horner, 1977a].

Pinacolization of 1,3-diketones gives cyclopropanediols which can be stabilized by acetylation [Curphey, 1969]. Bis-imines undergo cyclization. Chiral piperazines available by this method are valuable ligands for asymmetric synthesis [Shono, 1991b].

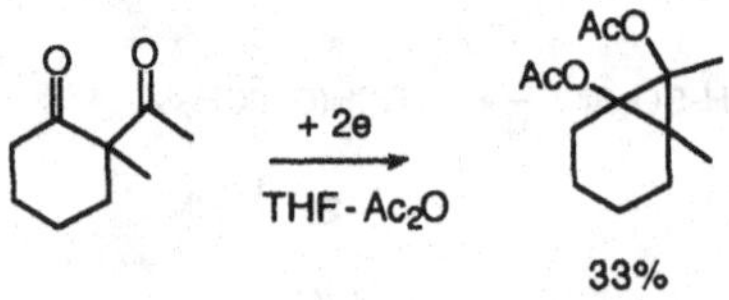

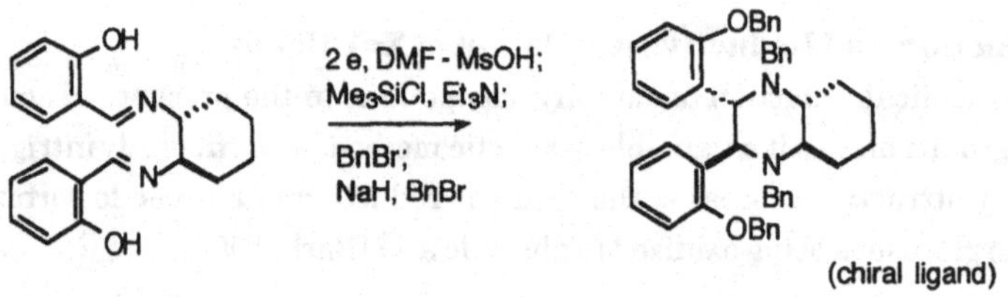

The reduction of carbonyl compounds in acidic media generates hydroxyalkyl radicals. In the presence of an alkene (preferably activated, although allyl alcohol is adequate) addition occurs [Bukhtiarov, 1980]. The intramolecular version of such a process is much more efficient, leading to cyclic alcohols [Shono, 1976d].

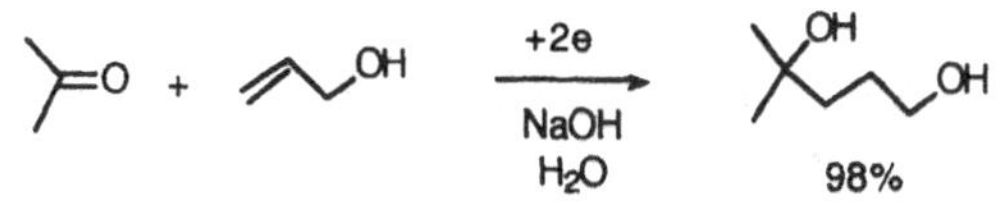

The electroreductive cyclization of keto nitriles is of significant importance in synthesis [Shono, 1992]. Thus convenient routes to dihydrojasmone, valeranone, $\Delta^{9(12)}$-capnellene and hirsutene are cleared.

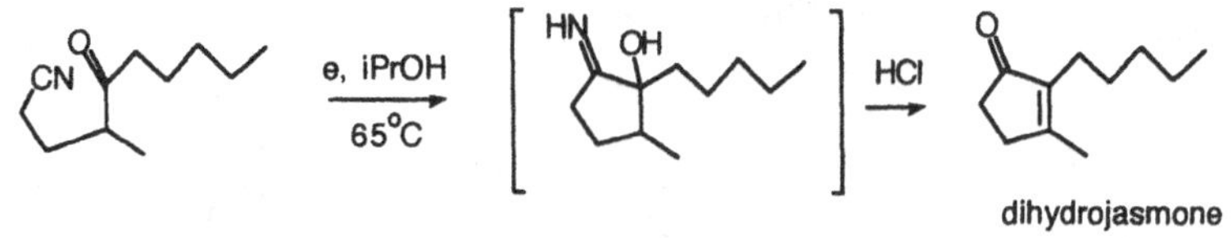

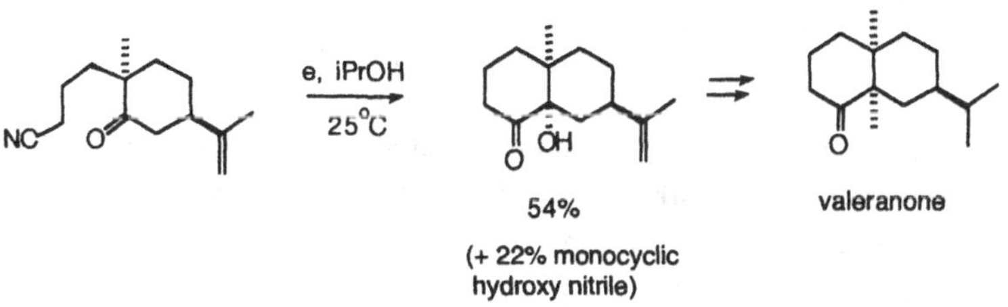

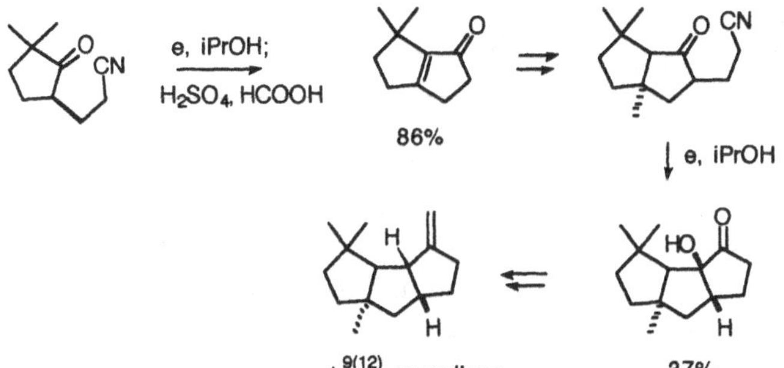

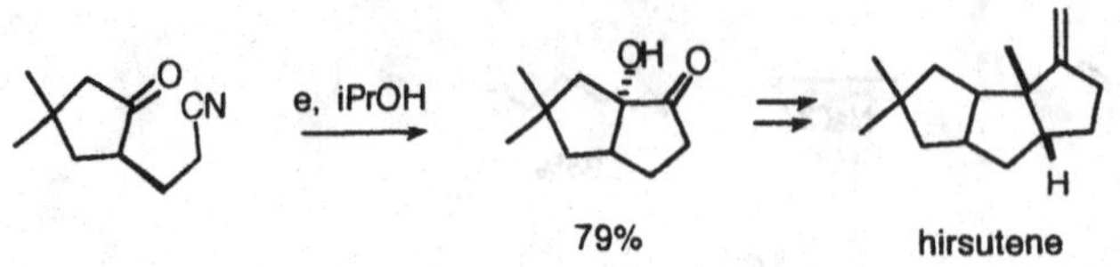

e, iPrOH

79% hirsutene

Remarkably, intramolecular attack of ketyl species on an aromatic ring has been observed, leading to condensed ring systems [Kise, 1994]. Analogously, *N*-oxoalkylpyridinium salts give quinolizidine and indolizidine derivatives on a mercury cathode [Gorny, 1995].

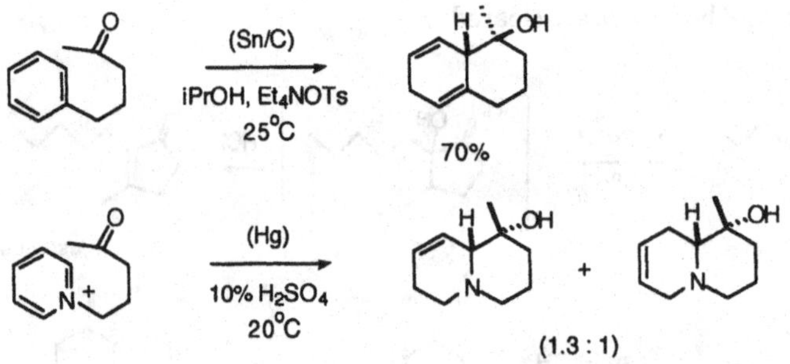

(Sn/C)
iPrOH, Et_4NOTs
25°C
70%

(Hg)
10% H_2SO_4
20°C
(1.3 : 1)

The tandem cyclization of nonconjugated dienones represents an efficient way to assemble angular bicycloalkanols [Kariv-Miller, 1989]. However, there seems to be a structural limitation that the second ring to be formed must be five-membered.

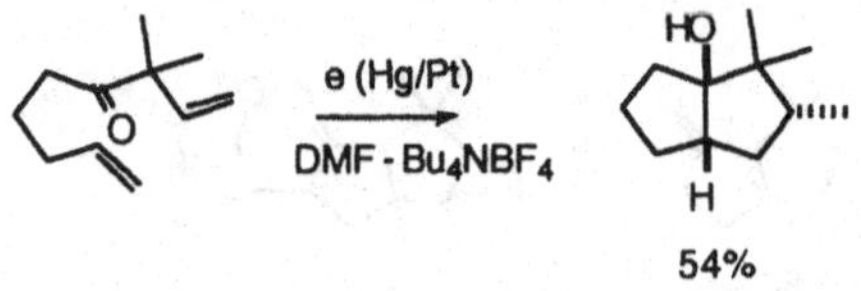

e (Hg/Pt)
DMF - Bu_4NBF_4
54%

In situ conversion of the carboxyl group to an acylphosphine is implicated in the reductive cyclization of δ-keto acids in an undivided cell in the presence of tri-*n*-butylphosphine and methanesulfonic acid [Maeda, 1992, 1994]. Cyclic α-ketols are readily available by this simple procedure.

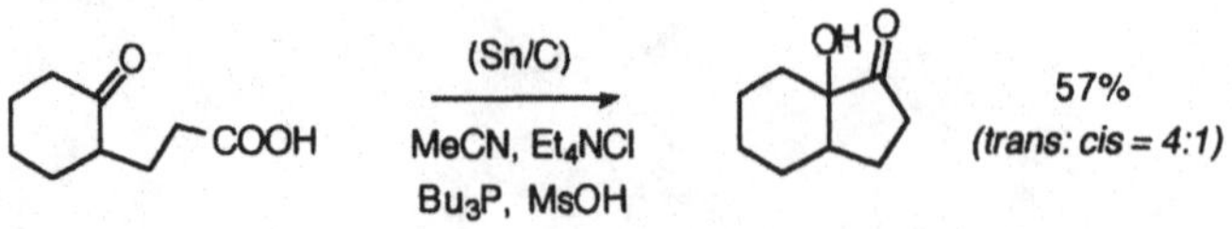

(Sn/C)
MeCN, Et_4NCl
Bu_3P, MsOH
57%
(trans: cis = 4:1)

On a lead cathode the amide group of pyroglutamic acid is reduced [Viscontini, 1966]. The electrolytic medium is acidic.

COOH $\xrightarrow[H_2SO_4]{+ 4e\ (Pb)}$ COOH 35%

An aldehyde-primary amine mixture is converted to the secondary amine [Pienemann, 1987], apparently via the Schiff base. Oximes and nitriles are also susceptible to cathodic reduction. The preparation of allylamine from acrylonitrile by electroreduction in a diaphragm cell is significant because conventional catalytic hydrogenation shows the opposite chemoselectivity.

+ $MeNH_2$ $\underset{H_2O}{\overset{-H_2O}{\rightleftharpoons}}$ NMe $\xrightarrow[(Hg)]{+ 2e}$ NHMe 90%

CN $\xrightarrow[30\%\ H_3PO_4\ (Pb)]{+ 4e}$ NH_2

α-Amino carbanions derived from reduction of iminium ions are nucleophilic; the tandem alkylation process is adaptable to alkaloid synthesis [Shono, 1978b]. Electrocarboxylation of certain imines uses carbon dioxide as external electrophile [N.L. Weinberg, 1971].

$\xrightarrow[DMF]{+ 2e}$ 60%

PhCH=N-Ph + CO_2 $\xrightarrow[Et_4N]{+ 2e}$ PhCH(COOH)-NHPh 60%

The behavior of 4-cyanopyridine in accordance with the pH and potential of the mercury cathode [Volke, 1972] is noteworthy.

CN + e → (pyridine) + (NH_2-substituted pyridine)

pH <7	-	++
pH ~7	more negative potential	more positive potential
pH >>7	++	-

The nitro group has a low reduction potential and it is reduced with extreme ease to either the hydroxylamine or the amine at controlled potential [Wawzonek, 1971]. The reduction of nitroarenes is more complicated, but it is possible to obtain one product selectively (among arylhydroxylamine, azoxyarene, azoarene, hydrazoarene, arylamine) by adjusting the cathode material, pH, and electric potential.

HO–…–NO_2 + 6e, H_2SO_4, 65-85°C → HO–…–NH_2 88%

+ 4e, H_2SO_4, 25-30°C → HO–…–NHOH 79%

6.3. REACTIONS WITH ELECTROCHEMICALLY GENERATED ACIDS AND BASES

Certain supporting electrolytes form acids near a platinum anode on electrolysis in organic solvents. Under such conditions benzylic cations are formed from alkyl benzenes, allylic alcohols are ionized to initiate cyclization [Uneyama, 1984a], and epoxides undergo isomerization to carbonyl compounds [Uneyama, 1983]. Strongly acidic catalysts used in acetalization can be replaced by electrogenerated acid [Torii, 1983a]. The Mukaiyama aldol condensation occurs when a mixture of a silyl enol ether and a dimethyl acetal is electrolyzed at low temperatures in the presence of the proper electrolytes [Torii, 1984; Inokuchi, 1988].

OH … (acetone), $LiClO_4$, (Pt) → HO … + …

Cl⁄⁄Cl
NaOTs, Et_4N OTs
(Pt)
79%

Me_3SiO + Me_3SiO
-e
$LiClO_4$-CH_2Cl_2
90%

Anions and radical anions derived from cathodically reduced organic compounds can act as bases. They are useful if they show strong base character while being weak nucleophiles and weak reductants. Electrochemically generated trichloromethide anion from carbon tetrachloride can be used to deprotonate malonic esters, and the anion of pyrrolidone has been employed to promote the Dieckmann condensation. Stabilized phosphonium ylides are created from the conjugate acids (salts) as the split carbanion components from the latter act as the base.

Due to the mild conditions for electrogeneration of the pyrrolidinone anion the esterification of unstable carboxylic acids including hydroxy acids and amino acids (e.g., gibberellic acid, penicillins) as well as sterically hindered acids can be safely performed in excellent yields [Shono, 1986]. Furthermore, macrolactonization of ω-bromo carboxylic acids under these conditions is feasible. A virtually quantitative yield of the methoxymethyl ester of 3-phenylpropanoic acid is obtainable even the electrophile ($ClCH_2OMe$) is unstable (has very high and nonselective reactivity).

$$PhCH_2CH_2COOH + ClCH_2OMe \longrightarrow PhCH_2CH_2COOCH_2OMe$$

The same base can deprotonate trifluoromethane to initiate trifluoromethylation of benzaldehyde [Shono, 1991a]. The reaction is most efficient when the counterion is tetra-*n*-octylammonium.

On electroreduction azobenzene gives rise to a sufficiently strong base to perform many functions. Thus a nitroalkane is readily deprotonated to be used as a Michael donor. With subsequent conversion of the nitro group to a carbonyl it provides an expedient alternative to the preparation of γ-keto esters and related compounds [Monte, 1983].

NO_2 + COOMe $\xrightarrow[\text{Bu}_4\text{NBr, PhN=NPh}]{\text{e, MeCN}}$ NO_2 COOMe $\xrightarrow[\text{Bu}_4\text{NBr, } O_2]{\text{e, MeCN}}$ O COOMe 60%

Monographs

Baizer, M.M.; Lund, H. (1983) *Organic Electrochemistry*, 2nd Ed., M. Dekker: New York.

Fry, A.J. (1989) *Synthetic Organic Electrochemistry*, 2nd Ed., Wiley: New York.

Shono, T. (1984) *Electroorganic Chemistry as a New Tool in Organic Synthesis*, Springer: Berlin.

Volke, J.; Liska, F. (1994) *Electrochemistry in Organic Synthesis*, Springer: Berlin.

References

Baizer, M.M. (1964) *J. Electrochem. Soc.* **111:** 215.

Baizer, M.M. (1980) *Chemtech* **10:** 161.

Becking, L.; Schäfer, H.J. (1988) *Tetrahedron Lett.* **29:** 2801.

Benkeser, R.A.; Tincher, C.A. (1968) *J. Org. Chem.* **33:** 2727.

Benkeser, R.A.; Mels, S.J. (1969) *J. Org. Chem.* **34:** 3970.

Berube, D.; Caza, J.; Kimmerle, F.M.; Lessard, J. (1975) *Can. J. Chem.* **53:** 3060.

Bewick, A.; Edwards, G.J.; Jones, S.R.; Mellor, J.M. (1977) *J. Chem. Soc. Perkin Trans. 1* 1831.

Birch, A.J.; Slobbe, J. (1978) *Aust. J. Chem.* **31:** 2559.

Blum, Z.; Ekström, M.; Wistrand, L.-G. (1984) *Acta Chem. Scand.* **B38:** 297.

Bode, H.E.; Sowell, C.G.; Little, R.D. (1990) *Tetrahedron Lett.* **31:** 2525.

Breslow, R.; Goodin, R. (1976) *Tetrahedron Lett.* 2675.

Brinkhaus, K.H.G.; Steckhan, E.; Schmidt, W. (1983) *Acta Chem. Scand.* **B37:** 499.

Bukhtiarov, A.V.; Tomilov, A.P. (1980) *Usp. Chimii* **49:** 470.

Campbell, K.N.; Young, E.E. (1943) *J. Am. Chem. Soc.* **65:** 965.

Caroll, W.F.; Peters, D.G. (1980) *J. Am. Chem. Soc.* **102:** 4127.

Casanova, J.; Rogers, H.R. (1974) *J. Org. Chem.* **39:** 3803.

Clauson-Kaas, N.; Limborg, F.; Glens, K. (1952) *Acta Chem. Scand.* **6:** 531.

Coleman, J.P.; Lines, R.; Utley, J.H.P.; Weedon, B.C.L. (1974) *J. Chem. Soc. Perkin Trans. II* 1064.

Coleman, J.P.; Hallcher, R.C.; McMackins, D.E.; Rogers, T.E.; Wegenknecht, J.H. (1991) *Tetrahedron* **47:** 809.

Corey, E.J.; Sauers, R.R. (1959) *J. Am. Chem. Soc.* **81:** 1739, 1743.

Curphey, T.J.; Amelotti, C.W.; Layloff, T.P.; McCartney, R.L.; Williams, J.H. (1969) *J.*

Am. Chem. Soc. **91:** 2817.
Danly, D.E. (1984) *J. Electrochem. Soc.* **131:** 435C.
Eberson, L. (1963) *Acta Chem. Scand.* **17:** 2004.
Eberson, L.; Radner, F. (1980) *Acta Chem. Scand.* **B34:** 739.
Eberson, L.; Oberrauch, E. (1981) *Acta Chem. Scand.* **B35:** 193.
Elming, N.; Clauson-Kaas, N. (1955) *Acta Chem. Scand.* **9:** 23.
Elofson, R.M.; Gadallah, F.F.; Laidler, J.K. (1985) *Can. J. Chem.* **63:** 1170.
Engels, R.; Schäfer, H.J. (1978) *Angew. Chem. Int. Ed. Engl.* **17:** 460.
Evans, D.A.; Tanis, S.P.; Hart, D.J. (1981) *J. Am. Chem. Soc.* **103:** 5813.
Fritz, H.P.; Kornrumpf, W. (1978) *Liebigs Ann. Chem.* 1416.
Fritz, H.P.; Kornrumpf, W. (1979) *J. Electroanal. Chem.* **100:** 217.
Fry, A.J.; Moore, R.H. (1968) *J. Org. Chem.* **33:** 1283.
Fry, A.J.; Mitnick, M.; Reed, R.G. (1970) *J. Org. Chem.* **35:** 1232.
Fry, A.J.; Reed, R.G. (1972) *J. Am. Chem. Soc.* **94:** 8475.
Fry, A.J.; Susla, M.; Weltz, M. (1987) *J. Org. Chem.* **52:** 2496.
Fuchigami, T.; Nonaka, T.; Iwata, K. (1976) *Chem. Commun.* 951.
Fuchigami, T.; Fujita, T. (1994) *J. Org. Chem.* **59:** 7190.
Fuchigami, T.; Konno, A. (1997) *J. Synth. Org. Chem. Jpn.* **55:** 301.
Gassman, P.G.; Rasmy, O.M.; Murdock, T.O.; Saito, K. (1981) *J. Org. Chem.* **46:** 5457.
Gassman, P.G.; Lee, C. (1989a) *J. Am. Chem. Soc.* **111:** 739.
Gassman, P.G.; Lee, C. (1989b) *Tetrahedron Lett.* **30:** 2175.
Gora, J.; Smigielski, K.; Kula, J. (1982) *Synthesis* 310.
Gorny, R.; Schäfer, H.J.; Fröhlich, R. (1995) *Angew. Chem. Int. Ed. Engl.* **34:** 2007.
Gribble, G.W.; Sanstead, J.K.; Sullivan, J.W. (1973) *Chem. Commun.* 735.
Grimshaw, J.; Trocha-Grimshaw, J. (1975) *J. Chem. Soc. Perkin Trans. 2* 215.
Grujic, Z.; Tabakovic, I.; Trkovnik, M. (1976) *Tetrahedron Lett.* 4823.
Hammerich, O.; Parker, V.D. (1974) *Chem. Commun.* 245.
Harada, J.; Sakakibara, Y.; Kunai, A.; Sasaki, K. (1984) *Bull. Chem. Soc. Jpn.* **57:** 611.
Horner, L.; Singer, R.J. (1969) *Liebigs Ann. Chem.* **723:** 1.
Horner, L.; Hönl, H. (1977a) *Liebigs Ann. Chem.* 2036.
Horner, L.; Röder, H. (1977b) *Phosphorus* 2067.
Husstedt, U.; Schäfer, H.J. (1979) *Synthesis* 964.
Hutchinson, J.H.; Pattenden, G.; Myers, P.L. (1987) *Tetrahedron Lett.* **28:** 1313.
Ihara, M.; Katsumata, A.; Setsu, F.; Tokunaga, Y.; Fukumoto, K. (1996) *J. Org. Chem.* **61:** 677.
Inokuchi, T.; Takagishi, S.; Ogawa, K.; Kurokawa, Y; Torii, S. (1988) *Chem. Lett.* 1347.
Inokuchi, T.; Matsumoto, S.; Torii, S. (1991) *J. Org. Chem.* **56:** 2416.

Iwasaki, T.; Nishitani, T.; Horikawa, H.; Inoue, I. (1982) *J. Org. Chem.* **47:** 3799.
Jäger, P.; Hannebaum, H.; Nohe, H. (1978) *Chem. Ing. Techn.* **50:** 787.
Junghans, K. (1973) *Chem. Ber.* **106:** 3465.
Junghans, K. (1974) *Chem. Ber.* **107:** 3191.
Kariv-Miller, E.; Maeda, H.; Lombardo, F. (1989) *J. Org. Chem.* **54:** 4022.
Kaulen, J.; Schäfer, H.J. (1982) *Tetrahedron* **38:** 3299.
Kerr, J.B.; Jempty, T.C. Miller, L. (1979) *J. Am. Chem. Soc.* **101:** 7338.
Kise, N.; Suzumoto, T.; Shono, T. (1994) *J. Org. Chem.* **59:** 1407.
Klehr, M.; Schäfer, H.J. (1975) *Angew. Chem. Int. Ed. Engl.* **14:** 247.
Klünenberg, H.; Schäfer, H.J. (1978) *Angew. Chem. Int. Ed. Engl.* **17:** 47.
Knolle, J.; Schäfer, H.J. (1975) *Angew. Chem. Int. Ed. Engl.* **14:** 758.
Koch, V.R.; Miller, L.L. (1973) *J. Am. Chem. Soc.* **95:** 8631.
Kuhn, A.T.; Birkett, M. (1979) *J. Appl. Electrochem.* **9:** 777.
Laurent, E.; Tardivel, R. (1976) *Tetrahedron Lett.* 2779.
Lenoir, D.; Kornrumpf, W.; Fritz, H.P (1983) *Chem. Ber.* **116:** 2390.
Leonard, N.J.; Swann, S.; Figueras, J. (1952) *J. Am. Chem. Soc.* **74:** 4620.
Lines, R.; Parker, V.D. (1980) *Acta Chem. Scand.* **B34:** 47.
Maeda, H.; Maki, T.; Ohmori, H. (1992) *Tetrahedron Lett.* **33:** 1347.
Maeda, H.; Maki, T.; Eguchi, K.; Koide, T.; Ohmori, H. (1994) *Tetrahedron Lett.* **35:** 4129.
Mandell, L.; Daley, R.F.; Day, R.A. (1976) *J. Org. Chem.* **41:** 4087.
Masui, M.; Ueshima, T.; Ozaki, S. (1983) *Chem. Commun.* 479.
Michaelis, R.; Müller, U.; Schäfer, H,J. (1987) *Angew. Chem. Int. Ed. Engl.* **26:** 1026.
Miller, L.L.; Watkins, B.F. (1976) *J. Am. Chem. Soc.* **98:** 1515.
Miller, L.L.; Hoffmann, A.K. (1977) *J. Am. Chem. Soc.* **99:** 593.
Miller, L.L.; Stewart, R.F., Gillespie, J.P.; Ramachandran, V.; So, Y.H.; Stermitz, F.R. (1978) *J. Org. Chem.* **43:** 1580.
Moëns, L.; Baizer, M.M.; Little, R.D. (1986) *J. Org. Chem.* **51:** 4497.
Monte, W.T.; Baizer, M.M.; Little, R.D. (1983) *J. Org. Chem.* **48:** 803.
Mori, M.; Ban, Y. (1985) *Heterocycles* **23:** 317.
Mori, M.; Kagechika, K.; Tohjima, K.; Shibasaki, M. (1988) *Tetrahedron Lett.* **29:** 1409.
Moyer, B.A.; Thompson, M.S.; Meyer, T.J. (1980) *J. Am. Chem. Soc.* **102:** 2310.
Nilsson, A.; Palmquist, U.; Pettersson, T.; Ronlan, A. (1978) *J. Chem. Soc. Perkin Trans. I* 696.
Nokami, J.; Matsuura, N.; Sueoka, T.; Kusumoto, J.; Kawada, M. (1978) *Chem. Lett.* 1283.
Nokami, J.; Yamamoto, T.; Kawada, M.; Izumi, M.; Ochi, N.; Okawara, R. (1979) *Tetrahedron Lett.* 1047.

Nugent, S.T.; Baizer, M.M.; Little, R.D. (1982) *Tetrahedron Lett.* **23:** 1339.
Ohmori, H.; Nakai, S.; Masui, M. (1979) *J. Chem. Soc. Perkin Trans. I* 2023.
Okita, M.; Wakatmatsu, T.; Ban, Y. (1983) *Heterocycles* **20:** 401.
Osa, T.; Kashiwagi, Y.; Yanagisawa, Y.; Bobbitt, J.M. (1994) *Chem. Commun.* 2535.
Park, K.; Pintauro, K.V.; Baizer, M.M.; Nohe, K. (1985) *J. Electrochem. Soc.* **132:** 1850.
Pickup, P.G.; Osteryoung, R.A. (1984) *J. Am. Chem. Soc.* **106:** 2294.
Platen, M.; Steckhan, E. (1984) *Chem. Ber.* **117:** 1679.
Pletcher, D.; Razaq, M. (1980) *J. Appl. Electrochem.* **10:** 575.
Ponsold, K.; Kasch, H. (1979) *Tetrahedron Lett.* 4463.
Qiu, W.M.; Wang, Z.Q. (1989) *Chem. Commun.* 356.
Radlick, P.; Klem, R.; Spurlock, S.N.; Sims, J.J.; van Tamelen, E.E.; Whitesides, T. (1968) *Tetrahedron Lett.* 5117.
Rifi, M.R. (1971) *J. Org. Chem.* **36:** 2017.
Ruholl, H.; Schäfer, H.J. (1987) *Synthesis* 408.
Robertson, P.M.; Berg, P.; Reimann, H.; Schleich, K.; Seiler, P. (1983) *J. Electrochem. Soc.* **130:** 591.
Satoh, S.; Itoh, M.; Tokuda, M. (1978) *Chem. Commun.* 481.
Schäfer, H.; Koch, D. (1972) *Angew. Chem. Int. Ed. Engl.* **11:** 48.
Schäfer, H.J. (1985) *Dechema Monographie* 112: 399.
Scheeren, J.W.; Goossens, H.J.M.; Top, A.W.H. (1978) *Synthesis* 283.
Scheffold, R.; Rytz, G.; Walder, L.; Orlinski, R.; Chilmonczyk, Z. (1983) *Pure Appl. Chem.* **55:** 1791.
Semmelhack, M.F.; Heisohn, G.E. (1972) *J. Am. Chem. Soc.* **94:** 5139.
Semmelhack, M.F.; Chou, C.S.; Cortes, D.A. (1983) *J. Am. Chem. Soc.* **105:** 4492.
Shono, T.; Ikeda, A.; Kimura, Y. (1971) *Tetrahedron Lett.* 3599.
Shono, T.; Natsumura, Y.; Hibino, K.; Miyawaki, S. (1974) *Tetrahedron Lett.* 1295.
Shono, T.; Ikeda, A.; Hayashi, J.; Hakozaki, S. (1975a) *J. Am. Chem. Soc.* **97:** 4261.
Shono, T.; Matsumura, Y.; Hashimoto, T.; Hibino, K.; Hamaguchi, H.; Aoki, T. (1975b) *J. Am. Chem. Soc.* **97:** 2546.
Shono, T.; Matsumura, Y. (1976a) *Tetrahedron Lett.* 1363.
Shono, T.; Matsumura, Y.; Hamaguchi, H.; Nakamura, K. (1976b) *Chem. Lett.* 1249.
Shono, T.; Nishiguchi, I.; Nitta, M. (1976c) *Chem. Lett.* 1319.
Shono, T.; Nishiguchi, I.; Okawa, M. (1976d) *Chem. Lett.* 573.
Shono, T.; Hayashi, T.; Omoto, H.; Matsumura, Y. (1977b) *Tetrahedron Lett.* 2667.
Shono, T.; Matsumura, Y.; Hamaguchi, H. (1977c) *Chem. Commun.* 712.
Shono, T.; Nishiguchi, I.; Ohmizu, H. (1977d) *Chem. Lett.* 1021.
Shono, T.; Matsumura, Y.; Kashimura, S. (1978a) *Chem. Lett.* 69.
Shono, T.; Yoshida, K.; Ando, K.; Usui, Y.; Hamaguchi, H. (1978b) *Tetrahedron Lett.*

4819.

Shono, T.; Matsumura, Y.; Hayashi, J.; Mizoguchi, M. (1979) *Tetrahedron Lett.* 165.

Shono, T.; Matsumura, Y.; Tsubata, K. (1981) *Tetrahedron Lett.* **22:** 3249.

Shono, T.; Matsumura, Y.; Inoue, K.; Ohmizu, H.; Kashimura, S. (1982a) *J. Am. Chem. Soc.* **104:** 5753.

Shono, T.; Matsumura, Y.; Tsubata, K.; Sugihara, Y.; Yamane, S.; Kanazawa, T.; Aoki, T. (1982b) *J. Am. Chem. Soc.* **104:** 6697.

Shono, T.; Ohmizu, H.; Kise, N. (1982c) *Tetrahedron Lett.* **23:** 4801.

Shono, T.; Kashimura, S. (1983a) *J. Org. Chem.* **48:** 1939.

Shono, T.; Matsumura, Y.; Hayashi, J.; Usui, M. (1983b) *Denki Kagaku* **51:** 131.

Shono, T.; Matsumura, Y.; Kanazawa, T. (1983c) *Tetrahedron Lett.* **24:** 4577.

Shono, T. (1984a) *Tetrahedron* **40:** 811.

Shono, T.; Matsumura, Y.; Onomura, O.; Kanazawa, T.; Habuka, M. (1984b) *Chem. Lett.* 1101.

Shono, T.; Matsumura, Y.; Uchida, K.; Tsubata, K.; Makino, A. (1984c) *J. Org. Chem.* **49:** 300.

Shono, T.; Ishige, O.; Uyama, H.; Kashimura, S. (1986) *J. Org. Chem.* **51:** 546.

Shono, T.; Matsumura, Y. (1987) *Synthesis* 1099.

Shono, T.; Kise, N.; Uematsu, N.; Morimoto, S.; Okazaki, E. (1990) *J. Org. Chem.* **55:** 5037.

Shono, T.; Ishifune, M.; Okada, T.; Kashimura, S. (1991a) *J. Org. Chem.* **56:** 2.

Shono, T.; Kise, N.; Shirakawa, E.; Matsumoto, H.; Okazaki, E. (1991b) *J. Org. Chem.* **56:** 3063.

Shono, T.; Kise, N.; Fujimoto, T.; Tominaga, N.; Morita, H. (1992) *J. Org. Chem.* **57:** 7175.

Sibille, S.; Ratovelomanana, V.; Nedelec, J.Y.; Perichon, J. (1993) *Synlett* 425.

Slobbe, J. (1977) *Chem. Commun.* 82.

So, Y.H.; Miller, L.L. (1976) *Synthesis* 468.

Sowell, C.G.; Wolin, R.L.; Little, R.D. (1990) *Tetrahedron Lett.* **31:** 485.

Stork, G.; Meisels, A.; Davies, J.E. (1963) *J. Am. Chem. Soc.* **85:** 3419.

Swenton, J.S. (1983) *Acc. Chem. Res.* **16:** 74.

Takahashi, J.; Mori, K.; Matsui, M. (1979) *Agric. Biol. Chem.* **43:** 1605.

Takahashi, Y.; Yuasa, K.; Tokuda, M.; Suzuki, A. (1978) *Bull. Chem. Soc. Jpn.* **51:** 339.

Tanaka, H.; Kobayasi, Y.; Torii, S. (1976) *J. Org. Chem.* **41:** 3482.

Torii, S.; Tanaka, H.; Mandai, T. (1975) *J. Org. Chem.* **40:** 2221.

Torii, S.; Inokuchi, T.; Mizoguchi, K.; Yamazaki, M. (1979a) *J. Org. Chem.* **44:** 2303.

Torii, S.; Okumoto, H.; Tanaka, H. (1980a) *Chem. Lett.* 617.

Torii, S.; Uneyama, K.; Ono, M. (1980b) *Tetrahedron Lett.* **21:** 2653.
Torii, S.; Tanaka, H.; Saitoh, N.; Siroi, T.; Sasaoka, M.; Nokami, J. (1981a) *Tetrahedron Lett.* **22:** 3193.
Torii, S.; Uneyama, K.; Nakai, T.; Yasuda, T. (1981b) *Tetrahedron Lett.* **22:** 2291.
Torii, S.; Inokuchi, T.; Oi, R. (1982) *J. Org. Chem.* **47:** 47.
Torii, S.; Inokuchi, T. (1983a) *Chem. Lett.* 1349.
Torii, S.; Inokuchi, T.; Oi, R. (1983b) *J. Org. Chem.* **48:** 1944.
Torii, S.; Inokuchi, T.; Kobayashi, T. (1984) *Chem. Lett.* 897.
Torii, S.; Inokuchi, T.; Sugiura, T. (1986) *J. Org. Chem.* **51:** 155.
Torii, S.; Inokuchi, T.; Takagishi, S.; Sato, E.; Tsujiyama, H. (1987) *Chem. Lett.* 1469.
Torii, S.; Okumoto, H.; Rashid, M.A.; Mohri, M. (1992) *Synlett* 721.
Tourillon, G.; Garnier, F. (1982) *J. Electroanal. Chem.* **135:** 173.
Uneyama, K.; Isimura, A.; Fujii, K.; Torii, S. (1983) *Tetrahedron Lett.* **24:** 2857.
Uneyama, K.; Masatsuga, Y.; Ueda, T.; Torii, S. (1984a) *Chem. Lett.* 529.
Uneyama, K.; Matsuda, H.; Torii, S. (1984b) *Tetrahedron Lett.* **25:** 6017.
Vargas, R.R.; Pardini, V.L.; Vierter, H. (1989) *Tetrahedron Lett.* **30:** 4037.
Vincent, F.; Tardivel, R.; Mison, P. (1976) *Tetrahedron* **32:** 1681.
Viscontini, M.; Bühler, H. (1966) *Helv. Chim. Acta* **49:** 2524.
Volke, J.; Skala, V. (1972) *J. Electroanal. Chem.* **36:** 393.
Wada, M.; Akiba, K.; Oki, H. (1987) *Chem. Commun.* 708.
Warning, K.; Mitzlaff, M.; Jensen, H. (1978) *Liebigs Ann. Chem.* 1707.
Wegenknecht, J.H. (1983) *J. Chem. Educ.* **60:** 271.
Weinberg, N.L.; Hoffmann, A.K.; Reddy, T.B. (1971) *Tetrahedron Lett.* 2271.
Weinreb, S.M. Epling, G.A.; Comi, R.; Reitano, M. (1975) *J. Org. Chem.* **40:** 1356.
Westberg, H.H.; Dauben, H.J. (1968) *Tetrahedron Lett.* 5123.
Wharton, P.S.; Hiegel, G.A.; Coombs, R.V. (1963) *J. Org. Chem.* **28:** 3217.
Wuts, P.G.M.; Cheng, M.-C. (1986) *J. Org. Chem.* **51:** 2844.
Yadav, A.K.; Tissot, P. (1984) *Helv. Chim. .Acta* **67:** 1698.
Yamamura, S.; Shizuri, Y.; Shigemori, H.; Okuno, Y.; Ohkubo, M. (1991) *Tetrahedron* **47:** 635.
Yang, T.-K.; Yeh, S.-T.; Lay, Y.-Y. (1994) *Heterocycles* **38:** 1711.
Yoshida, J.; Yamamoto, M.; Kawabata, N. (1985) *Tetrahedron Lett.* **26:** 6217.
Yoshida, J.; Ishichi, Y.; Isoe, S. (1992) *J. Am. Chem. Soc.* **114:** 7594.
Yoshida, K.; Nagase, S. (1979) *J. Am. Chem. Soc.* **101:** 4268.

7 ORGANIC PHOTOCHEMISTRY

Photochemistry has a fairly long history. Record shows that chemists in Italy were curious about the effect of sunlight on various compounds and they took samples to rooftops for exposure. However, investigations of organic photochemistry on an extensive scope came to the fore only in the 1960s, which were further stimulated by the formulation of the Woodward-Hoffmann rules.

Photochemically induced reactions often differ from thermal reactions because they proceed from excited and ground states, respectively. Many reactions can only be accomplished photochemically, therefore a synthetic chemist must always keep an eye on the potential of the methodology.

When a molecule absorbs ultraviolet or visible light, one of its valence electrons is promoted to an antibonding orbital. For example, excitation of a π electron to the π^* state results in the ($\pi\pi^*$) transition. If the excited electron of a singlet molecule retains its spin of antiparallelity to its original paired partner which remains in the bonding orbital, the molecule is in a singlet excited state (S). On the other hand, if the exciited electron has undergone a change in its spin to align in parallel to its original twin, then the molecule is in the triplet excited state (T). Since elecytronic transitions are governed by spin transition rules which maintain identical multiplicities for the initial and final states, the change is made possible only by the spin-orbit coupling mechanism. Transition after light absorption is a rapid process, and if vertical excitation of a ground state molecule is held (Franck-Condon principle), the potential energy minima of S_1 and T_1 might not be reached immediately.

Most of the common molecules are in the siinglet (S_0) ground state, a rare exception being dioxygen which has a lower energy triplet. Every molecule is associated with a series of energy levels in the excited state, represented by S_1, S_2, S_3,... T_1, T_2, T_3... in which the larger number of the subscript indicates a higher energy. Organic compounds containing multiple bonds but without heteroatoms will undergo ($\pi\pi^*$) excitation. Those devoid of extended conjugated chromophores while containing heteroatoms and having a nonbonding n-orbital as the highest occupied molecular orbtial (HOMO), their lowest excited state S_1 may be ($n\pi^*$) or ($n\sigma^*$).

Even if a molecule reaches a higher energy state on excitation, its chemical reaction often occurs in S_1 or T_1 state because the lifetime of an excited molecule is only in the range of 10^{-12} to 10^{-3} sec. The demotion from S_2 to S_1 (and from S_1 to S_0) is called *internal conversion*, whereas the process from S_x to T_x, and vice versa, is known as *intersystem crossing*, resulting from mixing of spin-magnetic moment and orbital electron angular moment. This spin-orbit coupling is enhanced by either heavy atoms or paramagnetic compounds. Note that triplet excited states are rarely reached by direct photoexcitation, they are more commonly accessible from the energy transfer method. In other words, one can use another substance (e.g., an aryl ketone) whose excited singlet state shows a great tendency to undergo intersystem crossing to absorb photon and then deliver the light energy from its excited triplet to the compound which is intended to be put in the T_1 state.

The energy released from an excited state molecule may be dissipated as heat or in the form of light. Fluorescence is emission of radiation usually accompanies the relaxation of S_1 to S_0^v and S_0 states with $\tau_f \sim 10^{-8}$ s. Phosphorescence is spin-forbidden emission between two vibronic states of different multiplicities, most commonly T_1 to S_0^v, therefore it has a much longer lifetime ($\tau_f \sim 10^{-4}$ -10 s). Photochemical reactions must compete with these various processes. Unimolecular photophysical transformations can be summarized by the Jablonski diagram.

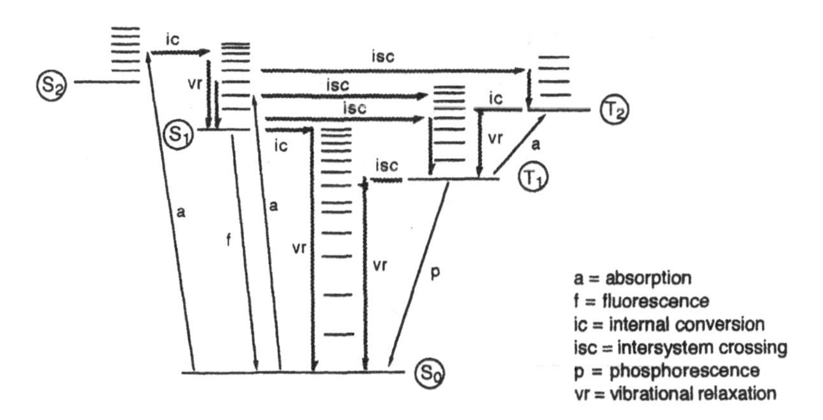

Jablonski Diagram.

7.1. GENERAL CONSIDERATIONS

Photochemical reactions usually occur in low-lying excited states. Their efficiency is measured by quantum yield, Φ, which indicates the percentage of excited molecules to undergo reactions. Quantum yields can exceed unity in cases of chain reactions. Thus the absorption of one proton by a molecule generates a reactive species which triggers a reaction cascade, as exemplified by photoinitiated polymerization. For ordinary photochemical reactions of organic compounds, low quantum yields do not preclude their applications although such situations are less desirable. It means a longer irradiation period is needed for high conversion of the substrate. A necessary condition is that it affords high chemical yields. For a photochemical reaction to be considered useful it must also fulfill the requirements of readily availability of the substrate which can be excited selectively (the product being photostable at the applied wavelength), good regio- and stereoselectivity, ease in product isolation.

The common light sources are low-pressure and medium-pressure mercury lamps. Within the lamp tube of a low-pressure lamp the mercury vapor is about 10^{-5} atm. and about 90% of the radiation is due to resonance bands at 185 nm and 254 nm. If the inside wall of the low-pressure lamp is coated with a phosphor, the 254 nm emission can be shifted to longer wavelength, and the band width can be increased from 20 nm to 100 nm. On the other hand,the radiation of the medium-pressure lamp is distributed as continuous spectral lines, and while containing 1-10 atm of mercury vapor, the exact pressure depending on operation temperature. High pressure lamps may be filled with mercury or rare gases (e.g., xenon), or a mixture of both. It supplies wide and continuous light waves.

Table 1. Solvent for Photochemical Reactions

Solvent	Cut-offWavelength(nm)
water	185
acetonitrile	190
n-hexane	195
ethanol	204
methanol	205
cyclohexane	215
diethyl ether	215
dioxane	230
dichloromethane	230
chloroform	245
tetrahydrofuran	245
acetic acid	250
ethyl acetate	255
carbon tetrachloride	265
dimethyl sulfoxide	277
benzene	280
toluene	285
pyridine	305
acetone	330

In operation one can select a certain wavelength by monochromatic filters. The most convenient procedure is to employ various sorts of glass to block out the undesirable light. The cut-off wavelengths for several filters are: quartz (< 200 nm),

Vycor 7910 (210-240 nm), Corex D (270-310 nm), pyrex (290-330 nm), Nonex 7720 (320-360 nm). Solutions of certain individual inorganic salts or their combinations can serve as liquid filters.

Since many photoreactions are carried out in solvents, the latter must not contain photoreactive impurities and do not interfere by absorbing the excitation radiation for the substrate, and by reaction with the substrate in the ground state or the excited state, unless the solvent is also a designated reactant. The cut-off wavelengths for several solvents are listed in Table 1 above.

The following sections describe synthetically useful photochemical behaviors according to functional groups, except the last section which deals with a common theme of photoinduced rearrangements.

7.2. CLEAVAGE AND RECOMBINATION

The (n,π^*) transition of the carbonyl group, through absorption in the 280-330 nm region to promote a nonbonding 2p electron of the oxygen atom to the antibonding π-orbital, usually initiates photochemistry of many aldehydes and ketones. In the (n,π^*) state the carbonyl group is less strongly polarized, consequently the oxygen atom of an excited carbonyl group exhibits radical-like character. Five types of reactions are often encountered:

(1) Norrish Type-I reaction (α-cleavage)

(2) Norrish Type-II reaction (intramolecular photoelimination)

(3) Photoreduction

(4) Paterno-Büchi reaction (in the presence of an alkene)

(5) Cyclobutanol formation (Yang reaction).

The Norrish Type-I reaction involves cleavage of the weakest bond. Only the (n,π^*) excited states but not the (π,π^*) can weaken the α-bond by overlap with the half-vacant n-orbital of the oxygen. The α-cleavage reactivity of triplet (n,π^*) excited states is about 100 times greater than that of the singlet with the same configuration. The acyl radical formed by this reaction can be linear or bent. Generally, the Norrish Type-I process is faster in aliphatic ketones than in alkyl aryl ketones. When a cyclic ketone undergoes this scission it generates a biradical which may lead to an unsaturated aldehyde or ketene by way of intramolecular hydrogen transfer, to a cyclic α-oxacarbene on reclosure of the ring (for cyclobutanone substrates), or to an alkene or a contracted cycloalkane after decarbonylation. In the presence of an alcohol solvent, the ketene and carbene intermediates as alluded to above are converted into esters and cyclic acetals, respectively.

An interesting reaction involving α-cleavage and recombination of the acyl/alkyl biradical accomplishes the photoisomerization of macrocyclic 2-phenylcycloalkanones to paracyclophanones [Lei, 1986]. The α-cleavage is favored by the formation of benzyl radical. Strain relief is also conducive to the process.

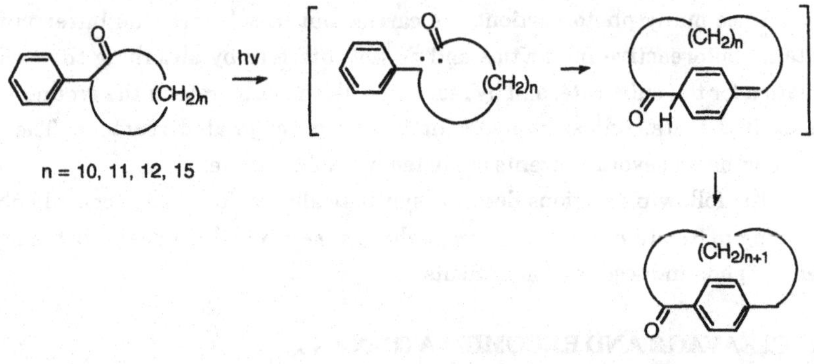

n = 10, 11, 12, 15

Decarbonylation is involved in the synthesis of tetra-*t*-butyltetrahedrane [Maier, 1978], [4.5]coronane [Fitjer, 1987], and the 9,9'-dehydrodimer of anthracene [Weinshenker, 1968] from ketonic precursors. A [3]rotane is readily produced photochemically from the 1,3,5-cyclohexanetrione by threefold extrusion of carbon monoxide [Krapcho, 1972]. *m*-Benzyne has been prepared in a matrix by photolysis of a diketo[2.2]metaparacyclophane [Marquardt, 1996].

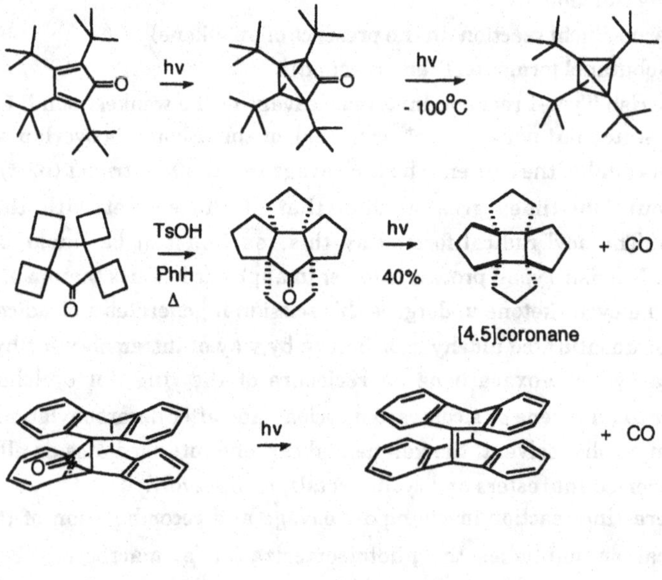

[4.5]coronane

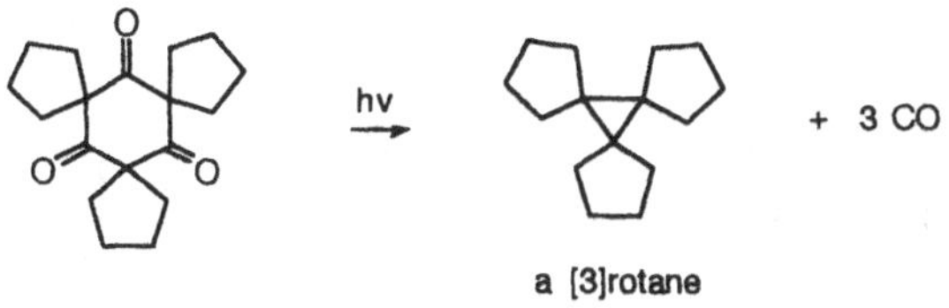

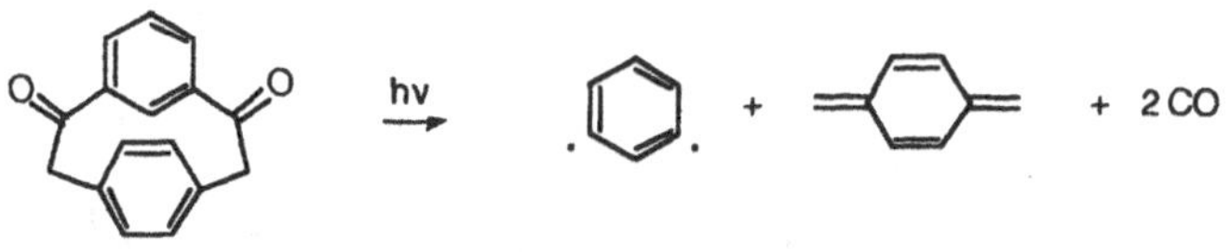

a [3]rotane

Photodecarbonylation initiated decomposition of a Diels-Alder adduct derived from cyclooctatetraene to give barrelene [Dauben, 1976].

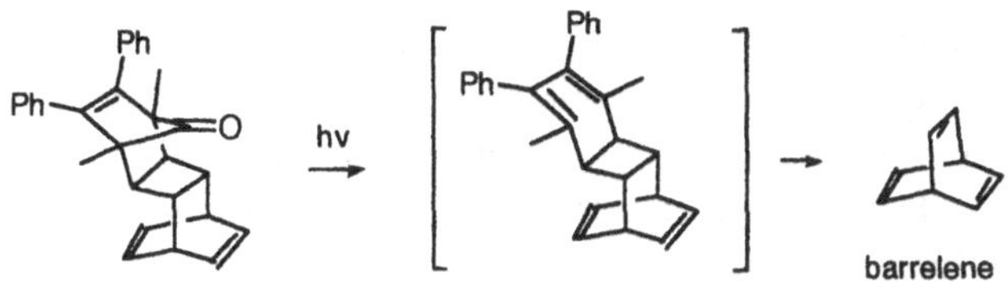

barrelene

It is not known whether the last step of the above reaction is photoinduced. Parenthetically, the highly efficient formation of bullvalene [Schröder, 1964] by photolysis of a cyclooctadiene dimer is most interesting.

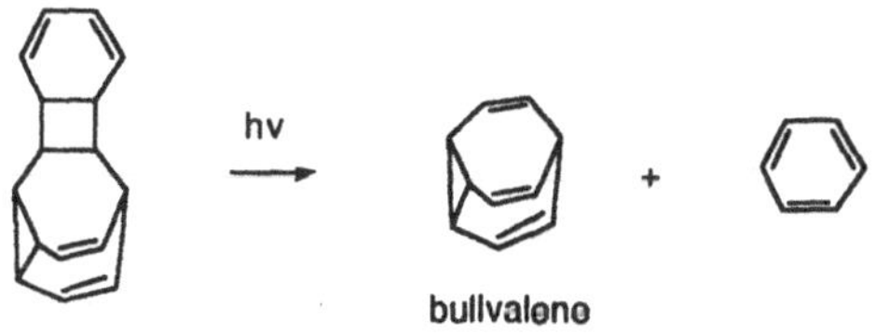

bullvalene

A pathway available to the biradical derived from Norrish Type-I cleavage of cyclobutanones is to complete the separation into two fragments (by a retro[2+2]-cycloaddition). A bicyclic substrate with a strategically positioned hydroxyl group can be transferred to lactone product by self-interception of the ketene moiety [Davies, 1985]. By this method intermediates of eldanolide and leukotriene-B4 can be constructed.

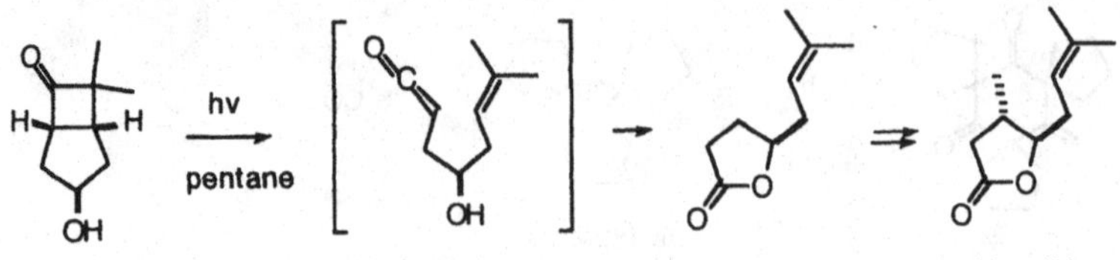

eldanolide

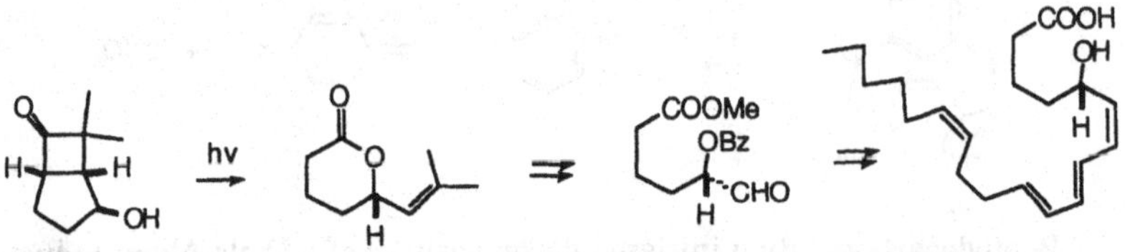

leukotriene-B_4

Cyclic lactol (acetal) formation is characteristic of cyclobutanone photolysis, particularly at low temperatures. Its exploitation in the synthesis of prostaglandins [Howard, 1980] is facilitated by the availability of the bicyclo[3.2.0]heptenone intermediates from the [2+2]adduct of cyclopentadiene and dichloroketene. The major side-product of the photochemical step is a cyclopentene, arising from the initial biradical by ketene elimination.

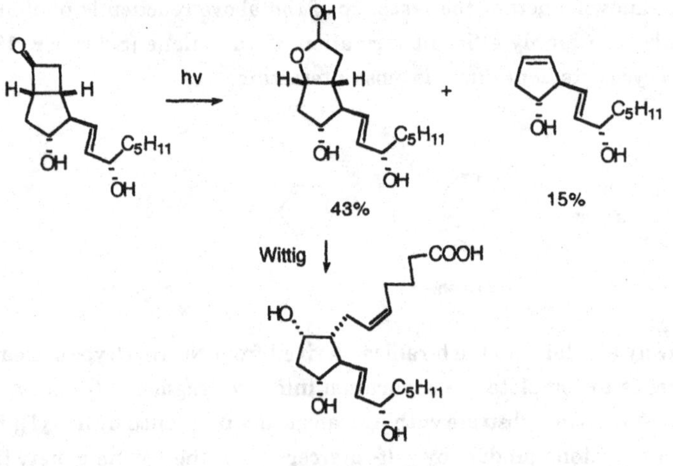

prostaglandin-$F_{2\alpha}$

Prostaglandin-$F_{2\alpha}$. [Howard, 1980]

A degassed solution of the fused cyclobutanone (378 mg, 1.5 mmol) in acetonitrile and water (68 mL) is irradiated with a Hanovia medium pressure mercury arc through quartz for 4.5 h under nitrogen. The resulting mixture is extracted with ethyl acetate (5x100 mL). An oil obtained from the extracts is a lactol precursor of $PGF_{2\alpha}$ (conversion by Wittig reaction).

Another application is in the synthesis of muscarine [Pirrung, 1988b].

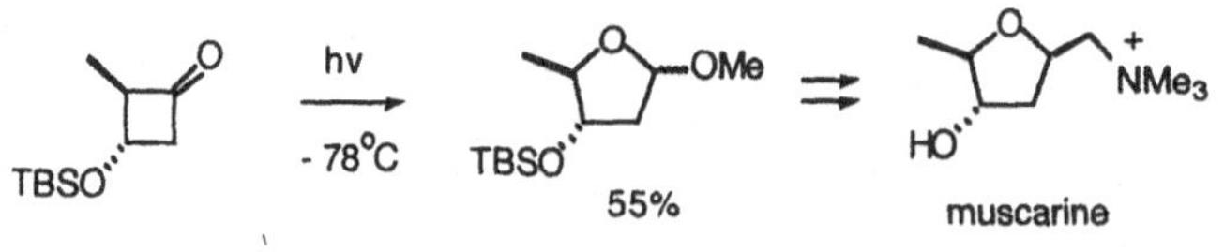

hv, - 78°C; 55%; muscarine

The Norrish Type-I reaction of cyclic ketones followed by intramolecular hydrogen transfer to give unsaturated aldehydes constitutes a useful transformation sequence. Thus a dienal employed in a preparation of propylure [Kossanyi, 1973] is accessible from a 2-substituted cyclopentanone. Syntheses of grandisol [Hobbs, 1976], prostaglandin-C_2 [Crossland, 1979], boschnialactone and teucriumlactone-C [Callant, 1983], and indolizidine-223AB [Muraoka, 1994] relied on photolysis of bridged ketones.

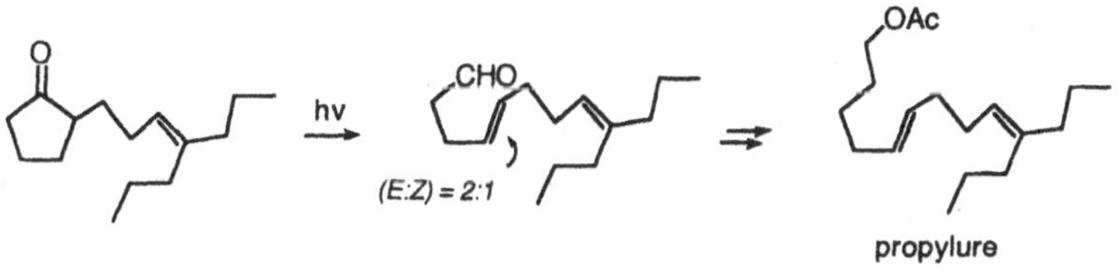

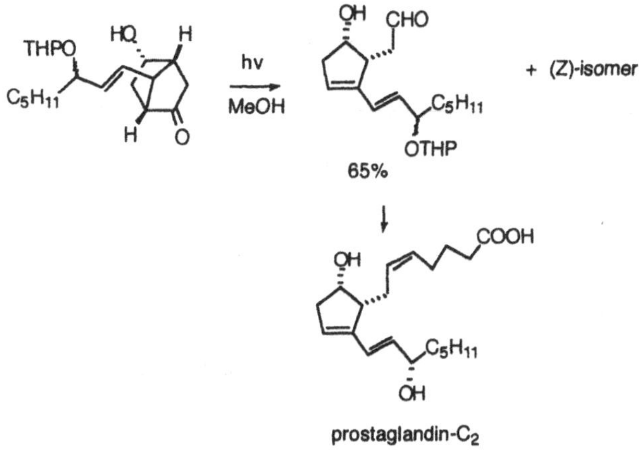

hv; (E:Z) = 2:1; propylure

hv, MeOH; 65%; + (Z)-isomer; prostaglandin-C_2

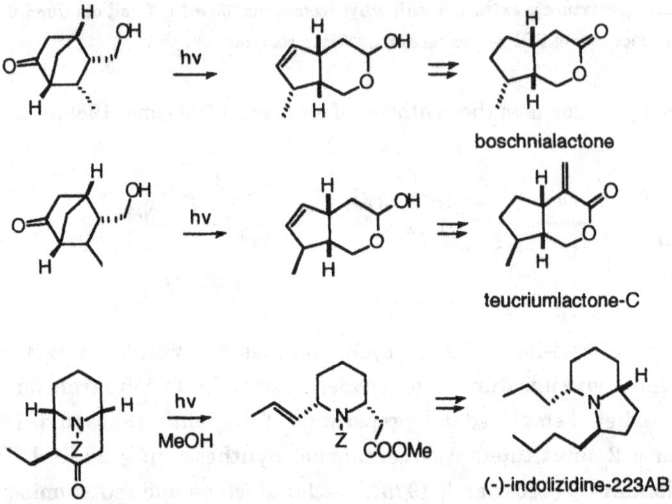

hv

boschnialactone

hv

teucriumlactone-C

hv MeOH

COOMe

(-)-indolizidine-223AB

For certain β,γ-unsatured ketones the α-cleavage may be followed by recombination of the diradical at the allylic position to result in 1,3-acyl shift. This change in carbon connectivity provides a very useful access channel to different ring size and constitution. Thus it has been exploited for the synthesis of sesqicarene [Uyehara, 1983], isosesquicarene [Uyehara, 1985b], pinguisone [Uyehara, 1986], ptilocaulin [Uyehara, 1988], methyl ether of mussaenoside aglucone [Hsu, 1993].

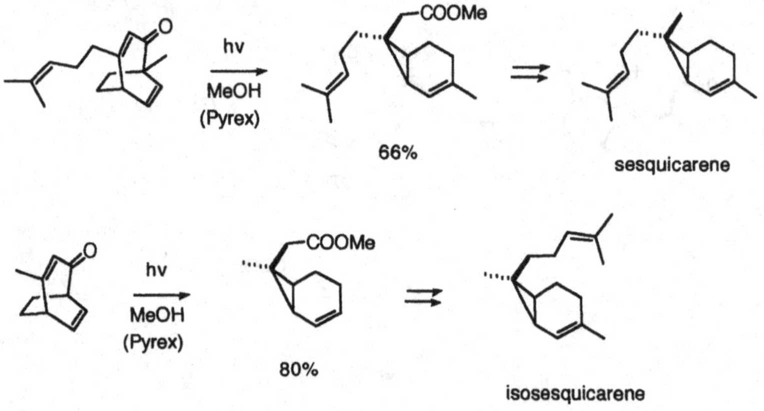

hv MeOH (Pyrex)

COOMe

66%

sesquicarene

hv MeOH (Pyrex)

COOMe

80%

isosesquicarene

Methyl 7-*exo*-Methyl-7-*endo*-bicyclo[4.1.0]hept-2-en-7-ylacetate. [Uyehara, 1985b]
A solution of 4-methylbicyclo[3.2.1]nona-3,6-dien-2-one (1.14 g, 7.69 mmol) in methanol (750 mL) is irradiated with a 300W mercury lamp for 1.5 h. The residue oil obtained upon evaporation of the solvent is chromatographed (10:1 hexane/EtOAc) to give the ester (1.11 g, 80%).

OMe hv THF MeO 59% pinguisone

hv THF 64% HN NH $^+NH_2$ NO_3^- ptilocaulin

COOMe hv >300 nm PhH MeOOC HO OH MeOOC mussaenoside aglucone

A key step of a hibaene synthesis [Wenkert, 1973] is the formation of the bridged ring moiety by irradiation of an enol lactone. This photochemical variant of a Claisen condensation might have involved homolysis of the C-O bond which is allylic.

hv hibaene

Certain N-acylindoles undergo 1,3-shift to afford the 3-acyl isomers. A very useful application of this process involves formation of a nine-membered lactam

which is a versatile intermediate for several types of indole alkaloids [Ban, 1983], including quebrachamine, aspidospermidine, eburnamine, strempeliopine, tubifoline, and condyfoline.

Photoisomerization of 6-Oxo-(2-aminoethyl)-6,7,8,9-tetrahydropyrido[1,2a]indole. [Ban, 1983]

A solution of the indole (1.84 g, 8.07 mmol) in ether (500 mL) is irradiated with a 300W high pressure Hg lamp under N_2 for 21 h. Solvent is removed in vacuo and the residue is chromatographed over silica gel (EtOAc-acetone as eluent) to afford the macrolactam (1.71 g, 89.3%).

condyfoline tubifoline

hv, MeOH or Et_2O 90% quebrachamine

strempeliopine dehydroaspidospermidine eburnamine

***O*-Alkyl *S*-(alkoxycarbonyl)dithiocarbonates suffer homolysis on exposure to light. With a juxtaposed double bond to intercept the acyl radical lactone formation results [Saicic, 1996]. Pyrolysis of such a product gives the α-methylenelactone. A synthesis of methylenolactocin by this method is highly efficient.**

hv 63% Cu / FVP; HCl methylenolactocin

The S_1 (n,π*) state of 2,4-cyclohexadienones yields ketenes, formally initiated by α-cleavage of the carbonyl group. Such species thermally revert to the starting compounds in the absence of nucleophiles, but they can also be trapped. This behavior led to the development of a convenient synthesis of crocetin dimethyl ester [Quinkert, 1977] and (+)-aspicilin [Quinkert, 1987], in which inter- and intramolecular interception of the ketene intermediates were accomplished, respectively.

Using a two-laser two-color technique benzyl phenyl ether and phenyl acetate can be induced to undergo sequential Claisen or Fries rearrangment, followed by fragmentation to the dienic ketenes [Jimenez, 1997].

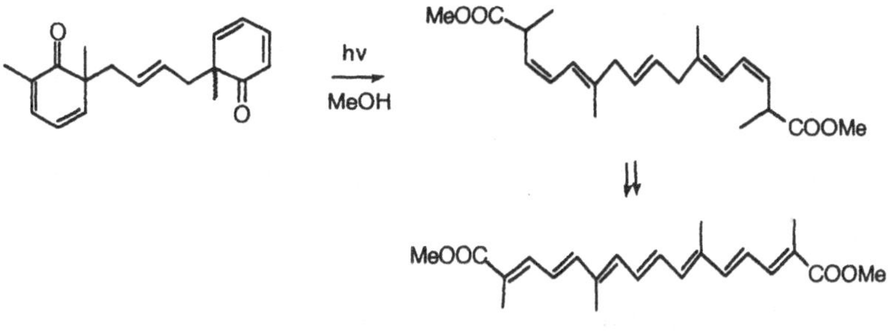

crocetin dimethyl ester

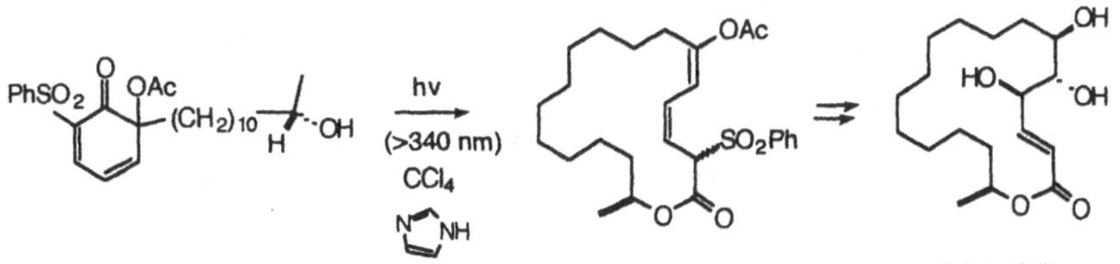

(+)-aspicilin

Photoreduction of ketones occurs in the triplet state. The reduction of benzophenone is very efficient, which results from hydrogen abstraction from solvent. Hydrogen abstraction can be rendered intramolecular and regioselective when the benzophenone is attached to a steroid skeleton. It serves to accomplish remote functionalization [Breslow, 1973, 1980].

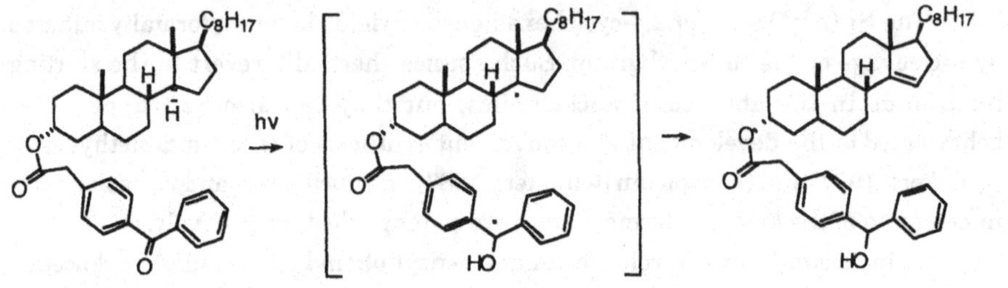

Reductive cyclization is the key to the skeletal construction of actinidine [Cossy, 1988]. The excited carbonyl group pursues normal chemistry of free radicals in the addition onto a multiple bond.

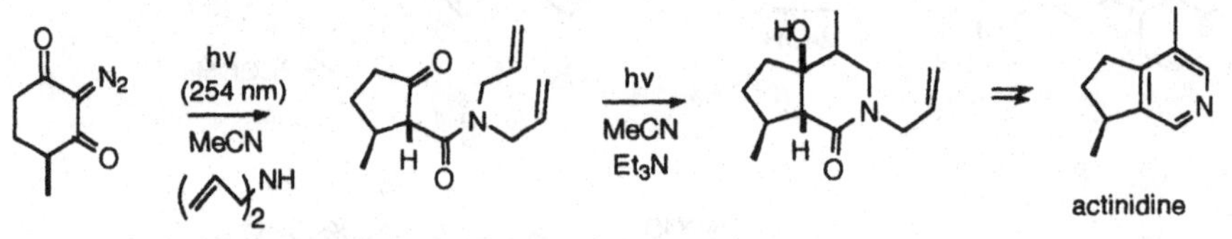

An excited ketone can abstract a hydrogen from solvent. In methanol it can lead to formation of *vic*-diols as illustrated in a synthesis of frontalin [Mitani, 1983]. If external hydrogen sources are not readily available, intramolecular γ-hydrogen abstraction (via a six-centered transition state) becomes a favorable pathway. The subsequent transformation, either β-cleavage or cyclization to give a cyclobutanol (Yang reaction) is dependent on the conformation of the 1,4-diradical. If the orbitals of the radical centers are parallel to the β-bond, Norrish Type-II cleavage predominates.

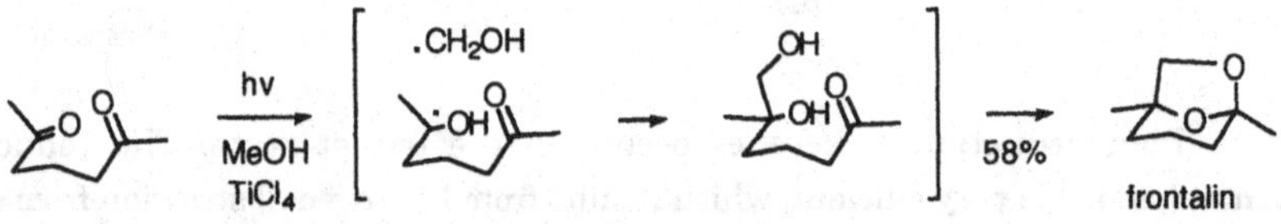

The cyclobutanol formation process has not found extensive use in synthesis. However, it is a key step in an approach to punctatin-A [Sugimura, 1987] and in the functionalization of C-18 in a synthesis of aldosterone acetate [Miyano, 1981].

<u>3-(Ethylenedioxy)-11α,20α-dihydroxy-18,20-cyclopregn-5-ene</u>. [Miyano, 1981]

A solution of 11α-hydroxy-3-(ethylenedioxy)-20-oxopregn-5-ene (10 g) in ethanol (950 mL) is irradiated with a 200W medium pressure lamp at 20° for 7 h under a nitrogen stream. On solvent removal the residue is chromatographed over silica gel (250 g) with toluene-ethyl acetate gradient to afford the cyclobutane (4.5 g) which is recrystallized from a cyclohexane-ethyl acetate mixture. The pure product (3.2 g) shows m.p. 205°.

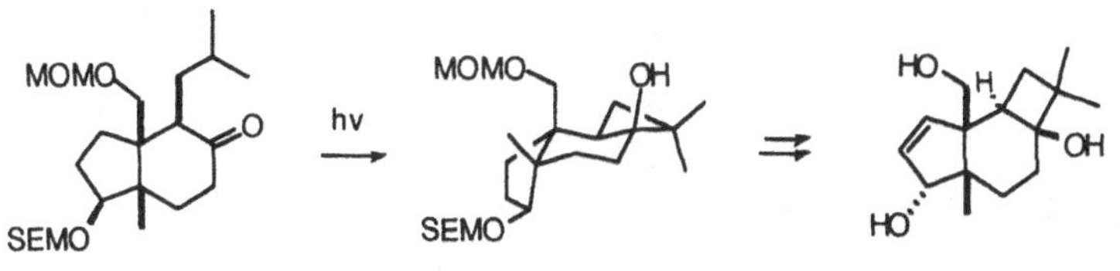

punctatin-A

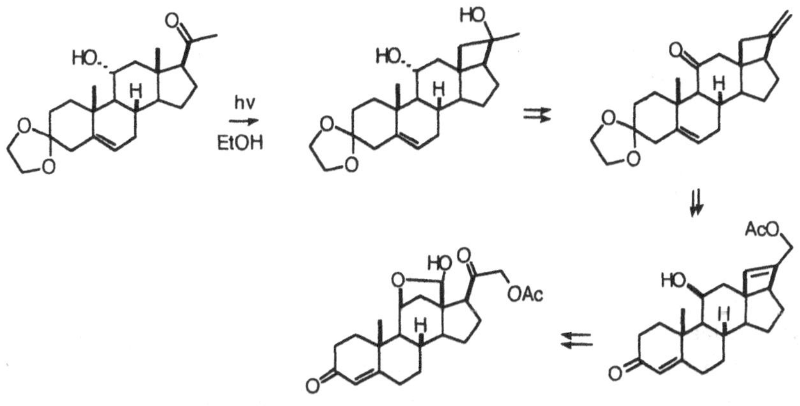

aldosterone acetate

2-Hydroxy-2-methylcyclobutanone can be obtained by photolysis of 2,3-pentanedione. In a few steps the compound is transformed into *cis*-jasmone [Weinreb, 1972]. Ring expansion of the photocyclized products of 1-bromo-2,3-alkanediones leads to 1,3-cyclopentanediones [Hamer, 1986].

Because the realm of solid state photochemistry is still being developed and applications of which are limited, it is rarely mentioned in this book. However, surprisingly different behaviors of certain molecules in solution and in the solid state have been noticed which will have syntheic implications. For example, among photoproducts emerged from 1,2-cyclodecanedione [Olovsson, 1977] the keto-aldehyde has been attributed to the consequence of conformational restricted excited ketone in the crystals. Hydrogen abstraction would be favored.

hν

cis-jasmone

hν Ag+

hν >340 nm

in PhH 7 : 1 : 0
solid 2 : 0 : 3

o-Alkylphenones undergo photoenolizations by way of γ-hydrogen abstraction. The photoenols can be intercepted intra- or intermolecularly, as shown in syntheses of estrone [Quinkert, 1981, 1982] and 7-deoxyaklavinone [Kraus, 1991]. It should be noted that 2,6-dialkylphenyl ketones give benzocyclobutenols under similar conditions [Sammes, 1976; Wagner, 1991].

hν

major

minor

estrone methyl ether

δ-Hydrogen abstraction occurs when a γ-hydrogen is inaccessible due to its absence or because of steric reasons. This reaction features prominently in a synthesis of dodecahedrane [Paquette, 1983] and paulownin [Kraus, 1990].

Secododecahedranol. [Paquette, 1983]

Disecododecahedranone (28 mg, 0.1 mmol) in a deoxygenated solvent system of benzene-*t*-butanol (4:1, 10 mL) containing triethylamine (2 drops) is irradiated with a 450W Hanovia lamp through Pyrex for 10 h under a nitrogen atmosphere. Solvent removal in vacuo provides the tertiary alcohol (28 mg, 100%) as a crystalline solid, m.p. >350°.

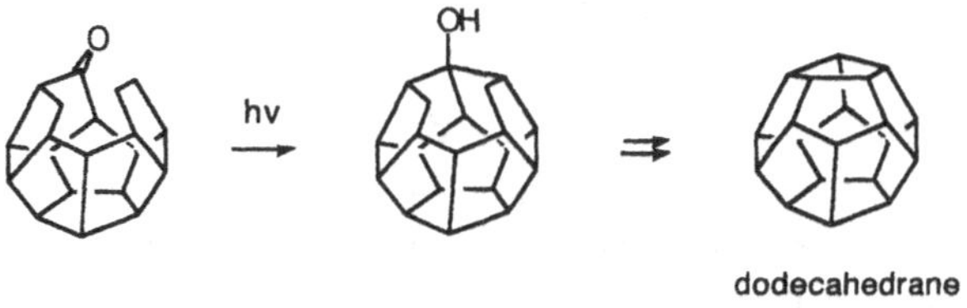

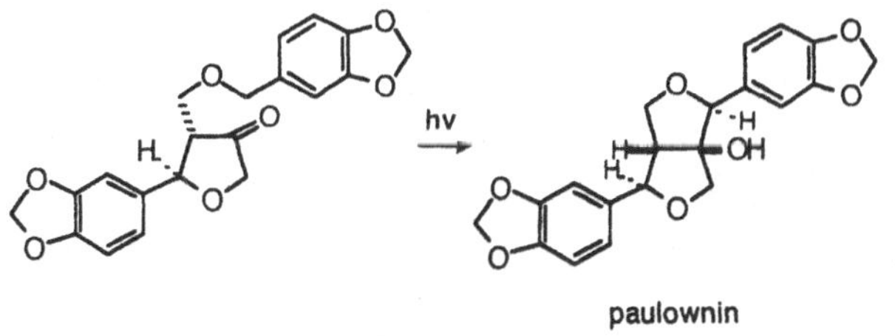

dodecahedrane

paulownin

The construction of the chilinene ring system [Mazzocchi, 1987] demonstrates hydrogen abstraction by an imide carbonyl at longer range followed by cyclization.

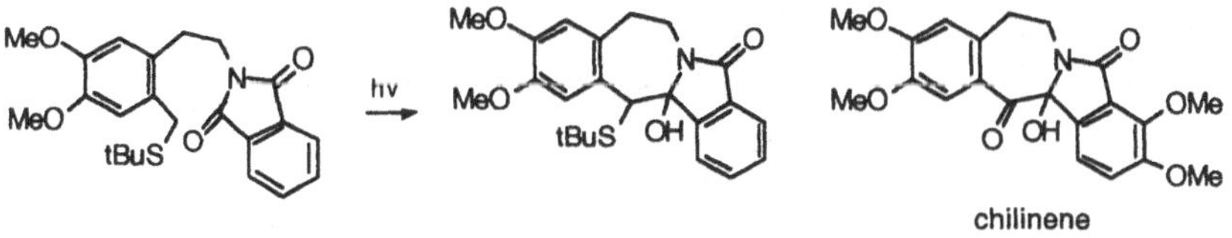

chilinene

An organic nitrite ester would undergo photolytic scission of the O-N bond to generate an alkoxy radical and NO. An accessible hydrogen at a δ-carbon is then abstracted by the alkoxy radical (via a six-center transition state), and the resulting carbon radical recombines with nitric oxide. If the δ-carbon still bears a hydrogen atom, tautomerization of the nitroso compound then gives an oxime. Experimental results have suggested that the recombination step is not confined in a solvent cage.

The transformation, known as Barton reaction, constitutes a useful method for remote functionalization within a conformationally rigid system. Two cases may be presented to illustrate its value in complex synthesis. Cross-ring functionalization within a spirocyclic system is responsible to a successful elaboration of perhydrohistrionicotoxin [Corey, 1975], and the exemplary transformation of the 18-methyl group in a steroid derivative into an oxime, and thence into aldosterone [Barton, 1961]. The original route to aldosterol is short but it suffers from a competing reaction at C-19. The problem can be removed by using the 11-nitrite of 1,2,6,7-tetradehydrocortisone 21-acetate [Barton, 1975].

Photolysis of 1,2,6,7-Tetradehydrocortisone 21-Acetate 11-Nitrite. [Barton, 1975]
1,2,6,7-Tetradehydrocortisone 21-acetate 11-nitrite (3.8 g) and triethylamine (1 mL) is dissolved in THF (160 mL) and irradiated under argon for 45 min. Evaporation and chromatography over florisil (eluent CH_2Cl_2-MeOH) furnishes the 18-oximino-11β-ol (1.41 g, 38%).

ONO=O hv py OH HO N OH NH O OH NH

perhydrohistrionicotoxin

ON-O OAc O H hv HO HON OAc O H HO OAc O O H O O

aldosterone acetate

Alkoxy radicals generated from alkyl hypohalites [Kalvoda, 1971] behave similarly. However, due to the involvement of halogen transfer in the subsequent step, the overall transformation differs. Furthermore, structural modification such as α-cyano substitution [Meystre, 1961] serves to extend the scope of synthetic use.

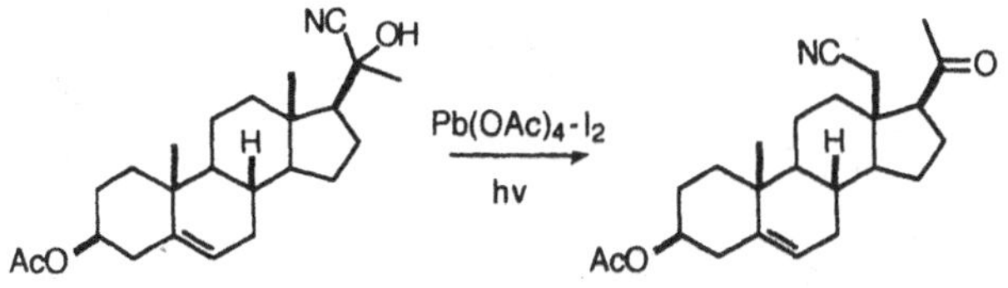

The α,β-bond of an alkoxy radical is also prone to cleavage. The process has been exploited in the synthesis of sesamin [Orito, 1991], taiwanin-C and justicidin-E [Kobayashi, 1992], muscone [Suginome, 1987], α-himachalene [Suginome, 1994] and caryophyllene [Suginome, 1995]. In the route to taiwanin-C/justicidin-E the lactone formation is accompanied by an ejection of the ethyl goup.

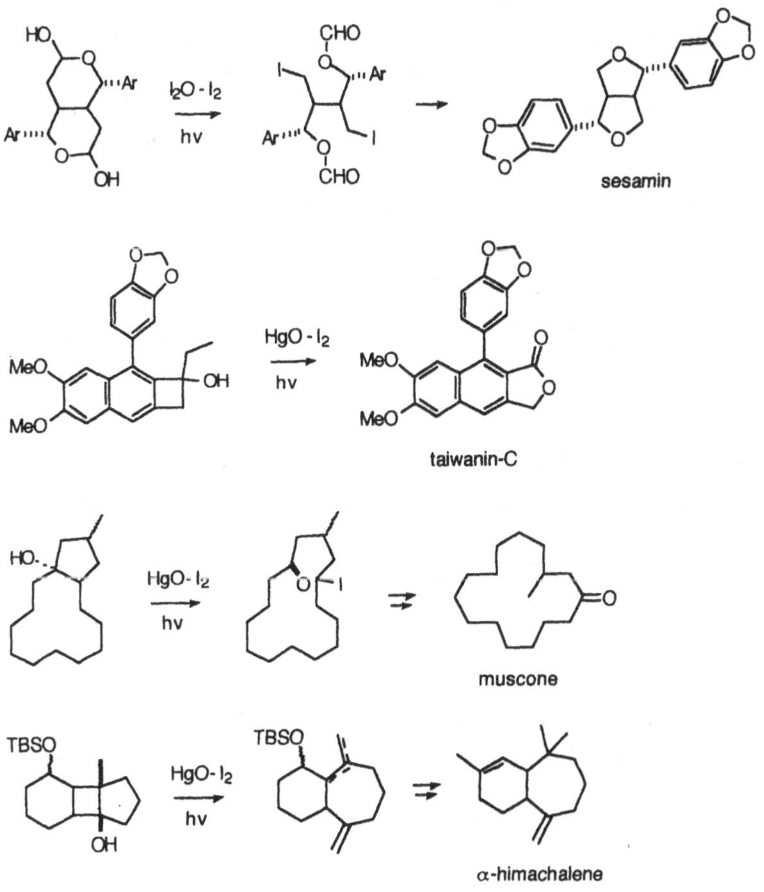

sesamin

taiwanin-C

muscone

α-himachalene

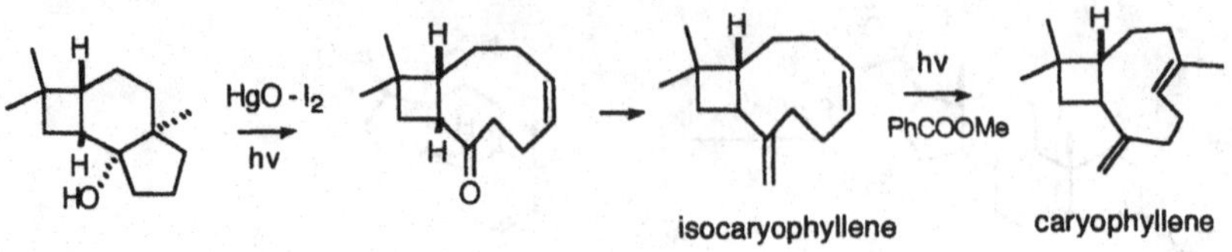

HgO - I_2 hv → hv PhCOOMe

isocaryophyllene caryophyllene

7.3. CYCLOADDITIONS

7.3.1. [2+2]Cycloadditions

7.3.1.1. Paterno-Büchi Reaction

The [2+2]-photocycloaddition of a carbonyl group to an alkene to give an oxetane is known as the Paterno-Büchi reaction. Its usefulness is due to the potential of oxetanes in being converted to many other functions. Generally, the reaction is clean, particularly when the alkene is present in high concentrations, because oxetanes do not compete with carbonyl compounds for light absorption. Generation of a stable biradical intermediate from attack of the triplet (nπ*) oxygen atom of an aldehyde or aryl ketone on the alkene accounts for the regioselectivity. (The less efficient intersystem crossing of aliphatic ketones permits the reaction of singlet (nπ*) state as well as the triplete state.) Interestingly, excitation of α-dicarbonyl compounds with visible light enables their participation in the cycloaddition because the products no longer absorb light.

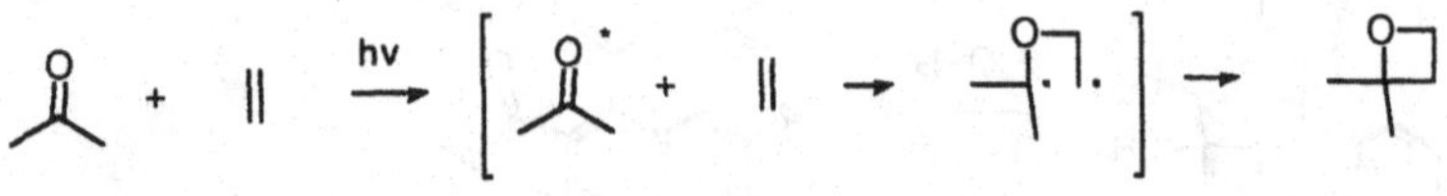

hv

A "metathetic" pathway to *(E)*-6-nonenol, a sex attractant of the Mediterranean fruit fly, involves Paterno-Büchi reaction to assemble all the skeletal elements [Jones, 1975].

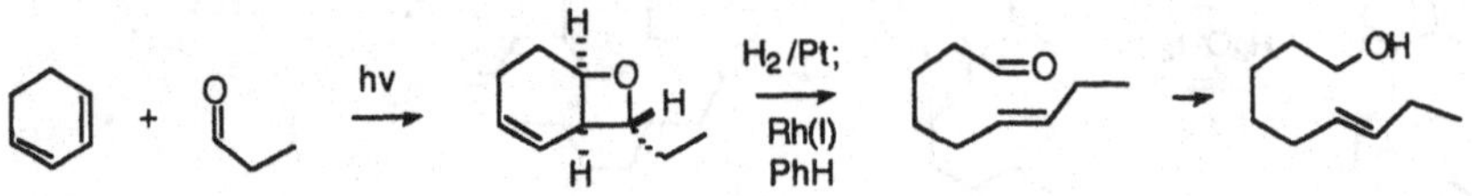

hv H_2/Pt; Rh(I) PhH

Vinylene carbonate adds to excited carbonyl group to form an adduct which is a protected glyceraldehyde. A synthetic approach to apiose [Araki, 1981] illustrates the utility of this process.

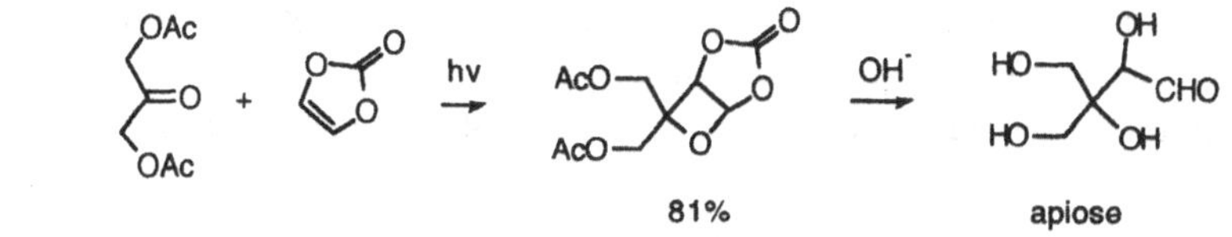

81%

apiose

Pyrrole and furan also participate in the Paterno-Büchi reaction. Of significance are the designs for synthesis of avenaciolide [Schreiber, 1984a] and asteltoxin [Schreiber, 1984c] on the basis of the process. Note the excellent stereoselectivity is fully exploited.

6-Octyl-2,7-dioxabicyclo[3.2.0]hept-3-ene. [Schreiber, 1984a]

Nonaldehyde (32.66 g, 0.23 mol) and furan (200 mL, 2.75 mol) are mixed and photolyzed with quartz immersion well containing a Vycor filter and a 450W Hanovia lamp. The system is kept at -20° while N_2 is bubbled through the solution. During irradiation additional amount of furan (150 mL) is introduced. After 20 h, excess furan is evaporated giving the desired product (48.7 g, 100%) as an oil.

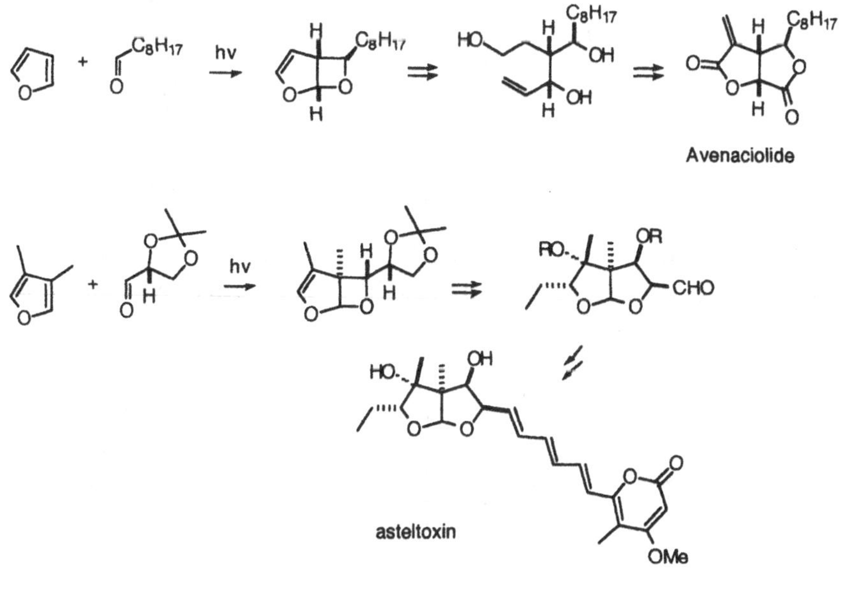

Avenaciolide

asteltoxin

The synthesis of isocomene [Rawal, 1994] involves opening of the oxetane derived from an intramolecular Paterno-Büchi reaction of a norbornene and further fragmentation of the resulting bridged ring system to provide a diquinane intermediate. Acess to triquinanes by the same strategy has also been demonstrated by a route to the endo isomer of hirsutene [Rawal, 1995], which is convertible to hirsutene.

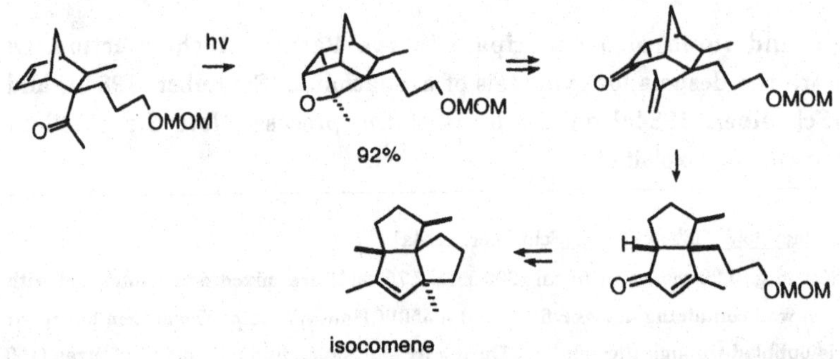

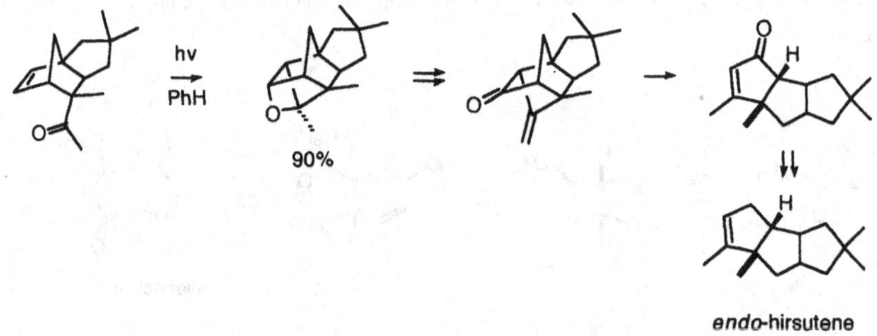

hv OMOM 92% OMOM OMOM isocomene OMOM

hv PhH 90% *endo*-hirsutene

When ring strain is too high for an impending cycloadduct, the precursorial diradical may seek other pathways to stabilize itself. Thus such a diversion provides an expedient access to *exo*-brevicomin [Chaquin, 1977].

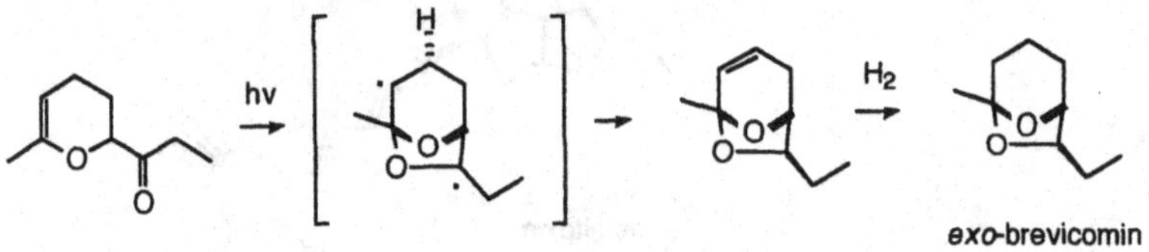

hv H_2 *exo*-brevicomin

7.3.1.2. [2+2]Photocycloaddition of Alkenes

While the Paterno-Büchi reaction involves attack of an excited C=O to ground state C=C linkage, a (π,π^*) excited state alkene also undergoes cycloaddition with another ground state alkene. The two alkene species may be the same (resulting in photodimerization) or different. Dimerization, best carried out by sensitization with aliphatic ketones and with copper triflate, usually leads to sterically less hindered *exo*-dimers. The Cu(I) salt forms complexes with alkenes which absorb at ca. 250 nm, but only the 1:2-complexes would undergo photocycloaddition. The complexation involves overlap of the alkene π-electrons with the vacant σ-orbital (4s) of copper and $3d_{\pi}$-orbital of copper with the antibonding π^*-orbital of the alkene.

The photodimerization of tetrabenzoheptafulvene forms a cage compound [Schönberg, 1968]. Intriguing molecules such as garudane [Mehta, 1987], a bridged prismane [Kostermans, 1987b], and a double tetraasterane [V.T. Hoffmann, 1987] have been acquired by means of photochemical method.

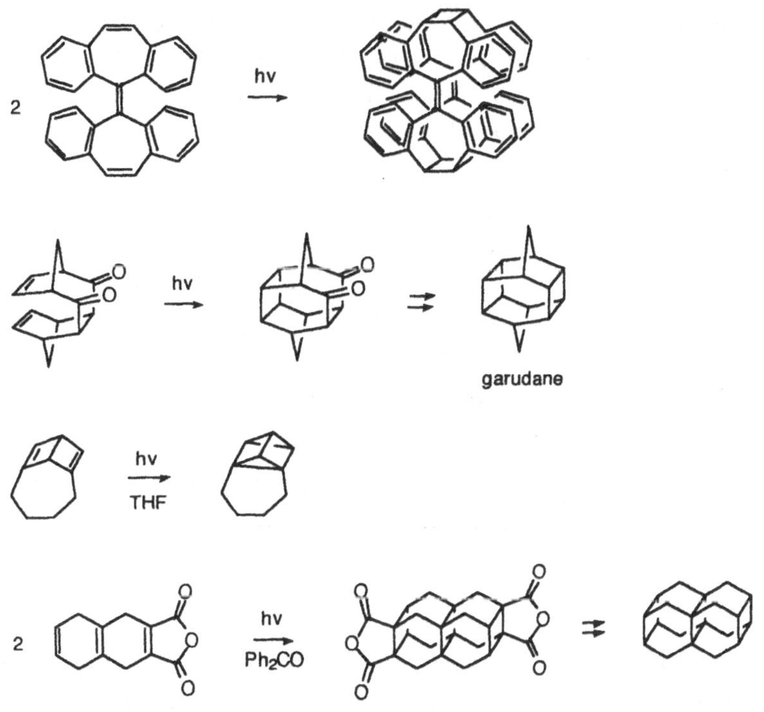

Perhaps more remarkable is the observation of a cubane derivative [Gleiter, 1988] in the photostationary state approached from the open diene. The behavior of this diene must be attributed to its inflexible juxtaposition as the parent tricyclo-octadiene does not adopt the same photochemical course.

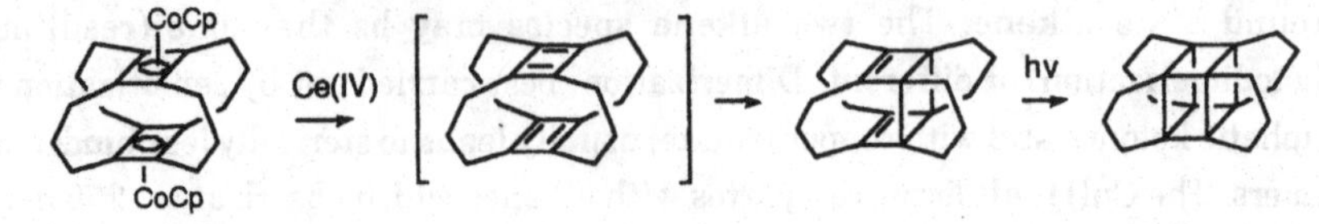

One of the sensitized photodimers of isoprene is a 1,1,2-trisubstituted cyclobutane delivered fragranol upon hydroboration [Katzenellenbogen, 1976]. For a synthesis of grandisol, the intramolecular photocycloaddition of a diene is a readily recognizable theme. Its development, exploiting the beneficial effect of Cu(I) salts, has advanced to an access of a chiral product [Langer, 1995].

<u>*(1R, 2S, 5R)- / (!S, 2S, 5S)-* 5-Methylbicyclo[3.2.0]heptan-2-ol</u>. [Langer, 1995]

(S)-6-Methylhepta-1,6-dien-3-ol (900 mg, 7.1 mmol) is dissolved in dry ether (30 mL) and divided in three portions to be irradiated in quartz vessels separately, after adding CuOTf (25 mg, 0.1 mmol) to each solution. After completion of the photoreaction (monitored by GLC) the combined organic mixture is evaporated and the crude product chromatographed to furnish two diastereomeric bicyclic alcohols (630 mg, 70%).

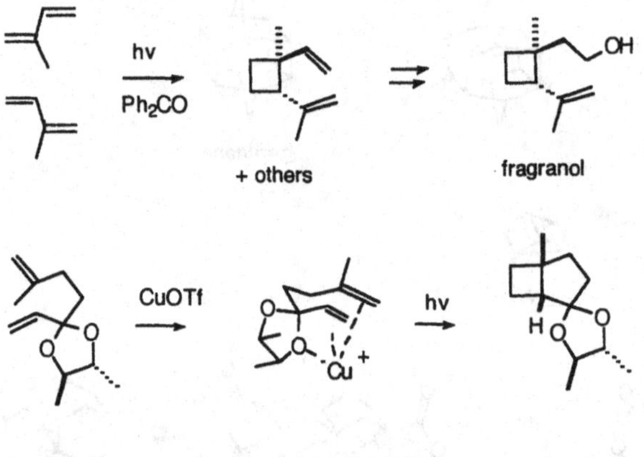

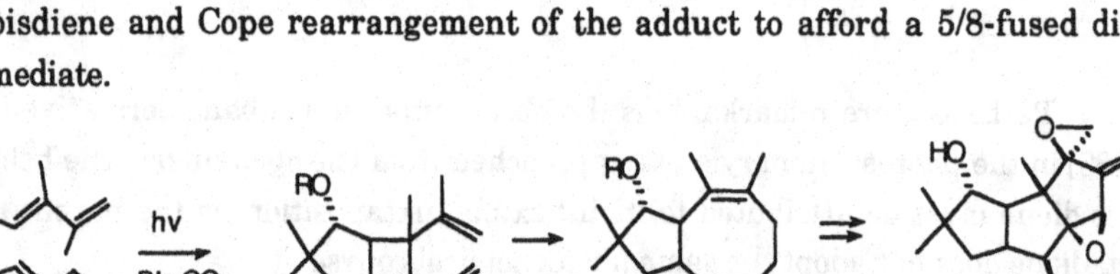

A novel approach to coriolin [Wender, 1987] involves [2+2]photocycloaddition of a bisdiene and Cope rearrangement of the adduct to afford a 5/8-fused diene intermediate.

The synthesis of α-bourbonene [Brown, 1968] featuring intramolecular photocycloaddition to form a bicyclo[3.2.0]heptane nucleus was a redirected effort because the original plan for elaboration of the copaenes/ylangenes via the head-to-tail isomer was thwarted. The conversion of (-)-germacrene-D into (-)-β-bourbonene [Yoshihara, 1969] follows the same pathway.

EtOOC EtOOC hv PhCOMe EtOOC EtOOC H ⇒ H

α-bourbonene

hv 254 nm H

β-bourbonene

It has been observed that intramolecular photocycloaddition of diene in the head-to-tail manner is effective when one of the double bonds is conjugated to a nonparticipating alkene. This aspect is evident in the synthetic routes to α-*trans*-bergamotene [Corey, 1971] and α-/β-longipinene [Miyashita, 1974]. (Note also the formation of a copaene intermediate [Heathcock, 1968].)

Preparation of 2,6,6-Trimethyl-9-methylenetricyclo[5.3.0.0^{2,8}]decane by Photocycloaddition.

[Miyashita, 1974]

A solution of 1,7,7-trimethyl-4-methylene-*(E)*-1,*(E)*-5-cyclodecadiene (700 mg, 3.7 mmol) and β-acetonaphthone (100 mg) in dry ether (400 mL) is irradiated through Pyrex with a 400W high pressure Hg lamp at 5-10° under nitrogen. After 19 h, two new products appear besides the starting material, in a ratio of 75:10:15. After the solvent is distilled through a Vigreux column the residue is chromatographed on a neutral alumina column and further purified by preparative GLC.

O O hv Ph_2CO O O ⇒

α-*trans*-bergamotene

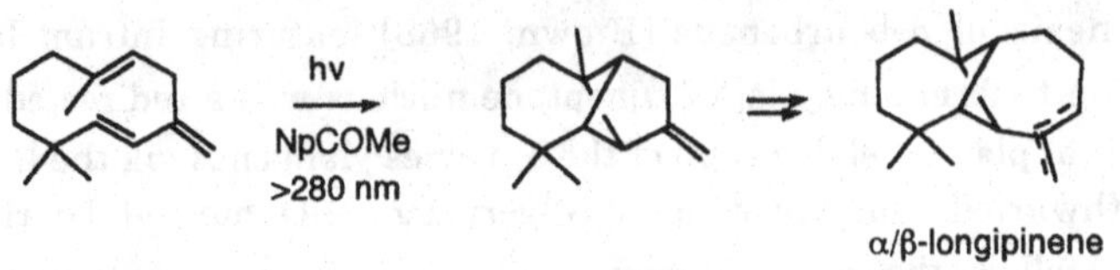

hν
NpCOMe
>280 nm

α/β-longipinene

7.3.1.3. [2+2]Photocycloaddition of Polarized Alkenes

Substituted alkenes such as enol ethers, enol esters, conjugated ketones, esters, including those containing a chiral component (e.g., diastereomeric menthoxybutenolides for the synthesis of grandisol enantiomers [N. Hoffmann, 1991]), participate in photocycloaddition. These reactions continue to enjoy the most popularity among various photochemical transformations [Dilling, 1977]. For example, a method for cyclopentanone synthesis which is further developed into a formal synthesis of cedrene [Ghosh, 1993] is based on an intramolecular cycloaddition and pinacol rearrangement.

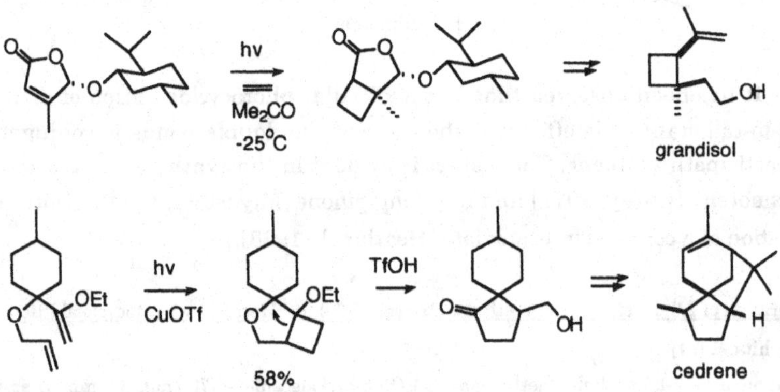

hν
Me_2CO
-25°C

grandisol

hν
CuOTf

58%

TfOH

cedrene

Acetylenes are also reactive as photocycloaddends, and the 1:1 adducts may undergo another photocycloaddition. An efficient synthesis of pterodactyladiene [H.-D. Martin, 1981] is based on the adduct of acetylene with two equivalents of maleic anhydride.

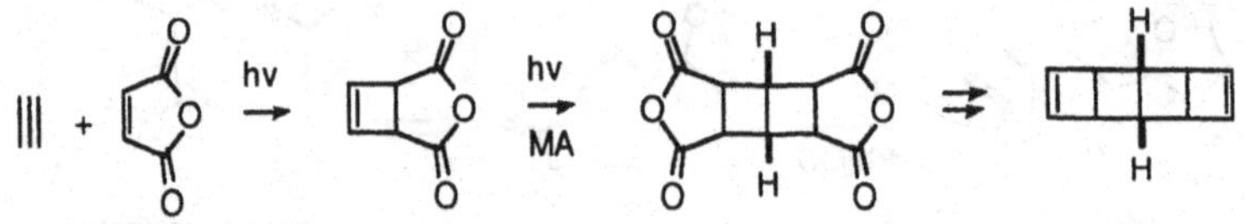

hν

hν
MA

Regioselectivity is of major concern in synthesis. In this respect the usefulness of photocycloaddition is attributed to high degree of control by judicious selection of reacting components. Regiochemical preferences in accordance of electronically directed alignment are shown in the photocycloaddition of 5,5-dimethyl-(5*H*)-furan-4-one with isobutene and with 1,1-bis(chlorodifluoromethyl)ethene [Magaretha, 1974].

hv hv F_2ClC CF_2Cl CF_2Cl CF_2Cl

Cycloaddition involving excited cycloalkenones or conjugated lactones and alkenes is the most widely explored photochemical method for synthesis [Crimmins, 1993]. The general consensus regarding the process is that $\pi\pi^*$ triplet states of the conjugated carbonyl systems are involved and products are derived from 1,4-diradical intermediates [Bauslaugh, 1970]. This is also the case of the cyclohexenone cycloaddition to C_{60} fullerene in benzene induced by a high pressure mercury lamp of with a XeCl excimer laser [Wilson, 1993]. As already stated, a tribute to its popularity is predictable regiochemistry (mainly in the cyclohexenone series), which suits well the construction of many natural products: caryophyllene and isocaryophyllene [Corey, 1964b], quadrone [Takeda, 1983], ormosanine [H.-J. Liu, 1969], hibaene [Duc, 1978], illudol [Kagawa, 1969; Matsumoto, 1971], sativene and helminthosporal [Yanagiya, 1979], khusimone [H.-J. Liu, 1979], paniculide-A [Smith, 1981b], daunomycinone [Boeckman, 1983], $\Delta^{9(12)}$-capnellene [H.-J. Liu, 1985], a pentalenolactone intermediate [Demuth, 1988b], 12,13-epoxytrichothec-9-ene [Masuoka, 1974, 1976], α-himachalene [Suginome, 1994], (-)-β-bourbonene [Tomioka, 1982], and spatol [Salomon, 1991]. It is apparent that the four-membered ring of the photoadducts is not always preserved, actually the various transformations of that structural unit enlarge the versatility of the method.

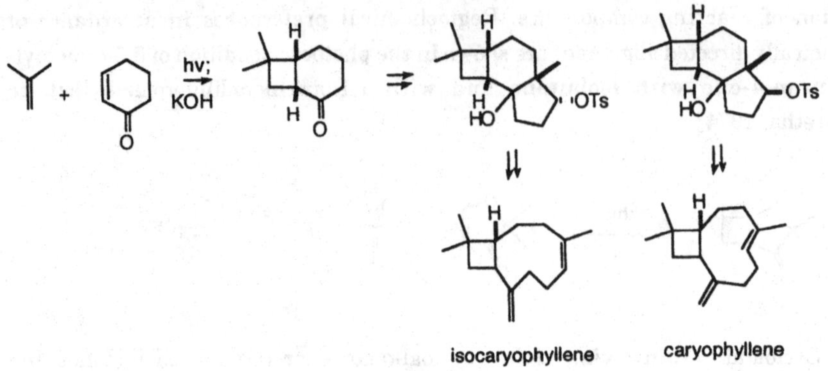

isocaryophyllene caryophyllene

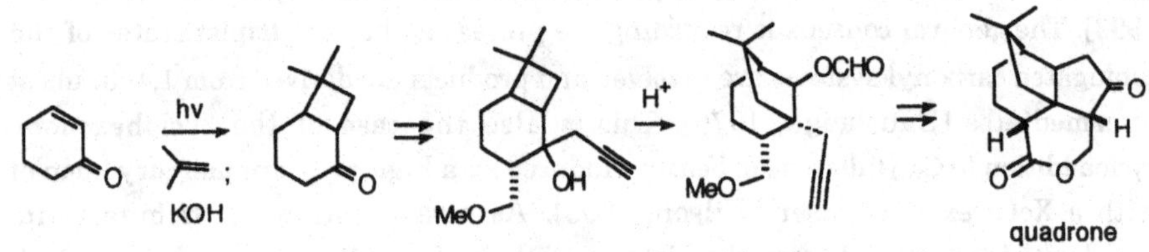

quadrone

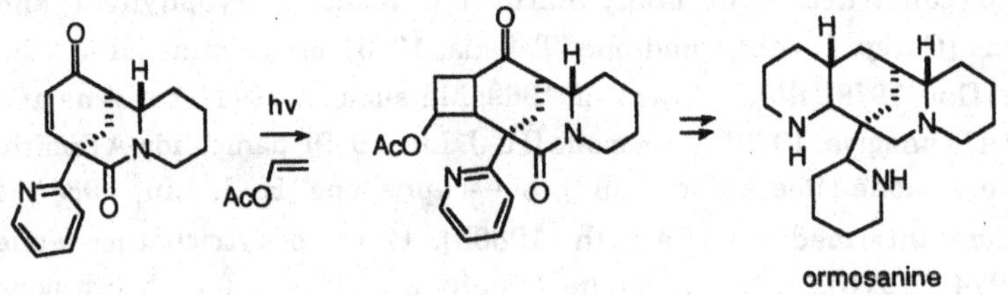

ormosanine

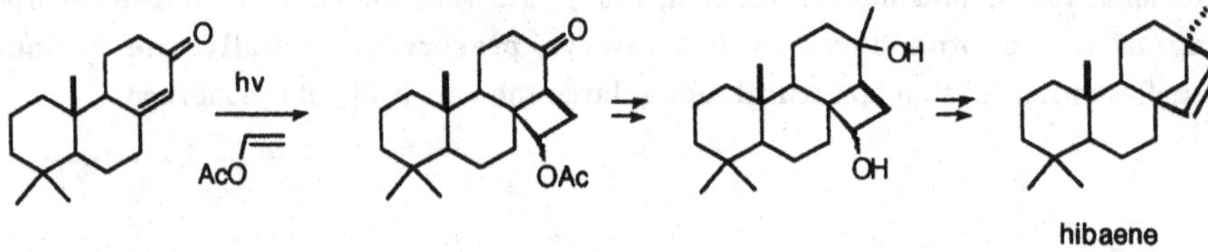

hibaene

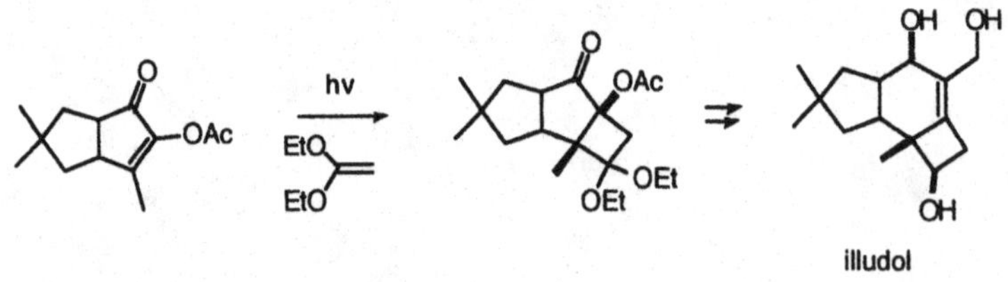

illudol

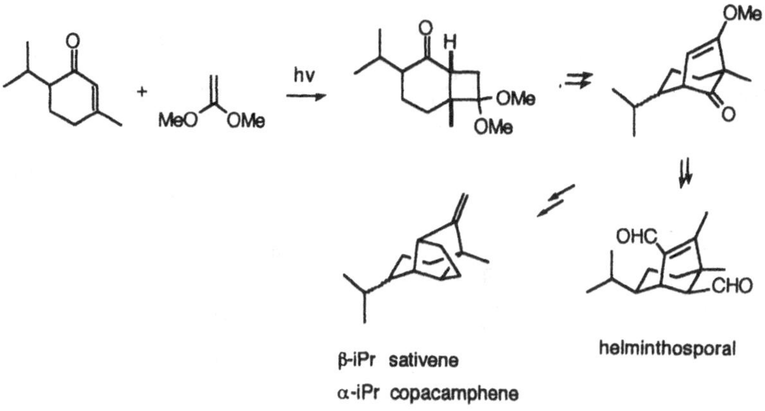

O + MeO OMe —hν→ O H OMe OMe ⇉ OMe O

OHC CHO

helminthosporal

β-iPr sativene

α-iPr copacamphene

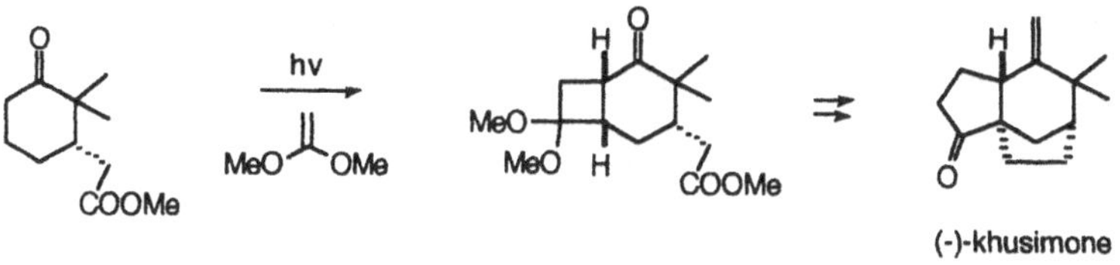

O COOMe —hν→ (MeO OMe) H O MeO MeO H COOMe ⇉ H O

(-)-khusimone

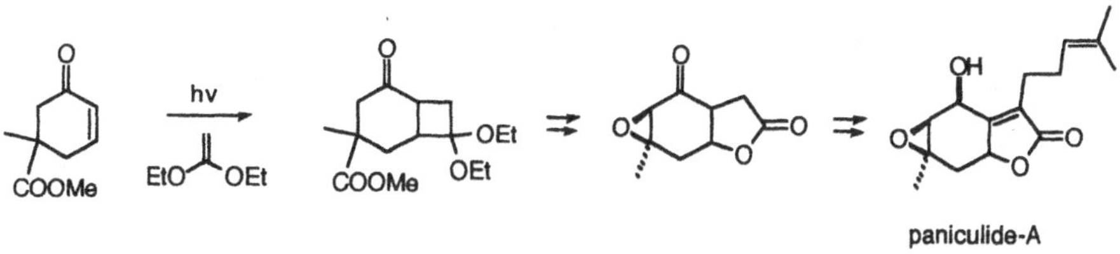

O COOMe —hν→ (EtO OEt) O OEt OEt COOMe ⇉ O O O O ⇉ OH O O O

paniculide-A

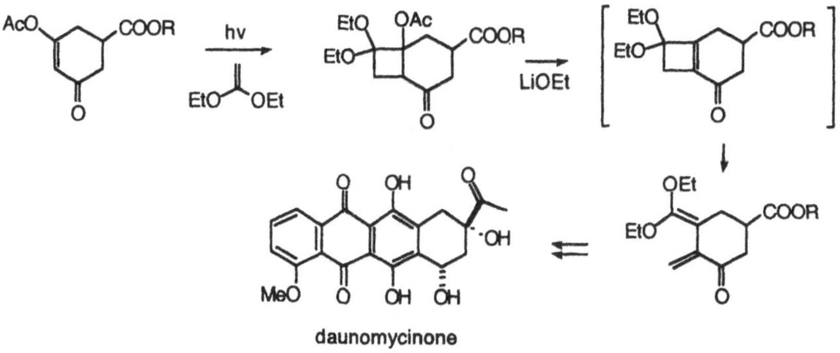

AcO COOR O —hν→ (EtO OEt) EtO OAc EtO COOR O —LiOEt→ [EtO EtO COOR O]

OEt EtO COOR O ⇉ O OH O OH MeO O OH OH

daunomycinone

EtO OEt hv EtO EtO O O COOEt OBn H

$\Delta^{9(12)}$-capnellene

O Me$_3$SiO hv O H H OSiMe$_3$ H H

α-himachalene

EtOOC Me$_3$SiO hv EtOOC OSiMe$_3$ O H O O O O H

O O O hv O O H O O O

12,13-epoxytrichothec-9-ene

O O OCOCMe$_3$ O O hv OCOCMe$_3$ O O H O H

β-bourbonene

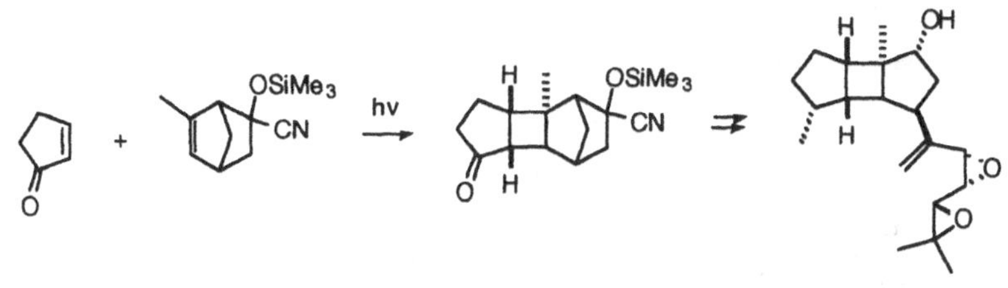

stoechospermol

The photochemical route to acorenone [Lange, 1978] which features spiroannulation is dependent on diastereofacially selective approach of the excited enone to the alkene.

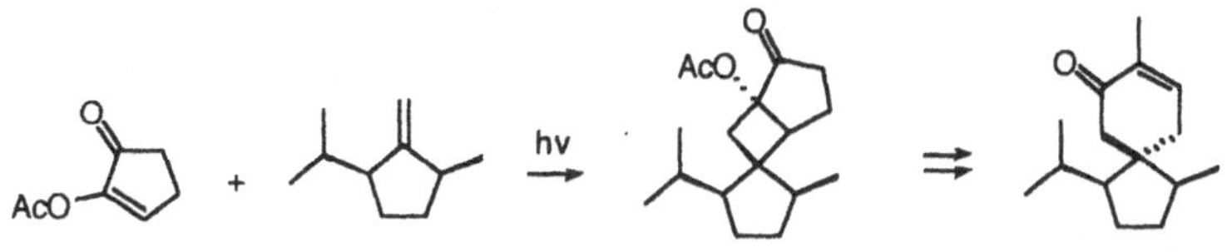

acorenone

Synthetically, stereoselectivity is also of great importance. For an excited cyclohexenone the most stable conformation may be defined by a substantial pyramidalized β-carbon atom [Wiesner, 1975]. This notion may be used to rationalize the bond formation process with the alkene.

When allene is employed as the alkene addend, it allows construction of intermediates for annotinine [Wiesner, 1969], modhephene [Tobe, 1984], atisine [Guthrie, 1966], ishwarane [Kelly, 1972], trachylobane [Kelly, 1973], talatisamine [Wiesner, 1974], phyllocladene [Duc, 1975], steviol methyl ester [Ziegler, 1977], pentalenolactone-E [Exon, 1981], periplanone-B [Schreiber, 1984b], aphidicolin [Marini-Bettelo, 1983], stemodin [Piers, 1985], shyobunone [Weyerstahl, 1987], and many others. The cycloadducts are individually modified in that the cyclobutane ring is either retained, cleaved, or rearranged. Dihydro-4-pyrones react in the same way as cycloalkenones, as demonstrated in an annulated derivative by its elaboration into forskolin [Ziegler, 1987].

Photocycloaddition of 2-Cyclohexenone with Allene. [Corey, 1964a]

A solution of 2-cyclohexenone (11.0 g, 0.1145 mol), allene (100 g, 2.5 mol) in pentane (530 mL) is cooled in dry ice bath and irradiated with a Type L, 450W Hanovia mercury arc through a Corex filter for 4.5 h. At the end of the photo reaction cooling bath is slowly lowered and the excess allene

is blown off by a stream of nitrogen introduced via a long hypodermic needle. Removal of the solvent and distillation furnishes the adduct (8.6 g, 55%), b.p. 43-45°C/0.65 torr.

hv
=C=
annotinine

hv
=C=
CH_2Cl_2 -78°
modhephene

hv
=C=
atisine

hv
=C=
trachylobane
phyllocladene

hv
=C=
ishwarane

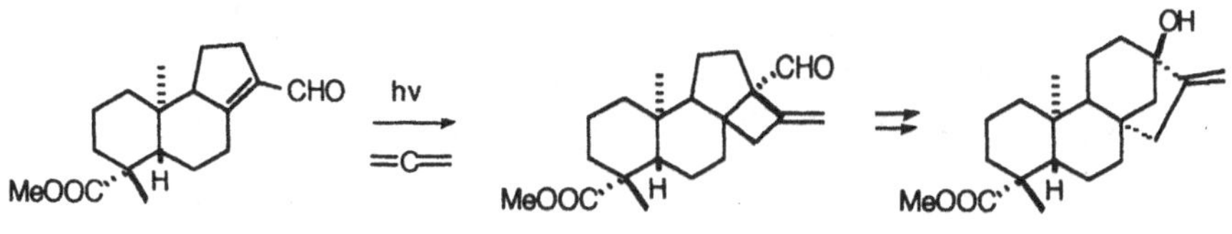

steviol methyl ester

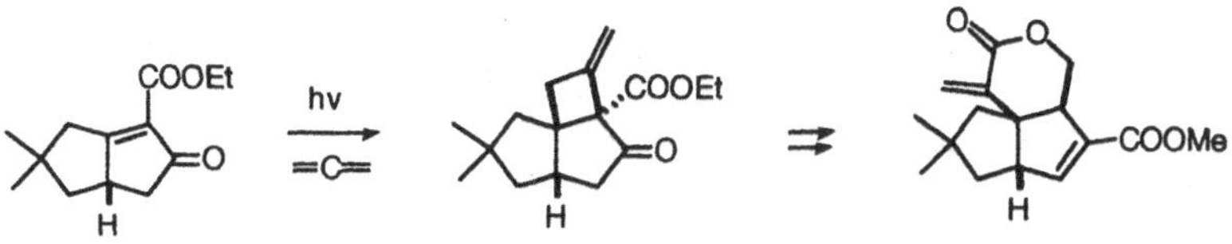

pentalenolactone-E
methyl ester

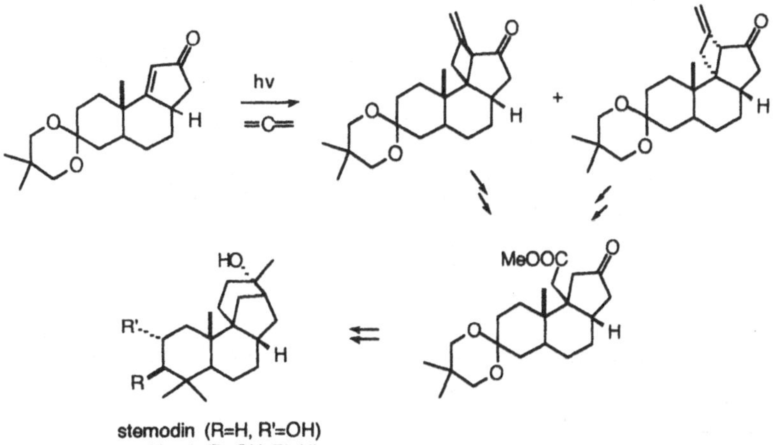

stemodin (R=H, R'=OH)
maritimol (R=OH, R'=H)

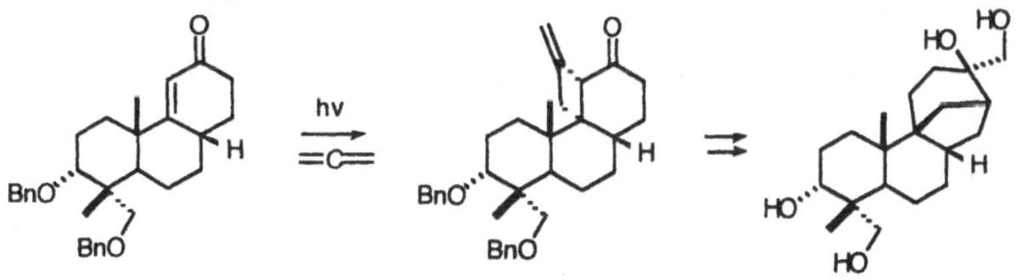

aphidicolin

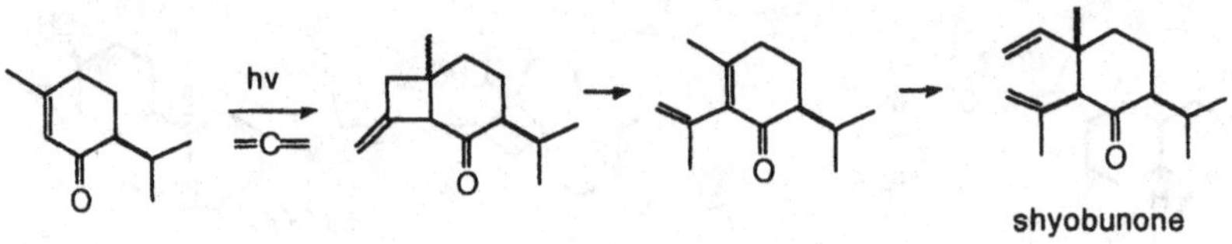
hv
=C=
shyobunone

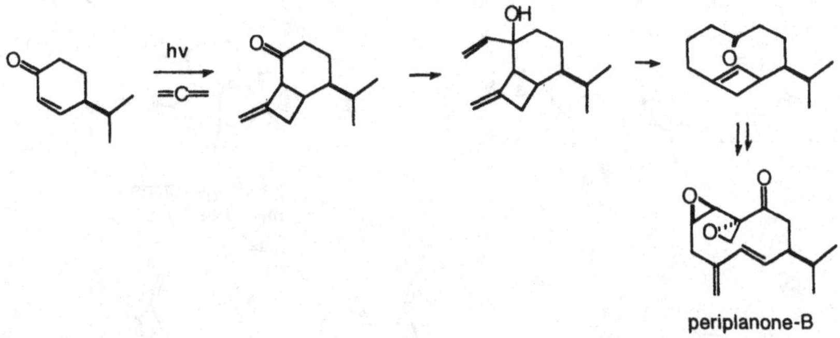
hv
=C=
OH
periplanone-B

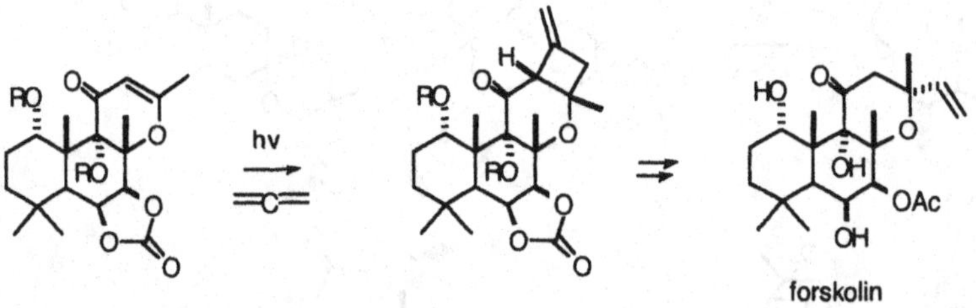
RO
RO
hv
=C=
H
OH
OAc
forskolin

Methyl 1-cyclobutenecarboxylate also enjoys an excellent reputation as an addend for photocycloaddition with conjugated cycloalkenones. The cycloadducts, while possessing a bicyclo[2.2.0]hexane moiety, are susceptible to thermal cleavage at the central C-C bond owing to high strain of the ring system and stabilization of the 1,4-diradical at one end by the ester group. Actually, the discovery of a remarkable transannular ene reaction of the thermolysis products enables the design of efficient routes to calameon [Wender, 1980a; Williams, 1980b], warburganal [Wender, 1980b], β-eudesmol, atractylon, and isoalantolactone [Wender, 1980c].

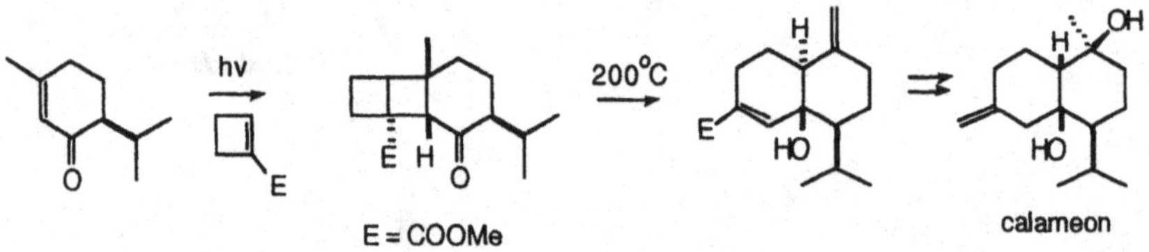
hv
E
E H
E = COOMe
200°C
HO
OH
calameon

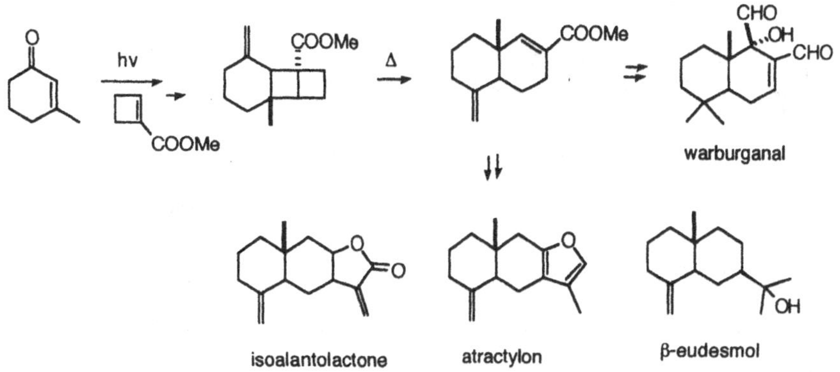

COOMe hv COOMe Δ COOMe CHO OH CHO warburganal

O O isoalantolactone atractylon OH β-eudesmol

The secondary reaction (transannular ene reaction) of the thermolytic step can be suppressed by structural modification, as is required for synthesis of 10-epijunenol [Wender, 1978] and isabelin [Wender, 1980b]. Note that the adduct from piperitone and 1-methylcyclobutene undergoes thermal cleavage to release *(ent)*-shyobunone, an elemane sesquiterpene, as well as a cyclodecadienone which is convertible to isoacoragermacrone [Williams, 1980a]. Another synthetic utility of the photocycloadduct is its conversion to zonarene [Williams, 1983].

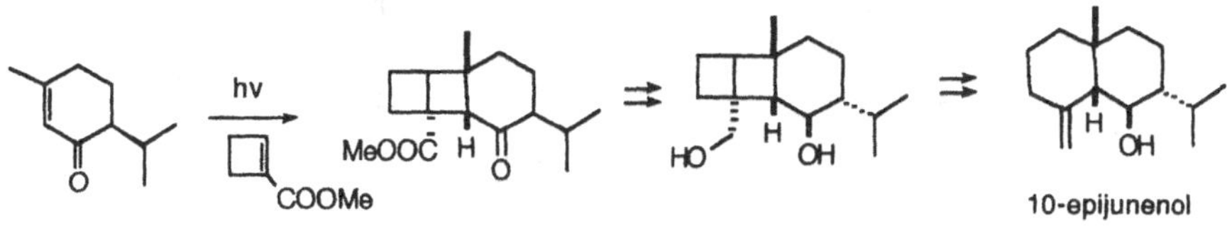

O hv COOMe MeOOC H O HO H OH H OH 10-epijunenol

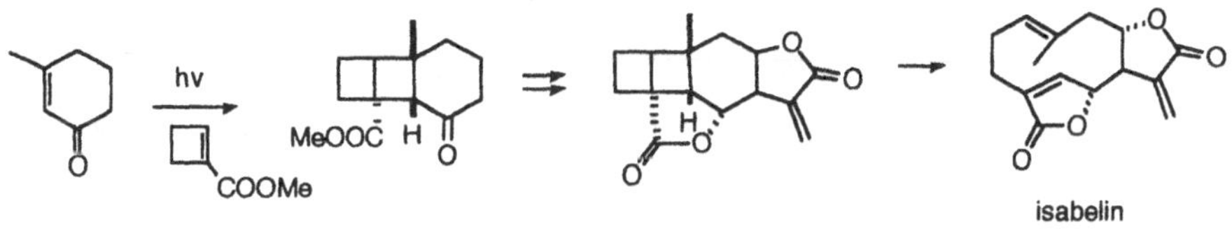

O hv COOMe MeOOC H O O H O O O O O O isabelin

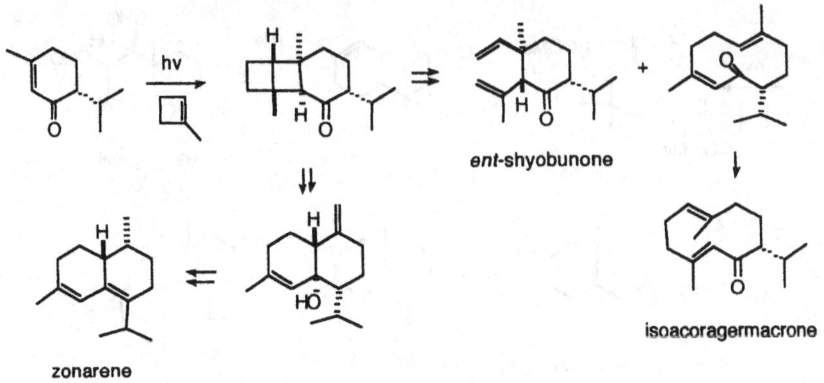

hv

ent-shyobunone

zonarene

isoacoragermacrone

Certain [2+2]photocycloadditions, such as those involving cyclopentenone, are plagued by low or nonexistent regioselectivity. The formation of equal amounts of the tricyclic products from 2-cyclopentenone and 1-methyl-3-isopropylcyclopentene, apparently due to insufficient perturbation power of the methyl substituent on the double bond, is a distraction of the synthesis of α-/β-bourbonene [White, 1968]. Only the head-to-tail isomer is useful for the purpose.

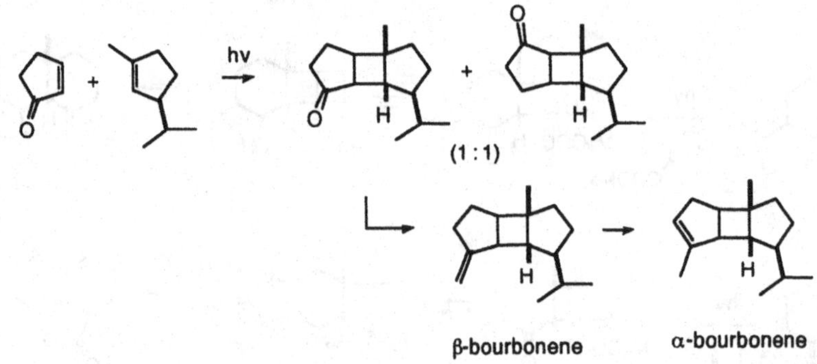

hv

(1 : 1)

β-bourbonene α-bourbonene

The same regiochemical problem appeared during a synthesis of lineatin [Mori, 1979] which was initiated by a photocycloaddition of 2,4,4-trimethyl-2-cyclopentenone with vinyl acetate.

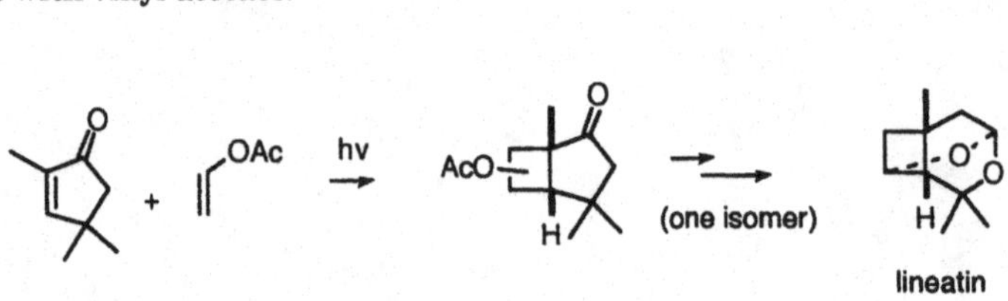

OAc hv AcO (one isomer)

lineatin

Fortunately on many other occasions, the issue never arises because of the symmetrical nature of at least one addend, such as concerning synthesis of grandisol [Zurflüh, 1970; Gueldner, 1972; Cargill, 1975; Rosini, 1979; Meyers, 1986], sterpurene [Moëns, 1986], sterpuric acid [Paquette, 1987], filiformin and aplysin [Nath, 1996], modhephene [Smith, 1981a], caryophyllene alcohol [Corey, 1964c], biotin [Whitney, 1983], dictamnine [Naito, 1984], and xanthocidin [Smith, 1983]. Interestingly, in the filiformin synthesis oxetane formation follows the photocycloaddition, which is of little consequence to the overall operation. However, the cycloaddition of the bicyclic enone with 1,2-dichloroethene en route to modhephene the stereoselectivity is poor.

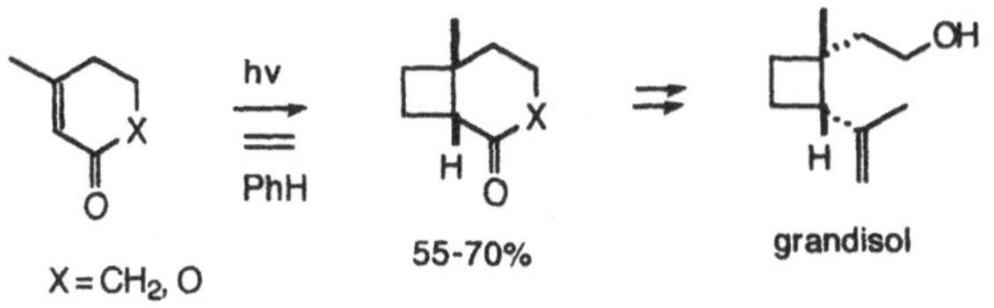

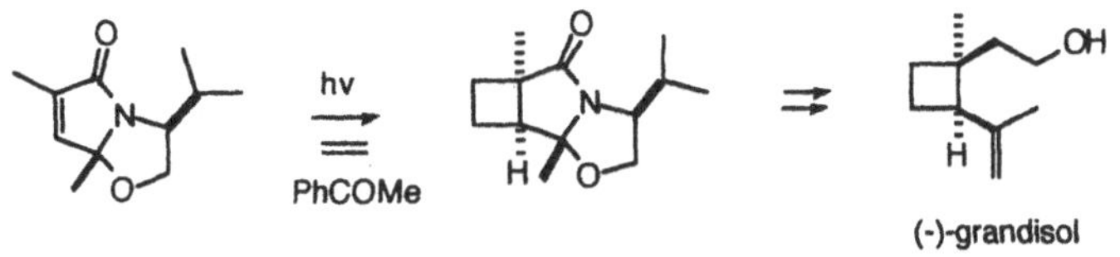

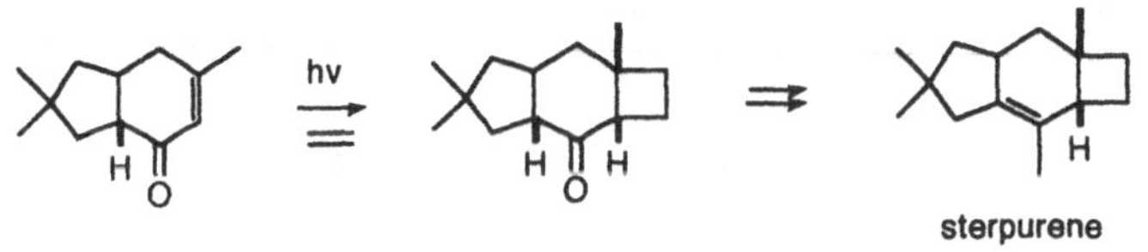

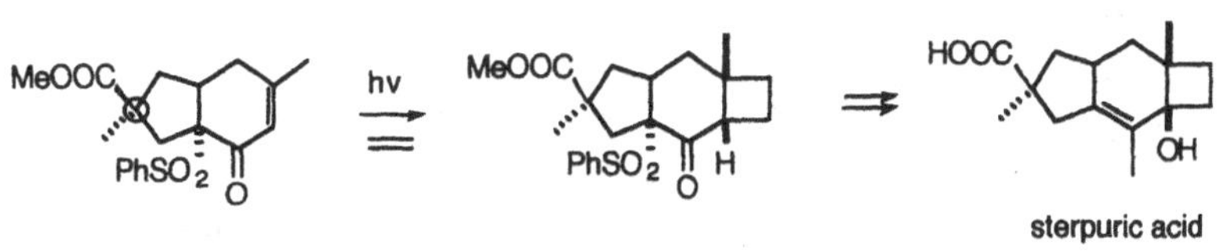

hv, PhH; X = CH_2, O; 55-70%; grandisol

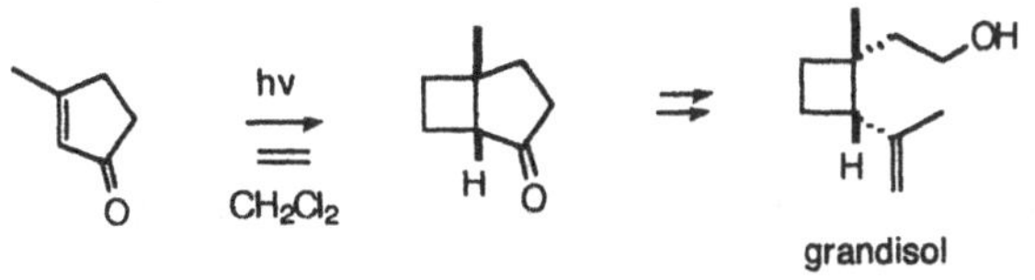

hv, CH_2Cl_2; grandisol

hv, PhCOMe; (-)-grandisol

hv; sterpurene

MeOOC, $PhSO_2$; hv; HOOC, OH; sterpuric acid

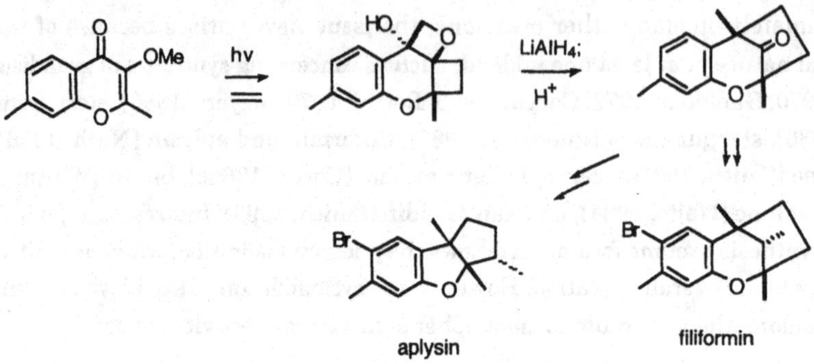

hν; LiAlH₄; H⁺

aplysin

filiformin

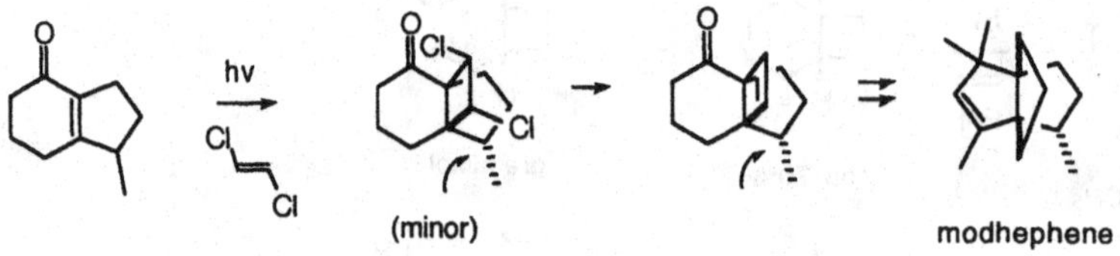

hν

(minor)

modhephene

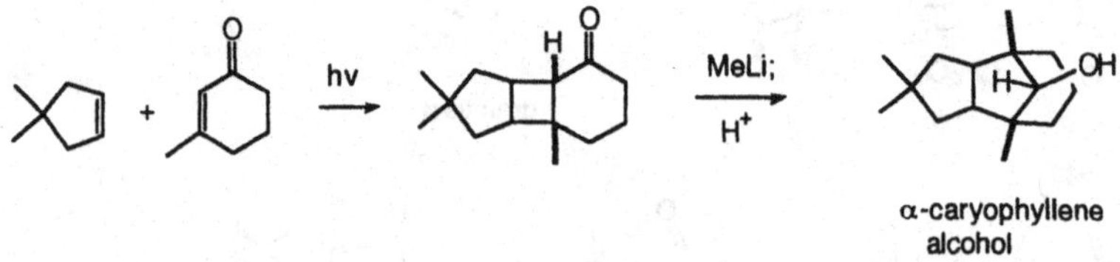

hν; MeLi; H⁺

α-caryophyllene alcohol

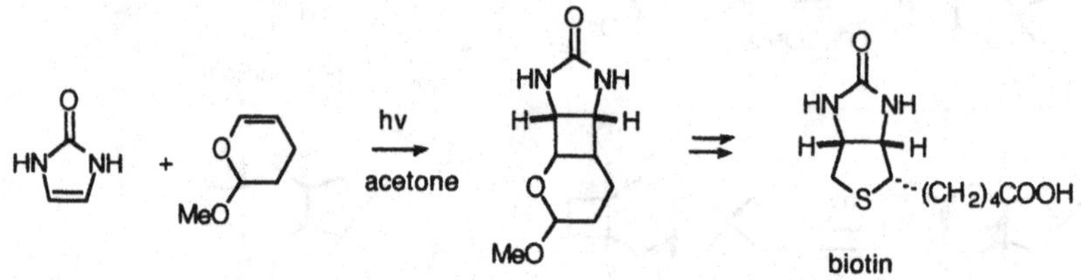

hν acetone

biotin

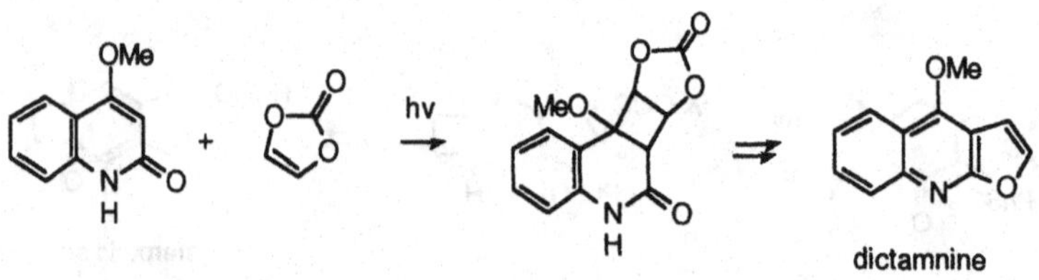

hν

dictamnine

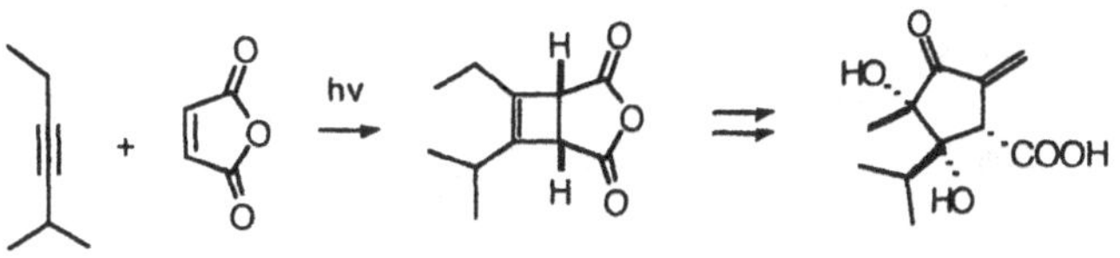

xanthocidin

Annulation of cyclopentenones with six- and seven-membered rings that possess carbonyl groups at the two subangular positions can be achieved via photocycloaddition with 1,2-bistrimethylsiloxycyclobutene and cyclopentene, followed by desilylation of the photoadducts and α-glycol cleavage [Termont, 1977]. The accessibility of the siloxy compounds by the modified acyloin condensation renders the method widely applicable, allowing the synthesis of damsin [De Clercq, 1977], hysterin [Demuynck, 1979], carpesiolin [Kok, 1979], compressanolide [Devreese, 1980], estafiatin [Demuynck, 1982], 1-oxocostic acid [van Hijfte, 1982], maritimin and α-santonin [van Hijfte, 1984], balanitol [Anglea, 1987], and daucene [Audenaert, 1987], among others.

Preparation of 6,10-Bis(trimethylsiloxy)tricyclo[5.3.0.0^{6,10}]decan-2-ones. [Termont, 1977]

A solution of cyclopentenone (8.2 g, 0.1 mol) in pentane is added to 1,2-bis(trimethylsiloxy)-cyclopentene (50.4 g, 0.2 mol) in five equal portions and regular intervals while being irradiated in a Rayonet photoreactor (366 nm) for 24 h. After distillation of the solvent and recovery of unreacted 1,2-bis(trimethylsiloxy)cyclopentene the cycloadducts are obtained (65% yield).

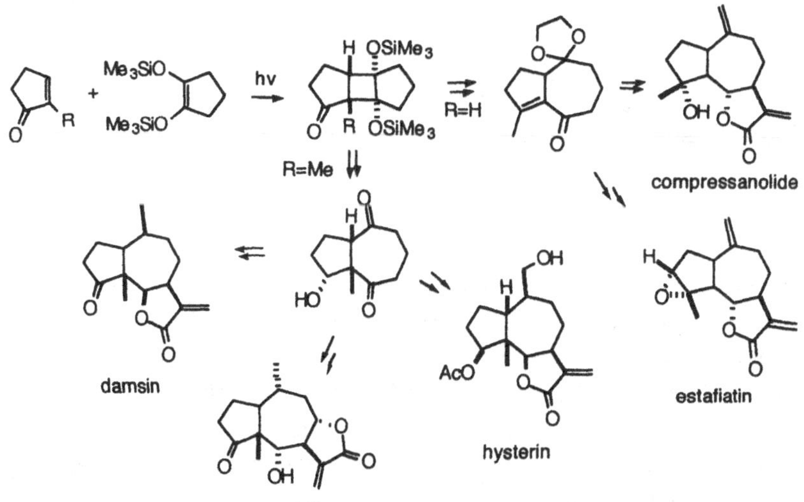

compressanolide

damsin

estafiatin

hysterin

carpesiolin

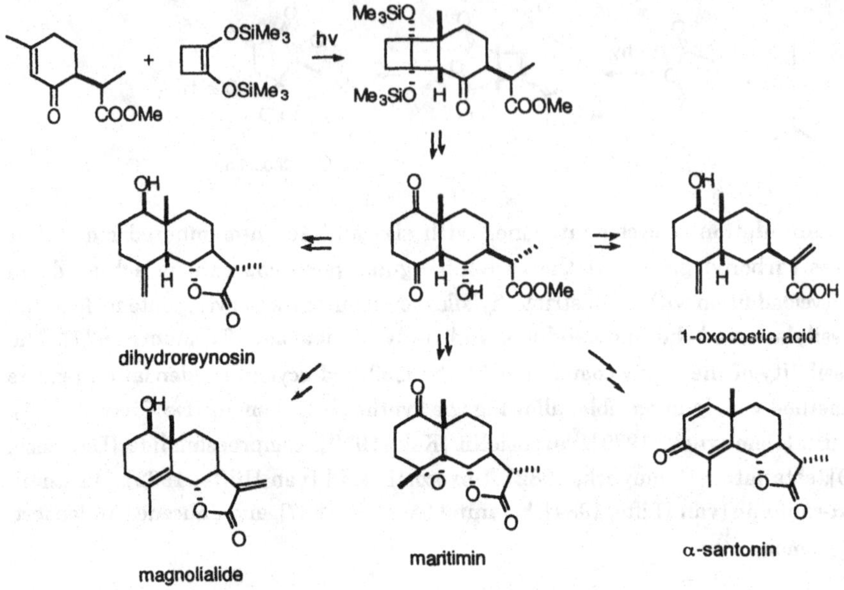

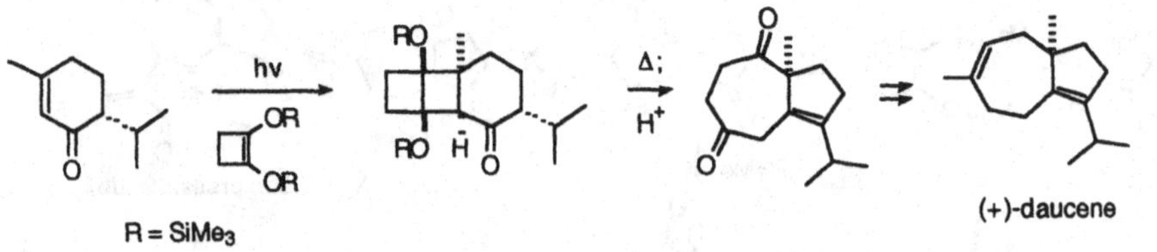

dihydroreynosin

1-oxocostic acid

magnolialide

maritimin

α-santonin

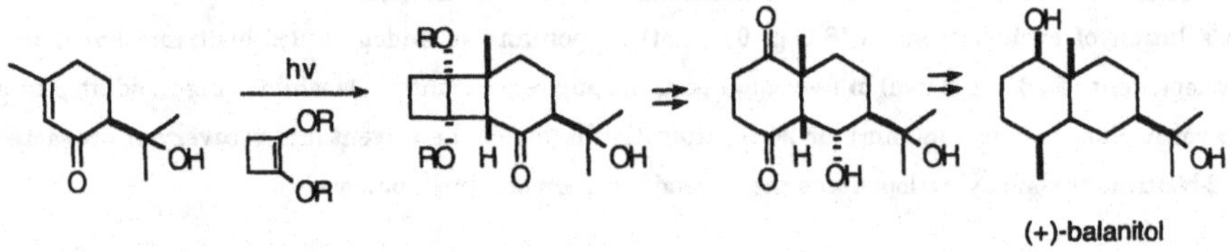

(+)-balanitol

$R = SiMe_3$

(+)-daucene

Intramolecular versions of photocycloaddition constitute on divers occasions unique solutions to synthetic problems, particularly regarding regiochemistry: cubane [Eaton, 1964], reserpine [Pearlman, 1979], isocomene [Pirrung, 1981], hibiscone-C [Koft, 1982], α-acoradiene [Oppolzer, 1983], pentalenic acid [Crimmins, 1984], trihydroxydecipiadiene [Dauben, 1984], italicene/isoitalicene [Leimner, 1984], silphinene [Crimmins, 1986], hirsutene and $\Delta^{9(12)}$-capnellene [Mehta, 1986], pentalenolactone-G [Pirrung, 1986, 1988a], laurenene [Crimmins, 1987], bilobalide

[Crimmins, 1992], paeoniflorin [Hatakeyama, 1994]. In the pentalenolactone-G synthesis there is an autoselection for one of the two butadienyloxy units in the photocycloaddition to produce the more stable and desired stereoisomer, in the paeoniflorin synthesis the proximal double bond of a conjugated diene undergoes cycloaddition, and in the bilabolide synthesis a double bond of the furan ring serves as an addend.

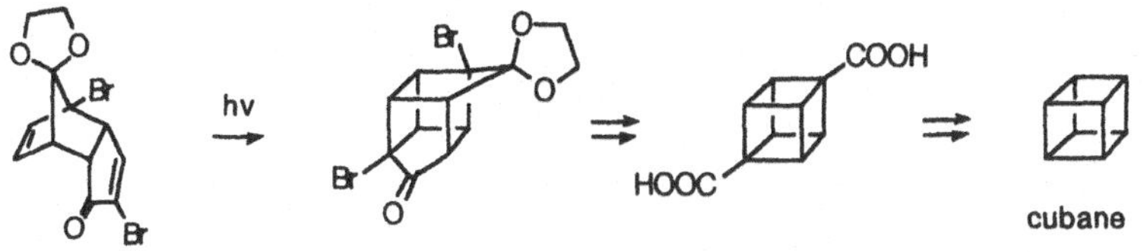

cubane

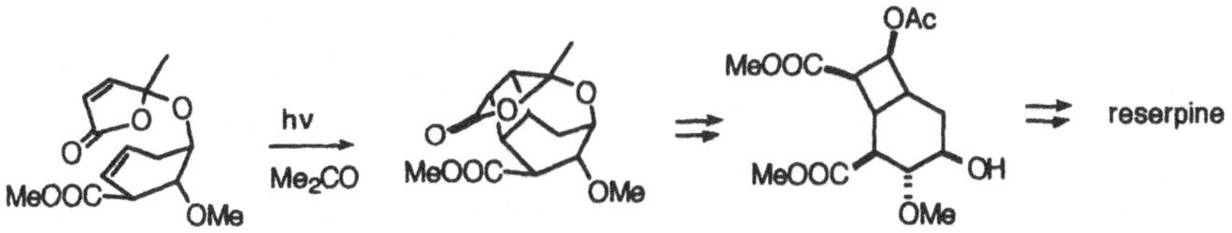

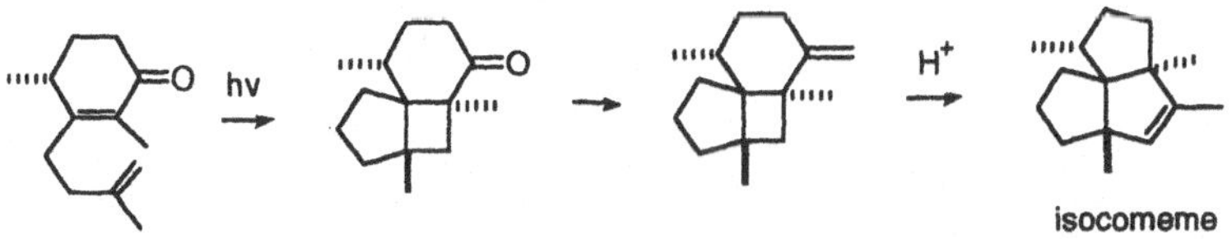

reserpine

isocomeme

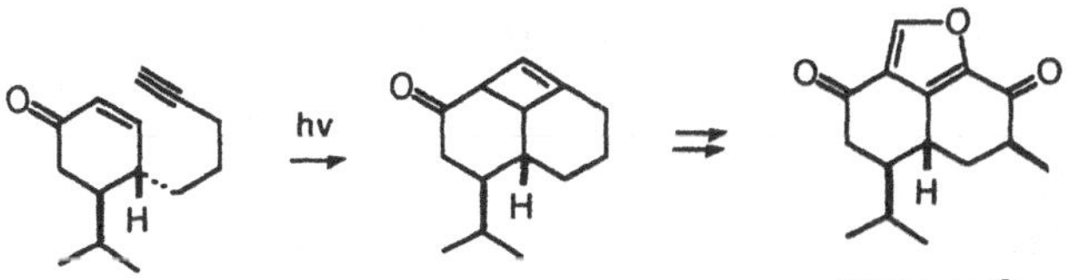

hibiscone-C

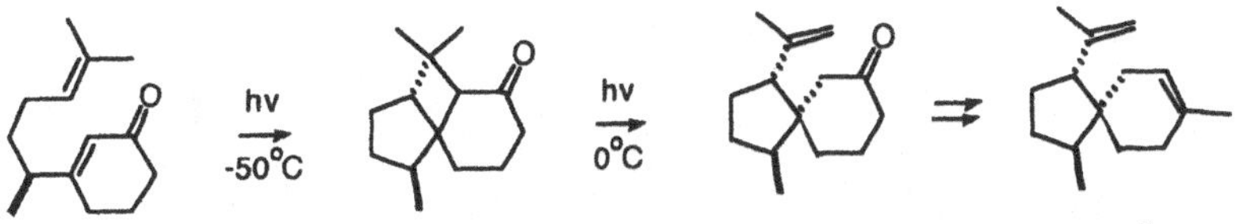

α-acoradiene

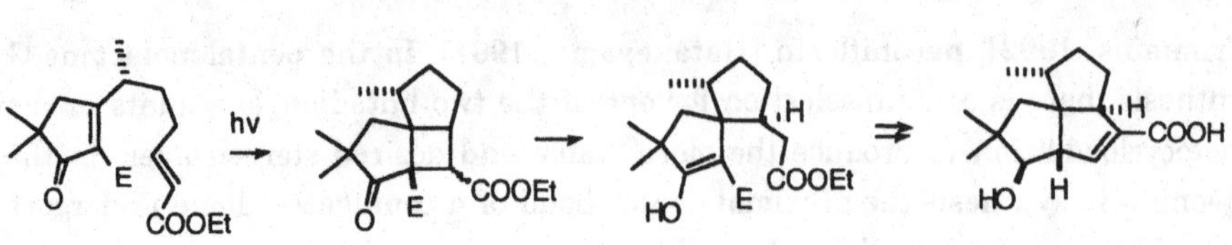

hv

COOEt

E = COOMe

COOEt

COOEt

H

HO

E

H

COOH

pentalenic acid

OH

hv

OH

HO

OH

OH

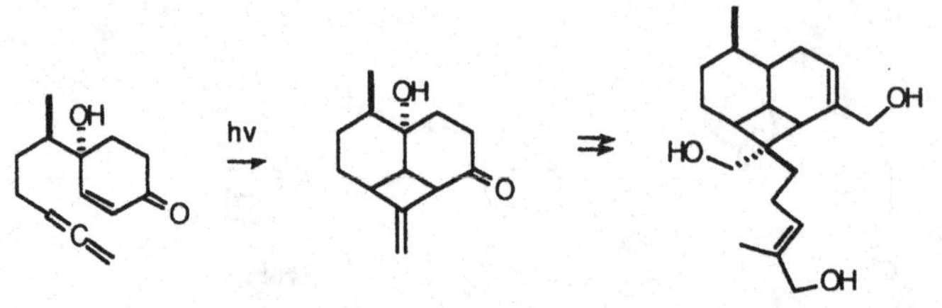

trihydroxydecipiadiene

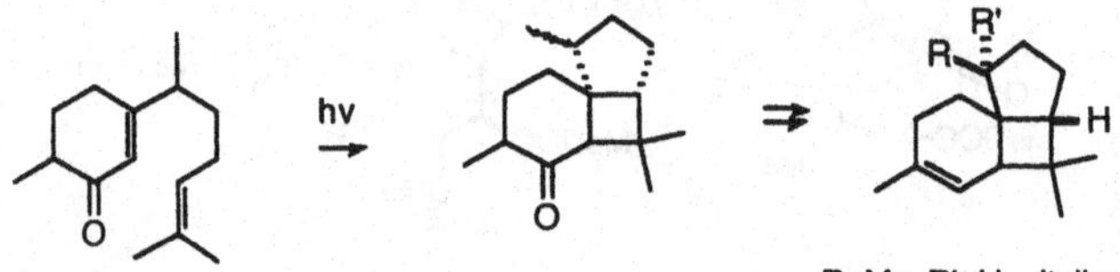

hv

R'

R

H

R=Me, R'=H italicene

R=H, R'=Me isoitalicene

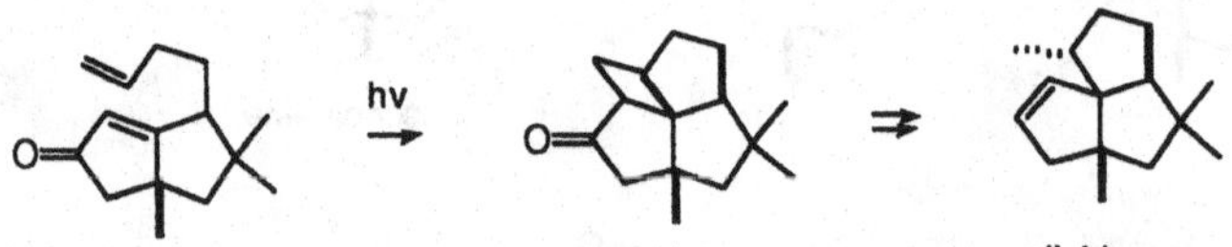

hv

silphinene

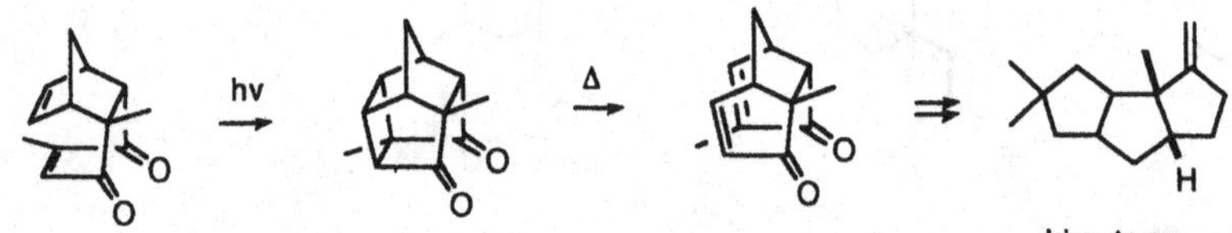

hv

Δ

H

hirsutene

hv

Δ

H

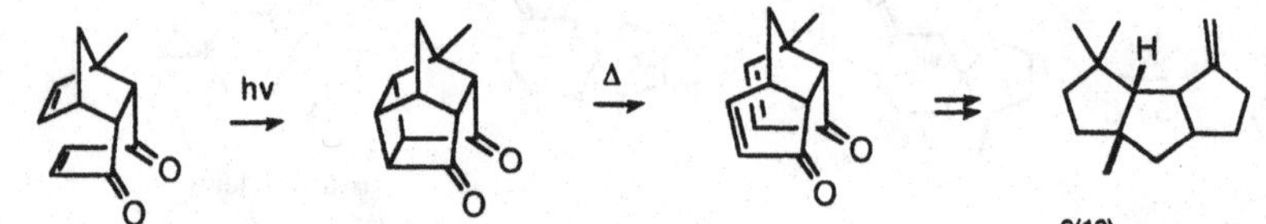

$\Delta^{9(12)}$-capnellene

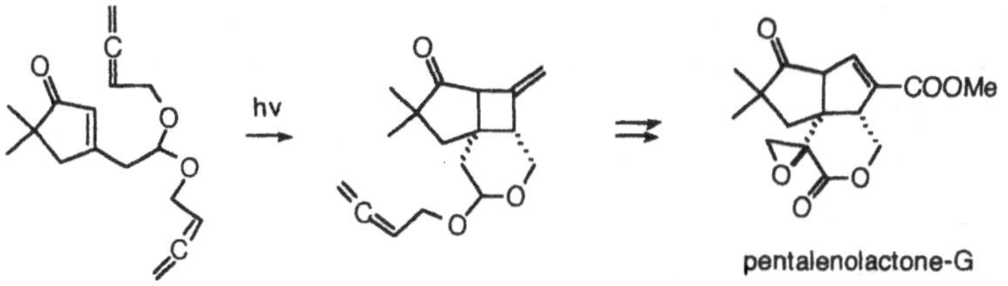

hv

pentalenolactone-G

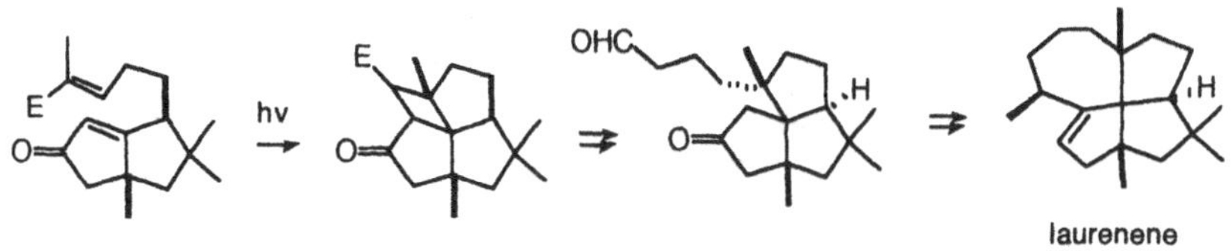

E = COOMe

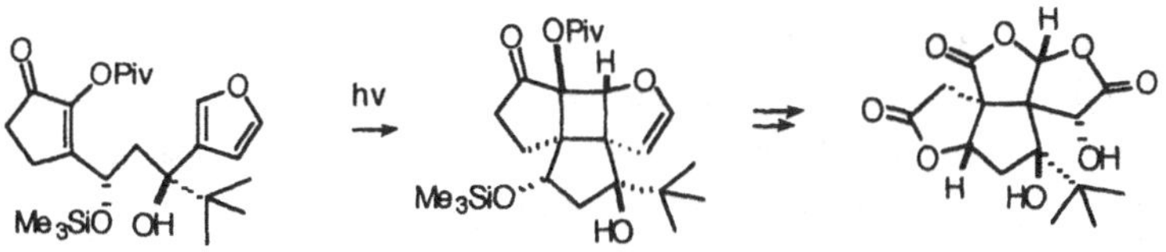

hv

laurenene

hv

bilobalide

hv

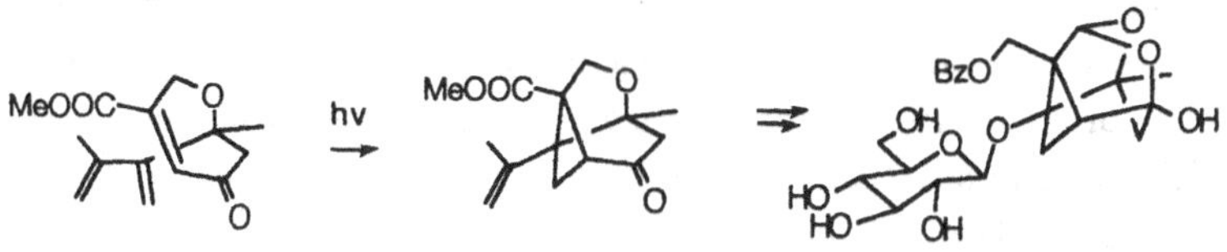

(-)-paeoniflorin

Carvonecamphor is a tricyclic intramolecular adduct derived from carvone. It is shown that the photocycloaddition becomes very efficient (90% yield) in chlorobenzene at 130°C. Pyrolysis of carvone camphor leads to a cyclopentenone which is useful for elaboration of methylenomycin-A [Sternbach].

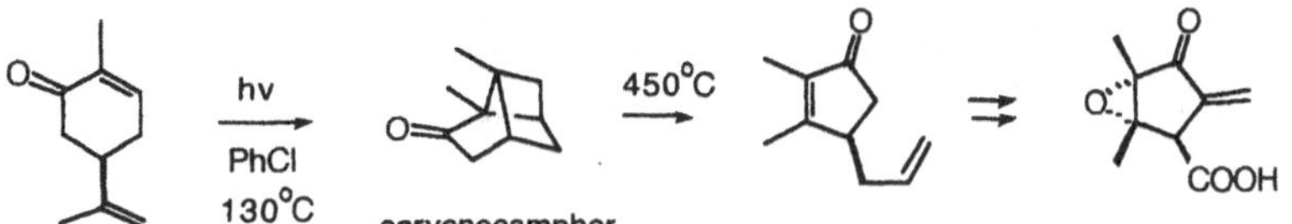

hv PhCl 130°C

carvonecamphor

450°C

methylenomycin-A

Regarding the photochemical synthesis of α-/β-panasinsene [McMurry, 1980], probably for steric reasons, the intermolecular pathway was found untenable.

hν 67%

β-panasinsene

It should be emphasized that regioselectivity may be manipulated through tethering, although steps for cleavage of the extraneous ring must be involved. Accordingly, the design of the tether is important. The structure of the potential diradical intermediate should be considered, as formation of a 5-membered ring is generally favored [Beckwith, 1981].

A remarkable method for β-lactam formation involves irradiation of a mixture of an imine and a Fischer carbene complex [Hegedus, 1984], the latter being transformed into a ketene prior to undergoing a [2+2]cycloaddition. Apparently such cycloadducts are prone to ring expansion rearrangment and followed by reductive elimination of the metallic species. The photochemical behavior of the system is very different from the ground state chemistry.

<u>1,3-Dimethyl-3-methoxy-4-phenyl-β-azetidinone.</u>

The methoxycarbene complex (1.25 g, 5 mmol) and N-methylbenzylideneimine (595 mg, 5.0 mmol) are combined in an Erlenmeyer flask containing petroleum ether (200 mL) and placed in direct sunlight at 24°. After 3 h, the mixture becomes brown and heterogeneous. The β-lactam (760 mg, 75%) is isolated and recrystallized, m.p. 74-75°.

OMe $Cr(CO)_5$ + Ph N Me hν MeO Ph NMe O

75%

7.3.1.4. De Mayo Reaction

The [2+2]photocycloaddition of an enolized β-dicarbonyl compound with an alkene gives an adduct embodying a 2-acylcyclobutanol unit is known as the de Mayo reaction. Quenching experiments indicated the enolized diketone triplet excited state energy is about 70 kcal/mol. The usefulness of the process lies in the fact that the adducts fragment by a retroaldol pathway very readily (ring strain relief) to generate two carbonyl-containing side chains. Acyclic β-hydroxy enones which can form internal hydrogen bond are reactive, whereas their derivatives are not, because the latter tend to deactivate by undergoing *(E,Z)*-isomerization. The reaction is best conducted in nonpolar solvents so as not to disturb the intramolecular hydrogen bond. It should be noted that derivatives of common-sized β-hydroxycycloalkenones do not suffer from this defect.

Photocycloaddition of Cyclohexene with Acetylacetone. [de Mayo, 1963]

A solution of acetylacetone (15.02 g) in cyclohexene (135 mL) is irradiated with a water-cooled apparatus (80W lamp) under nitrogen for 45 h. After removal of the excess cyclohexene the product is distilled through a 40-inch spinning band column to give the adduct (20.2 g, 78%), b.p. 89°C/1.7 torr.

The de Mayo reaction plays a key role in the synthesis of loganin [Büchi, 1973; Partridge, 1973], secologanin [Hutchinson, 1978], sweroside [Nakane, 1980], sarracenin [Baldwin, 1982], protoillud-7-ene [Takeshita, 1980], chrysomelidial [Takeshita, 1982], deoxytrisporone [Takeshita, 1984], agarospirol and hinesol [Hatsui, 1995].

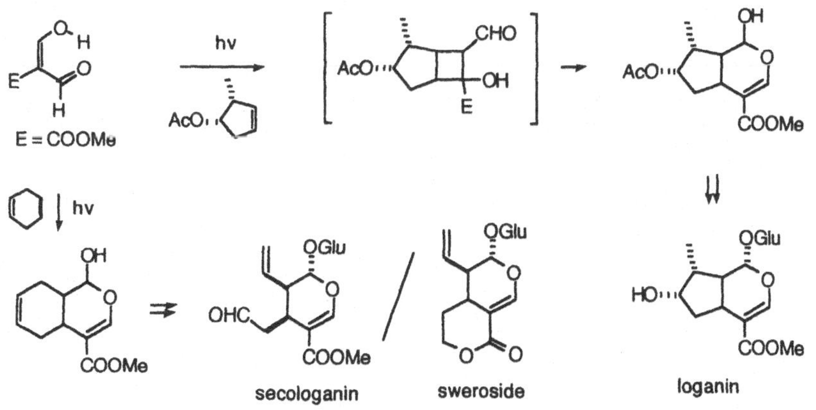

E = COOMe

secologanin sweroside loganin

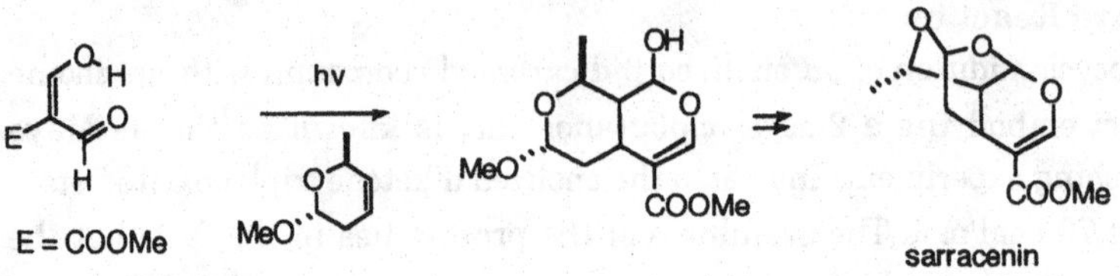

E = COOMe

hv

sarracenin

hv

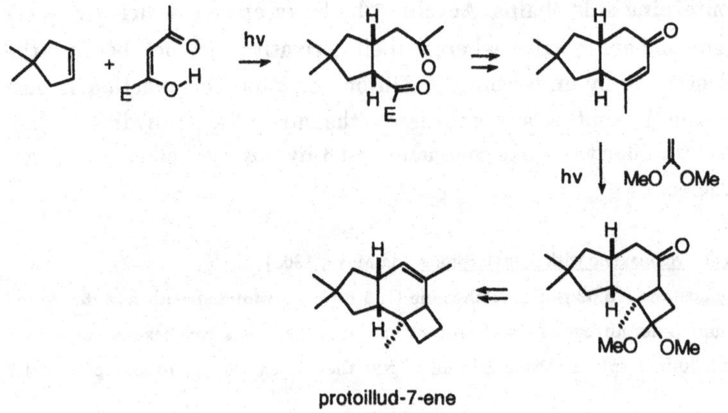

hv

protoillud-7-ene

hv

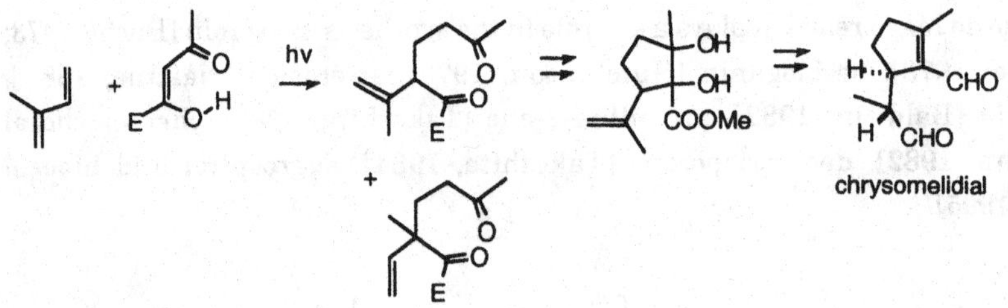

chrysomelidial

hv

deoxytrisporone

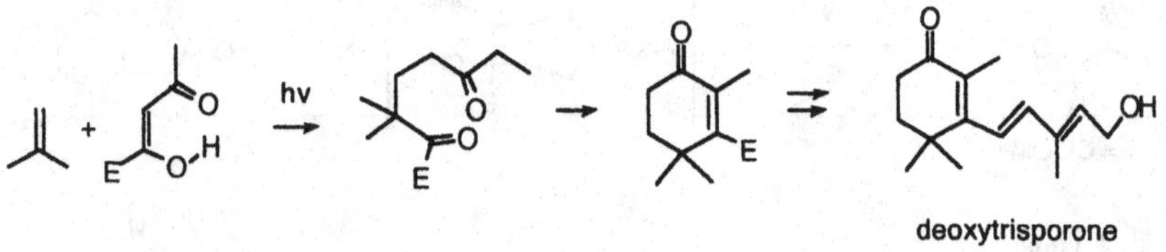

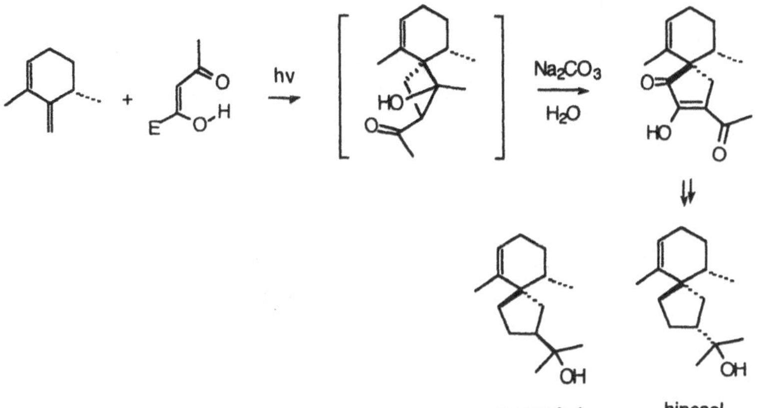

agarospirol hinesol

Many syntheses utilize the enol derivatives of 1,3-cycloalkanediones to engage in a deMayo reaction, in either an intermolecular or intramolecular fashion: β-himachalene and stipitatonic acid [Challand, 1969], hirsutene [Tatsuta, 1979], β-bulnesene [Oppolzer, 1980], zizaene [Baker, 1981], longifolene and sativene [Oppolzer, 1984], phoracantholide-M [Ikeda, 1983], pentalenene [Pattenden, 1984], daucene [Seto, 1985], isoamijiol [Pattenden, 1986]. A rare instance in which an unprotected (enolized) 1,3-cyclohexanedione is involved pertains to an approach to hirsutene [Disanayaka, 1985].

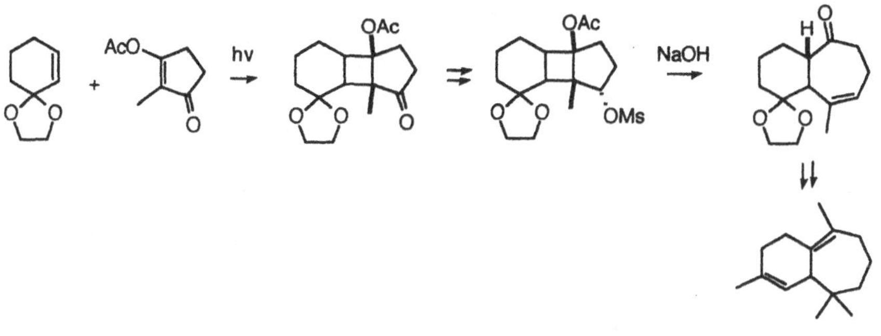

β-himachalene

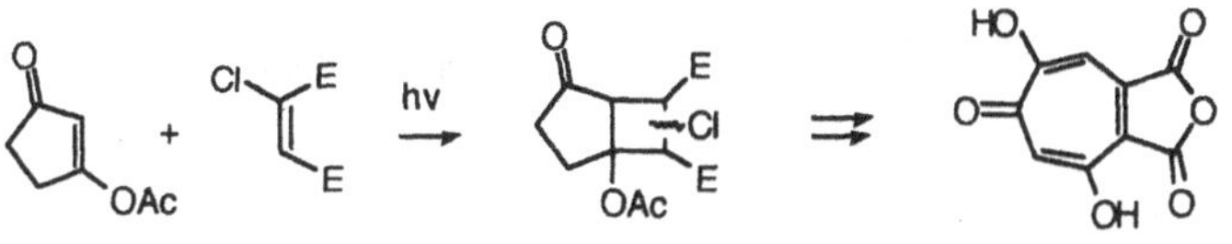

stipitatonic acid

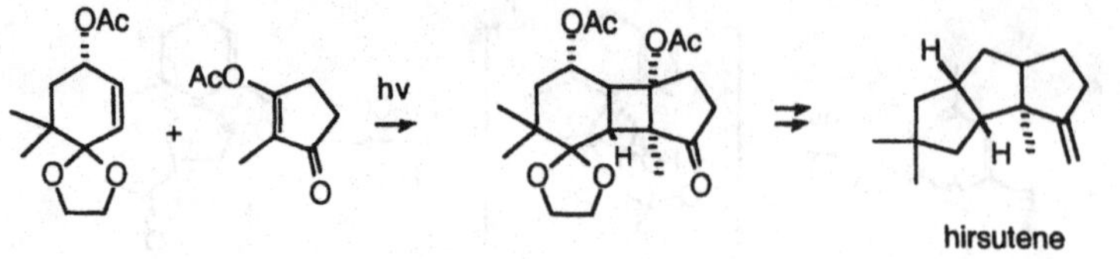
OAc
AcO
hv
OAc
OAc
H
H
H
hirsutene

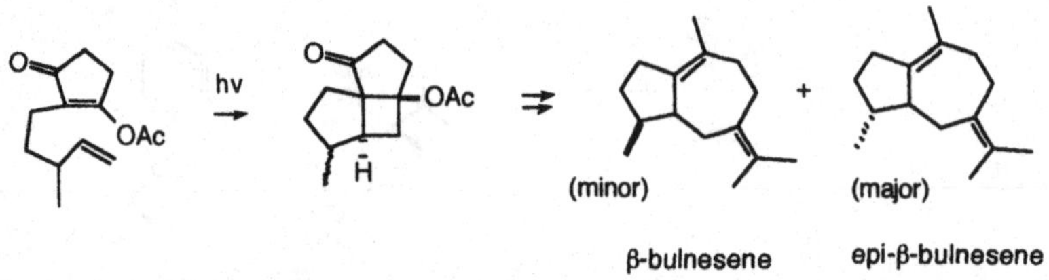
OAc
hv
OAc
H
(minor)
(major)
β-bulnesene
epi-β-bulnesene

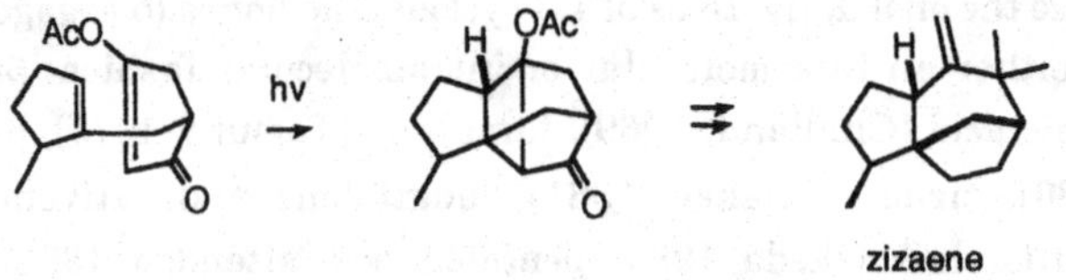
AcO
hv
H
OAc
H
zizaene

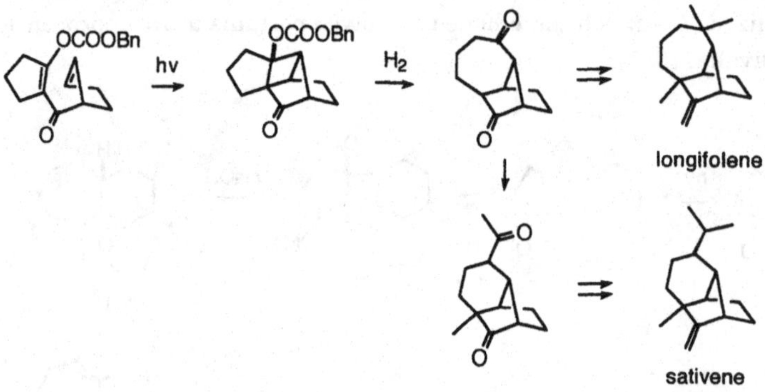
OCOOBn
hv
OCOOBn
H2
longifolene
sativene

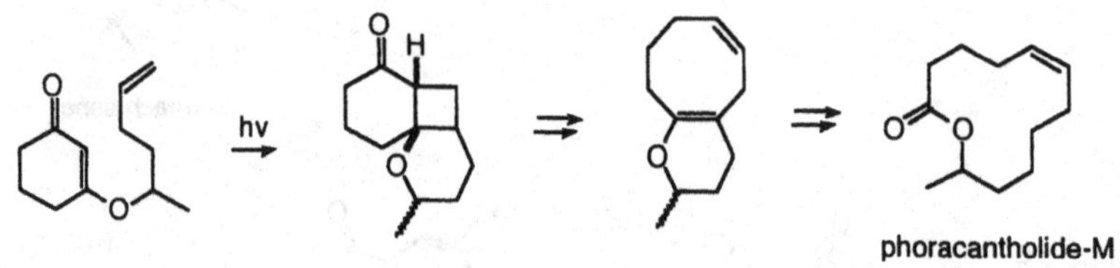
hv
H
phoracantholide-M

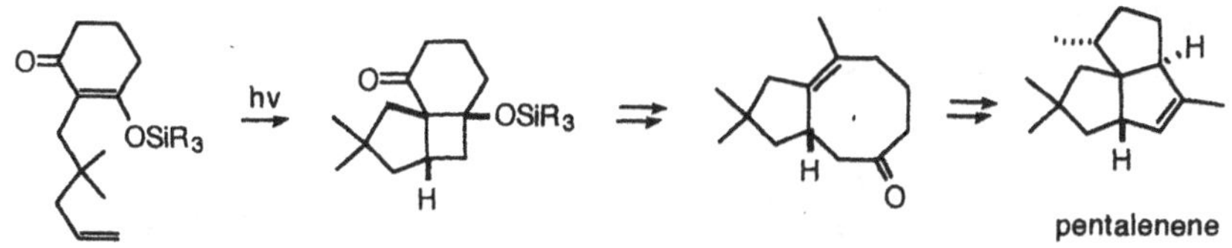

pentalenene

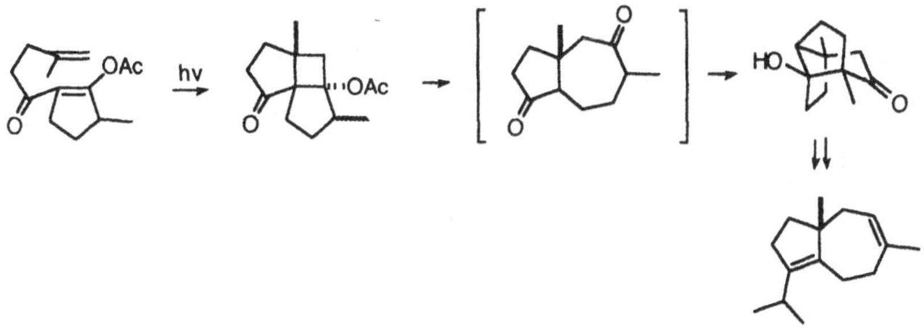

daucene

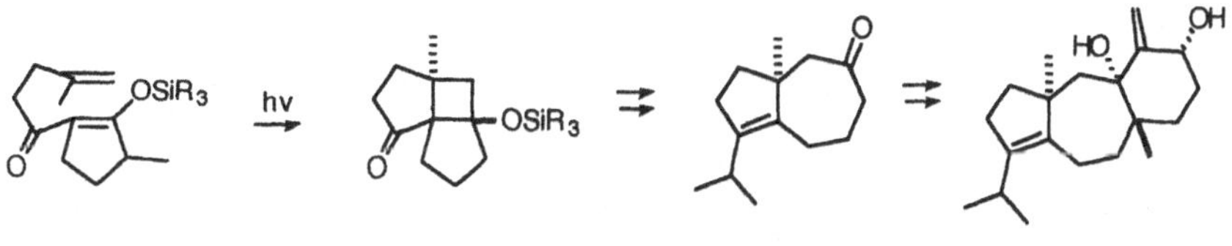

isoamijiol

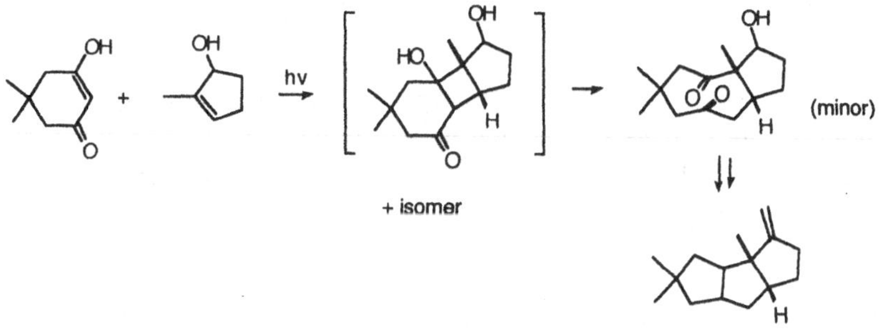

hirsutene

5,5-Dimethyl-(5*H*)-furan-3-one and 2,2,6-trimethyl-(4*H*)-1,3- dioxin-4-one have been employed as variant of the enolized acetoacetic ester. By alternative processing it is possible to achieve chemoselectivity of the functionalities. Demonstration of the methodology is imparted in the synthesis of acorenone [Baldwin, 1982a], occidentalol

[Baldwin, 1982b], and elemol [Baldwin, 1985]. From a chiral dioxinone (+)-grandisol has been prepared [Demuth, 1986].

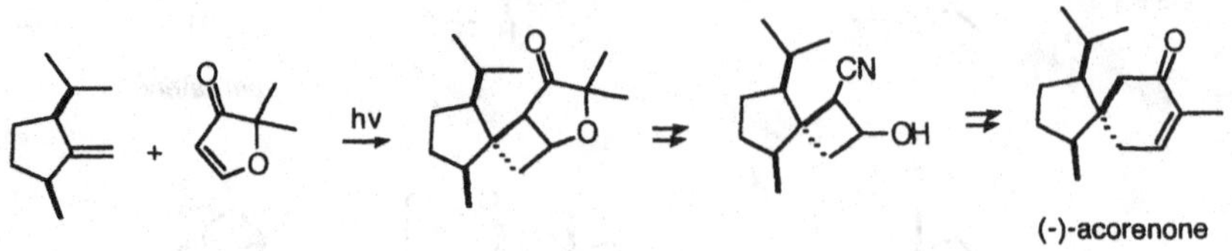

(-)-acorenone

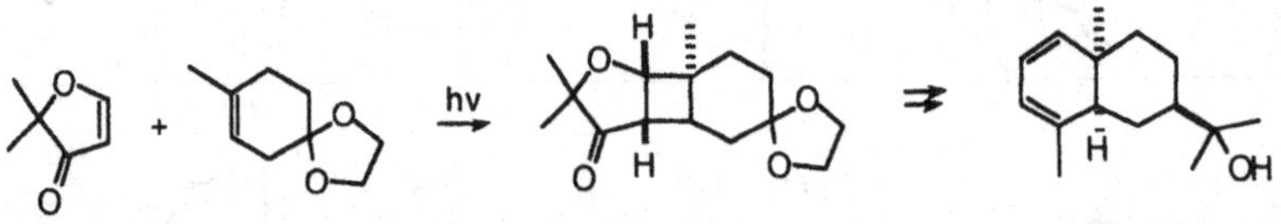

occidentalol

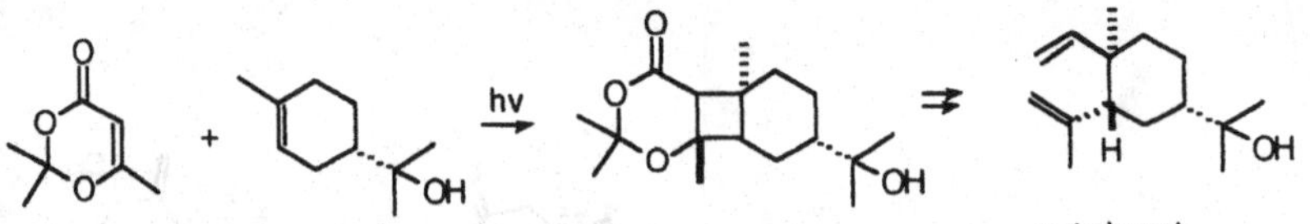

ent-elemol

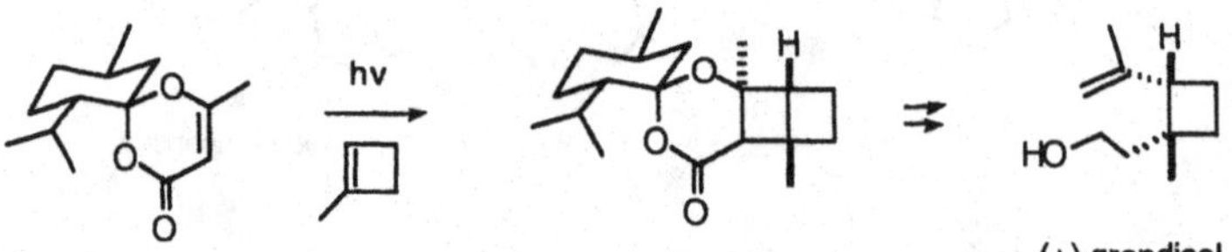

43%

(+)-grandisol

It is remarkable that an intramolecular version opens a pathway to inside-outside bridged ring systems such as that present in ingenol [Winkler, 1987]. The photocycloadduct of a simple model [Winkler, 1991] underwent fragmentation to unravel the intriguing skeleton.

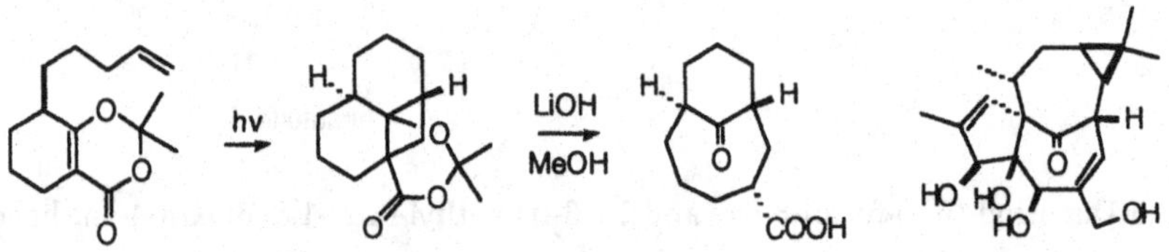

ingenol

(1RS,3RS,7RS)-3-Carboxybicyclo[5.3.1]undecan-11-one. [Winkler, 1991]

A degassed solution of the dioxinone (360 mg, 1.44 mmol) in 9:1 acetonitrile/acetone (300 mL) is cooled to 0°C and irradiated through a Pyrex filter sleeve using a 450 W Hanovia lamp in an immersion well for 2 h. Concentration of the solution and flash chromatography (10% EtOAc in petrolether) provides the photoadduct (325 mg, 90%).

The tricarbocyclic dioxanone (982 mg, 4.38 mmol) is stirred with LiOH (480 mg) in methanol (50 mL) at 25°C for 10 h. The solution is diluted with ether, acidified to pH 1 with 6N hydrochloric acid, and extracted with CH_2Cl_2. From the dried organic solution the carboxylic acid product (737 mg, 80% yield; mp.p. 108-109°C) is obtained by evaporation and flash chromatography (20% EtOAc in petrolether).

Intramolecular versions of this reaction serve to assemble histrionicotoxin [Winkler, 1989]. In this case both photoaddend units are β-heteroenones therefore its is difficult to determine which is being excited.

COOMe HN O O O hv MeOOC HN O O O H H O ⇒ HN HO

histrionicotoxin

A very concise synthesis of mesembrine [Winkler, 1988] is based on intramolecular aza-version of the deMayo reaction. The preparation of an intermediate of vindorosine [Winkler, 1990] by the same process is particularly interesting in view of asymmetric induction.

OMe OMe N H O hv Ar N H H O → Ar N O ⇒ OMe OMe O N H Me

mesembrine

vindorosine

7.3.2. Photocycloaddition of Arenes

For obvious reasons, aromatic compounds do not undergo dimerization. However, a steric enforcement makes photocycloaddition of two benzene rings possible and such is a key step of an elegant synthesis of pagodane [Fessner, 1987]. Equally of interest is the formation of dibenzoequinene (25-50%) by irradiation of [2.2](1,4)-naphthalenophane [Wasserman, 1969].

pagodane

dibenzoequinene

The achievement of photocycloaddition that unites an arene and an alkene molecule has a relatively short history. The novel reaction has been subjected to theoretical treatment [Bryce-Smith, 1980; Houk, 1982].

The *ortho*-cycloaddition was discovered accidentally when attempts were made to intercept fulvene, which was expected to form on photolysis of benzene, with maleic anhydride. Instead, a 1:2 adduct whose structure clearly indicated a new photochemical process which was followed by a thermal Diels-Alder reaction [Angus, 1960].

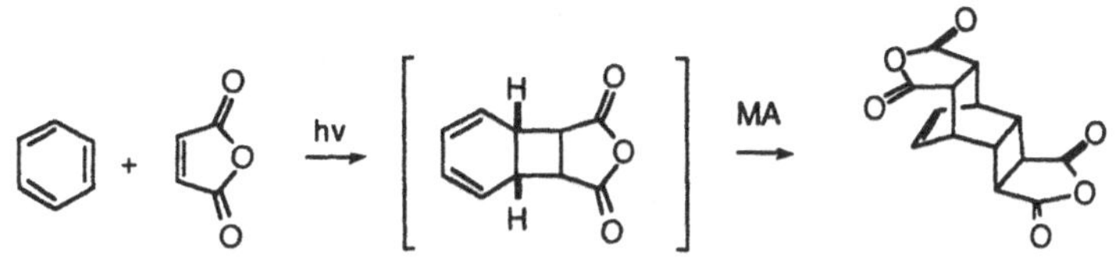

In subsequent studies many other 1:1 cycloadducts have been isolated. Interestingly, with alkynes as reactants cyclooctatetraenes are generated [Grovenstein, 1961]. Here the alkynes (but not the arenes) are the excited species [Bryce-Smith, 1970].

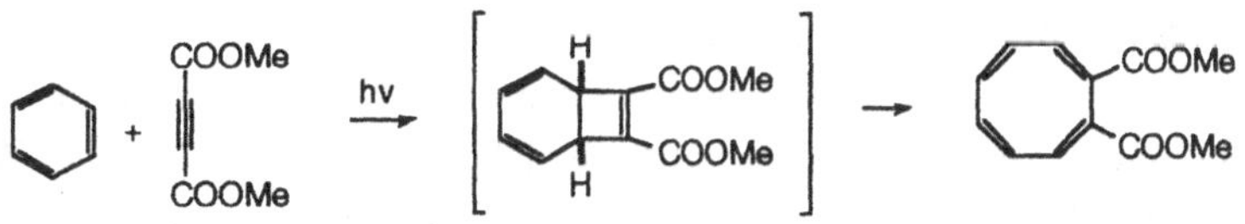

Because of poor orbital overlap between an excited arene and a ground state alkene, *para*-cycloaddition is more difficult and less commonly observed. Conjugated dienes are better reaction partners for arenes. There is very limited synthetic value of *p*-photocycloaddition.

meta-Cycloaddition gives rise to tricyclic adducts [Bryce-Smith, 1966; Wilzbach, 1966]. By virtue of its generality, formation of three C-C bonds in one step, and extensive reorganization of the carbon framework (of the arene) which is favorable to manipulation into several different skeletons, this process makes an outstanding contribution to synthetic methodology. The *m*-cycloaddition involves singlet excited arenes. Generally, it is a concerted but synchronous process. An exciplex model proposed for the reaction is consistent with all the evidence. Exciplex is referred to stoichiometric complex of two molecular species which shows appreciable stability only in the excited state. It is different from the donor-acceptor or charge-transfer complex which is readily formed from ground state molecules. The exciplex then turns into a diradical, and eventually a collapse into the cycloadduct terminates the reaction. As indicated below, such photocycloadducts present themselves many opportunities for structural modification, accordingly they are extremely versatile building blocks for synthesis.

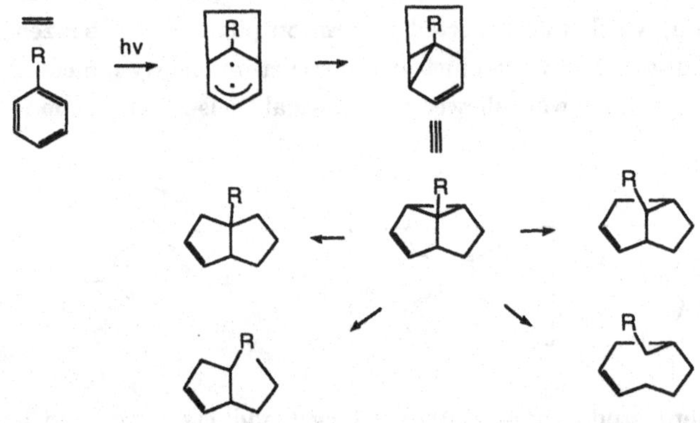

Based on an incisive analysis of the synthetic potential of *m*-cycloaddition, particularly the intramolecular version, Wender and his group have executed many elegant and efficient syntheses of terpenes: cedrene [Wender, 1981b], isocomene [Wender, 1981a], hirsutene [Wender, 1982b], coriolin [Wender, 1983b], silphiperfol-6-ene [Wender, 1985a], retigeranic acid [Wender, 1990c], subergorgic acid [Wender, 1990a], silphinene [Wender, 1985b], rudmollin [Wender, 1986], laurenene [Wender, 1988]. (Using intermolecular *m*-cycloaddition it has been possible to acquire modhephene [Wender, 1982a] and isoiridomyrmecin [Wender, 1983a].) Generally in these synthetic applications the benzylic (and also homobenzylic) substitution pattern is determinative of stereoselectivity of the photocycloaddition. It is also remarkable that remote steric interactions also contribute to the observed and agreeable steric outcome. For example, in the synthetic route of silphiperfol-6-ene the photocycloaddition avoided the more encumbered transition state which brought the methyl groups at the allylic and the ortho positions close to one another.

It should be noted that there are some structural limitations to the generality of the intramolecular *m*-cycloaddition. Thus 1,2,3-trisubstituted benzenes seem to lack photoreactivity, as elaboration of the entire skeleton of ceratopicanol cannot be achieved [Baralotto, 1996]. It requires an angular methylation step to complete the synthesis.

Photocycloaddition of *o*-((*Z*)-5-Methylhept-5-en-2-yl)toluene. [Wender, 1981a]

A solution of the substituted toluene (1.688 g, 8.356 mmol) in cyclohexane (130 mL) is purged with nitrogen while being irradiated (Vycor filter) for 4.5 h. After evaporation of the solvent, silica gel chromatography of the residue furnishes a product (1.391 g) consisting mostly (85-90%) of an equal mixture of two tetracyclic compounds.

Thermolysis of the mixture at 230-244°C for 14.5 h returns 61% of the mass which contains about 90% of dehydroisocomene.

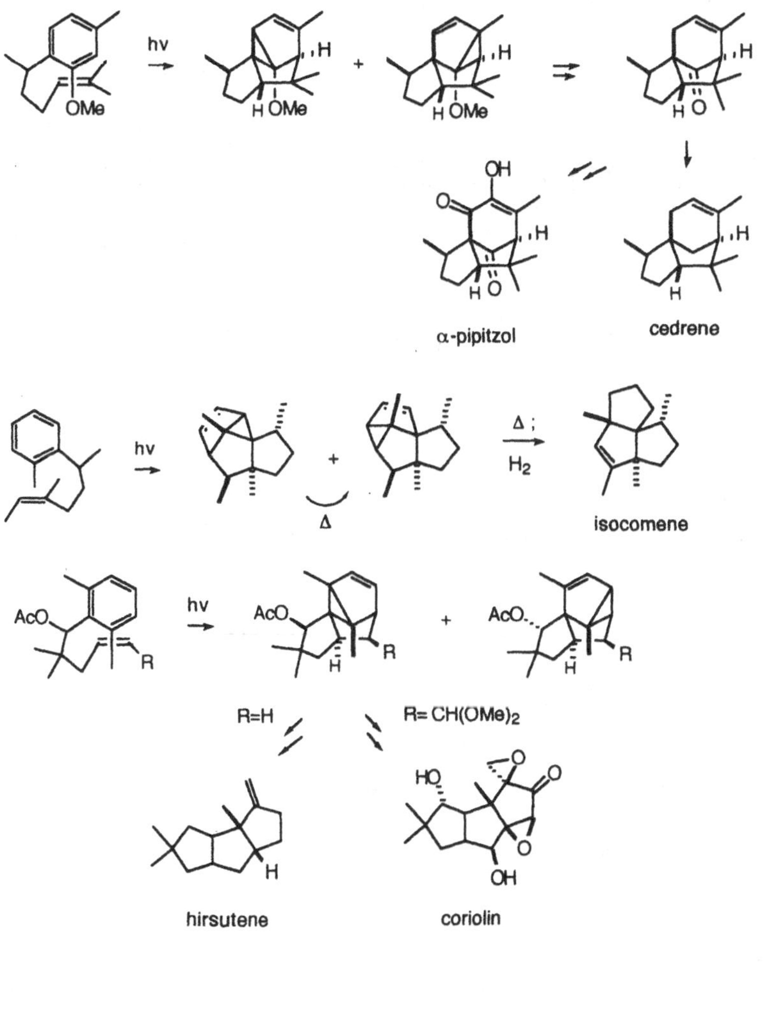

hv
H OMe
H OMe
H O
OH
α-pipitzol
cedrene

hv
Δ
Δ ;
H_2
isocomene

AcO
R
hv
H
R=H
R= $CH(OMe)_2$
HO
OH
hirsutene
coriolin

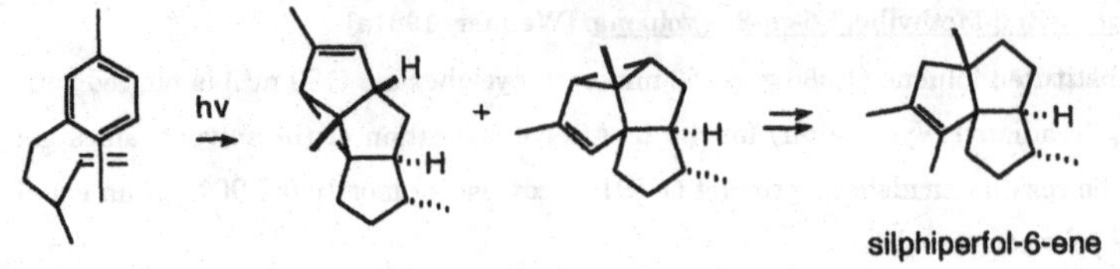

hv

silphiperfol-6-ene

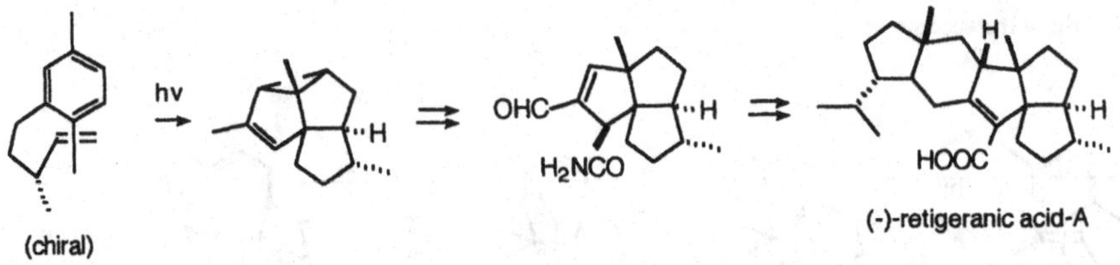

hv

(chiral)

OHC

H_2NCO

HOOC

(-)-retigeranic acid-A

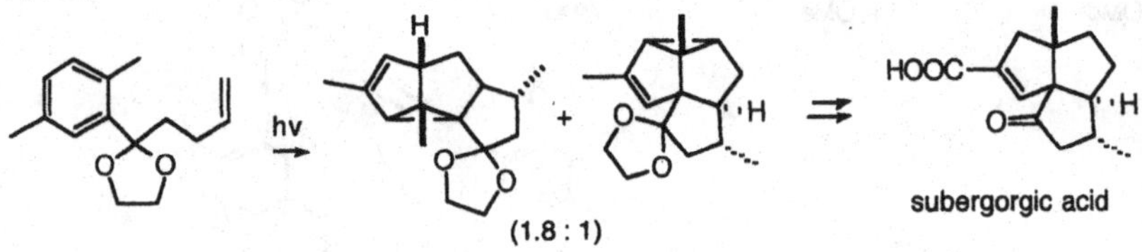

hv

(1.8 : 1)

HOOC

subergorgic acid

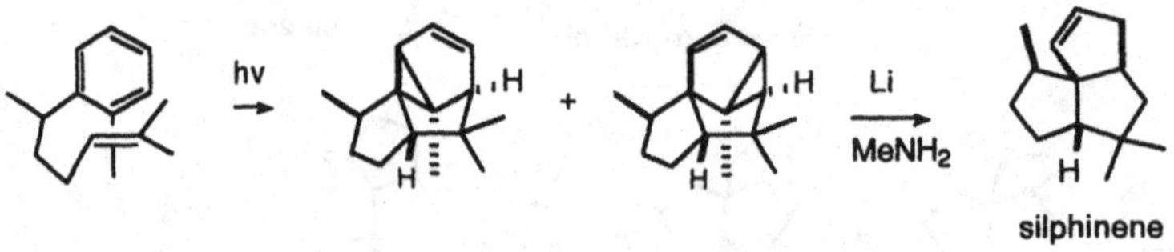

hv

Li

$MeNH_2$

silphinene

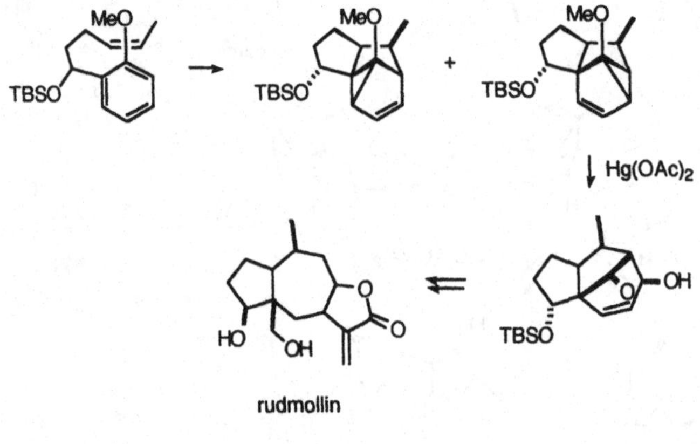

MeO

TBSO

$Hg(OAc)_2$

HO

OH

rudmollin

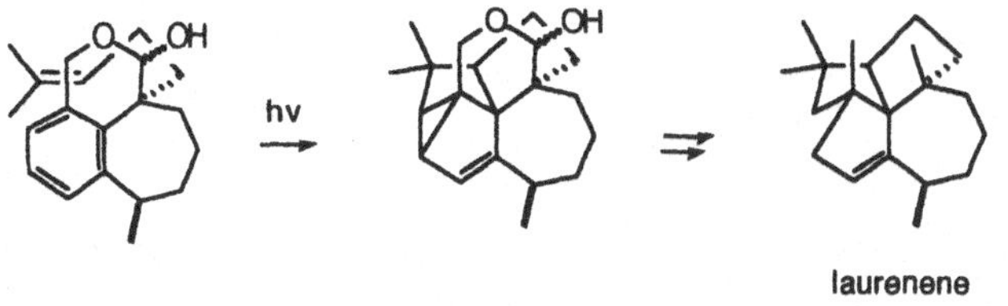
OH
hν
OH
laurenene

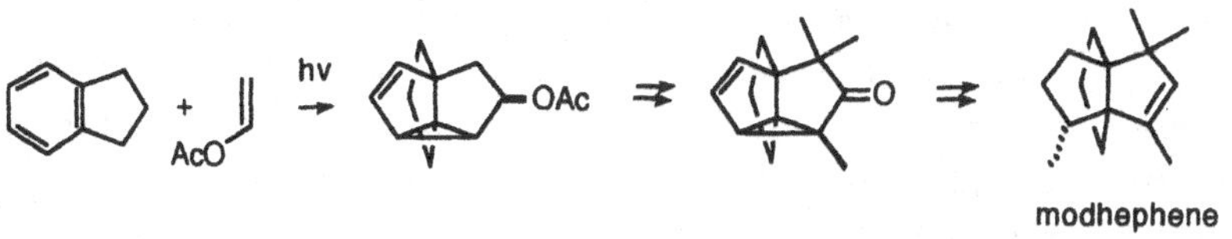
AcO
hν
OAc
O
modhephene

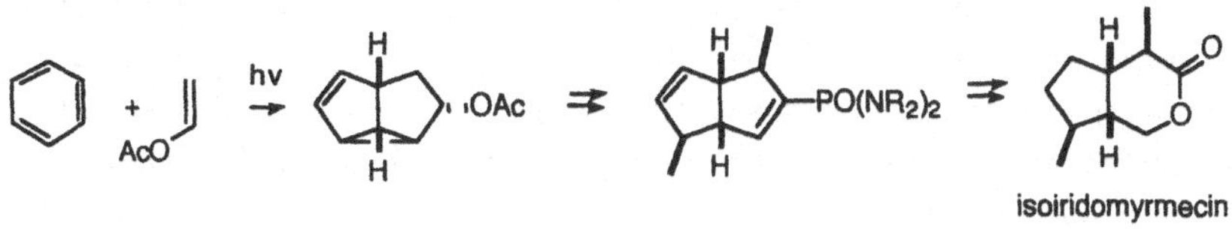
AcO
hν
H
OAc
PO(NR2)2
O
isoiridomyrmecin

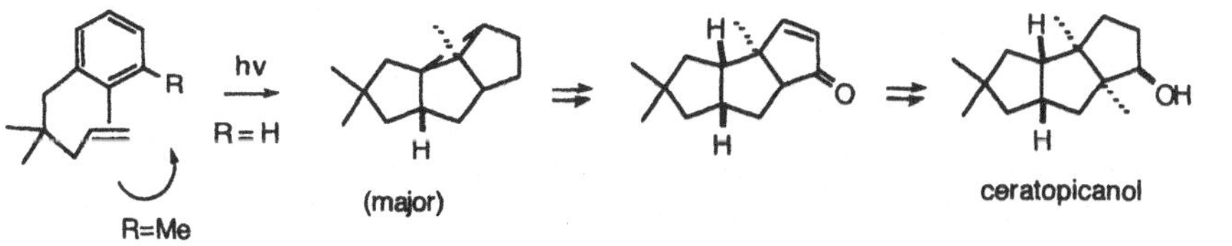
R
hν
R = H
R=Me
(major)
H
O
OH
ceratopicanol

7.3.3. Other Photocycloadditions

Other photocycloadditions are much less developed, and accordingly their synthetic utility is much rarer. The BC-core structure of phorbol has been acquired from an intramolecular [3+2]cycloaddition of a 4-pyrone which involves silatropic shift to generate a 3-oxidopyrylium zwitterion, but its efficiency is inferior (15% vs. 77%) to the thermal process [Wender, 1990b].

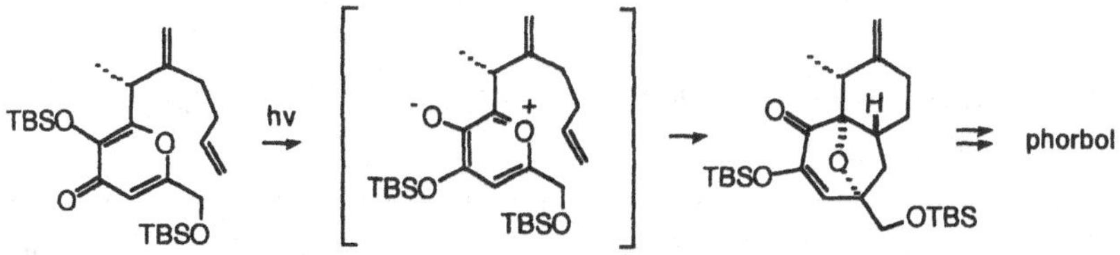
TBSO
O
TBSO
hν
TBSO
TBSO
H
OTBS
phorbol

[6+2]Cycloadditions involving tropolone are known and an intramolecular reaction forms the basis of a synthesis of dactylol [Feldman, 1990].

hv
>350 nm
dactylol

Photoinduced cycloadditions involving cycloheptatrienechromium tricarbonyl complexes proceed in the [6+4] and [6+2] modes, with conjugated dienes and electron-deficient alkenes, respectively [Rigby, 1990, 1991]. Synthetic applications of these efficient processes are eagerly awaited.

Singlet dioxygen has a higher potential energy and its photochemical accessibility is only by sensitized excitation. Contrary to many other chemical species, excited dioxygen participates in the Diels-Alder reaction as a dienophile. A classical example for this reactivity is in the generation of an endoperoxide en route to cantharidin [Schenck, 1953]. Its use in synthesis of stemolide [van Tamelen, 1980] has also been demonstrated.

hv
O_2
sens.
cantharidin

hv
O_2
sens.
Δ
xylene
α : β peroxide 2:1
stemolide

7.4. PHOTOISOMERIZATION OF ALKENES AND ARENES

7.4.1. (*E*,*Z*)-Isomerizations

Simple alkenes require very short wavelength ultraviolet light to be excited and little synthetic value is attached to their singlet state photoreactions. In spite of the accessibility to their triplet states by sensitization which amends the situation somewhat, relatively scant attention has been paid to the photochemistry of isolated alkenes regarding its use in synthesis in comparison with that of conjugated polyenes.

Perhaps this status reflects the intrinsic character of excited alkenes which have limited modes of reaction. It must be emphasized that alkene addends for cycloaddition with enones are in the ground state.

In both mechanistic and synthetic terms the most important photoreaction of an alkene is isomerization. In the (π,π^*) excited state the π-bond of an alkene is broken, and free rotation allows *(E,Z)*-isomerization. In fact, the lowest energy species is a biradical with orthogonal configuration and its decay to ground state may result in the geometrical isomer. Thus a convenient approach to betweenanenes [Nakazaki, 1977; Marshall, 1980] by photoisomerization of macrobicyclic alkenes have been developed.

$(CH_2)_{10}$ $(CH_2)_8$ hv xylene (2.4 : 1) $(CH_2)_{10}$ $(CH_2)_8$

[8.10]betweenanene

For synthetic expediency, humulene was prepared [Corey, 1967] from the *(E,E,Z)*-cycloundecatriene isomer by photoisomerization in the presence of diphenyl disulfide.

hv (350 nm) PhSSPh

humulene

The acyclic precursor for the humulene synthesis was assembled by a *(Z)*-selective Wittig reaction. Analogously, an industrial preparative method for vitamin-A [Pommer, 1977] also involves photoisomerization of the 11-*cis* isomer of retinyl acetate [Fischer, 1978].

PPh_3 + CHO OAc → 11 12 OAc hv → OAc

Photoisomerization is of industrial significance in the production of vitamin-D and related substances [Braun, 1991]. For synthesis of an A-ring synthon [Posner,

1990] the geometrical correction of an *(E)*-dienoic ester was accomplished by photochemical means.

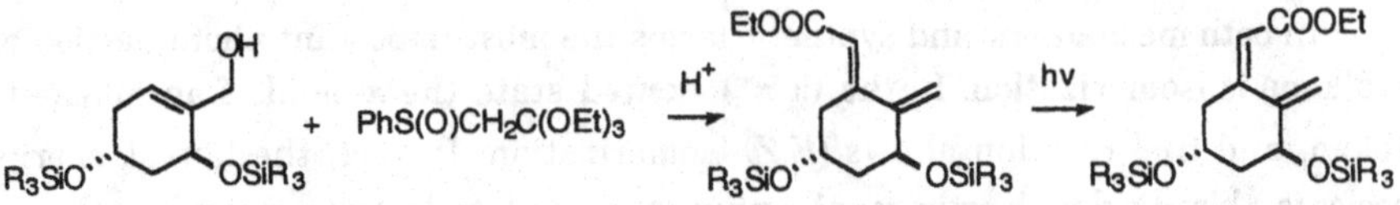

The conversion of stilbenes to phenanthrenes is most efficient when the former are restricted to the *(Z)*-configuration, and this feature was exploited in a synthesis of cannithrene-II [Ben, 1985]. For *(E)*-stibenes the transformation necessarily involves *(E)*->*(Z)* isomerization, electrocyclization, and oxidation, all of which can be carried out in one operation. It has been effectively applied to the synthesis of helicenes [Laarhoven, 1974; R.H. Martin, 1974] and kekulene [Diederich, 1978]. Recent improvement of the process involves the use of stoichiometric amount of iodine in the presence of propene oxide while excluding air [L. Liu, 1991].

Photocyclization of Stilbenes. [L. Liu, 1991]

A solution of the stilbene and iodine in benzene is purged with argon for 20-30 min before an excess of propene oxide is added and irradiation begins. The progress of reaction (Ar flow being maintained throughout) is monitored by NMR of aliquots or disappearance of the iodine color. When excess iodine remains at the end, workup includes washing with 15% sodium thiosulfate and brine, drying, filtering, and concentrating to dryness. Otherwise the solution is evaporated directly.

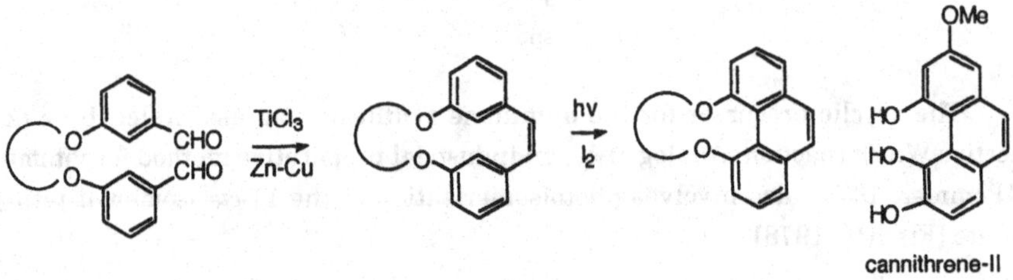

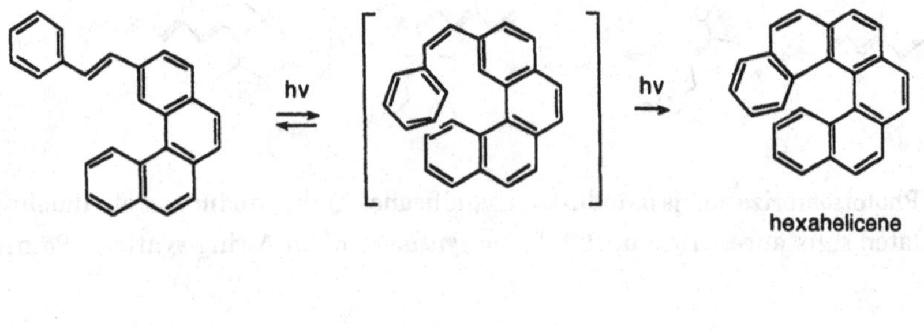

cannithrene-II

hexahelicene

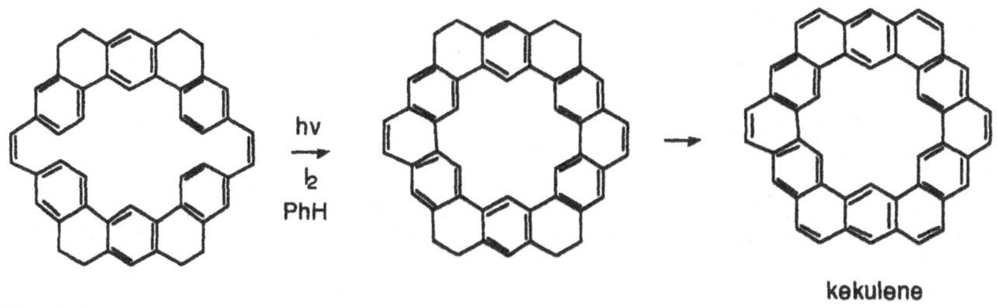

hv
I_2
PhH

kekulene

Hexahelicene.

To a 10^{-3}M solution of 2-styrylbenzo[*c*]phenanthrene (single isomer or *(E,Z)*-mixture) in toluene is added some iodine (ca. 25 mg). Irradiation continues for 8 h using a high-pressure mercury lamp (Philips HPK 125) which is surrounded by a pyrex tube and immersed in the solution. Evaporation of the solvent is followed by addition of a small volume of carbon tetrachloride which is again removed by distillation. The procedure is repeated until the violet color of iodine is not longer seen. The residue is purified by silica gel chromatography to give hexahelicene (80% yield. M.p. 240-242°C).

Relevant to stilbene isomerization is that involving azobenzenes. Molecular systems containing an azobenzene unit show remarkable photoresponsive behavior, for example the cation complexation by a crown ether portion in the same molecule [Shinkai, 1980].

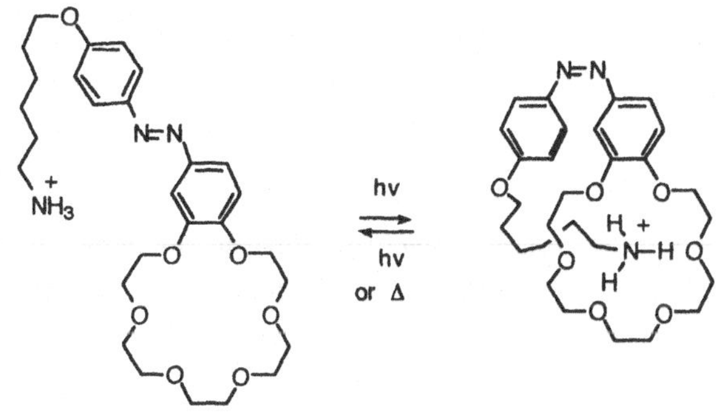

hv
hv
or Δ

On excitation at 254 nm benzene reaches the S_1 state which decays to give benzvalene and fulvene. On the other hand, Dewar benzene and prismane are formed (from the S_2 state) when benzene is irradiated with 205 nm light.

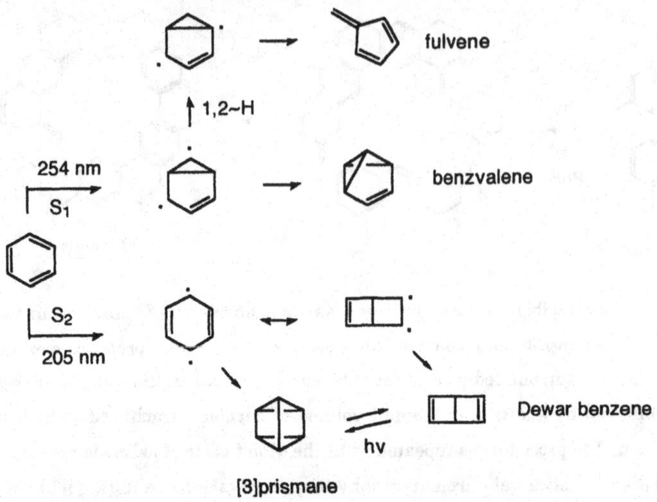

fulvene

1,2~H

254 nm

S_1

benzvalene

S_2

205 nm

Dewar benzene

hv

[3]prismane

Benzene derivatives with bulky substituents (e.g., *t*-butyl) are particularly prone to undergo valence isomerization. The isomers revert to the Kekule structures on further irradiation, leading to mixtures of positional isomers. The photoisomerization of [2.2]metaparacyclophane to [2.2]metacyclophane [Delton, 1971] is one such case.

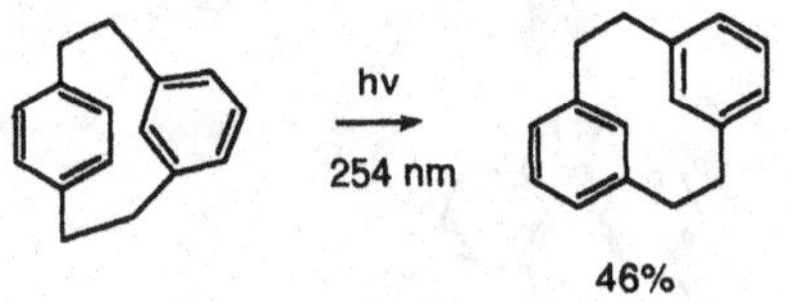

hv

254 nm

46%

Forced nonplanarity of the aromatic ring (e.g., [6]paracyclophane) also favors such reaction [Liebe, 1982; Kostermans, 1987a]. Aromatic heterocyles behave similarly.

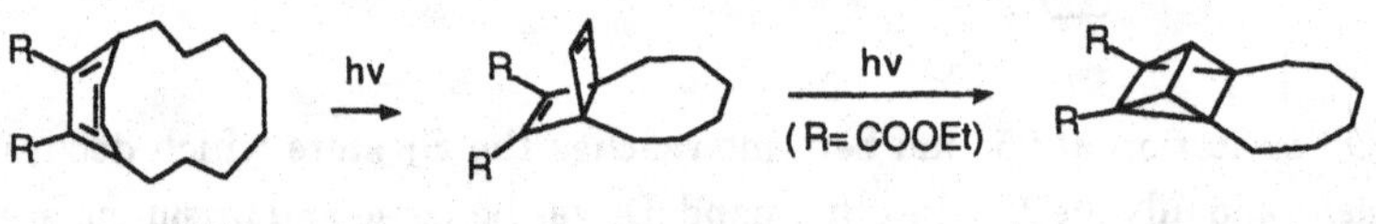

R R hv R R hv (R= COOEt) R R

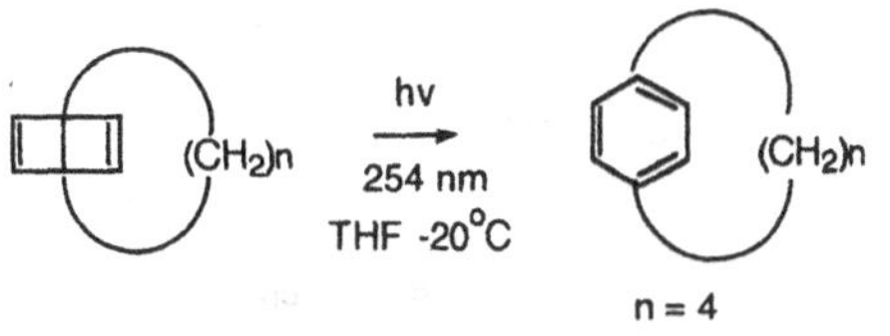

$(CH_2)n$ hv 254 nm THF -20°C $(CH_2)n$

n = 4

7.4.2. Electrocyclic Reactions

The photochemical formation of phenanthrenes alluded to above actually involves electrocyclization and dehydrogenation. Stilbenes and more common conjugated dienes and polyenes undergo photochemically induced pericyclic reactions by steric courses according to the Woodward-Hoffmann rules [Woodward, 1970]. Thus the stilbene-to-phenanthrene transformation is initiated by conrotatory cyclization before aromatization (oxidation).

It is interesting to note that ergosterol and isopyrocalciferol behave differently on uv irradiation [Dauben, 1959a; Havinga, 1961]. Ring opening of ergosterol is favored because electrocyclization would generate a highly strained product, due to the relative configuration of C-9 and C-10.

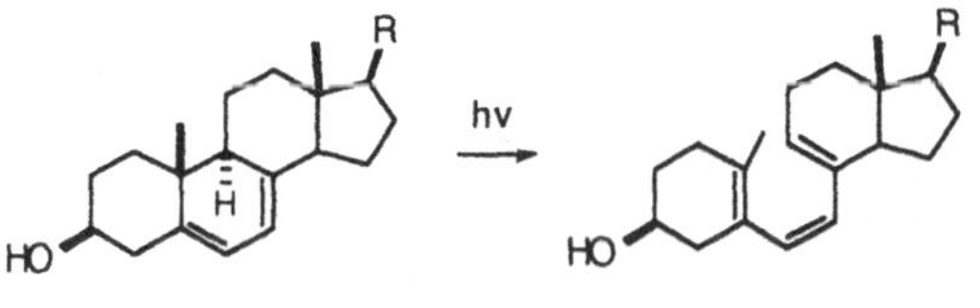

R H HO hv R HO

ergosterol

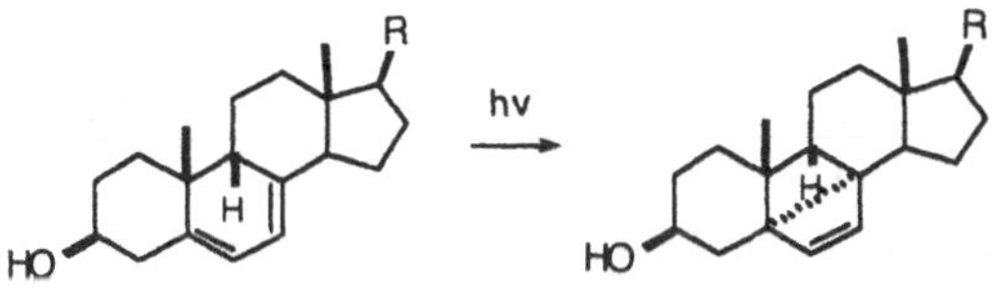

R H HO hv R H HO

isopyrocalciferol

The divergent pathways have been exploited in synthesis, for example, of dihydrocostunolide [Corey, 1965] and Dewar benzene [van Tamelen, 1971].

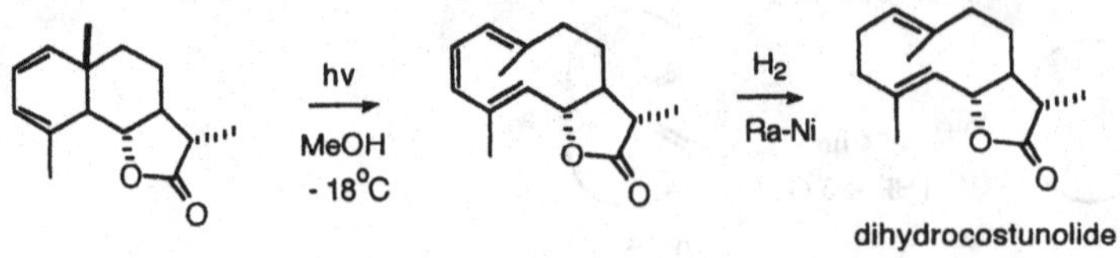
hv
MeOH
- 18°C
H2
Ra-Ni
dihydrocostunolide

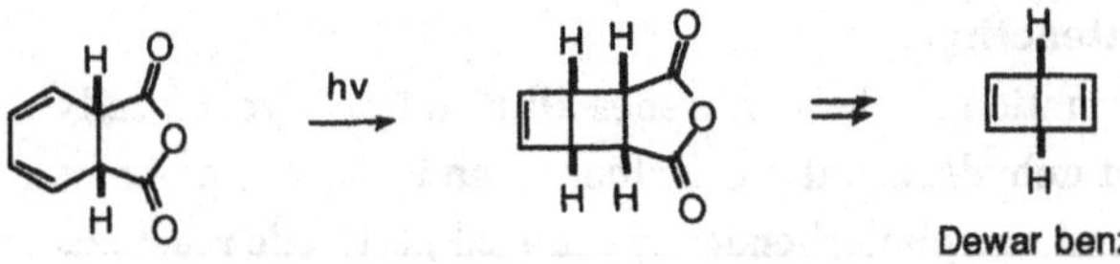
hv
Dewar benzene

α-Pyrone undergoes cyclization when irradiated in pyrex at room temperature. Further irradiation in quartz vessel leads to cyclobutadiene [Corey, 1964d]. An analogous bicyclic lactam obtained from α-pyridone is a useful synthon for β-lactam antibiotics [Brennan, 1981].

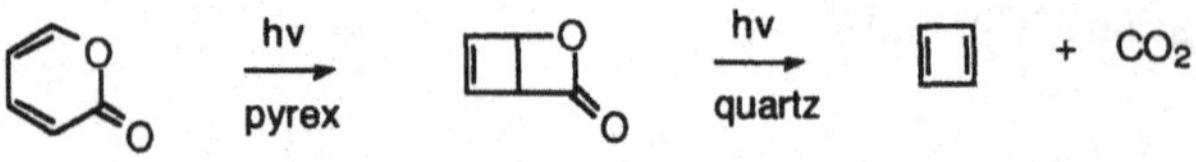
hv
pyrex
hv
quartz
+ CO2

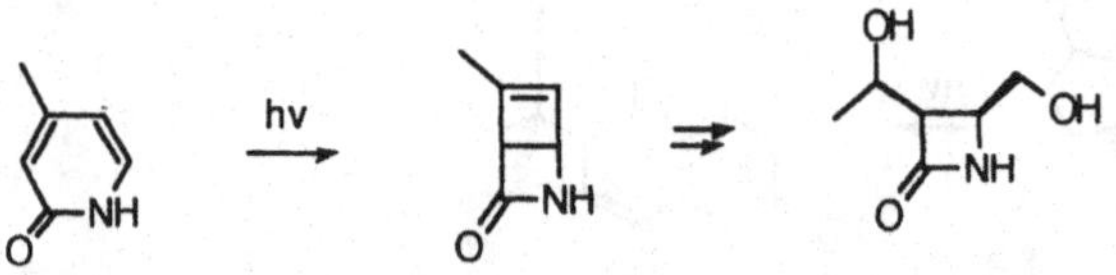
hv
NH
OH

Eucarvone is useful for synthesis of grandisol [Ayer, 1974] on account of its ready conversion to the bicyclic enone. Tropolone methyl ether gives a bicyclic dienone on irradiation [Dauben, 1959b, 1963]. The agreeable features of this compound to the elaboration of brefeldin-A [Baudouy, 1977] has been duly noted. The dienone actually arises from rearrangement of the primary electrocyclized product.

5-Methoxybicyclo[3.2.0]hepta-3,6-dien-2-one. [Dauben, 1959]

A solution of α-tropolone methyl ether (5.0 g) in distilled water (2.5 L) is flushed with nitrogen, sealed in a Pyrex flask in vacuo and exposed to sunlight (8 days). After centrifugation of the turbid solution to remove a dark-red amorphous powder (1.45 g) the liquid is extracted thrice with CH_2Cl_2. The dried extracts are concentrated and then distilled to give the bicyclic dienone (1.19 g), b.p. 34°C/0.1 torr.

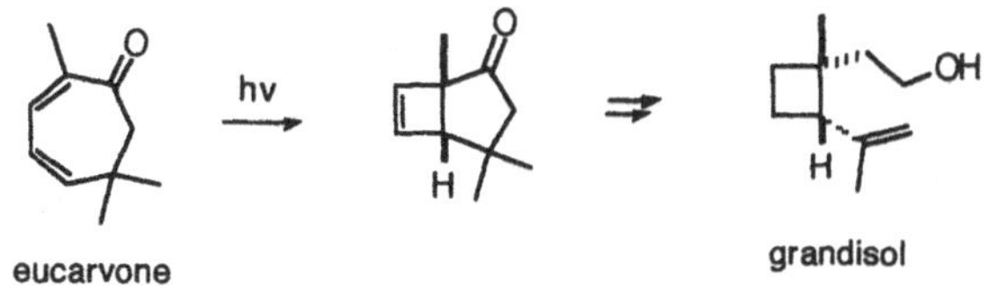

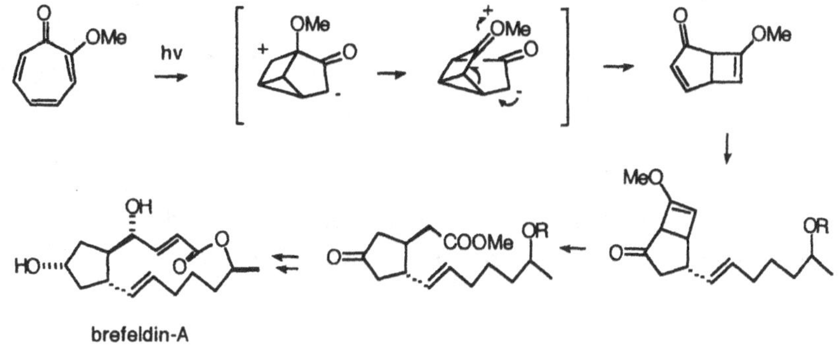

eucarvone

grandisol

brefeldin-A

By exploiting long-range steric interactions to bias the ground state conformation it is possible to favor the formation of the desired tetracarbocyclic intermediate for ikarugamycin [Whitesell, 1989] by photocyclization of the triene precursor.

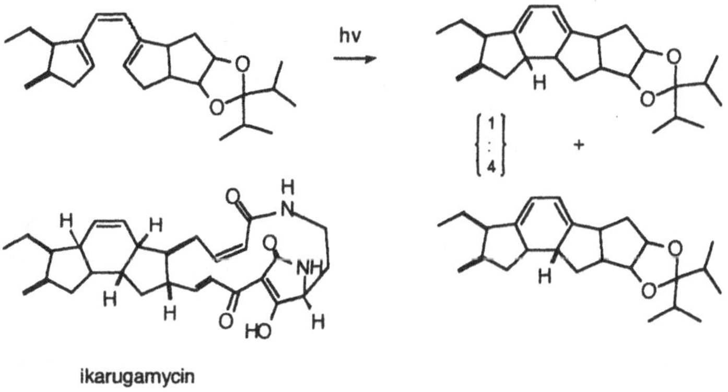

ikarugamycin

Carbomethoxylated β-ionone cyclizes on irradiation in basic solutions [White, 1978; Kim, 1983] thus providing a valuable synthon for terpenes of the drimane and related types.

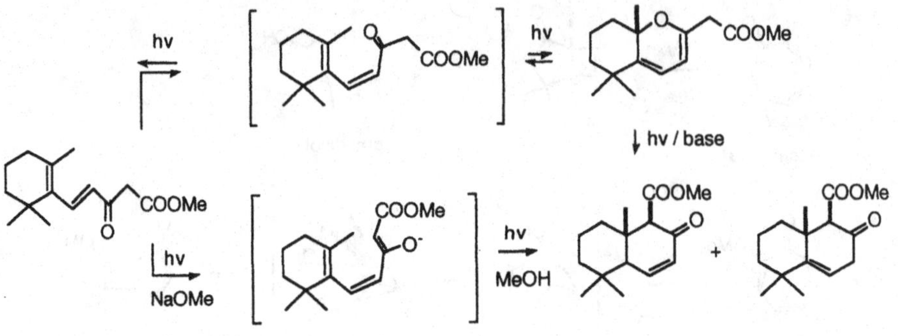

The phototransformation of 2-vinylbiphenyls to 9,10-dihydrophenanthrenes presumably consists of electrocyclization and thermal [1,5]-hydrogen migration. Two slightly different syntheses of juncusol [McDonald, 1978; Kende, 1979] are based on this method.

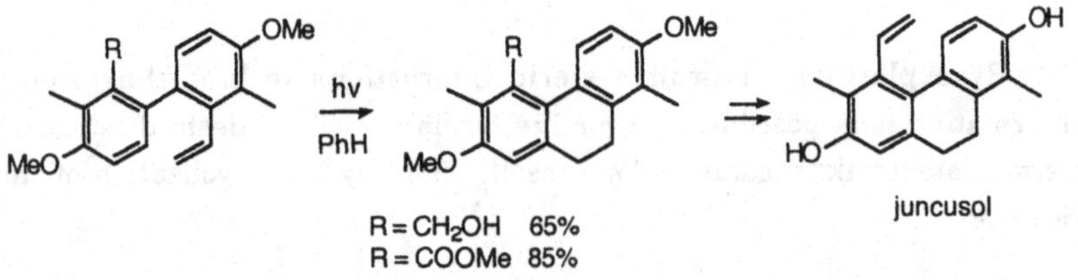

R = CH_2OH 65%
R = COOMe 85%

juncusol

Aporphine alkaloids such as nuciferine are readily synthesized using photochemical reaction of 1-benzylidene-1,2,3,4-tetrahydroisoquinolines [Yang, 1966]. A precursor of methoxatin is similarly accessible from a diarylethene in which the two existing rings are a pyrrole and a pyridine [Hendrickson, 1982].

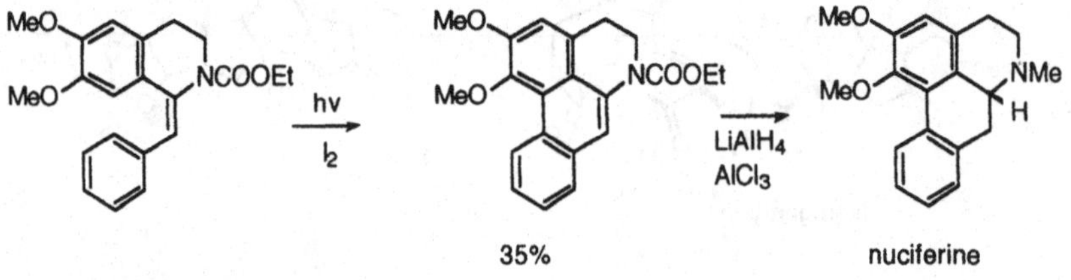

35%

nuciferine

E = COOMe

hν, $(PhSe)_2$

44%

methoxatin

The presence of a removable group or atom (OR, Br,...) in the terminal position of the "triene" unit facilitates the process, thus ensuring high conversion of the substrates and good regioselectivity. Cepharanone-B was synthesized [Castedo, 1982] in such a manner.

hν (-HI)

cepharanone-B

The successful photocyclization of *N*-benzalanilines, anilides, and enamides in which the acyl group is conjugated prompted intenive studies with respect to their synthetic utility [Ninomiya, 1984; Campbell, 1987]. As illustrated by a simple approach to xylopinine [Ninomiya, 1973], the enamides probably undergo cyclization in their zwitterionic form.

hν

5% + 40%

xylopinine

It may be advantageous to effect photocyclization of enamides under reductive conditions (e.g., in the presence of $NaBH_4$) to furnish products more amenable to elaboration into synthetic targets. Thus it has been possible to acquire an intermediate for lysergic acid [Kiguchi, 1982] rather effortlessly. Actually, by a systematic extension of this methodology the access to a great number of alkaloid families has been realized. For example, the indole alkalods yohimbine [Naito, 1988], deserpidine [Miyata, 1984], ajmalicine [Naito, 1985], corynantheine [Naito, 1986a], hirsuteine and geissoschizine [Naito, 1987], and the ipecac alkaloid emetine [Naito, 1985, 1986b].

<u>16,17,19,20-Tetradehydro-17-methoxyyohimban-21-one</u>. [Naito, 1988]

To a solution of *N*-(4-methoxybenzoyl)-1-methylene-1,2,3,4-tetrahydro-β-carboline (318 mg) in acetonitrile (150 mL) are added sodium borohydride (350 mg) and methanol (15 mL) successively at room temperature. After dissolution of the borohydride the mixture is cooled to 5-10° and irradiated through a Pyrex filter with a 300W high pressure mercury lamp for 30 min. Evaporation of solvent in vacuo, and addition of water to the residue cause the deposition of a solid which is recrystallized from methanol to afford the pentacyclic precursor of yohimbine and deserpidine in 90% yield.

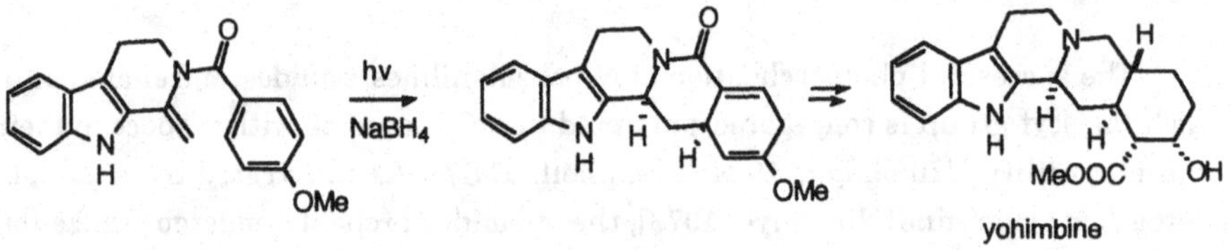

hv, $NaBH_4$

OMe

yohimbine

MeOOC OH

At least one of the double bonds in the six-electron system for photochemical electrocyclization can be replaced by a heteroatom which bears an electron lone-pair. Diarylamines and aryl vinyl ethers are among such photoactive substances, proofs for their reactivity are found in syntheses of glycozoline [Carruthers, 1968] and lycoramine [Schultz, 1977]. Similarly, cyclization of β-(*N*-phenylamino)acrylic esters with proper substituents provides indolines which can be converted into spirotricyclic compounds for elaboration of certain indole alkaloids [Ibrahim-Ouali, 1996].

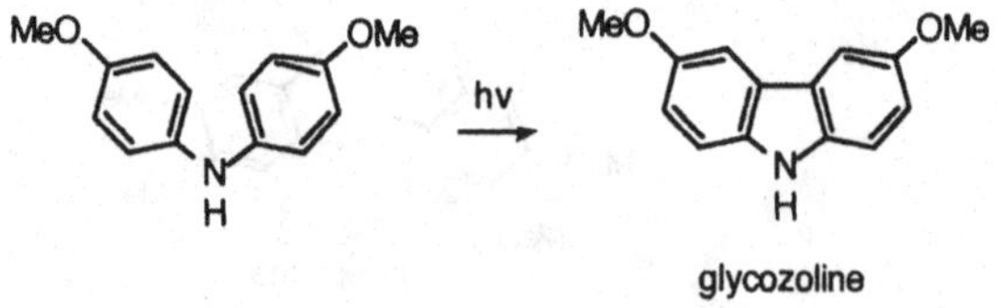

MeO OMe

hv

glycozoline

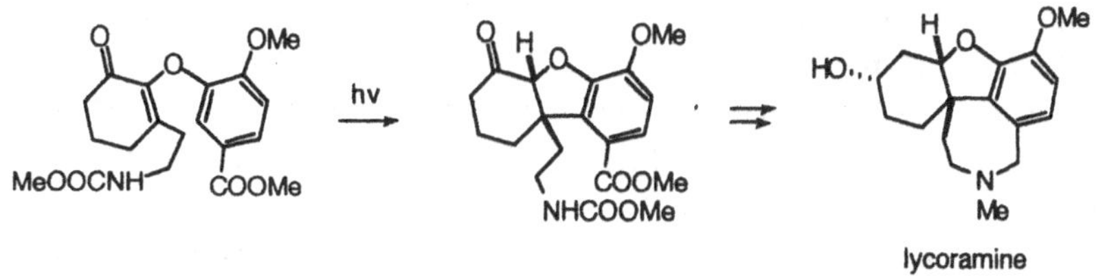

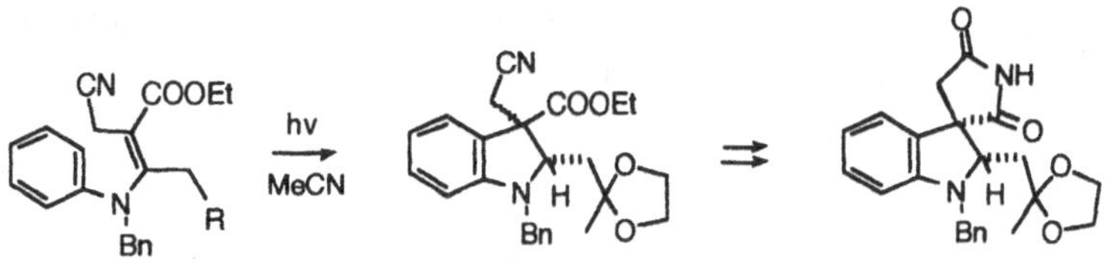

One example of electrocyclic opening (dihydrocostunolide synthesis) was alluded to above. Several other significant results include the syntheses of *N*-carbethoxyaza[17]annulene [Röttele, 1978].

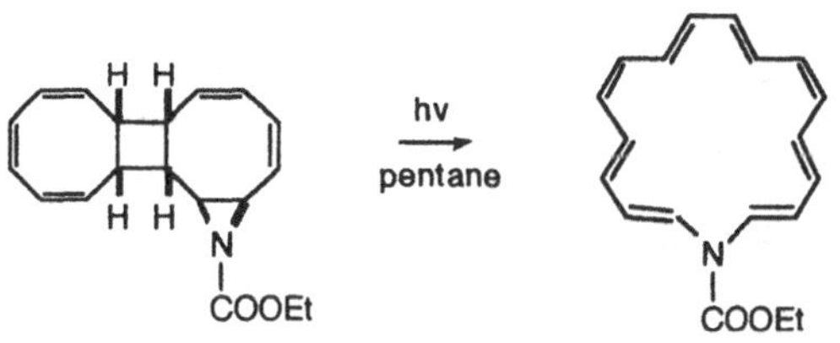

Photolysis of *trans*-10b,10c-dimethyldihydropyrene gives the [2.2]metacyclophane isomer [Blattmann, 1965]. The reverse reaction occurs thermally in the dark.

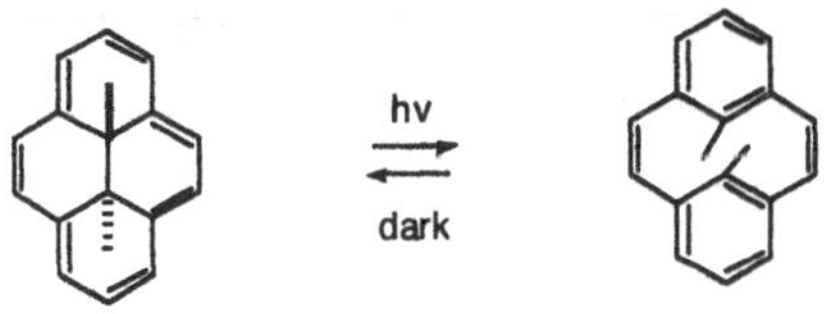

7.5. REARRANGEMENTS

7.5.1. Of Conjugated Ketones

The photochemistry of conjugated enones is dominated by *(E,Z)*-isomerization and cycloaddition. However, rearrangement pathways are competitive for 4,4-

disubstituted 2-cyclohexenones and 2,5-cyclohexadienones. In fact, the photochemistry of α-santonin has fascinated chemists for one and half centuries (although in the olden days the structure of α-santonin was unknown to the investigators) and the various rearrangement products were clarified only in recent times [Barton, 1958].

Depending on the solvent, several types of photochemical products may be formed. In dioxane, lumisantonin is obtained, but it undergoes further transformation to mazdasantonin. The latter gives rise to photosantonic acid on irradiation in water. In aqueous acidic solutions α-santonin is rearranged to isophotosantonic lactone which has a hydrazulene skeleton.

hv, dioxane
α-santonin
lumisantonin
mazdasantonin
hv, H_2O
hv, H_3O^+
isophotosantonic lactone
HOOC
photosantonic acid

The cyclomutation mode (6:6- to 5:7-ring system) has inspired the formulation of synthetic routes for compounds such as cyclocolorenone [Caine, 1972], oplopanone [Caine, 1973], α-cadinol [Caine, 1977], grayanotoxin-II [Hamanaka, 1972; Gasa, 1976]. On the other hand, since the lumi-type products can be cleaved to give spirocyclic systems, novel routes become available to natural products including β-vetivone [Marshall, 1970; Caine, 1976].

Preparation of the Lumiketone Precursor for β-Vetivone. [Kropp, 1965]

A solution of the dimethylhexalindione (808 mg) freshly distilled dioxane (115 mL) is vigorously agitated with a stream of nitrogen and irradiated with a Hanau NK 6/12 low pressure Hg lamp for 2.5 h. The solvent is evaporated and the residue chromatographed over silica gel to afford the lumiketone (479 mg, 59%).

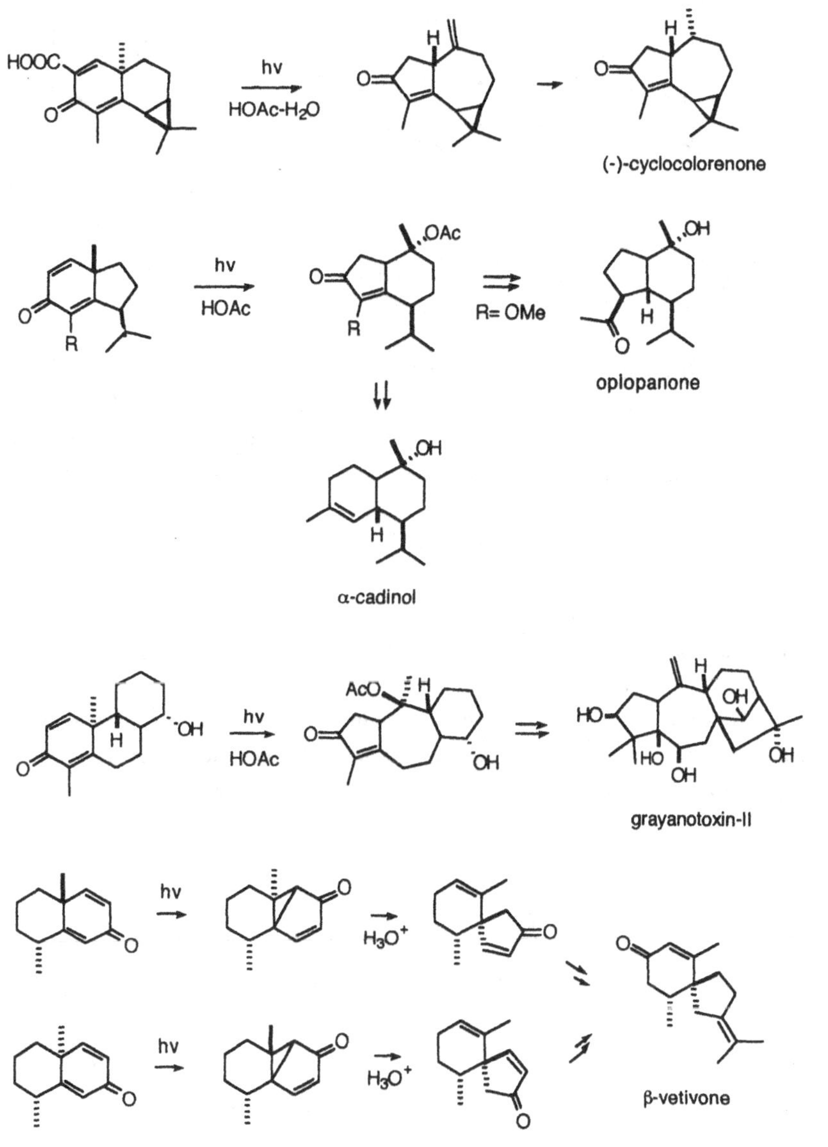

A very interesting observation pertains to the steric courses for the photorearrangement of orientalinone and roemerialinone [Gözler, 1986]. These two compounds differ only in one A-ring substituent (OMe vs. OH) which is in close proximity to the cyclohexadienone unit.

hv (sunlight)

isoboldine   isothebaine

R = H orientalinone (major) (minor)
R = Me roemerialinone (minor) (major)

The ring contraction of a 3-hydroxy-4-pyrone derived from kojic acid on irradiation in water, followed by direct reduction of the product led to two diastereomeric dihydroxycyclopentenones, the minor component is terrein [Barton, 1977]. This photochemical contruction of the cyclopentenone is clearly superior to many other methods in view of mild reaction conditions and the sensitivity of the target molecule to acids and bases.

Terrein and Isoterrein. [Barton, 1977]

An aqueous solution of 3-hydroxy-6-(1-propenyl)-4-pyrone (62 mg, 0.41 mmol) in water is irradiated at 0° in a Pyrex vessel for 33 min. with a 125W high pressure Hg lamp. During the first 15 min, $NaBH_3CN$ (2 mol. equiv.) is introduced. Workup by extraction with ethyl acetate (3x and then continuously for 24 h) gives a crude product which is separated by preparative thin layer chromatography. Terrein (12.3%) and isoterrein (24.7%) are obtained.

kojic acid   hv $NaBH_3CN$

25%   8%
terrein

7.5.2. Di-π-methane and Oxadi-π-methane Rearrangements

1,4-Pentadienes in the excited state rearrange to alkenylcyclopropanes, formally involving a 1,2-shift of a double bond and cyclization. A structural requirement for the substrates is that the two double bonds must be prohibited to move into conjugation for example by full substitution at the central atom or placement of the 1,4-diene chromophore in a bridged ring system. The two situations are exemplified in the preparation of methyl chrysanthemate [Bullivant, 1976] and the conversion of barrelene to semibullvalene.

COOMe hv COOMe

methyl chrysanthemate

hv

barrelene semibullvalene

The di-π-methane rearrangement [Hixson, 1973] as shown above is also possible for compounds in which one of the double bonds is part of an aromatic ring or replaced by a carbonyl group. The photorearrangement of β,γ-unsaturated ketones via (π,π*) triplet, and referred to as oxadi-π-methane rearrangement, is actually a much more versatile synthetic reaction in view of the high chemical yields often obtained. Acetone frequently serves the dual role of solvent and sensitizer. Note the alternative 1,3-acyl shift pathway due to ketone excitation is mentioned in a prvious section.

Synthetic applications of the oxadi-π-methane rearrangement are found in the skeletal elaboration of iridodial [Ritterskamp, 1984], loganin aglucone acetate [Demuth, 1984], coriolin [Demuth, 1986b], α-/β-biotol [Yates, 1987], pentalenolactone-P [Paquette, 1992], silphiperfol-6-en-5-one [Demuth, 1988a], and by two different routes, modhephene [Mehta, 1991; Uyehara, 1996], when the proper bicyclo[2.2.2]-octenones are irradiated.

Tricyclo[3.3.0.0^{2,8}]octan-3-one. [Demuth, 1980]

A solution of bicyclo[2.2.2]oct-5-en-2-one (10 g) in acetone (1 L) is purged with argon and irradiated in a water-cooled quartz vessel placed in a Rayonet photoreactor equipped with RUL-3000A lamps until 96-98% of the enone is consumed (72 h). Acetone is distilled off and residue is

chromatographed over silica gel to give the tricyclic ketone (8.6 g) which can be purified by distillation (8.1 g, 81%).

iridodial

loganin aglucone acetate

coriolin

α-/ β-biotol

pentalenolactone-P

silphiperfol-6-en-5-one

modhephene

modhephene

The bicyclo[3.2.1]octenone analogs also undergo rearrangement, as shown in a synthesis of α-*trans*-bergamotene [Larsen, 1977], α-santalene [Monti, 1978].

71%

α-santalene

7.5.3. Sigmatropic Rearrangements

By far the most remarkable corroboration of the Woodward-Hoffmann rules is the observation of a photoinduced sigmatropic migration of a hydrogen atom in a secocorrin Cd complex [Eschenmoser, 1976], resulting in cyclization with a stereoselectivity of 95%.

7.5.4. Other Rearrangements

The Fries rearrangement of aryl esters is commonly promoted by Lewis acids. With the advent of the photochemical version [Stenberg, 1967; Bellus, 1971; R. Martin, 1992] corrosive and sometimes deleterious catalysts are avoided. Furthermore, this method has been extended to aryl carbonates, carbamates, sulfonates, sulfamates. Anilides and analogues also undergo the rearrangement. As this reaction is known to proceed via a caged radical pair which is formed by homolysis of the O-C(acyl) bond, the recombination is favored by physical constraints such as those provided by cyclodextrins [Ohara, 1975; Syamala, 1988]. A particularly interesting report describes the virtually quantitative formation of 4-chromanones in the benzene-sodium hydroxide two phase system [Miranda, 1982].

MeO R hv PhH- NaOH MeO C(R)Me O MeO R O

R = H, Me

There are many applications of the photo-Fries rearrangement to synthesis, including routes to griseofulvin [Taub, 1963], bikaverin [Katagiri, 1981]. It also entails design of lithography based on poly(*p*-acetoxystyrene) [Frechet, 1985] as the light-exposd sections of the polymer are soluble in alkali.

<u>Phto-Fries Rearrangement of 2-Chloro-3,5-dimethoxyphenyl 4-Acetoxy-2-methoxy-6-methylbenzoate.</u>
[Taub, 1963]

A solution of the ester (250 mg) in ethanol (2.5 mL) in a quartz tube is exposed to a Hanovia low-pressure lamp at 40° for 66 h. Chromatography of the concentrate on florisil and crystallization gives the benzophenone (42 mg) which can be purified by recrystallization.

OMe MeO Cl O OMe MeO OH hv OMe O OMe MeO Cl OH MeO OH OMe O OMe MeO Cl O

griseofulvin

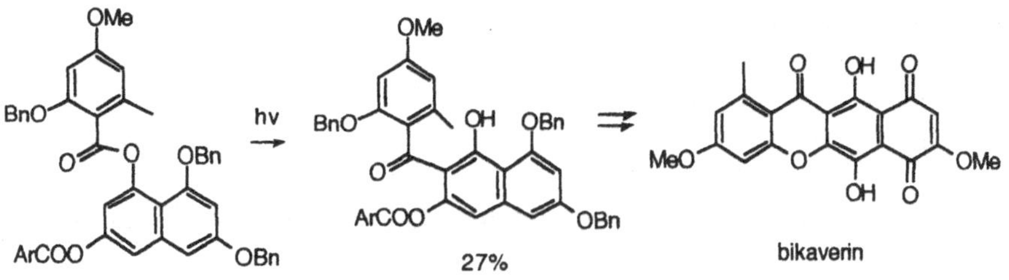

OMe BnO O O OBn ArCOO OBn hv OMe BnO O OH OBn ArCOO OBn 27% O OH O MeO O OMe OH O bikaverin

The photoinduced transformation of verbenone to chrysanthenone which was known previously now features in an efficient approach to taxol [Wender, 1995]. The α-bromo derivative forms a nucelophile on bromine-lithium exchange.

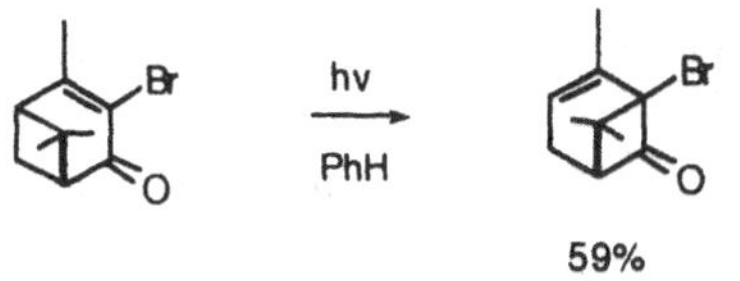

Br O hv PhH Br O 59%

2-Carene and derivatives undergo photoreaction in analogy to the thermal vinylcyclopropane-to-cyclopentene rearrangement, apparently via the cycloheptene biradicals. The racemic bicyclo[3.2.0]heptene isomer from chiral 2-carene has been converted into grandisol whereas a synthesis of (-)-$\Delta^{9(12)}$-capnellene can be realized by manipulation of one of the two diastereomers derived from (+)-2-caren-4-ol [Sonawane, 1991].

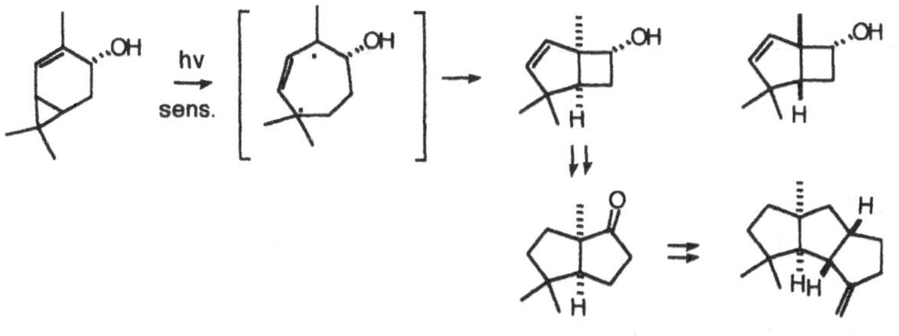

OH hv sens. OH OH H OH H O H H HH

$\Delta^{9(12)}$-capnellene

cis-1,4,4-Trimethylbicyclo[3.2.0]hept-2-ene. [Sonawane, 1991]

A solution of 2-carene (4.80 g, 30 mmol) in petroleum ether (500 mL containing 3.1 mL toluene) is irradiated with a 450 W Hanovia immersion-type lamp (Vycor filter) for 50 h. Distillation removes the solvent and the product (3.04 g, 75%) is obtained by further purification using preparative GC.

The photoisomerization of cyclooctatetraene to semibullvalene [Turro, 1980] is quite efficient.

hv

semibullvalene

Semibullvalene.

In a Rayonet apparatus equipped with 12 RPR 350 nm lamps is placed at its center a cyclindrical pyrex vessel (25x11 cm) with a cold finger (10 mm long). The cold finger is charged with cyclooctatetraene (0.27 g, 2.6 mmol), cooled with dry ice/acetone while the vessel is sealed under vacuum (0.05 torr) and irradiated for 60 h. The product condensed onto the cold finger is removed. It contains >98% semibullvalene and weighs 0.24 g.

Fischer carbene complexes undergo photochemical rearrangement to produce species with ketene-like reactivity. In situ trapping of the latter gives rise to carboxylic acid derivatives (with alcohols and amines), β-lactams (with imines), and cyclobutanones (with alkenes). Asymmetric synthesis of amino acids and peptides have also been demonstrated [Hegedus, 1995].

Ph Ph OH HN $Cr(CO)_5$ Ar — hv, MeCN, DMAP → [Ph Ph OH HN C=O Ar $Cr(CO)_4$] → Ph Ph O HN O Ar

7.6. PHOTOELIMINATIONS AND PHOTOADDITIONS

Photoinduced cleavage of one or more carbon-heteroatom bonds is a rather common occurrence. Photodecomposition of diazo compounds leads to carbenes, and of α-diazo ketones to ketenes through a Wolff rearrangement. These species are very reactive therefore their fate depends on the environment (and solvent). Nevertheless, the photoreaction has offered service to the generation of tetra-*t*-butyltetrahedrane [Maier, 1991], benzvalene and Dewar benzene [Turro, 1976], [3.5.4]fenestrane [Brinker, 1990], cyclocopacamphene [Piers, 1975], and a key intermediate of longifolene [Schultz, 1985]. 1,3-Diradical generation from a 2,3-diazabicyclo[2.2.1]-hept-2-ene can be performed thermally or photolytically, and with a juxtaposed diylophile to intercept the reactive species the reaction affords a tricyclic product. The photolytic method was used to construct the triquinane core of coriolin [Van Hijfte, 1985].

It should be noted that nitrogen extrusion and Wolff rearrangement proceed better by photolysis than thermolysis. For example, 3-diazocamphor gives rise to the bicyclo[2.1.1]hexanecarboxylic acid only under photolytic conditions [Horner, 1955].

Methyl bicyclo[2.1.1]hexane-5-carboxylate. [Wiberg, 1961].

A solution of diazonorcamphor (73.6 g, 0.54 mol) in methanol (4.5 L) is irradiated with a water-cooled Hanovia immersion quartz lamp which is surrounded by a Corex filter. After 24 h (ca. 95% reaction) most of the solvent is removed by distillation, the residue is treated with water (1 L) and crushed ice, and extracted with pentane (4x 1 L). The extracts are dried over Na_2SO_4, evaporated, and the product is distilled to give the ester (41 g, 54% yield).

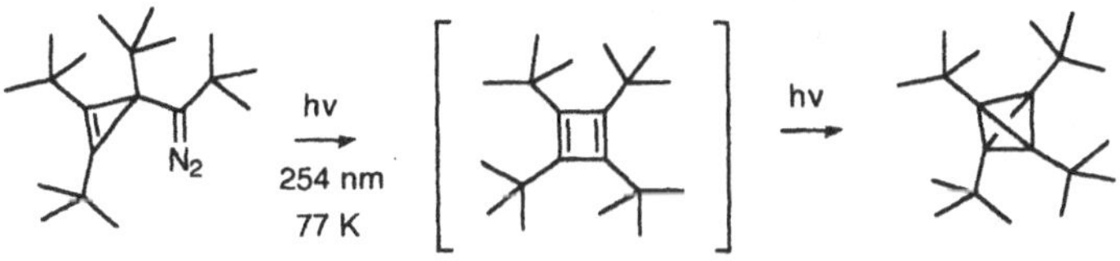

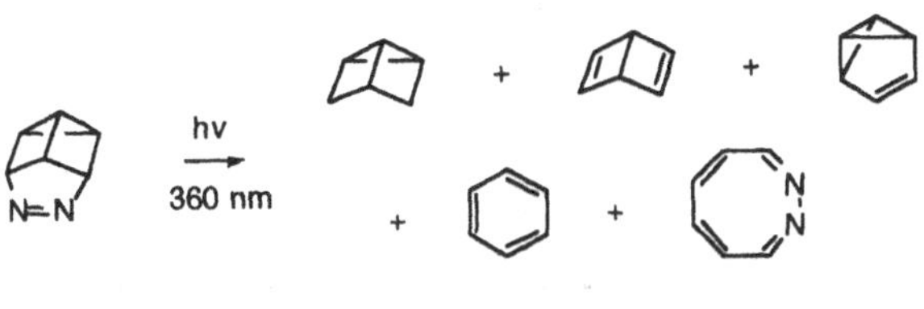

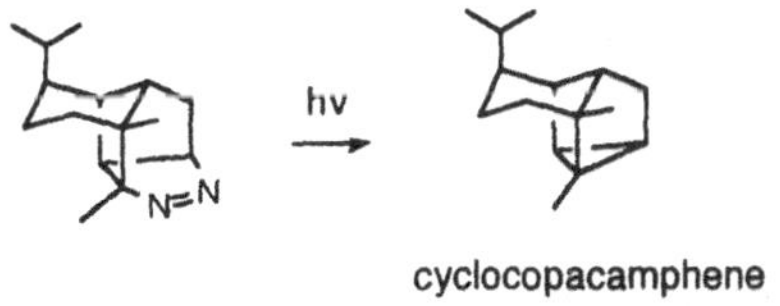

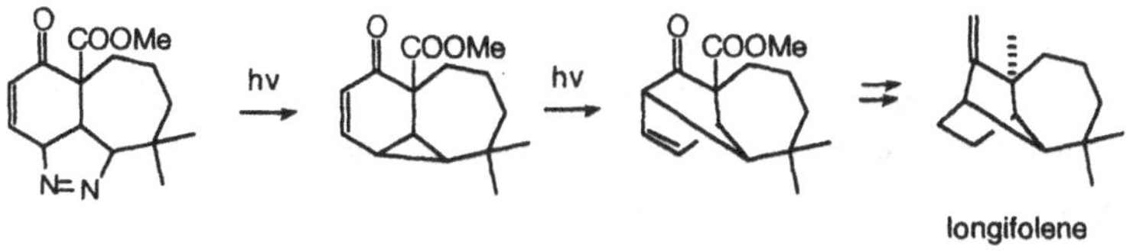

hv 254 nm 77 K

hv

hv 360 nm

hv

cyclocopacamphene

hv

hv

longifolene

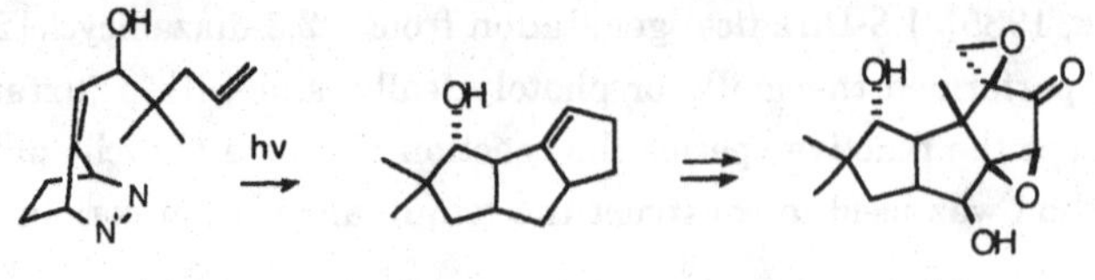

coriolin

Photolysis of organic azides generates nitrenes. A synthetically useful method for carbazole formation is by exposure of *o*-azidobiphenyl (and various ring-substituted analogues) to light [Swenton, 1970].

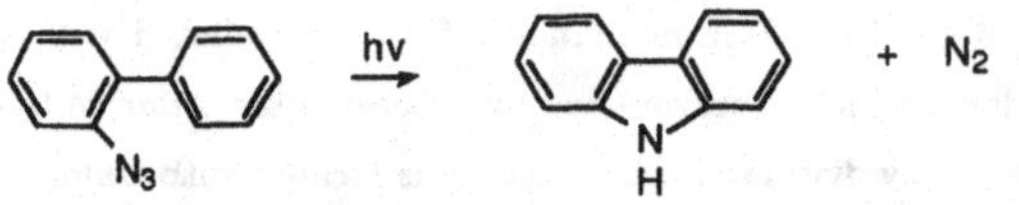

As mentioned previously, the photolysis of α-pyrone leads to cyclobutadiene. Photodecarboxylation has also been employed in in the conversion of α-santonin to α-cyperone [Murai, 1982], and a preparation of [2.2]paracyclophane from a diolide [Kaplan, 1976].

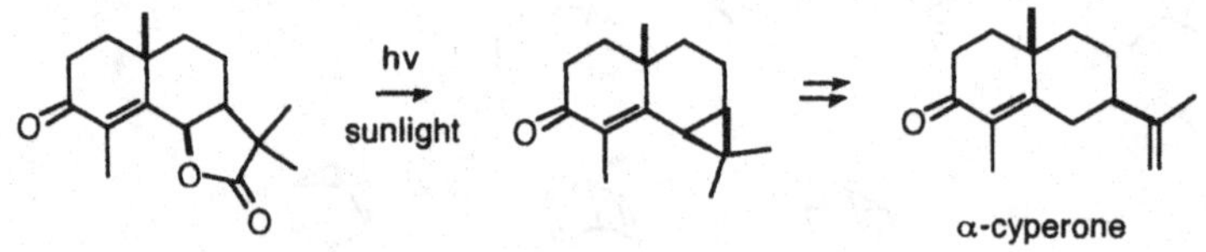

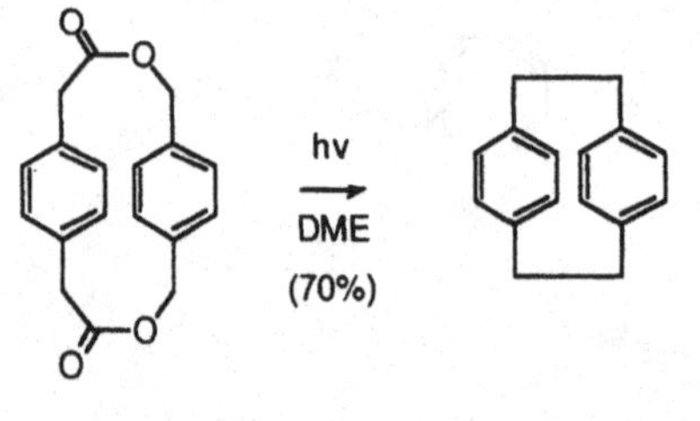

α-cyperone

Reductive degradation of an unactivated carboxylic acid can be performed by photolyzing the corresponding *N*-(acyloxy)phthalimide in the presence of *N*-methylcarbazole and *t*-butanethiol [Okada, 1988]. This method proved to be of critical importance to the assembly of a molecular segment of rapamycin [Fisher, 1991], as it contributed to the realization of merging two chiral building blocks by esterification and an Ireland-Claisen rearrangement.

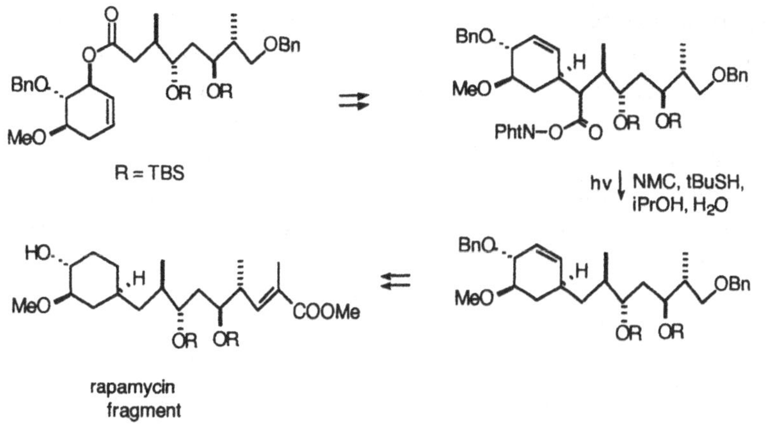

R = TBS

hv, NMC, tBuSH, iPrOH, H_2O

rapamycin fragment

Photodeoxygenation of alcohols via the acetates [Pete, 1977] also has synthetic utility, for example in delivering dactylol [Feldman, 1990]. Deoxygenation via the xanthate or mesylate failed.

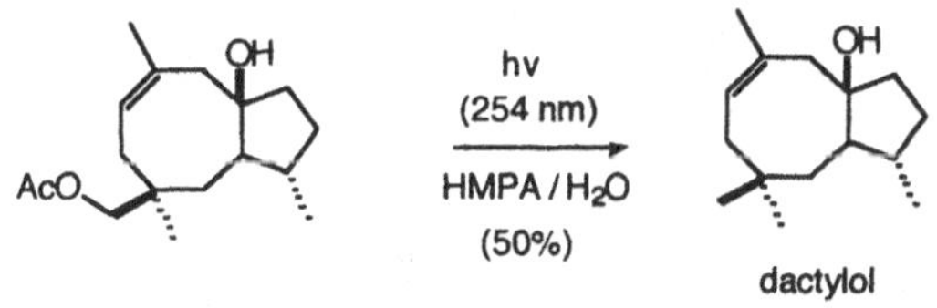

hv (254 nm), HMPA / H_2O (50%)

dactylol

2-Chloroallyl anions readily lose chloride ion on photoexcitation because of the increased charge density at C-2. Accordingly, strained cyclic allenes such as 1,3-diphenyl-1,2-cyclohexadiene can be generated (with in situ trapping as furan adduct) in good yields [Tolbert, 1990].

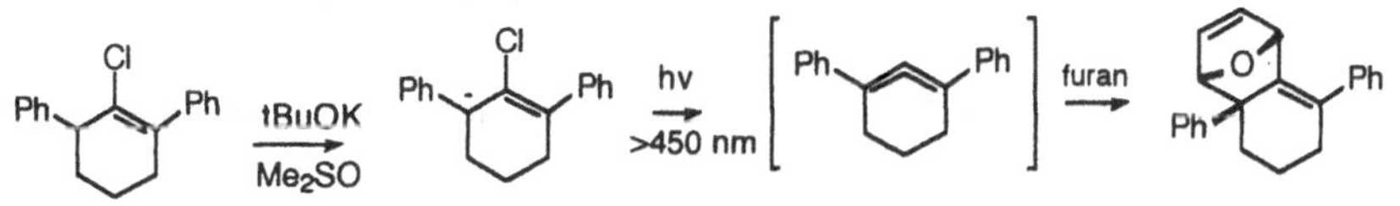

tBuOK, Me_2SO; hv >450 nm; furan

Singlet dioxygen has many applications [Balci, 1981; Wasserman, 1981], besides its ability to under Diels-Alder reaction with conjugated dienes, such as forming ene reaction products and 1,2-dioxetanes. Thus the elaboration of the trioxabicyclic unit of qinghaosu [Schmid, 1983] and the dearomatization of the C-ring with a highly stereoselective introduction of a hydroxyl group at C-6 of a tetracycline precursor [Scott, 1962; Schach v. Wittenau, 1964] were initiated by such a reaction.

Enamines such as the indole ring either suffer oxidative cleavage (in an alcohol solvent) via a dioxetane intermediate or undergo an ene-type reaction (in an aprotic solvent). A biomimetic transformation of a tetracyclic indole derivative to the 4-quinolone system and thence to a precursor of camptothecin [Kametani, 1981] is most interesting.

It should be mentioned that in the total synthesis of chlorophyll-*a* [Woodward, 1990] a purpurin intermediate bearing an extra five-membered ring was degraded by photooxidation.

hv, O_2, sens. — (MeO)$_2$CH, HOO — qinghaosu

hv / O_2 ; H / Pd-C — H_2, Pd — tetracycline

The method of ketene dithioacetal cleavage with singlet dioxygen [Adam, 1982; Geller, 1983] has been applied in a synthesis of (+)-castanospermine [Miller, 1990].

It should be mentioned that the deliberate introduction of some additives in a photoxygenation medium may achieve desired transformation of the primary products. For example, a cycloalkene is converted to open chain chloroketone(s) [Sato, 1994] or keto nitrile [Shimizu, 1997] in the presence of $FeCl_3$ and $Cu(OTf)_2$-NaN_3, respectively.

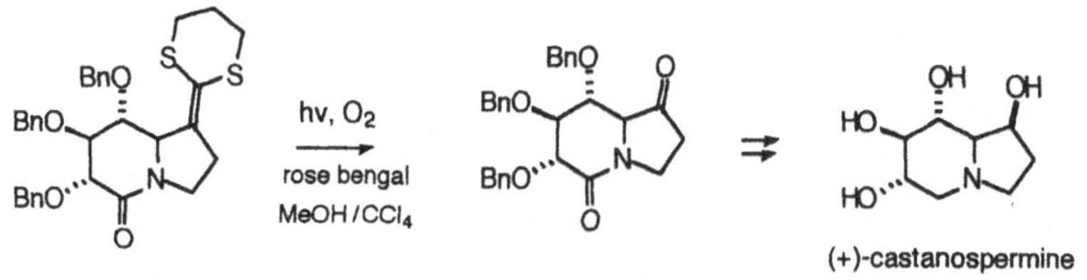

hv, O_2
rose bengal
MeOH / CCl_4

(+)-castanospermine

7.7. MISCELLANEOUS PHOTOCHEMICAL PROCESSES OF SYNTHETIC SIGNIFICANCE

The *o*-nitrobenzyl residue is an excellent protective device [Patchornik, 1970] because of the photoreactivity of the nitro group. The intramolecular redox transformation introduces a hydroxyl group to the benzylic position which can then trigger the release of the protected alcohol or an amine. In a synthesis of (-)-furodysinin [T.-L. Ho, 1996] it rendered an indispensable service, and in the context of generating a nitrosoarene for an intramolecular Diels-Alder reaction to establish a tricyclic 1,2-oxazine which underwent rearrangement to provide a viable precursor of mitomycin-K [Benbow, 1993].

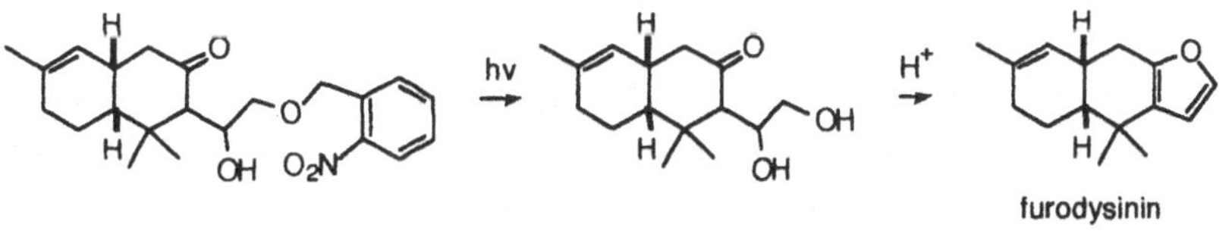

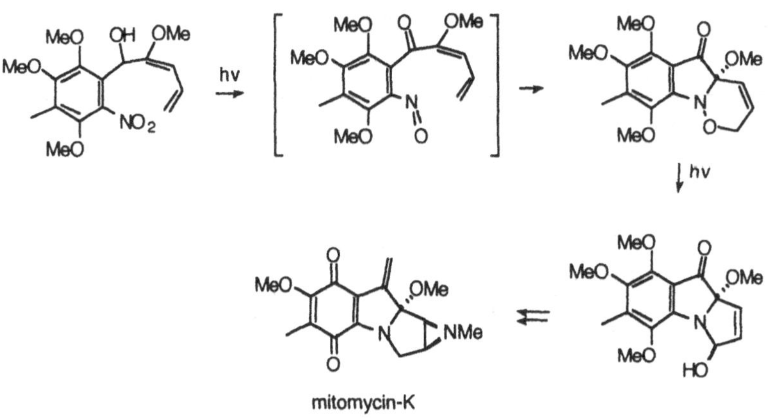

furodysinin

mitomycin-K

Exciplexes are stabilized by charge transfer, and in highly polar solvents such species tend to collapse by the single electron transfer (SET) mode. Iminium ions participate in various SET-promoted reactions with neutral electron donors which can lead to C-C bond formation. Photocyclization based on this process has been realized in the synthetic courses toward stylopine [G.D. Ho, 1987] and the cyclic array of the Cephalotaxus alkaloids [Kavash, 1989].

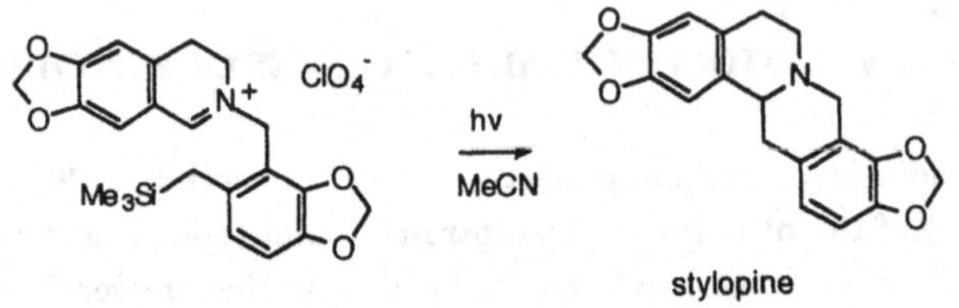

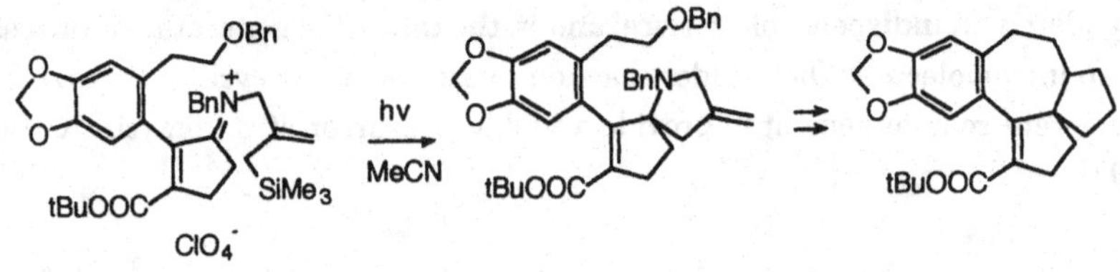

ClO_4^- hv MeCN Me_3Si

stylopine

OBn BnN $SiMe_3$ tBuOOC ClO_4^- hv MeCN OBn BnN tBuOOC tBuOOC

Nucleophilic substitution of haloarenes is promoted often by irradiation. While the true nature of the reaction is still unclear, the process (photoinduced SET?) has found considerable synthetic uses, including the direct formation of 2-substituted indoles [Bard, 1980; Beugelmans, 1980] and photocyclization leading to cephalotaxine [Semmelhack, 1975] and cherylline [Kessar, 1981].

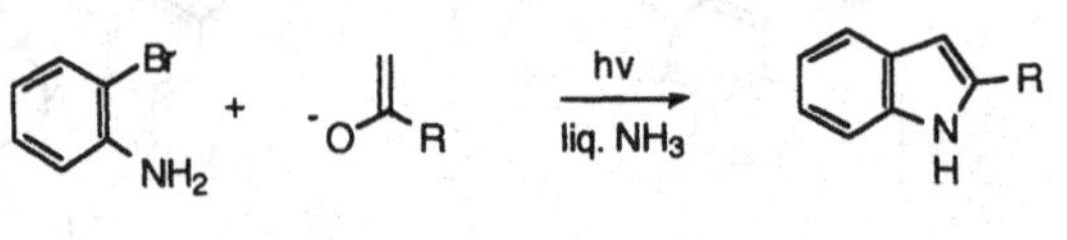

Br NH_2 + R hv liq. NH_3 R N H

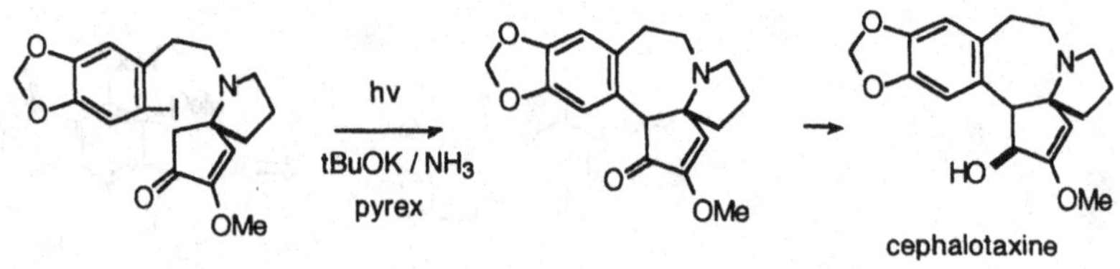

I N O OMe hv tBuOK / NH_3 pyrex N O OMe → N HO OMe

cephalotaxine

hν

KNH_2/NH_3

cherylline

Monographs and Reviews

Coyle, J.D. (ed.) (1986) *Photochemistry in Organic Synthesis*, Royal Society of Chemistry: Cambridge.

Demuth, M.; Mikhail, G. (1989) *Synthesis* 145.

Dilling, W.L. (1977) *Photochem. Photobiol.* **25:** 605.

Hoorspool, W.M. (ed.) (1984) *Synthetic Organic Photochemistry*, Plenum Press: New York.

Hoorspool, W.M.; Armesto, D. (1992) *Organic Photochemistry*, Ellis Horwood: New York.

Hoorspool, W.M.; Song, P.-S. (eds.) (1994) *CRC Handbook of Organic Photochemistry and Photobiology*, CRC Press: Boca Raton.

Kopecky, J. (1992) *Organic Photochemistry: A Visual Approach*, VCH: Weinheim.

Mattay, J.; Griesbeck, A. (eds.) (1994) *Photochemical Key Steps in Organic Synthesis*, VCH: Weinheim.

Ninomiya, I.; Naito, T. (1989) *Photochemical Synthesis*, Academic Press: London.

Sammes, P.G. (1970) *Synthesis* 636.

References

Adam, W.; Arias, L.; Schetzow, D. (1982) *Tetrahedron Lett.* **23:** 2835.

Anglea, T.A.; Pinder, A.R. (1987) *Tetrahedron* **43:** 5537.

Angus, H.J.F.; Bryce-Smith, D. (1960) *J. Chem. Soc.* 4791.

Araki, Y.; Nagasawa, J.; Ishido, Y. (1981) *J. Chem. Soc. Perkin Trans. I* 12.

Audenaert, F.; De Keukeleire, D.; Vandewalle, M. (1987) *Tetrahedron* **43:** 5593.

Ayer, W.A.; Browne, L.M. (1974) *Can. J. Chem.* **52:** 1352.

Baker, R.; Sims, R.J. (1981) *J. Chem. Soc. Perkin Trans. I* 3087.

Balci, M. (1981) *Chem. Rev.* **81:** 91.

Baldwin, S.W.; Crimmins, M.T. (1982a) *J. Am. Chem. Soc.* **104:** 1132.

Baldwin, S.W.; Fredericks, J.E. (1982b) *Tetrahedron Lett.* **23:** 1235.

Baldwin, S.W.; Landmesser, N.G. (1982c) *Tetrahedron Lett.* **23:** 4443.

Baldwin, S.W.; Martin, G.F.; Nunn, D.S. (1985) *J. Org. Chem.* **50:** 5720.
Ban, Y.; Yoshida, K.; Goto, J.; Oishi, T.; Takeda, E. (1983) *Tetrahedron* **39:** 3657.
Baralotto, C.; Chanon, M.; Julliard, M. (1996) *J. Org. Chem.* **61:** 3576.
Bard, R.R.; Bunnett, J.F. (1980) *J. Org. Chem.* **45:** 1546.
Barton, D.H.R.; De Mayo, P.; Shafiq, M. (1958) *J. Chem. Soc.* 140.
Barton, D.H.R.; Beaton, J.M. (1961) *J. Am. Chem. Soc.* **83:** 4083.
Barton, D.H.R.; Basu, N.K.; Day, M.J.; Hesse, R.H.; Pechet, M.M.; Starratt, A.N. (1975) *J. Chem. Soc. Perkin Trans. I.* 2243.
Barton, D.H.R.; Hulshof, L.A. (1977) *J. Chem. Soc. Perkin Trans. I.* 1103.
Baudouy, R.; Crabbe, P.; Greene, A.E.; Le Drian, C.; Orr, A.F. (1977) *Tetrahedron Lett.* 2973.
Bauslaugh, P.G. (1970) *Synthesis* 287.
Beckwith, A.L.J. (1981) *Tetrahedron* **37:** 3073.
Bellus, D. (1971) *Adv. Photochem.* **8:** 109.
Ben, I.; Castedo, L.; Saa, J.M.; Seijas, J.A.; Suau, R.; Tojo, G. (1985) *J. Org. Chem.* **50:** 2236.
Benbow, J.W.; McClure, K.F.; Danishefsky, S.J. (1993) *J. Am. Chem. Soc.* **115:** 12305.
Beugelmans, R.; Boudet, B.; Quintero, L. (1980) *Tetrahedron Lett.* **21:** 1943.
Blattmann, H.-R.; Meuche, D.; Heilbronner, E.; Molyneux, R.J.; Boekelheide, V. (1965) *J. Am. Chem. Soc.* **87:** 130.
Boeckman, R.K.; Cheon, S.H. (1983) *J. Am. Chem. Soc.* **105:** 4112.
Braun, A.M.; Maurette, M.-T.; Oliveros, E. (1991) *Photochemical Technology*, Wiley: Chichester.
Brennan, J. (1981) *Chem. Commun.* 880.
Breslow, R.; Baldwin, S.; Flechtner, T.; Kalicky, P.; Liu, S.; Washburn, W. (1973) *J. Am. Chem. Soc.* **95:** 3251.
Breslow, R. (1980) *Acc. Chem. Res.* **13:** 170.
Brinker, U.H.; Schrievers, T.; Xu, L. (1990) *J. Am. Chem. Soc.* **112:** 8609.
Brown, M. (1968) *J. Org. Chem.* **33:** 162.
Bryce-Smith, D.; Gilbert, A.; Orger, B.H. (1966) *Chem. Commun* 512.
Bryce-Smith, D.; Gilbert, A.; Grzonka, J. (1970) *Chem. Commun* 498.
Bryce-Smith, D.; Foulger, B.; Forrester, J.; Gilbert, A.; Orger, B.H.; Tyrrell, H.M. (1980) *J. Chem. Soc. Perkin Trans. I* 55.
Büchi, G.; Carlson, J.A.; Powell, J.E.; Tietze, L.-F. (1973) *J. Am. Chem. Soc.* **95:** 540.
Bullivant, M.J.; Pattenden, G. (1976) *J. Chem. Soc. Perkin Trans. I* 256.
Caine, D.; Ingwalson, P.F. (1972) *J. Org. Chem.* **37:** 3751.
Caine, D.; Tuller, F.N. (1973) *J. Org. Chem.* **38:** 3663.
Caine, D.; Boucugnani, A.A.; Chao, S.T.; Dawson, J.B.; Ingwalson, P,F. (1976) *J. Org.*

Chem. **41:** 1539.
Caine, D.; Frobese, A.S. (1977) *Tetrahedron Lett.* 3107.
Callant, P.; Storme, P.; Van der Eycken, E.; Vandewalle, M. (1983) *Tetrahedron Lett.* **24:** 5797.
Campbell, A.L.; Lenz, G.R. (1987) *Synthesis* 421.
Cargill, R.L.; Wright, B.W. (1975) *J. Org. Chem.* **40:** 120.
Carruthers, W. (1968) *J. Chem. Soc. (C)* 2244.
Castedo, L.; Guitian, E.; Saa, J.M.; Suau, R. (1982) *Heterocycles* **19:** 279.
Challand, B.D.; Hikino, H.; Kornis, G.; Lange, G.; de Mayo, P. (1969) *J. Org. Chem.* **34:** 794.
Chaquin, P.; Morizur, J.-P.; Kossanyi, P. (1977) *J. Am. Chem. Soc.* **99:** 903.
Corey, E.J.; Bass, J.D.; LeMahieu, R.; Mitra, R.B. (1964a) *J. Am. Chem. Soc.* **86:** 5570.
Corey, E.J.; Mitra, R.B.; Uda, H. (1964b) *J. Am. Chem. Soc.* **86:** 485.
Corey, E.J.; Nozoe, S. (1964c) *J. Am. Chem. Soc.* **86:** 1652.
Corey, E.J.; Streith, J. (1964d) *J. Am. Chem. Soc.* **86:** 950.
Corey, E.J.; Hortmann, A.G. (1965) *J. Am. Chem. Soc.* **87:** 5736.
Corey, E.J.; Hamanaka, E. (1967) *J. Am. Chem. Soc.* **89:** 2758.
Corey, E.J.; Cane, D.E.; Libit, L. (1971) *J. Am. Chem. Soc.* **93:** 7016.
Corey, E.J.; Arnett, J.F.; Widiger, G.N. (1975) *J. Am. Chem. Soc.* **97:** 430.
Cossy, J.; Belotti, D. (1988) *Tetrahedron Lett.* **29:** 6113.
Crimmins, M.T.; DeLoach, J.A. (1984) *J. Org. Chem.* **49:** 2076.
Crimmins, M.T.; Mascarella, S.W. (1986) *J. Am. Chem. Soc.* **108:** 3435.
Crimmins, M.T.; Gould, L.D. (1987) *J. Am. Chem. Soc.* **109:** 6199.
Crimmins, M.T.; Jung, D.K.; Gray, J.L. (1992) *J. Am. Chem. Soc.* **114:** 5445.
Crimmins, M.T. (1993) *Org. React.* **44:** 297.
Crossland, N.M.; Roberts, S.M.; Newton, R.F. (1979) *J. Chem. Soc. Perkin Trans. I,* 2397.
Dauben, W.G.; Fonken, G.J. (1959a) *J. Am. Chem. Soc.* **81:** 4060.
Dauben, W.G.; Koch, K.; Thiessen, W.E. (1959b) *J. Am. Chem. Soc.* **81:** 6087.
Dauben, W.G.; Koch, K.; Smith, S.L.; Chapman, O.L. (1963) *J. Am. Chem. Soc.* **85:** 2616.
Dauben, W.G.; Rivers, G.T.; Twieg, R.J.; Zimmerman, W.T. (1976) *J. Org. Chem.* **41:** 887.
Dauben, W.G.; Shapiro, G. (1984) *J. Org. Chem.* **49:** 4252.
Davies, H.G.; Roberts, S.M.; Wakefield, B.J.; Winders, J.A. (1985) *Chem. Commun.* 1166.
De Clercq, P.; Vandewalle, M. (1977) *J. Org. Chem.* **42:** 3447.

Delton, M.H.; Gilman, R.E.; Cram, D.J. (1971) *J. Am. Chem. Soc.* **93:** 2329.

de Mayo, P.; Takeshita, H. (1963) *Can. J. Chem.* **41:** 440.

Demuth, M.; Raghavan, P.R.; Carter, C.; Nakano, K.; Schaffner, K. (1980) *Helv. Chim. Acta* **63:** 2434.

Demuth, M.; Chandrasekhar, S.; Schaffner, K. (1984) *J. Am. Chem. Soc.* **106:** 1092.

Demuth, M. (1986a) *Pure Appl. Chem.* **58:** 1233.

Demuth, M.; Ritterskamp, P.; Weigt, E.; Schaffner, K. (1986b) *J. Am. Chem. Soc.* **108:** 4149.

Demuth, M.; Hinsken, W. (1988a) *Helv. Chim. Acta* **71:** 569.

Demuth, M.; Pandey, B.; Wietfeld, B.; Said, H.; Viader, J. (1988b) *Helv. Chim. Acta* **71:** 1392.

Demuynck, M.; DeClercq, P.J.; Vandewalle, M. (1979) *J. Org. Chem.* **44:** 4863.

Demuynck, M.; Devreese, A.A.; DeClercq, P.J.; Vandewalle, M. (1982) *Tetrahedron Lett.* **23:** 2501.

Devreese, A.A.; DeClercq, P.J.; Vandewalle, M. (1980) *Tetrahedron Lett.* **21:** 4767.

Diederich, F.; Staab, H.A. (1978) *Angew. Chem. Int. Ed. Engl.* **17:** 372.

Dilling W.L. (1977) *Photochem. Photobiol.* **25:** 605.

Disanayaka, B.W.; Weedon, A.C. (1985) *Chem. Commun.* 1282.

Duc, D.K.M.; Fetizon, M.; Lazare, S. (1975) *Chem. Commun.* 282.

Duc, D.K.M.; Fetizon, M.; Lazare, S. (1978) *Tetrahedron* **34:** 1207.

Eaton, P.E.; Cole, T.W. (1964) *J. Am. Chem. Soc.* **86:** 3157.

Eschenmoser, A. (1976) *Chem. Soc. Rev.* **5:** 377.

Exon, C.; Nobbs, M.; Magnus, P. (1981) *Tetrahedron* **37:** 4515.

Feldman, K.S.; Wu, M.-J.; Rotella, D.P. (1990) *J. Am. Chem. Soc.* **112:** 8490.

Fessner, W.-D.; Sedelmeier, G.; Spurr, P.R.; Rihs, G.; Prinzbach, H. (1987) *J. Am. Chem. Soc.* **109:** 4626.

Fischer, M. (1978) *Angew. Chem. Int. Ed. Engl.* **17:** 16.

Fisher, M.J.; Myers, C.D.; Joglar, J.; Chen, S.-H.; Danishefsky, S.J. (1991) *J. Org. Chem.* **56:** 5826.

Fitjer, L.; Quabeck, U. (1987) *Angew. Chem. Int. Ed. Engl.* **26:** 1023.

Frechet, J.M.J.; Tessier, T.G.; Willson, C.G.; Ito, H. (1985) *Macromolecules* **18:** 317.

Gasa, S.; Hamanaka, N.; Matsunaga, S.; Okuno, T.; Takeda, N.; Matsumoto, T. (1976) *Tetrahedron Lett.* 553.

Geller, G.G.; Foote, C.S.; Pechman, D.B. (1983) *Tetrahedron Lett.* **24:** 673.

Ghosh, S.; Patra, D.; Saha, G. (1993) *Chem. Commun.* 783.

Gleiter, R.; Karcher, M. (1988) *Angew. Chem. Int. Ed. Engl.* **27:** 840.

Gözler, B.; Guinaudeau, H.; Shamma, M.; Sariyar, G. (1986) *Tetrahedron Lett.* **27:** 1899.

Grovenstein, E.; Rao, D.V. (1961) *Tetrahedron Lett.* 148.
Gueldner, R.C.; Thompson, A.C.; Hedin, P.A. (1972) *J. Org. Chem.* **37:** 1854.
Guthrie, R.W.; Valenta, Z.; Wiesner, K. (1966) *Tetrahedron Lett.* 4645.
Hamanaka, N.; Matsumoto, T. (1972) *Tetrahedron Lett.* 3087.
Hamer, N.K. (1986) *Tetrahedron Lett.* **27:** 2167.
Hatakeyama, S.; Kawamura, M.; Takano, S. (1994) *J. Am. Chem. Soc.* **116:** 4081.
Hatsui, T.; Wang, J.-J.; Takeshita, H. (1995) *Bull. Chem. Soc. Jpn.* **68:** 2393.
Havinga, E.; Schlatmann, J.L.M.A. (1961) *Tetrahedron* **16:** 146.
Heathcock, C.H.; Badger, R.M. (1968) *Chem. Commun.* 1510.
Hegedus, L.S.; McGuire, M.A.; Schultze, L.M.; Yijun, C.; Anderson, O.P. (1984) *J. Am. Chem. Soc.* **106:** 2680.
Hegedus, L.S. (1995) *Acc. Chem. Res.* **28:** 299.
Hendrickson, J.B.; de Vries, J.G. (1982) *J. Org. Chem.* **47:** 1148.
Hixson, S.S.; Mariano, P.S.; Zimmerman, H.E. (1973) *Chem. Rev.* **73:** 531.
Ho, G.D.; Mariano, P.S. (1987) *J. Org. Chem.* **52:** 704.
Ho, T.-L.; Chein, R.-J. (1996) *Chem. Commun.* 1147.
Hobbs, P.D.; Magnus, P.D. (1976) *J. Am. Chem. Soc.* **98:** 4594.
Hoffmann, N.; Scharf, H.-D. (1991) *Liebigs Ann. Chem.* 1273.
Hoffmann, V.T.; Musso, H. (1987) *Angew. Chem. Int. Ed. Engl.* **26:** 1006.
Horner, L.; Spietschka, E. (1955) *Chem. Ber.* **88:** 934.
Houk, K.N. (1982) *Pure Appl. Chem.* **54:** 1633.
Howard, C.C.; Newton, R.F.; Reynolds, D.P.; Wadsworth, A.H.; Kelly, D.R.; Roberts, S.M. (1980) *J. Chem. Soc. Perkin Trans. I* 852.
Hsu, L.-F.; Chang, C.-P.; Li, M.-C.; Chang, N.-C. (1993) *J. Org. Chem.* **58:** 4756.
Hutchinson, C.R.; Mattes, K.C.; Nakane, M.; Partridge, J.J.; Uskokovic, M.R. (1978) *Helv. Chim. Acta* **61:** 1221.
Ibrahim-Ouali, M.; Sinibaldi, M.-E.; Troin, Y.; Gramain, J.-C. (1996) *Tetrahedron Lett.* **37:** 37.
Ikeda, M.; Ohno, K.; Takahashi, M.; Homma, K.-I. (1983) *Heterocycles* **20:** 1005.
Jimenez, M.C.; Miranda, M.A.; Scaiano, J.C.; Tormos, R. (1997) *Chem. Commun.* 1487.
Jones, G., II; Acquadro, M.A.; Carmody, M.A. (1975) *Chem. Commun.* 206.
Kagawa, S.; Matsumoto, S.; Nishida, S.; Yu, S.; Morita, J.; Ichihara, A.; Shirahama, H.; Matsumoto, T. (1969) *Tetrahedron Lett.* 3913.
Kalvoda, J.; Heusler, K. (1971) *Synthesis* 501.
Kametani, T.; Ohsawa, T.; Ihara, M. (1981) *J. Chem. Soc. Perkin Trans. I* 1563.
Kaplan, M.L.; Truesdale, E.A. (1976) *Tetrahedron Lett.* 3665.
Katagiri, N.; Nakano, J.; Kato, T. (1981) *J. Chem. Soc. Perkin Trans. I* 2710.

Katzenellenbogen, J.A. (1976) *Science* **194:** 139.

Kavash, R.W.; Mariano, P.S. (1989) *Tetrahedron Lett.* **30:** 4185.

Kelly, R.B.; Zamecnik, J.; Beckett, B.A. (1972) *Can. J. Chem.* **50:** 3455.

Kelly, R.B.; Eber, J.; Hung, H.-K. (1973) *Can. J. Chem.* **51:** 2534.

Kende, A.S.; Curran, D.P. (1979) *J. Am. Chem. Soc.* **101:** 1857.

Kessar, S.V.; Singh, P.; Chawla, R.; Kumar, P. (1981) *Chem. Commun.* 1074.

Kiguchi, T.; Hashimoto, C.; Naito, T.; Ninomiya, I. (1982) *Heterocycles* **19:** 2279.

Kim, T.H.; Hayase, Y.; Isoe, S. (1983) *Chem. Lett.* 651.

Kobayashi, K.; Kanno, Y.; Seko, S.; Suginome, H. (1992) *J. Chem. Soc. Perkin Trans. I* 3111.

Koft, E.R; Smith, A.B., III (1982) *J. Am. Chem. Soc.* **104:** 5568.

Kok, P.; DeClercq, P.J.; Vandewalle, M. (1979) *J. Org. Chem.* **44:** 4553.

Kossanyi, J.; Furth, B.; Morizur, J.P. (1973) *Tetrahedron Lett.* 3459.

Kostermans, G.B.M.; Bobeldijk, M.; de Wolf, W.H.; Bickelhaupt, F. (1987a) *J. Am. Chem. Soc.* **109:** 2471.

Kostermans, G.B.M.; Hogenbirk, M.; Turkenburg, L.A.M.; de Wolf, W.H.; Bickelhaupt, F. (1987b) *J. Am. Chem. Soc.* **109:** 2855.

Krapcho, A.P.; Waller, F.J. (1972) *J. Org. Chem.* **37:** 1079.

Kraus, G.A.; Chen, L. (1990) *J. Am. Chem. Soc.* **112:** 3464.

Kraus, G.A.; Chen, L. (1991) *J. Org. Chem.* **56:** 5098.

Kropp, P.J. (1965) *J. Am. Chem. Soc.* **87:** 3914.

Laarhoven, W.H.; Cuppen, T.J.H.M.; Nivard, R.F.J. (1974) *Tetrahedron* **30:** 3343.

Lange, G.L.; Neidert, E.E.; Orrom, W.J.; Wallace, D.J. (1978) *Can. J. Chem.* **56:** 1628.

Langer, K.; Mattay, J. (1995) *J. Org. Chem.* **60:** 7256.

Larsen, S.D.; Monti, S.A. (1977) *J. Am. Chem. Soc.* **99:** 8015.

Lei, X.-g.; Doubleday, C.E.; Zimmt, M.B.; Turro, N.J. (1986) *J. Am. Chem. Soc.* **108:** 2444.

Leimner, J.; Marschall, H.; Meier, N.; Weyerstahl, P. (1984) *Chem. Lett.* 1769.

Liao, C.C.; Wei, C.P. (1989) *Tetrahedron Lett.* **30:** 2255.

Liebe, J.; Wolff, C.; Tochstermann, W. (1982) *Tetrahedron Lett.* **23:** 2439.

Liu, H.-J.; Valenta, Z.; Wilson, J.S.; Yu, T.T.J. (1969) *Can. J. Chem.* **47:** 509.

Liu, H.-J.; Chan, W.H. (1979) *Can. J. Chem.* **57:** 708.

Liu, H.-J.; Kulkarni, M.G. (1985) *Tetrahedron Lett.* **26:** 4847.

Liu, L.; Yang, B.; Katz, T.J.; Pointdexter, M.K. (1991) *J. Org. Chem.* **56:** 3769.

McDonald, E.; Martin, R.T. (1978) *Tetrahedron Lett.* 4723.

McMurry, J.E.; Choy, W. (1980) *Tetrahedron Lett.* 2477.

Magaretha, P. (1974) *Helv. Chim. Acta* **57:** 1866.

Maier, G.; Pfriem, S.; Schäfer, U.; Matusch, R. (1978) *Angew. Chem. Int. Ed. Engl.* **17:**

520.

Maier, G.; Fleischer, F. (1991) *Tetrahedron Lett.* **32:** 57.

Marini-Bettolo, R.; Tagliatesta, P.; Lupi, A.; Bravetti, D. (1983) *Helv. Chim. Acta* **66:** 1922.

Marquardt, R.; Sander, W.; Kraka, E. (1996) *Angew. Chem. Int. Ed. Engl.* **35:** 746.

Marshall, J.A.; Johnson, P.C. (1970) *J. Org. Chem.* **35:** 192.

Marshall, J.A.; Black, T.H. (1980) *J. Am. Chem. Soc.* **102:** 7581.

Martin, H.-D.; Mayer, B.; Pütter, M.; Höchstetter, H. (1981) *Angew. Chem. Int. Ed. Engl.* **20:** 677.

Martin, R. (1992) *Org. Prep. Proc. Int.* **24:** 369.

Martin, R.H.; Eyndels, C.; Defay, N. (1974) *Tetrahedron* **30:** 3339.

Masuoka, N.; Kamikawa, T.; Kubota, T. (1974) *Chem. Lett.* 751.

Masuoka, N.; Kamikawa, T. (1976) *Tetrahedron Lett.* 1691.

Matsumoto, T.; Miyano, K.; Kagawa, S.; Yu, S.; Ogawa, J.; Ichihara, A. (1971) *Tetrahedron Lett.* 3521.

Mazzocchi, P.; King, C.R.; Ammon, H.L. (1987) *Tetrahedron Lett.* **28:** 2473.

Mehta, G.; Subrahmanyan, D. (1991) *J. Chem. Soc. Perkin Trans. I.* 395.

Mehta, G.; Murthy, A.N.; Reddy, D.S.K.; Reddy, A.V. (1986) *J. Am. Chem. Soc.* **108:** 3443.

Mehta, G.; Padma, S. (1987) *J. Am. Chem. Soc.* **109:** 7230.

Meyers, A.I.; Fleming, S.A. (1986) *J. Am. Chem. Soc.* **108:** 306.

Meystre, C.; Heusler, K.; Kalvoda, J.; Wieland, P.; Anner, G.; Wettstein, A. (1961) *Experientia* **17:** 475.

Miller, S.A.; Chamberlin, A.R. (1990) *J. Am. Chem. Soc.* **112:** 8100.

Miranda, M.A.; Primo, J.; Tormos, R. (1982) *Heterocycles* **19:** 1819.

Mitani, M.; Yamamoto, Y.; Koyama, K. (1983) *Chem. Commun.* 1446.

Miyano, M. (1981) *J. Org. Chem.* **46:** 1846.

Miyashita, M.; Yoshikoshi, A. (1974) *J. Am. Chem. Soc.* **96:** 1917.

Miyata, O.; Hirata, Y.; Naito, T.; Ninomiya, I. (1984) *Heterocycles* **22:** 1041.

Moëns, L.; Baizer, M.M.; Little, R.D. (1986) *J. Org. Chem.* **51:** 4497.

Monti, S.A.; Larsen, S.D. (1978) *J. Org. Chem.* **43:** 2282.

Murai, A.; Abiko, A.; Ono, M.; Masamune, T. (1982) *Bull. Chem. Soc. Jpn.* **55:** 1191.

Muraoka, O.; Okumura, K.; Maeda, T.; Tanabe, G.; Momose, T. (1994) *Tetrahedron: Asymmetry* **5:** 317.

Naito, T.; Kaneko, C. (1984) *J. Synth. Org. Chem. Jpn.* **42:** 51.

Naito, T.; Kojima, N.; Miyata, O.; Ninomiya, I. (1985) *J. Chem. Soc. Perkin Trans. 1* 1611.

Naito, T.; Kojima, N.; Miyata, O.; Ninomiya, I. (1986a) *Heterocycles* **24:** 2117.

Naito, T.; Kojima, N.; Miyata, O.; Ninomiya, I. (1986b) *Chem. Pharm. Bull.* **34:** 3530.

Naito, T.; Miyata, O.; Ninomiya, I. (1987) *Heterocycles* **26:** 1739.

Naito, T.; Hirata, Y.; Miyata, O.; Ninomiya, I. (1985) *J. Chem. Soc. Perkin Trans. 1* 2219.

Nakane, M.; Gollman, H.; Hutchinson, C.R.; Knutson, P.L. (1980) *J. Org. Chem.* **45:** 2536.

Nakazaki, M.; Yamamoto, K.; Yanagi, J. (1977) *Chem. Commun.* 346.

Nath, A.; Mal, J.; Venkateswaran, R.V. (1996) *J. Org. Chem.* **61:** 4391.

Ninomiya, I.; Naito, T. (1973) *Chem. Commun.* 137.

Ninomiya, I.; Naito, T. (1984) *J. Synth. Org. Chem. Jpn.* **42:** 225.

Okada, K.; Okamoto, K.; Oda, M. (1988) *J. Am. Chem. Soc.* **110:** 8736.

Ohara, M.; Watanabe, K. (1975) *Angew. Chem. Int. Ed. Engl.* **14:** 820.

Olovsson, G.; Scheffer, J.R.; Trotter, J.; Wu, C.-H. (1997) *Tetrahedron Lett.* **38:** 6549.

Oppolzer, W.; Wylie, R.D. (1980) *Helv. Chim. Acta* **63:** 1198.

Oppolzer, W.; Zutterman, F.; Bättig, K. (1983) *Helv. Chim. Acta* **66:** 522.

Oppolzer, W.; Godel, T. (1984) *Helv. Chim. Acta* **67:** 1154.

Orito, K.; Yorita, K.; Suginome, H. (1991) *Tetrahedron Lett.* **32:** 5999

Paquette, L.A.; Ternansky, R.J.; Balogh, D.W.; Kentgen, G. (1983) *J. Am. Chem. Soc.* **105:** 5446.

Paquette, L.A.; Lin, H.-S.; Coghlan, M.J. (1987) *Tetrahedron Lett.* **28:** 5017.

Paquette, L.A.; Kang, H.-J.; Ra, C.S. (1992) *J. Am. Chem. Soc.* **114:** 7387.

Partridge, J.J.; Chadha, N.K.; Uskokovic, M.R. (1973) *J. Am. Chem. Soc.* **95:** 532.

Patchornik, A.; Amit, B.; Woodward, R.B. (1970) *J. Am. Chem. Soc.* **92:** 6333.

Pattenden, G.; Teague, S.J. (1984) *Tetrahedron Lett.* **25:** 3021.

Pattenden, G.; Robertson, G.M. (1986) *Tetrahedron Lett.* **27:** 399.

Pearlman, B.A. (1979) *J. Am. Chem. Soc.* **101:** 6404.

Pete, J.P.; Portella, C. (1977) *Synthesis* 774.

Piers, E.; Geraghty, M.B.; Smillie, R.D.; Saucy, M. (1975) *Can. J. Chem.* **53:** 2849.

Piers, E.; Abeysekera, B.F.; Herbert, D.J.; Suckling, I.D. (1985) *Can. J. Chem.* **63:** 3418.

Pirrung, M.C. (1981) *J. Am. Chem. Soc.* **103:** 82.

Pirrung, M.C.; Thomson, S.A. (1986) *Tetrahedron Lett.* **27:** 2703.

Pirrung, M.C.; Thomson, S.A. (1988a) *J. Org. Chem.* **53:** 227.

Pirrung, M.C.; De Amicis, C.V. (1988b) *Tetrahedron Lett.* **29:** 159.

Pommer, H. (1977) *Angew. Chem. Int. Ed. Engl.* **16:** 423.

Posner, G.H.; Kinter, C.M. (1990) *J. Org. Chem.* **55:** 3967.

Quinkert, G.; Schmieder, K.R.; Dürner, G.; Hache, K.; Stegk, A.; Barton, D.H.R (1977) *Chem. Ber.* **110:** 3582.

Quinkert, G.; Weber, W.-D.; Schwartz, U.; Stark, H.; Baier, H.; Dürner, G. (1981) *Liebigs Ann. Chem.* 2335.

Quinkert, G.; Schwartz, U.; Stark, H.; Weber, W.-D.; Adam, F.; Baier, H.; Frank, G.; Dürner, G. (1982) *Liebigs Ann. Chem.* 1999.

Quinkert, G.; Heim, N.; Glenneberg, J.; Billhardt, U.-M.; Autze, V.; Bats, J.W.; Dürner, G. (1987) *Angew. Chem. Int. Ed. Engl.* **26:** 362.

Rawal, V.H.; Dufour, C.; Eschbach, A. (1994) *Chem. Commun.* 1797.

Rawal, V.H.; Fabré, A.; Iwasa, S. (1995) *Tetrahedron Lett.* **36:** 6851.

Rigby, J.H.; Ateeq, H.S. (1990) *J. Am. Chem. Soc.* **112:** 6442.

Rigby, J.H.; Henshilwood, J.A. (1991) *J. Am. Chem. Soc.* **113:** 5122.

Ritterskamp, P.; Demuth, M.; Schaffner, K. (1984) *J. Org. Chem.* **49:** 1155.

Rosini, G.; Salomoni, A.; Squarcia, F. (1979) *Synthesis* 942.

Röttele, H.; Heil, G.; Schröder, G. (1978) *Chem. Ber.* **111:** 84.

Saicic, R.N.; Zard, S.Z. (1996) *Chem. Commun.* 1631.

Salomon, R.G.; Sachinvala, N.D.; Roy, S.; Basu, B.; Raychaudhuri, S.R.; Miller, D.B.; Sharma, R.B. (1991) *J. Am. Chem. Soc.* **113:** 3085.

Sammes, P.G. (1976) *Tetrahedron* **32:** 405.

Sato, T.; Yonemochi, S. (1994) *Tetrahedron* **50:** 7375.

Schach v. Wittenau, M. (1964) *J. Org. Chem.* **29:** 2746.

Schenck, G.O.; Wirtz, R. (1953) *Naturwiss.* **40:** 581.

Schmid, G.; Hofheinz, W. (1983) *J. Am. Chem. Soc.* **105:** 624.

Schönberg, A.; Sodtke, U.; Praefcke, K. (1968) *Tetrahedron Lett.* 3669.

Schreiber, S.L.; Hoveyda, A.H. (1984a) *J. Am. Chem. Soc.* **106:** 7200.

Schreiber, S.L.; Santini, C. (1984b) *J. Am. Chem. Soc.* **106:** 4038.

Schreiber, S.L.; Satake, K. (1984c) *J. Am. Chem. Soc.* **106:** 4186.

Schröder, G. (1964) *Chem. Soc.* **97:** 3140.

Schultz, A.G.; Yee, Y.K.; Berger, M.H. (1977) *J. Am. Chem. Soc.* **99:** 8065.

Schultz, A.G.; Puig, S. (1985) *J. Org. Chem.* **50:** 915.

Scott, A.I.; Bedford, C.T. (1962) *J. Am. Chem. Soc.* **84:** 2271.

Semmelhack, M.F.; Chong, B.P.; Stauffer, R.D.; Rogerson, T.D.; Chong, A.; Jones, L.D. (1975) *J. Am. Chem. Soc.* **97:** 2507.

Seto, H.; Fujimoto, Y.; Tatsuno, T.; Yoshioka, H. (1985) *Synth. Commun.* **15:** 1217.

Shimizu, I.; Fujita, M.; Nakajima, T.; Sato, T. (1997) *Synlett* 887.

Shinkai, S.; Nakaji, T.; Nishida, Y.; Ogawa, T.; Manabe, O. (1980) *J. Am. Chem. Soc.* **102:** 5860, and later papers.

Smith, A.B., III; Jerris, P.J. (1981a) *J. Am. Chem. Soc.* **103:** 194.

Smith, A.B., III; Richmond, R.E. (1981b) *J. Org. Chem.* **46:** 4814.

Smith, A.B., III; Boschelli, D. (1983) *J. Org. Chem.* **48:** 1217.

Sonawane, H.R.; Naik, V.G.; Bellur, N.S.; Shah, V.G.; Purohit, P.C.; Kumar, M.U.; Kulkarni, D.G.; Ahuja, J.R. (1991) *Tetrahedron* **47:** 8259.

Stenberg, V.I. (1967) *Org. Photochem.* **1:** 127.

Sternbach, D.D.; Mane, M. priv. commun. to Crimmins, M.T. (1988) *Chem. Rev.* **88:** 1453.

Sugimura, T.; Paquette, L.A. (1987) *J. Am. Chem. Soc.* **109:** 3017.

Suginome, H.; Yamada, S. (1987) *Tetrahedron Lett.* **28:** 3963.

Suginome, H.; Nakayama, Y. (1994) *Tetrahedron* **50:** 7771.

Suginome, H.; Kondoh, T.; Gongonea, C.; Singh, V.; Goto, H.; Osawa, E. (1995) *J. Chem. Soc. Perkin Trans. I.* 69.

Swenton, J.C.; Ikele, T.K.; Williams, B.H. (1970) *J. Am. Chem. Soc.* **92:** 3103.

Syamala, M.S.; Rao, B.N.; Ramamurthy, V. (1988) *Tetrahedron* **44:** 7233.

Takeda, K.; Shimono, Y.; Yoshii, E. (1983) *J. Am. Chem. Soc.* **105:** 563.

Takeshita, H.; Kunno, I.; Iino, M.; Iwabuchi, H.; Nomura, D. (1980) *Bull. Chem. Soc. Jpn.* **53:** 3641.

Takeshita, H.; Hatsui, T.; Kato, N.; Masuda, T.; Tagoshi, H. (1982) *Chem. Lett.* 1153.

Takeshita, H.; Hatta, S.; Hatsui, T. (1984) *Bull. Chem. Soc. Jpn.* **57:** 619.

Tatsuta, K.; Akimoto, K.; Kinoshita, M. (1979) *J. Am. Chem. Soc.* **101:** 6116.

Taub, D.; Kuo, C.H.; Slates, H.L.; Wendler, N.L. (1963) *Tetrahedron* **19:** 1.

Termont, D.; De Clercq, P.; De Keukeleire, D.; Vandewalle, M. (1977) *Synthesis* 46.

Tobe, Y.; Yamashita, S.; Yamashita, T.; Kakiuchi, K.; Odaira, Y. (1984) *Chem. Commun.* 1259.

Tolbert, R.P.; Islam, M.N.; Johnson, R.P.; Loiselle, P.M.; Shakespeare, W.C. (1990) *J. Am. Chem. Soc.* **112:** 6416.

Turro, N.J.; Renner, C.A.; Waddell, W.H.; Katz, T.J. (1976) *J. Am. Chem. Soc.* **98:** 4320.

Turro, N.J.; Liu, J-M.; Zimmerman, H.E.; Factor, R.E. (1980) *J. Org. Chem.* **45:** 3511.

Uyehara, T.; Ogata, K.; Yamada, J.; Kato, T. (1983) *Chem. Commun.* 17.

Uyehara, T.; Yamada, J.; Kato, T.; Bohlmann, F. (1985) *Bull. Chem. Soc. Jpn.* **58:** 861.

Uyehara, T.; Kabasawa, Y.; Kato, T.; Furuta, T. (1986) *Bull. Chem. Soc. Jpn.* **59:** 2521.

Uyehara, T.; Furuta, T.; Kabasawa, Y.; Yamada, J.; Kato, T.; Yamamoto, Y. (1988) *J. Org. Chem.* **53:** 3669.

Uyehara, T.; Murayama, T.; Sakai, K.; Ueno, M.; Sato, T. (1996) *Tetrahedron Lett.* **37:** 7295.

van Hijfte, L.; Vandewalle, M. (1982) *Tetrahedron Lett.* **23:** 2229.

van Hijfte, L.; Vandewalle, M. (1984) *Tetrahedron* **40:** 4371.

van Hijfte, L.; Little, R.D. (1985) *J. Org. Chem.* **50:** 3940.
van Tamelen, E.E.; Pappas, S.P.; Kirk, K.L. (1971) *J. Am. Chem. Soc.* **93:** 6092.
van Tamelen, E.E.; Taylor, E.G. (1980) *J. Am. Chem. Soc.* **102:** 1202.
Wagner, P.J.; Subrahmanyam, D.; Park, B.-S. (1991) *J. Am. Chem. Soc.* **113:** 709.
Wasserman, H.H.; Keehn, P.M. (1969) *J. Am. Chem. Soc.* **91:** 2374.
Wasserman, H.H.; Ives, J.L. (1981) *Tetrahedron* **37:** 1825.
Weinreb, S.M.; Cvetovich, R.J. (1972) *Tetrahedron Lett.* 1233.
Weinshenker, N.M.; Greene, F.D. (1968) *J. Am. Chem. Soc.* **90:** 506.
Wender, P.A.; Lechleiter, J.C. (1978) *J. Am. Chem. Soc.* **100:** 4321.
Wender, P.A.; Hubbs, J.C. (1980a) *J. Org. Chem.* **45:** 365.
Wender, P.A.; Lechleiter, J.C. (1980b) *J. Am. Chem. Soc.* **102:** 6340.
Wender, P.A.; Letendre, L.J. (1980c) *J. Org. Chem.* **45:** 367.
Wender, P.A.; Dreyer, G.B. (1981a) *Tetrahedron* **37:** 4445.
Wender, P.A.; Howbert, J.J. (1981b) *J. Am. Chem. Soc.* **103:** 688.
Wender, P.A.; Dreyer, G.B. (1982a) *J. Am. Chem. Soc.* **104:** 5805.
Wender, P.A.; Howbert, J.J. (1982b) *Tetrahedron Lett.* **23:** 3983.
Wender, P.A.; Dreyer, G.B. (1983a) *Tetrahedron Lett.* **24:** 4543.
Wender, P.A.; Howbert, J.J. (1983b) *Tetrahedron Lett.* **24:** 5325.
Wender, P.A.; Singh, S.K. (1985a) *Tetrahedron Lett.* **26:** 5987.
Wender, P.A.; Ternansky, R.J. (1985b) *Tetrahedron Lett.* **26:** 2625.
Wender, P.A.; Fisher, K. (1986) *Tetrahedron Lett.* **27:** 1857.
Wender, P.A.; Correia, C.R.D. (1987) *J. Am. Chem. Soc.* **109:** 2523.
Wender, P.A.; van Geldern, T.W.; Levine, B.H. (1988) *J. Am. Chem. Soc.* **110:** 4858.
Wender, P.A.; de Long, M.A. (1990a) *Tetrahedron Lett.* **31:** 5429.
Wender, P.A.; McDonald, F.E. (1990b) *J. Am. Chem. Soc.* **112:** 4956.
Wender, P.A.; Singh, S.K. (1990c) *Tetrahedron Lett.* **31:** 2517.
Wender, P.A.; Wessjohann, L.A.; Peschke, B.; Rawlins, D.B. (1995) *Tetrahedron Lett.* **36:** 7181.
Wenkert, E.; Bindra, J.S.; Mylari, B.L.; Nussim, M.; Wilson, N.D.V. (1973) *Synth. Commun.* **3:** 431.
Weyerstahl, P.; Rilk, R.; Marschall-Weyerstahl, H. (1987) *Liebigs Ann. Chem.* 89.
White, J.D.; Gupta, D.N. (1968) *J. Am. Chem. Soc.* **90:** 6171.
White, J.D.; Skeean, R.W. (1978) *J. Am. Chem. Soc.* **100:** 6296.
Whitesell, J.K.; Minton, M.A.; Tran, V.D. (1989) *J. Am. Chem. Soc.* **111:** 1473.
Wiberg, K.B.; Lowry, B.R.; Coby, T.H. (1961) *J. Am. Chem. Soc.* **83:** 3998.
Winkler, J.D.; Henegar, K.E.; Williard, P.G. (1987) *J. Am. Chem. Soc.* **109:** 2850.
Wiesner, K.; Poon, L.; Jirkovsky, I. Fishman, M. (1969) *Can. J. Chem.* **47:** 433.
Wiesner, K.; Tsai, T.Y.R.; Huber, K.; Bolton, S.E.; Vlahov, R. (1974) *J. Am. Chem. Soc.*

96: 4990.

Wiesner, K. (1975) *Tetrahedron* **31:** 1655.

Williams, J.R.; Callahan, J.F. (1980a) *J. Org. Chem.* **45:** 4475.

Williams, J.R.; Callahan, J.F. (1980b) *J. Org. Chem.* **45:** 4479.

Williams, J.R.; Callahan, J.F.; Lin, C. (1983) *J. Org. Chem.* **48:** 3162.

Wilson, S.R.; Kaprinidis, Wu, Y.; Schuster, D.I. (1993) *J. Am. Chem. Soc.* **115:** 8495.

Wilzbach, K.E.; Kaplan, L. (1966) *J. Am. Chem. Soc.* **88:** 2066.

Winkler, J.D.; Muller, C.L.; Scott, R.D. (1988) *J. Am. Chem. Soc.* **110:** 4831.

Winkler, J.D.; Hershberger, P.M. (1989) *J. Am. Chem. Soc.* **111:** 4852.

Winkler, J.D.; Scott, R.D.; Williard, P.G. (1990) *J. Am. Chem. Soc.* **112:** 8971.

Winkler, J.D.; Hong, B.-C.; Hey, J.P.; Williard, P.G. (1991) *J. Am. Chem. Soc.* **113:** 8839.

Woodward, R.B.; Hoffmann, R. (1970) *The Conservation of Orbital Symmetry*, Verlag Chemie: Weinheim.

Woodward, R.B.; et al. (1990) *Tetrahedron* **46:** 7599.

Yanagiya, M.; Kaneko, K.; Kaji, T.; Matsumoto, T. (1979) *Tetrahedron Lett.* 1761.

Yang, N.C.; Shani, A.; Lenz, G.R. (1966) *J. Am. Chem. Soc.* **88:** 5369.

Yates, P.; Stevens, K.E. (1981) *Tetrahedron* **37:** 4401.

Yates, P.; Burnell, D.J.; Freer, V.J.; Sawyer, J.F. (1987) *Can. J. Chem.* **65:** 69.

Yoshihara, K.; Ohta, Y.; Sakai, T.; Hirose, Y. (1969) *Tetrahedron Lett.* 2663.

Ziegler, F.E.; Kloek, J.A. (1977) *Tetrahedron* **33:** 373.

Ziegler, F.E.; Jaynes, B.H. (1987) *Tetrahedron Lett.* **28:** 2339.

Zurflüh, R.; Dubham, L.L.; Spain, V.L.; Siddall, J.B. (1970) *J. Am. Chem. Soc.* **92:** 425.

8 EPILOG

Let us examine the following reaction sequence which was reported by Hermann Kolbe in 1845:

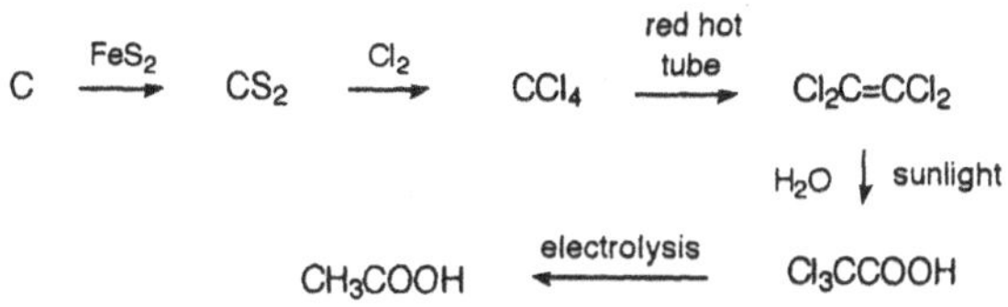

$$C \xrightarrow{FeS_2} CS_2 \xrightarrow{Cl_2} CCl_4 \xrightarrow[\text{tube}]{\text{red hot}} Cl_2C{=}CCl_2$$

$$Cl_2C{=}CCl_2 \xrightarrow{H_2O,\ \text{sunlight}} Cl_3CCOOH$$

$$CH_3COOH \xleftarrow{\text{electrolysis}} Cl_3CCOOH$$

This transformation of carbon into acetic acid is the very first multistep organic synthesis achieved by chemists in the laboratory. It constituted the bona fide and overwhelming evidence that rang the death knell of vitalism, as opposed to the often-cited accidental event pertaining to Wöhler's preparation of urea. It is significant that among the five steps employed in this work three reactions are the subject matters of the present volume.

Organic synthesis has indeed come a long way during the past 150 years, yet some of the historical methods, through much refinement, continue to retain their value along with new developments. Let us not forget the pioneers and be ever grateful to them for laying down the solid technical foundation.

INDEX